3702382772

THE PARK
LEARNING CENTRE

PARK
LEARNING CENTRE
The Park, Cheltenham
Gloucestershire GL50 2QF
Telephone: 01242 532721

UNIVERSITY OF
GLOUCESTERSHIRE

WITHDRAWN

WEEK LOAN

WEEK LOAN

12 FEB 1999
10 MAR 1999
17 MAR 1999
22 APR 1999
-4 MAY 1999
-6 MAY 1999
18 MAY 1999
-2 JUN 1999
11 JUN 1999
13 OCT 1999
11 NOV
27 JAN 2000
10th FEB
22 FEB 2000
03 APR 2000
23 OCT 2000
11 DEC 2000
11 JAN 2001
-7 FEB 2001
22 FEB 2001
10 DEC
21.1.2002
14 OCT 2002
16 JAN 2003
12 MAY 2003
-4 NOV 2004
-6 NOV 2006
01 FEB 2010

D1345729

SCIENTIFIC AND TECHNOLOGICAL CHALLENGES

Nathaniel I. Durlach and Anne S. Mavor, *Editors*

Committee on Virtual Reality Research and Development

Commission on Behavioral and Social Sciences and Education

Commission on Physical Sciences, Mathematics,
and Applications

National Research Council

PARK CAMPUS
LEARNING CENTRE
C & G C.H.E. P.O. Box 220
The Park
Cheltenham GL50 2QF
Tel: (01242) 532721

NATIONAL ACADEMY PRESS
Washington, D.C. 1995

NATIONAL ACADEMY PRESS • 2101 Constitution Ave., N.W. • Washington, DC 20418

NOTICE: The project that is the subject of this report was approved by the Governing Board of the National Research Council, whose members are drawn from the councils of the National Academy of Sciences, the National Academy of Engineering, and the Institute of Medicine. The members of the committee responsible for the report were chosen for their special competences and with regard for appropriate balance.

This report has been reviewed by a group other than the authors according to procedures approved by a Report Review Committee consisting of members of the National Academy of Sciences, the National Academy of Engineering, and the Institute of Medicine.

The work of the Committee on Virtual Reality Research and Development is supported by Department of the Army Contract No. DAAD05-92-C-0087 issued by the U.S. Aberdeen Proving Ground Support Activity. The views and opinions, and findings contained in this report are those of the author(s) and should not be construed as an official Department of the Army position, policy, or decision, unless so designated by other official documentation.

Library of Congress Cataloging-in-Publication Data

Virtual reality : scientific and technological challenges / Nathaniel I. Durlach and Anne S. Mavor, editors.

p. cm.

"Committee on Virtual Reality Research and Development, Commission on Behavioral and Social Sciences and Education, Commission on Physical Sciences, Mathematics, and Applications, National Research Council."

Includes bibliographical references and index.

ISBN 0-309-05135-5

1. Human-computer interaction. 2. Virtual reality. I. Durlach, Nathaniel I. II. Mavor, Anne S. III. National Research Council (U.S.). Committee on Virtual Reality Research and Development.

QA76.9.H85V6 1994

006—dc20 94-37695

CIP

Copyright 1995 by the National Academy of Sciences. All rights reserved.

Printed in the United States of America

COMMITTEE ON VIRTUAL REALITY RESEARCH AND DEVELOPMENT

NATHANIEL DURLACH (Chair), Department of Electrical Engineering and Computer Science, Research Laboratory of Electronics, Massachusetts Institute of Technology
STEVE BRYSON, NASA-Ames Research Center, Moffett Field, California
NORMAN HACKERMAN, Robert A. Welch Foundation, Houston, Texas
JOHN N. HOLLERBACH, Department of Biomedical Engineering, McGill University
JAMES R. LACKNER, Ashton Graybiel Spatial Orientation Lab, Brandeis University
J. MICHAEL MOSHELL, Institute for Simulation and Training, The University of Central Florida
RANDY PAUSCH, Department of Computer Science, University of Virginia
RICHARD W. PEW, BBN Laboratories, Inc., Cambridge, Massachusetts
WARREN ROBINETT, Virtual Reality Games, Inc., Chapel Hill, North Carolina
JOSEPH ROSEN, Dartmouth Hitchcock Medical Center, Lebanon, New Hampshire
MANDAYAM A. SRINIVASAN, Department of Mechanical Engineering, Research Laboratory of Electronics, Massachusetts Institute of Technology
JAMES J. THOMAS, Battelle Pacific Northwest Laboratory, Richland, Washington
ANDRIES VAN DAM, Computer Science Department, Brown University
ELIZABETH WENZEL, NASA-Ames Research Center, Moffett Field, California
ANDREW WITKIN, School of Computer Science, Carnegie Mellon University
EUGENE WONG, Department of Electrical Engineering and Computer Science, University of California, Berkeley
MICHAEL ZYDA, Department of Computer Science, Naval Postgraduate School

ANNE S. MAVOR, Study Director
HERBERT S. LIN, Senior Staff Officer, Computer Science and Telecommunications Board
CINDY S. PRINCE, Project Assistant

The National Academy of Sciences is a private, nonprofit, self-perpetuating society of distinguished scholars engaged in scientific and engineering research, dedicated to the furtherance of science and technology and to their use for the general welfare. Upon the authority of the charter granted to it by the Congress in 1863, the Academy has a mandate that requires it to advise the federal government on scientific and technical matters. Dr. Bruce M. Alberts is president of the National Academy of Sciences.

The National Academy of Engineering was established in 1964, under the charter of the National Academy of Sciences, as a parallel organization of outstanding engineers. It is autonomous in its administration and in the selection of its members, sharing with the National Academy of Sciences the responsibility for advising the federal government. The National Academy of Engineering also sponsors engineering programs aimed at meeting national needs, encourages education and research, and recognizes the superior achievements of engineers. Dr. Robert M. White is president of the National Academy of Engineering.

The Institute of Medicine was established in 1970 by the National Academy of Sciences to secure the services of eminent members of appropriate professions in the examination of policy matters pertaining to the health of the public. The Institute acts under the responsibility given to the National Academy of Sciences by its congressional charter to be an adviser to the federal government and, upon its own initiative, to identify issues of medical care, research, and education. Dr. Kenneth I. Shine is president of the Institute of Medicine.

The National Research Council was organized by the National Academy of Sciences in 1916 to associate the broad community of science and technology with the Academy's purposes of furthering knowledge and advising the federal government. Functioning in accordance with general policies determined by the Academy, the Council has become the principal operating agency of both the National Academy of Sciences and the National Academy of Engineering in providing services to the government, the public, and the scientific and engineering communities. The Council is administered jointly by both Academies and the Institute of Medicine. Dr. Bruce M. Alberts and Dr. Robert M. White are chairman and vice chairman, respectively, of the National Research Council.

Contents

Preface

The Committee on Virtual Reality Research and Development was established by the National Research Council in 1992 at the request of a consortium of federal government agencies: the Advanced Research Projects Agency, the Air Force Office of Scientific Research, the Human Research and Engineering Directorate of the Army Research Laboratory, the Crew Systems Directorate and the Human Resources Directorate of the Armstrong Laboratory, the Army Natick RD&E Center, the National Aeronautics and Space Administration, the National Science Foundation, the National Security Agency, and the Sandia National Laboratory. As a group, these agencies sought guidance and direction regarding the federal investment in research and development in the area of virtual reality.

This report constitutes the committee's response to the charge to "recommend a national research and development agenda in the area of virtual reality" to guide government research and development over the next generation. Although the charge refers only to virtual reality systems (systems in which the world the human operator interacts with is generated by a computer), the committee has also considered teleoperator systems. Such an extension is required not only for logical and scientific reasons, but also because many of the examples cited in the charge are teleoperator systems.

The purpose of the agenda is to provide a technical rationale to federal agencies for the allocation of their resources in support of research and technology development—and thereby to help define and shape the field. In keeping with this purpose, the recommendations in this report have been generated by a committee consisting primarily of computer scientists, engineers, and psychologists, some of whom have extensive experience in federal science and technology policy. The committee, in addition to drawing on the expertise of its own members, has made substantial use of advisers and consultants. Overall, this group's expertise is well matched to its task of surveying the scientific and technological state of the field, the potential of current and future technology for improving performance in various application areas, and the types of research and development required to realize this potential. In addition, it is important to note that most members of the committee have a vested interest in seeing the field of virtual reality and synthetic environments flourish because much of their professional work is related to it. Indeed, serious involvement in and knowledge about the field were among the main criteria used to form the committee.

Ultimately, however, the federal research and development agenda must also take account of societal needs, the potential of various research and development accomplishments for fulfilling these needs, and the kinds and magnitudes of attention and support various programs are likely to receive without special efforts. Furthermore, given a research and development program that receives high ratings according to these criteria, it is important to specify the kinds of infrastructure that will facilitate the carrying out of this program. We recognize that the full range of expertise required to determine and judge many of these factors goes well beyond the competencies represented on the committee. In particular, both the formulation and evaluation of societal needs and the specification of appropriate societal infrastructure to ensure efficient performance in carrying out federal research and development programs require inputs from individuals with expertise not only in computer science, engineering, and psychology, but also in sociology, economics, business, social policy, and government.

We have attempted, to the extent possible, to avoid basing our recommendations on assumptions concerning the relative importance of various societal needs. Also, to the extent possible, we have avoided basing our recommendations on assumptions concerning the extent to which and the manner in which given technological advances will influence society—the history of such predictions is extremely humbling. However, given the importance of assigning some sort of priorities to different possible research and development programs (a laundry list of all possible programs would be of little use to anyone), these issues cannot be avoided entirely. What we have tried to do is to make our assumptions explicit. We also suggest that individuals with additional expertise be asked to follow up this work. Finally, with regard to the problem of ensuring appropriate infrastructures to carry out the recommended research and development programs—a problem that is perhaps more difficult to solve than the problem of selecting the programs themselves—we point out some of the factors that we believe need to be considered in the follow-up work.

This research and development agenda for the federal investment in virtual environments and teleoperation is rooted in a careful review and analysis of the current state of research and technology and of the steps required to reach a point at which significant applications can be fully realized. Our perspective on infrastructure takes cognizance of the need for additional studies of issues related to market forces and societal design as this agenda is further shaped and implemented. In the committee's view, our scientific and technical assessment represents only the first step in a multistage process that would appropriately lead to a more fully developed federal research agenda based on considerations of societal as well as scientific and technical issues.

Although our report is directed primarily to the federal agencies who requested it, we have attempted to make it useful to a wide variety of readers, including students and professionals working in academia and industry. The material is structured in the following way. The executive summary provides a brief synopsis of the main points that can easily be read in a few minutes. The overview is intended to provide a general understanding of the field and of our recommendations; it can be read in a few hours. The remainder of the report provides the detailed analysis as well as numerous bibliographical citations.

As in most reports prepared by large committees, there has been a struggle to find an appropriate balance between coherence and uniformity of style on one hand and free and full expression by individual experts on the other. As is evident when reading the report, different chapters have been prepared by different people. In other words, with the exception of the executive summary and the overview, we have leaned toward full and free expression rather than uniformity. It should also be noted that the differences among the chapters reflect not only the different writing styles of the various contributors but also the different cultures associated with the topics being discussed. For example, even within the single domain of human-machine interfaces, discussions of the visual, auditory, and haptic channels would require rather disparate treatment even if written by the same individual.

Many individuals have made a significant contribution to the committee's thinking and to various sections of this report by serving as presenters, consultants, and advisers. A complete list of contributors and their affiliations is presented in Appendix B. Although all of these individuals provided us with valuable information, a few played a more direct role in the preparation of this manuscript and deserve special mention. We extend our gratitude to Kurt Akeley for his contribution on computer architecture; to Walter Aviles for his work on visual technology and remote, teleoperated vehicles; to Norman Badler for his contributions on modeling virtual worlds and virtual actors; to Paul DiZio for his work on full-body motion interfaces; to Blake Hannaford and Thomas Sheridan for their discussions of teleoperator systems; to Michael Macedonia for his work on networking; to John Makhoul and Kenneth Stevens for their contributions on speech communication; to Barbara Shinn-Cunningham for her work on the auditory channel; to Thomas Wiegand and Richard Held for their discussions of the status of research on the human visual system; to David Zeltzer for his work on visual technology and computer modeling; and to George Zweig for his sustained interest in our study and his extremely valuable critical comments on a wide variety of crucial issues.

In the course of preparing this report, each member of the committee took an active role in drafting sections of chapters, leading discussions, and reading and commenting on successive drafts. Even so, John Hollerbach and Michael Zyda deserve special acknowledgment for their heroic efforts throughout the study. John Hollerbach assumed major responsibility for the work on telerobotics as well as contributing extensively to several other sections of the report. Michael Zyda took responsibility for obtaining, summarizing, and integrating material on the computer generation of virtual environments and networking. Other committee members contributing work on the computer generation of virtual environments include Steve Bryson for scientific visualization, Randy Pausch for techniques for interacting in three-dimensional environments, Andrew Witkin for modeling software, and Andries van Dam for his overall review and insightful comments. Committee members who contributed their expertise to the human-machine interface sections of the report were James Lackner for whole-body motion and illusions; Elizabeth Wenzel for the auditory channel; and Mandayam Srinivasan for the haptic channel and for the discussion of hardware and software requirements for multimodal interfaces. Those who provided expertise in the various application domains of virtual environments include Norman Hackerman, Michael Moshell, Richard Pew, Warren Robinett, Joseph Rosen, James Thomas, and Eugene Wong.

Staff of the National Research Council made important contributions to our work in many ways. We would like to extend a special note of thanks to Harold Van Cott and Herbert Lin. Harold Van Cott was instrumental in the original thinking about this project and the formation of the committee; he has been of immeasurable assistance throughout the process in providing advice and draft materials. Herbert Lin, a senior staff officer to the committee from the Computer Science and Telecommunications Board, contributed extensively to the applications section of the report. Further, we would like to express our appreciation to Marjory Blumenthal, director of the board, for her contributions in the early stages of project development, and to Alexandra Wigdor, director of the Division on Education, Labor, and Human Performance, for her valuable insight and support. Thanks are also due to Cindy Prince, the committee's administrative assistant, who was indispensable in organizing meetings, arranging travel, and compiling agenda materials. We are also indebted to Carolyn Sax and Nora Luongo for their work on the manuscript, and to Christine McShane, who edited and significantly improved the report.

To our sponsors we are most grateful for their interest in the topic of this report and for their many useful contributions to the committee's deliberations. We are particularly appreciative of the early efforts of Bernard Corona, of the Human Research and Engineering Directorate, U.S. Army Research Laboratory, and James Jenkins, of the National Aeronautics and Space Administration, which led to the formation of the committee.

Finally, it is a deep pleasure to acknowledge the remarkable feelings that appear to have been generated by this cooperative, interdisciplinary study. These feelings can be best summarized by noting the following two ideas expressed at the last meeting:

- Whatever the reception of the report, the field of synthetic environments was significantly advanced by the education of the committee members, along with their consultants and advisers, that took place during the study;
- Many of the committee members intend to continue meeting on a regular basis to continue the spirited interdisciplinary dialogue that was initiated during the study.

Nathaniel I. Durlach, *Chair*
Anne S. Mavor, *Study Director*
Committee on Virtual Reality
Research and Development

Virtual REALITY

Executive Summary

At the request of a consortium of federal government agencies, the Committee on Virtual Reality Research and Development was established to provide guidance and direction on the allocation of resources for a coordinated federal program in the area of virtual reality. In responding to this charge, the committee has included both virtual environments and teleoperation in its assessment of the field.

This report includes recommendations and extensive background material concerning systems popularly referred to by such terms as *virtual reality*,[1] *cyberspace*, *virtual environments*, *teleoperation*, *telerobotics*, *augmented reality*, and *synthetic environments*. In all such systems, the basic components are a human operator, a machine, and a human-machine interface linking the human operator to the machine.

In a *teleoperator system*, the machine is an electromechanical tool containing sensors and actuators (i.e., a telerobot) that effectively extend the operator's sensorimotor system and thereby allow him or her to sense and manipulate the real environment in new ways. In a *virtual environment (VE) system*, the machine is an appropriately programmed computer

[1] Our use of the term *virtual reality* in the title of the book differs from that in the report itself. In the title it is intended, as is often the case in the popular press, to encompass the entire field by including both teleoperator and virtual environment systems. In the text, *virtual reality* and *virtual environment* are used synonymously and do not include teleoperation. The term we use to refer to systems of this general kind when we have no need to distinguish among different types is *synthetic environments*.

that generates or synthesizes virtual worlds with which the operator can interact. Whereas the purpose of a teleoperator system is to sense and transform the real world (as in removal of hazardous waste by teleoperation), the purpose of a virtual environment system is to alter the state of the human operator or the computer (as in the use of virtual environment systems for training, designing, marketing, or scientific modeling). In many systems, such as teleoperator systems that make use of virtual environment systems to help plan future actions, teleoperator systems and virtual environment systems are combined. In an *augmented-reality system*, the operator's interaction with the real world (either directly or via a teleoperator system) is enhanced by overlaying the associated real-world information with information stored in the computer (generated from models, derived previously from other sensing systems, etc.). In general, we refer to all systems of the types just described as *synthetic environment (SE) systems.*

Virtual environment systems differ from traditional simulator systems in that they rely much less on physical mock-ups for simulating objects within reach of the operator and are much more flexible and reconfigurable. Virtual environment systems differ from other previously developed computer-centered systems in the extent to which real-time interaction is facilitated, the perceived visual space is three-dimensional rather than two-dimensional, the human-machine interface is multimodal, and the operator is immersed in the computer-generated environment.

In recent years, synthetic environment systems, particularly virtual environment systems, have generated both great excitement and great confusion. These factors are evident in the extensive material published in the popular press; in the unrealistic expectations on the part of the public; in the inadequate terminology being used; in the deluge of conferences, articles, books, and demonstrations occurring; in the difficulties being experienced in communicating across disciplinary boundaries even by individuals whose professional work lies within the domain of synthetic environment systems; and in the frenetic pace at which most of the individuals concerned with synthetic environments are working.

In this book, we attempt to describe the current state of research and technology that is relevant to the development of synthetic environment systems, provide a summary of the application domains in which such systems are likely to make major contributions, and outline a series of recommendations that we believe are crucial to rational and systematic development of the synthetic environment field. Inasmuch as the "bottom line" of the committee's work is our recommendations (presented in the final section of the overview), the remainder of this executive summary focuses on these recommendations. They are summarized under the headings of Applications, Psychological Considerations, Technology, Evaluation, and Government Policy and Infrastructure.

In discussing these areas, it should be noted that the recommendations have not been prioritized in any detailed manner. This is due primarily to our judgment that successful development and application of SE systems depends on an entire matrix of interrelated factors. We nevertheless feel that it is important to stress the crucial need for improved hardware technologies to enable development of improved interface devices and improved computer generation of multimodal images. Unlike the situation in the area of teleoperation, in the area of VEs there are relatively few individuals who have primary interests or backgrounds in hardware; most individuals in the VE area are involved primarily in the software end of computer science, in communication or entertainment media, and in human perception and performance. Thus, the importance of adequate hardware, without which the VE field will never come close to realizing its potential, tends to be underplayed by the VE community. A somewhat similar comment concerns the issue of user comfort. To date, a very large fraction of VE usage has occurred in the context of short demonstrations, a context in which the degree of comfort is relatively unimportant. However, if the comfort of VE systems (particularly head-mounted displays) cannot be radically improved, the practical usage of these systems will be limited to emergency situations or to very short time periods. In other words, adequate comfort, as well as technically adequate hardware, are essential to realizing the potential of the SE field.

Finally, it should also be noted that our thoughts about government policy and infrastructure are stated as comments and suggestions rather than as recommendations. They are based solely on the experience and judgment of the committee members.

RECOMMENDATIONS

Applications

Significant research and development is taking place in a wide variety of application domains, and in some cases the results of this work are beginning to be applied on an experimental basis. Although it is not yet clear which tasks will eventually gain the most from the use of SE systems, the committee has identified four application domains that show particular promise: (1) design, manufacturing, and marketing; (2) medicine and health care; (3) hazardous operations; and (4) training.

Other important application domains that are assigned lower priority are education, information visualization, and telecommunications and teletravel. The application domain of education is of critical concern;

however, in our judgment, the overwhelming issues in this domain are social, political, and economic rather than technological. Because the domains of information visualization and telecommunications and teletravel cut across all others, we expect them to receive substantial attention in connection with work that is primarily addressed to the other domains. Training has the same cross-cutting property; however, it was judged to be so important and so well matched to the technology that it was nevertheless given high priority.

Psychological Considerations

Because human beings constitute an essential component of all SE systems, there are very few areas of knowledge about human behavior that are not relevant to the design, use, and evaluation of these systems.

The committee recommends that work in this area be organized around the following objectives: (1) development of a comprehensive review of theory and data on human performance characteristics from the viewpoint of synthetic environment systems; (2) development of a theory that facilitates quantitative predictions of human responses to alterations in sensorimotor loops; (3) development of cognitive models that facilitate effective design of VE systems for purposes of education, training, and information visualization; and (4) development of improved understanding of possible deleterious effects of spending substantial portions of time in synthetic environments.

Technology

Despite the enthusiasm and the "hype" surrounding the SE field, there is a substantial gap between the technology that is available and the technology that is needed to realize the potential of SE systems envisioned in the various application domains. Two partial exceptions are the virtual environment technology used in the entertainment industry and the teleoperator technology used for hazardous operations such as waste removal. For most applications to be truly successful, however, the development of substantially improved technology is a major requirement. In our review, we divide the relevant technology into four general categories: human-machine interfaces, the computer generation of virtual environments, telerobotics, and networks.

Human-Machine Interfaces

Human-machine interfaces for SE systems include all the devices used to present information to human users or to sense the human actions or responses that control the machine in question. In this book we examine the needs for human research studies and technology development for the visual, auditory, and haptic channels; for whole body motion and locomotion displays; for position tracking; and for speech communication, physiological, olfactory, and gustatory interfaces. After careful analysis we have determined that the most important research and development needs in the interface area concern the visual channel, the haptic channel, locomotion displays, and position tracking, and we make specific recommendations on these topics accordingly.

Computer Hardware and Software

It is computer hardware and software that produce virtual environments. Technology should be capable of generating such environments in a way that makes them appear convincingly real to human users and that allows them to interact with the environments in real time. With available technology, however, there is a major trade-off between realistic images and realistic interactivity.

Hardware requirements for virtual environments include very large physical memories, multiple high-performance scalar processors, high-bandwidth mass storage devices, and high-speed interface ports for various input and output peripherals. In the committee's judgment, commercial market forces, if they continue to grow at the current rate, will probably be sufficient to support the needed development. Therefore, the committee recommends no aggressive federal involvement in computer hardware development in the SE area at this time. Rather we conclude that hardware development remain largely a private-sector activity. Should serious lags in development occur, the government might then consider strategies for leveraging private-sector development efforts.

Software requirements are such that a major unified research program, focusing on the generation, implementation, and application of virtual environments, should be undertaken. The basic topics that need to be considered in such a program include: (1) multimodal human-computer interactions, (2) rapid specification and rendering of visual, auditory, and haptic images, (3) models and tools for representing and interacting with physical objects under multimodal conditions (including automated model acquisition from real data), (4) simulation frameworks (5) a new time-critical, real-time operating system suitable for virtual environments with relatively simple input/output requirements, (6) registration of real and virtual images in augmented reality applications, (7) navigational cues in virtual space, (8) the behavior of autonomous actors, and (9) computer generation of auditory and haptic images. Because the natural tendency of computer scientists to concentrate on graphics will focus sufficient attention on the visual channel, special attention must be given to the modeling and generation of auditory and haptic images and to the needs associated with integrating the different modalities in virtual environment systems.

Telerobotics

In many ways, research issues in teleoperation are similar to those in virtual environments. Independent of the purpose for which a system is being designed (e.g., to train an operator, to remove hazardous waste) and independent of whether the relevant environment is real or virtual, both are concerned with the design, construction, and application of multimodal, immersive systems that enable the operator to interact usefully with some structured environment. Because of these similarities and the relatively long history of research in the teleoperation area, results in teleoperation can be usefully exploited in the virtual environment area. Concerns unique to teleoperator systems relate to the design and performance of the complex electromechanical systems referred to as telerobots and the unavoidable time delays that arise in communicating between the human-machine interface and the telerobot when these subsystems are separated by large distances. Such communication delays can result in degraded or unstable teleoperator performance.

Improved teleoperator systems require improved control algorithms and methods for constructing and using predictive displays and for realizing effective supervisory control techniques as strategies for combatting communication time delays. Hardware requirements include: (1) multiaxis, high-resolution tactile sensors to provide telerobots with an adequate sense of touch; (2) robot proximity sensors for local guidance prior to grasping, (3) multiaxis force sensors to measure net force and torque exerted on end effectors, (4) improved actuator and transmission designs for high-performance joints, and (5) real-time computational architectures. Also, since many new problems arise when a human is interfaced to a microtelerobot, research is needed to capitalize on the advances now being made in the field of microelectromechanics. Similarly, research is needed to explore possibilities and problems associated with the development and application of distributed telerobots (macro and micro).

Networks

Communication networks have the potential to transform virtual environments into shared worlds in which individuals, objects, and processes interact without regard to their location. In the future, such networks will allow us to use virtual environments for such purposes as distance learning, group entertainment, distributed training, and communication among telerobots in diverse locations. Although the technology is becoming able to support the development of distributed virtual environments, it is currently insufficient to support multiple users and multiple modes of input in real time. Other problems to be resolved are network host-interface slowdowns caused by the multiple layers of operating system software and the high cost of purchasing time on high-speed, wide-area networks.

We anticipate that in the future most virtual environment applications will rely heavily on network hardware and software. Because several forces in the federal government and in the private sector are driving the major advances in hardware, we do not advise additional investment in network hardware development at this time. We do propose, however, that the federal government provide funding for a program (to be conducted with industry and academia in collaboration) aimed at developing network standards that support the requirements for implementing distributed virtual environments on a large scale. Furthermore, we propose funding of an open VE network that can be used by researchers, at a reasonable cost, to experiment with various VE network software developments and applications.

Evaluation

In general, SE technology and SE systems are not being adequately evaluated. Admittedly, the evaluation task is complex: it involves considerations of many disciplines, both whole systems and individual components, a wide variety of component technologies, and many different types of evaluation goals. Nevertheless, we believe that if the SE field is to progress beyond the stage of demonstrations, serious evaluations are crucial. They are needed not only for estimating overall cost-effectiveness, but also for analyzing performance in terms of the contributions made by different component features and thereby guiding the directions of future research and development.

Developers of SE systems should conduct evaluations using a variety of approaches throughout the development process, and the federal government should therefore encourage the developers whose work it supports to include a comprehensive evaluation plan in the design stages of their research and development projects. The federal government should also help coordinate the development of standardized testing procedures for use across studies, systems, and laboratories, particularly in areas in which the private sector has not been active.

Government Policy and Infrastructure

Because the field of synthetic environments is in its very early stages, the federal government has a rare and important opportunity to foster careful planning for its research and development. In this section, we discuss some mechanisms that we think federal agencies should consider as part of their strategic planning for the research and development agenda in the area of synthetic environments.

First, a national information system that provides comprehensive coverage of research activities and results on synthetic environments in a user-friendly way to a wide variety of users could be a useful tool for promoting cross-fertilization and integration of the research and development efforts. Such a system could serve as a repository of text, data, computational models, and software and could include effective subsystems for both retrieval and dissemination.

Second, federal agencies might fruitfully consider establishing a small number of national research and development teams, each focusing on a specific application. These teams could involve government, industry, and academia, as well as the various disciplines relevant to the given application. Funding could be provided jointly by both the federal government and the private sector.

Third, it might be useful for some federal agencies and offices to explore the use of synthetic environment technology to meet their own administrative and program needs. In addition to the application of synthetic environments to the defense and space programs already under way, other application domains, such as training, telecommunication and teletravel, and information visualization, are relevant to the activities of many agencies. One way for the government to facilitate the development of the SE field would be to select a few agencies to serve as test beds for synthetic environment technology in these general domains.

Fourth, although it is probably too early in the development of synthetic environment systems to establish standards and regulations, it is not too early for the federal government to begin to evaluate the work already under way in connection with the telecommunications and entertainment industries. Problems that are already of concern and are likely to increase as the field develops relate to technological compatibility issues, enforcement and control issues, and social and ethical issues.

Finally, in developing a funding strategy for specific research and development projects, it is essential that federal planners be aware of existing market forces, which are as likely to be shaped by the results of research and development as they are to shape the research and development that is performed. With strategic planning, it would be possible for the federal government to use its investment in research to leverage developments in the private sector. Similarly, it is essential that federal planners take account of the societal implications of the technology. As with most other technologies, the effects of the advances in synthetic environments are likely to be mixed: some effects will be positive and others negative. It cannot be assumed that all technological advances, even those that are likely to have substantial practical applications, will necessarily be beneficial.

Overall, the committee believes that synthetic environment systems have great potential for helping to satisfy various societal needs and stimulating advances in some important areas of science and technology. In pursuing the committee's recommendations for research and development, the federal government has an opportunity to make important contributions to the development of this exciting new field.

I

OVERVIEW

At the request of a consortium of federal government agencies, the Committee on Virtual Reality Research and Development was established to provide guidance and direction on the allocation of resources for a coordinated federal program in the area of virtual reality. In responding to this charge, the committee has included both virtual environments and teleoperation in its assessment of the field. Such an extension is required not only for logical and scientific reasons, but also because many of the examples cited in the charge feature the use of teleoperator systems.

In a synthetic environment (SE) system, the human operator is transported into a new interactive environment by means of devices that display signals to the operator's sense organs and devices that sense various actions of the operator. In teleoperator systems, the human operator is connected by means of such displays and controls to a telerobot that can sense, travel through, and manipulate the real world. In virtual reality (VR) or virtual environment (VE) systems, the human operator is connected to a computer that can simulate a wide variety of worlds, both real and imaginary. Simple remote manipulators are an example of the first type of system; video games of the second type.

Teleoperator systems effectively provide the operator with a transformed sensorimotor system that enables him or her to perform new types of actions in the real world. Virtual environment systems effectively provide the operator with controllable methods for generating new types of experiences. Using both teleoperator and virtual environment systems, one can (or will be able to) explore the ocean floor and outer space, visit Samarkand while staying in Elmira, try out products not yet manufactured, dig up a 10-ton container of hazardous waste, take a canoe trip through the human circulatory system, and have one's hair trimmed by a barber in Seville.

SCOPE OF THE SYNTHETIC ENVIRONMENT FIELD

The research and development required to realize the potential of SE systems is extremely challenging. The systems are complicated because they involve both complex artificial devices and a complex biological system (the human operator). There is a crucial need for cooperation among many disciplines, including computer science, electrical and mechanical engineering, sensorimotor psychophysics, cognitive psychology, and human factors. Also, the range of possible applications is exceedingly broad. Overall, the committee believes that the SE field has great potential, that the research and development required to realize this potential is just beginning, and that work in this area should be vigorously pursued by a wide variety of specialists in a wide variety of institutions.

There is currently a great deal of excitement, a great deal of "hype," and a great deal of confusion associated with the SE area. A major source of the confusion is the combination of rapid acceleration of interest in the area and the coming together of individuals from widely varying disciplines. In some cases, individuals are coming together because the problem to be solved requires expertise in diverse areas. In other cases, they are coming together because it has suddenly become apparent that essentially the same problems are being addressed by individuals in different fields who have never had the benefit of communicating with each other about them.

Associated with this interdisciplinary feature of the SE field is confusion over terminology: each discipline brings to the field its own language and its own biases. For example, whereas computer scientists naturally use the terms *input* and *output* in reference to the computer, psychologists use these terms in reference to the human user. Thus, in a virtual environment system, what is output to the psychologist is input to the computer scientist. Similar confusions often arise with the term *interface.* Whereas computer scientists frequently use this term to designate a component internal to the computer's hardware or software, many others use the term as a shorthand for human-computer interface devices external to the computer. Also, of course, in addition to the communication difficulties associated with the interdisciplinary nature of the field, there are communication difficulties associated with the tendency of different individuals, institutions, and countries to compete rather than to cooperate.

Another source of confusion results from political and public relations considerations. *Virtual reality* and *virtual environment* (two terms that we regard as equivalent) are such "hot" terms that many people tend to use them even when their use is logically inappropriate. Thus, for example, these terms are often used in a manner that implies that teleoperator systems are a special case of virtual reality systems. At the same time, however, when describing the origins of virtual reality systems, the history of teleoperator systems (in particular, the use of head-mounted displays in these systems) is entirely ignored. Similar distortions often occur in connection with simulator systems. Although simulator systems, like teleoperator systems, are closely related to virtual environment systems and have a long and distinguished history, past accomplishments in the simulation area are often inappropriately downplayed. Further discussion of the basic concepts and terminology is presented in the next section.

Generally speaking, virtual reality currently has an extremely high "talk-to-work" or "excitement-to-accomplishment" ratio. Between 1992 and 1994, roughly 12 new books have been published, 4 new journals or magazines have been started, and 200 new articles have been published on the topic of virtual reality. Major professional meetings and trade shows are occurring at a rate of roughly one per month. Over 10 government agencies have held conferences or written reports on VE during the same two-year period. And practically everyone in the field is spending substantial time traveling to other laboratories that are working on VE and providing demonstrations of their own facilities in their own laboratories.

Despite this high talk-to-work or excitement-to-accomplishment ratio, substantial efforts are, in fact, under way in various research and development areas and in various application domains. Significant research and development programs, as well as applications of currently available technology, are being pursued in government, in academia, and in industry. Also, some attempts are being made to develop adequate course material for educational programs in the SE area; however, it is likely to be some time before most academic departments recognize SE as a legitimate field of specialization (e.g., one in which faculty can achieve tenure).

Current research and development efforts directly relevant to the creation of useful SE technology are concerned with (1) computer generation of virtual environments, (2) design of telerobots, (3) improvement of human-machine interfaces, (4) study of relevant aspects of human behavior, and (5) development of communication systems that are adequate to support networking of SE systems. Items (3) and (4) are relevant to all the kinds of systems considered, item (1) to VE systems, item (2) to teleoperation systems, and item (5) to networked systems. An additional item of importance when augmented-reality systems are considered is (6) merging of computer-generated images with images derived directly from the real world.

The "SE Challenge" is related to the High Performance Computing and Communications (HPCC) Grand Challenge program initiated by the federal government through both the computer generation of VEs and networked systems. For many applications, adequate computer generation of the associated virtual worlds is going to require very high-performance computing. Similarly, the networking of SE systems is going to require very high-performance communications. In general, SE systems will provide both a major application area for HPCC and an important source of constraints for the design of HPCC systems.

Currently, the main commercial driving force for the development of VE systems is the entertainment application. There is no equivalent commercial driving force for the development of teleoperators or augmented-reality systems at this time.

Programs on SE technologies and applications are under way in almost every developed country (Thompson, 1993). Major players are the

United States, Japan, and the European Economic Community; other players include South Korea, Singapore, the Netherlands, and Sweden. Although each of these regions is engaged in a full range of research, development, and commercial activities, the work in each region bears the marks of its distinctive culture.

Today, more than 25 universities, at least 15 federal agencies, and more than 100 large and small companies throughout the United States are contributing to the growth of research and development in the SE field. In industry, research and development directed toward defense, space, scientific visualization, and medicine are more prominent in the United States than elsewhere. The European Economic Community and Japan have regional or national initiatives on SE, but such initiatives are still being debated in this country.

Although the recession of the early 1990s in Europe has slowed down investment, a variety of SE projects are under way in industry and, to a lesser extent, in universities. Interests in the United Kingdom are similar to those in the United States but place more emphasis on education, training, and entertainment. The United Kingdom may well be the world leader in SE entertainment systems. On the continent, work on SE applications is being conducted at the European Space Research Center in Noordwijk, the Netherlands. Research on computer-aided architectural software and a virtual railroad environment are also being supported in the Netherlands. In France, the university at Metz is developing an autonomous motor vehicle for people with disabilities that uses SE technology. In Lille, the University of Technology is exploring the use of teleoperation in surgery. At the University of Paderborn in Germany, a new method for walk-through animation in three-dimensional scenes is under way. In Italy, the University of Genoa is developing a knowledge-based simulation for production engineers.

Japan entered the VE part of the SE world later than the United States and Europe. Recently, however, that country has realized that VE, as well as teleoperation, is a logical extension of its strong national interest and background in robotics, automation, and high-definition television. Concern with haptic interfaces and force-feedback sensor display systems is also intense. As a consequence, Japan has established 10 national consortia for research and development in the SE area that, taken together, provide more funds per year than all SE investment in the United States (Larinaji, 1994). In 1992, the Japan Technology Transfer Association formed an Artificial Reality and Tele-Existence Research Committee of 90 participating companies from the SE industry. Knowledge and technology sharing among companies—generally a boon to Japanese industry—are extensive. These indicators, together with its typical long-range financial horizon, large targeted investments, and a national technology agenda, could give Japan a major competitive advantage in SE. The extent to which this advantage is actually realized will depend, at least in part, on the extent to which Japan can become a leader in the relevant computer software areas.

This overview begins by presenting some basic concepts and terminology that are important in talking about virtual environments and teleoperator systems. We then present some visions of where we think the technology may be leading. The visions section differs from the rest of this report in the speculative nature of the material and in the incorporation of societal issues into the scenarios. The overview then goes on to summarize the current state of the synthetic environment (SE) field, covering application domains, knowledge about human behavior and performance, technology issues, and evaluation issues. The committee's assessment of needs and priorities completes the overview. In making these recommendations, we include consideration of the extent to which various research goals are likely to be realized without special government funding efforts or are likely to require such efforts. Similarly, we consider issues related to the infrastructure required to carry out various research and development programs.

BASIC CONCEPTS AND TERMINOLOGY

There are currently no precise and generally accepted definitions of the terms being used in our area of interest. This is due in part, as already discussed, to the interdisciplinary nature of the field and to public relations matters. It is also due to fundamental problems of the type usually encountered in efforts to create language that faithfully reflects the structures and processes to which the language refers. For example, whereas language is fundamentally discrete, the evolutionary process by which virtual environment systems have developed from antecedent systems (such as desktop computing systems, simulators, teleoperator systems, etc.) is effectively continuous. Thus, either the definition of virtual environment systems must remain rather fuzzy, or one must set arbitrary thresholds on the complex, continuous evolutionary process.

Here, we outline some of the principal defining ideas and indicate how the terms *virtual environment*, *teleoperator*, and *augmented reality* are related to each other and to other closely related terms such as *simulator*, *telerobot*, and *robot*. Our purpose is to provide background on the meaning of the terms we use in order to permit readers to understand later sections of the report. The process of creating and defining terms in this area will of course continue for many years.

A *teleoperator* system consists of a human operator, a human-machine interface, and a telerobot (Figure 1). Environmental signals are sensed by sensors (cameras, microphones, etc.) located in the telerobot, transmitted to the human-machine interface, and presented to the human by means of display devices (e.g., cathode ray tubes, earphones) in the interface. Human responses, usually motor actions, are sensed by the interface and used to control the actions of the telerobot. Thus, a teleoperator system can be viewed as a system for extending the sensorimotor system of the human organism. The purpose of such a system is to facilitate the human operator's ability to sense, maneuver in, and manipulate the environment. Teleoperator systems vary along many dimensions, including the structure of the human-machine interface and the telerobot and the nature of the control algorithms.

Teleoperator systems have been used to conduct work in outer space and under the ocean; to perform a variety of tasks in connection with security, firefighting, nuclear plants, and hazardous waste removal; to assist in various types of military operations; to perform microsurgery; and to aid in the rehabilitation of individuals with severe physical disabilities. In some teleoperator systems, the human operator has direct and detailed control of all the telerobot's actions. In other systems, the human's control occurs only at a supervisory level and many of the telerobot's detailed actions are controlled locally and automatically. In the extreme, there is no human control, all actions of the telerobot are automatic and autonomous, and the telerobot is called simply a robot.

A *virtual environment* system (also illustrated in Figure 1) consists of a human operator, a human-machine interface, and a computer. The computer and the displays and controls in the interface are configured to immerse the operator in an environment containing three-dimensional objects with three-dimensional locations and orientations in three-dimensional space. Each virtual object has a location and orientation in the surrounding space that is independent of the operator's viewpoint, and the operator can interact with these objects in real time using a variety of motor output channels to manipulate them. The extent to which a virtual environment is designed to simulate a real environment depends on the specific application in mind.

As illustrated in Figure 1, teleoperator and virtual environment systems are similar in that they both involve human operators and elaborate human-machine interfaces. They differ however, with respect to what takes place on the nonhuman side of the interface. Whereas in a teleoperator system the interface is connected to a telerobot that operates in a master-slave or supervisory control mode in a real-world environment, in a VE system the interface is connected to a computer.

Consistent with this difference in structure is the difference in purpose between the two types of systems: whereas the purpose of a teleoperator system is to sense, manipulate, and transform the state of the

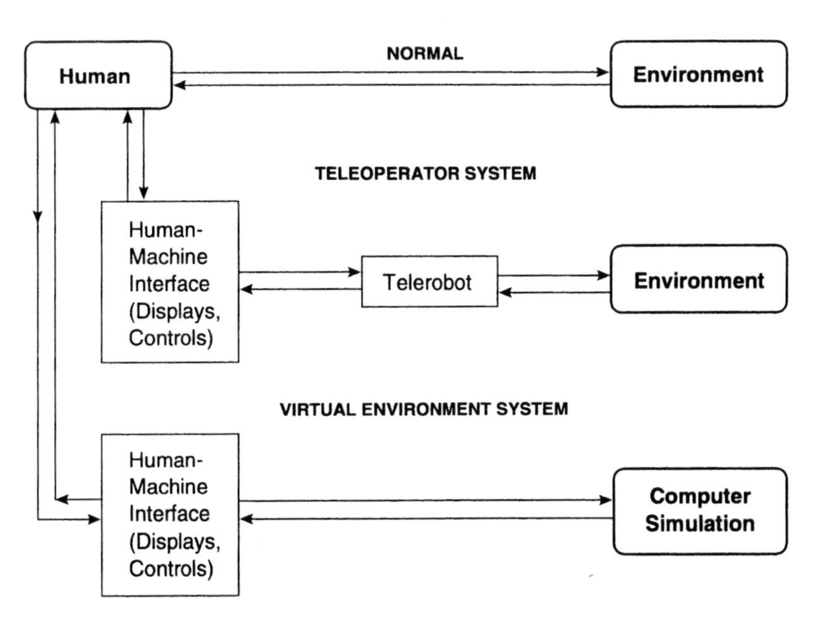

FIGURE 1 Schematic outline comparing a teleoperator system, a virtual environment system, and an unmediated (normal) system.

real-world environment, the purpose of a VE system is to sense, manipulate, and transform the state of the human operator (as in training or in scientific visualization) or to modify the state of the information stored in the computer (e.g., the virtual environment or some theoretical model represented in the computer software). Virtual environment systems are being used in the areas of telecommunication, information visualization, health care, education and training, product design, manufacturing, marketing, and entertainment. In the near future, such systems are likely to find further applications in various areas of psychology, including basic psychophysical research, biofeedback, and psychotherapy.

Many systems are now being developed that are mixtures or blends of teleoperator and virtual environment systems. Thus, for example, VE systems are now being introduced as subsystems of teleoperator systems in order to assist the human operator in controlling the telerobot. In particular, when the telerobot is sufficiently far removed from the human operator to cause significant time delays in the transmission of information between the telerobot and the human operator, virtual environments can be used to present computer-generated information derived from predictive models in the computer.

People are also designing systems in which virtual and real environments are combined (Figure 2). The use of such *augmented-reality* systems

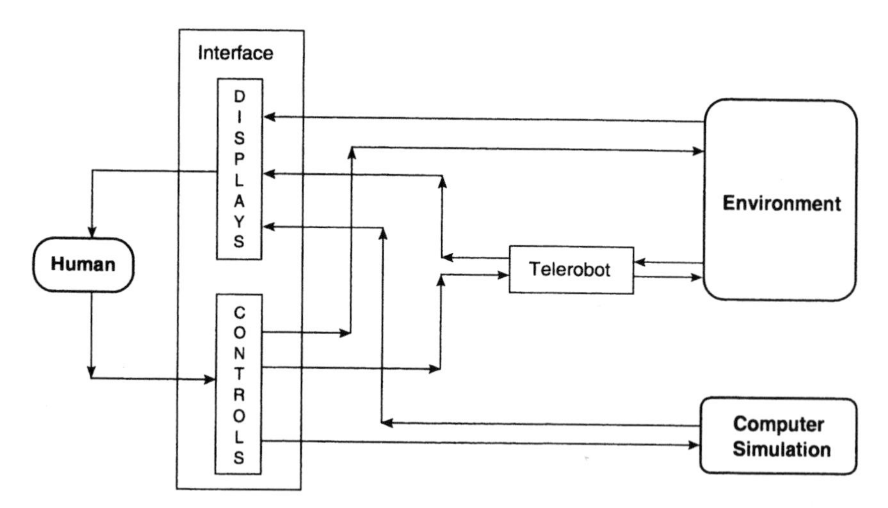

FIGURE 2 Schematic outline of some augmented reality systems. One kind of augmented-reality system combines images obtained from direct sensing of objects in the environment with images generated by a computer using see-through (or hear-through or feel-through) displays. A second kind combines images obtained by means of a telerobot with those generated by a computer. In principle, systems that combine all three channels of input information could be combined. Also, of course, and as mentioned in the text, it is possible to consider systems that merge output (control) information rather than, or as well as, input (display) information.

is being explored in medical applications, manufacturing applications, and driving applications (both airplanes and cars). In many such cases, information from the real environment is sensed directly by means of a see-through display, and the supplementary information from the virtual environment is overlaid on this display. In other cases, the real-environment information to be combined with the virtual-environment information is derived by means of a teleoperator system. Although currently receiving less attention in the SE community, it is also possible of course to consider augmented-reality systems in which, instead of combining input channels, output channels are combined. For example, speech sounds or commands uttered by the human operator might be combined with those uttered by an automatic speech-synthesis system, or physical objects in the environment might be manipulated by systems that include both the hand of the human operator and a telerobotic hand controlled by the operator. There are certainly many tasks in which it would be extremely useful to have a third hand (with special features perhaps) that could work cooperatively with one's own two hands and be controlled, perhaps, by simple speech commands. A further way of picturing some possible relations between teleoperator and virtual environment systems within an SE system is illustrated in Figure 3.

In all of these systems, the human operator is projected into a new interactive environment that is mediated by artificial electronic and electromechanical devices, and in all of these systems, the operator's performance and subjective experience in the new environments depend strongly on the human-machine interface and the associated environmental (real or virtual) interactions. In general, we refer to all of these systems (teleoperator systems, virtual environment systems, augmented-reality systems, etc.) as *synthetic environment* (SE) systems.

In considering these different kinds of systems, it should be noted that many of the problems now facing designers of VE systems have been studied previously in the field of telerobotics. This is the case, for example, in the area of human-machine interfaces. Although the constraints

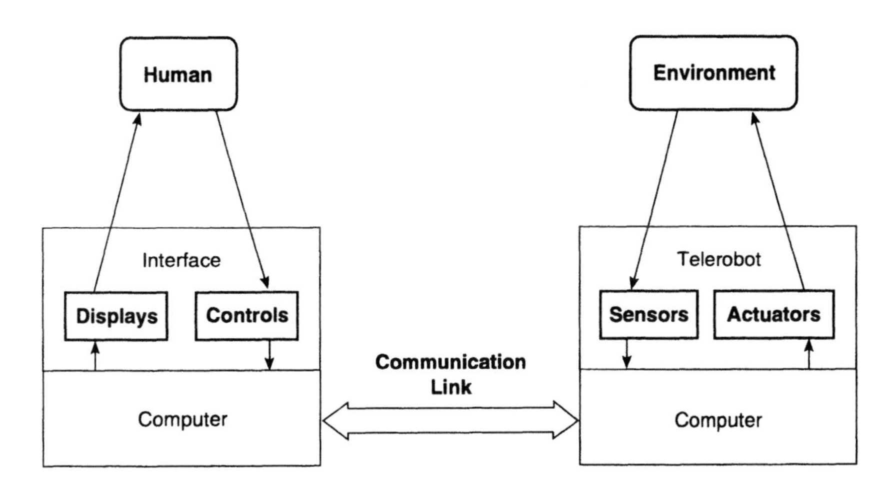

FIGURE 3 Schematic outline of a further configuration of SE system components. In this configuration, there are two computers, one on each side of a communication link. If all the components on the left side of this link are deleted and the computer on the right is used to control the telerobot, the system reduces to an autonomous robot. If the components to the right of the link are deleted and the computer on the left is used to generate a virtual environment, the system reduces to a pure virtual environment system. If all the components are included, but the computer on the left is not used to generate a virtual environment, then the system reduces to a pure teleoperator system (which would have supervisory control if local, low-level actions were controlled by one of the computers). If all the components are included, and the computer on the left is used to generate a virtual environment, then the system becomes an augmented-reality system of the second type described in the caption to Figure 2.

on such interfaces for VE systems and for teleoperator systems are not identical, there is considerable overlap. Similarly, many of the problems now facing VE designers in the area of autonomous agents (i.e., computer-generated entities with programmed-in behaviors that enable the entity to function without direct commands or supervision by the human operator) have been studied for many years by the designers of autonomous electromechanical robots.

One aspect of the subjective experience in SE systems that has received considerable attention is the extent to which the human operator loses his or her awareness of being present at the site of the interface and instead feels present in the artificial environment. This feature, often referred to under the headings of *telepresence, virtual presence,* or *synthetic presence,* is dependent on many factors, including the extent to which the interface is transparent and attenuates stimulation from the immediate environment, as well as the amounts and kinds of interaction that take place in the artificial environment.

The distinction between VE systems and simulator systems is more subtle than the distinction between VE systems and teleoperator systems. Also, there is a more or less continuous transition from simulator systems to VE systems. Generally speaking, the term *VE system* rather than *simulator system* is increasingly used as the following conditions are more fully satisfied:

(1) The system is easily reconfigurable by changes in software;

(2) the system can be used to create highly unnatural environments as well as a wide variety of natural ones;

(3) the system is highly interactive and adaptive;

(4) the system makes use of a wide variety of human sensing modalities and human sensorimotor systems; and

(5) the user becomes highly immersed in the computer-synthesized environment and experiences a strong sense of presence in the artificial environment.

It has also been suggested (Breglia, personal communication, 1992) that, whereas a simulator is most intimately tied to the given physical system with which the user is expected to interact (i.e., it is designed to simulate this physical system), a VE system is most intimately tied to the human operator (i.e., it is designed to include a general-purpose interface to match the human organism, as well as the capability for generating a large range of virtual worlds). Accordingly, it is not surprising that a large fraction of VE equipment constitutes a kind of high-tech clothing (head-mounted displays, gloves, body suits, etc.). A further suggestion (Allard, personal communication, 1994) is that simulators and VE systems differ in the extent to which the near field (i.e., the world within the user's reach) is real or simulated. A simulator ordinarily simulates only the far field and uses real physical mock-ups for the near field, whereas VEs can simulate the near field as well as the far field.

The distinction between VE systems and other types of highly flexible computer systems (e.g., conventional desktop computing systems) is based mainly on the extent to which the system is interactive, multimodal, and immersive (items 3, 4, and 5 listed above). When focused on the visual channel, the characteristics of three-dimensional rather than two-dimensional presentations, plus a large field of view, are often cited as distinguishing characteristics.

Most current systems involve visual and auditory displays; very few involve olfactory or gustatory displays (one exception is discussed in Chapter 7). Often the displays are presented by means of devices mounted on the operator's head in a *head-mounted display* system. Control signals are usually derived from the human operator's motor behavior—actions of the head, hands, feet, or speech production mechanism. The use of control signals derived from neural behavior (e.g., electroencephalogram signals) is still rare. In the case of head-mounted displays, the interface usually includes a system for monitoring head position, and the visual images displayed to the eyes and the auditory images displayed to the ears are modified in real time according to the measured head position. By monitoring head position, the visual image seen by the user can be continually modified so that, no matter how he or she moves, the objects in the virtual world remain in stable locations, just as they would in the real world. The user is given the impression that she or he is moving about in a stable world even though the stable world is created artificially. (In a teleoperator system, the position of the human's head is used to control the position of the telerobot's optical and acoustic sensors and thereby the images presented on the displays; in a VE system, the human's head position is used to control the characteristics of the synthesized visual and auditory images.)

The term *haptic interfaces* refers to interfaces involving the human hand and to manual sensing and manipulation. One common type of haptic interface currently in use is a device consisting of a glove or manual exoskeleton that monitors hand position and posture (i.e., finger joint angles). These devices, like head-position trackers, provide no feedback and are used solely for control. Other devices, such as force-reflecting joysticks, act as tool handles and serve not only a control function but also a display function because they are capable of providing force feedback. Haptic interfaces in which hand position and posture are tracked and object properties such as texture and temperature are displayed to the hand (as well as simple force information) are not yet commercially available.

The availability of force feedback is a powerful addition to a virtual environment. By sensing the position of the fingers relative to a virtual object, such as a simulated rubber ball, the system can introduce force cues as the user closes his or her hand around the virtual object. With suitable sensors and actuators, the object can be made to feel stiff or spongy by systematically manipulating the characteristics of the force cues as a function of the position and motion of the fingers relative to the position of the object. In this manner, it is possible to create haptic images of virtual objects (a further defining characteristic of VEs).

VISIONS

In this section, we attempt to provide the reader with a glimpse of the future that we foresee if SE systems continue to develop at the current rapid rate. To indicate the special nature of this discussion, presented in the form of speculative vignettes, we have used a different typography in the sections that follow.

For specificity, we have chosen to convey this picture of the future in terms of the activities of a family of two adults and two children in their home. Although many of these activities clearly require advances in certain components of the technology, we believe that most such advances will take place within the next 5 to 10 years. In cases in which there is substantial uncertainty about the achievability of a given hypothesized development, we have indicated such uncertainty by referring to it as a research project. (Perhaps the most unrealistic aspect of the picture we paint is that of a traditional nuclear family at home together: current statistics indicate both a decrease in the incidence of traditional two-parent families and an increase in the incidence of multigenerational families.) For convenience, we have chosen to focus on activities inside rather than outside the home, despite the fact that most of the activities considered in our discussion of applications in Chapter 12 take place outside the home.

The envisioned family, the Roberts family, includes a mother, Jennifer, a father, Henry, a 12-year-old daughter, Samantha, and a 16-year-old son, Peter. Henry does not appear until the end of the sketch because he is not actively involved in any SE activity; he is suffering from "SE overdose."

Finally, it should be noted that in writing this section we have not hesitated to interweave images based on assumed future technology with images based on assumed future social and psychological phenomena. We have included the latter not because we have any particular expertise in predicting such phenomena, but because we believe that technology must be considered in the light of such phenomena. It is our hope that those who follow up on this report will have the expertise appropriate to serious consideration of these issues.

Surgical Training

Jennifer Roberts, the mother, is training to become a surgeon and is at her SE station studying past heart operations.

She previously spent many hours familiarizing herself with the structure and function of the heart by working with the virtual-heart system she acquired after deciding to return to medical school and to specialize in heart surgery. This system includes a special virtual-heart computer program obtained from the National Medical Library of Physical/Computational Models of Human Body Systems and a special haptic interface that enables her to interact manually with the virtual heart. Special scientific visualization subroutines enable her to see, hear, and feel the heart (and its various component subsystems) from various vantage points and at various scales. Also, the haptic interface, which includes a special suite of surgical tool handles for use in surgical simulation (analogous to the force-feedback controls used in advanced simulations of flying or driving), enables her to practice various types of surgical operations on the heart. As part of this practice, she sometimes deliberately deviates from the recommended surgical procedures in order to observe the effects of such deviations. However, in order to prevent her medical school tutor (who has access to stored versions of these practice runs on his own SE station) from thinking that these deviations are unintentional (and therefore that she is poor material for surgical training), she always indicates her intention to deviate at the beginning of the surgical run.

*Her training also includes studying heart action in real humans by using see-through displays (augmented reality) that enables the viewer to combine normal visual images of the subject with images of the beating heart derived (in real time) from ultrasound scans. Although there are still some minor imperfections in the performance of the subsystem used to align the two types of visual images, the overall system provides the user with what many years ago (in Superman comics) was called X-*ray vision. *In this portion of her training, Jennifer examines the effect of position, respiration, exercise, and medication on heart action using both the see-through display and the traditional auditory display of heart sounds.*

Today, Jennifer is studying recordings of a number of past real heart operations that had been recorded at the Master Surgical Center in Baltimore. In all of these operations, the surgery was performed by means of a surgical teleoperator system. Such systems not only enable remote surgery to be performed, but also increase surgical precision (e.g., elimination of hand tremor) and decrease need for immobilization of the heart during surgery (the surgical telerobot is designed to track the motion of the heart and to move the scalpel along with the heart in such a way that the relative position of the scalpel and the target can be precisely controlled even when the heart is beating).

The human operator of these surgical teleoperator systems generally has access not only to real-time visual images of the heart via the telerobotic cameras employed in the system, but also to augmented-reality information derived from other forms of sensing and overlaid on the real images. Some of these other images, like the ultrasound image mentioned above, are derived in real time; others summarize information obtained at previous times and contribute to the surgeon's awareness of the patient's heart history.

All the operations performed with such telerobotic surgery systems are recorded and stored using visual, auditory, and mechanical recording and storage systems. These operations can then be replayed at any time (and the operation felt as well as seen and heard) by any individual, such as Jennifer, who has the appropriate replay equipment available. Recordings are generally labeled "master," "ordinary," and "botched," according to the quality of the operation performed. As one might expect, the American Medical Association initially objected to the recording of operations; however, they agreed to it when a system was developed that guaranteed anonymity of the surgeon and the Supreme Court ruled that patients and insurance companies would not have access to the information. This particular evening, Jennifer is examining two master double-bypass operations and one botched triple-bypass operation.

During her training time on the following day, she is going to monitor a heart operation in real time being performed by a surgeon at the Master Surgical Center in Baltimore on a patient in a rural area of Maryland roughly 200 miles away. Although substantial advances have been made in combatting problems of transport delay in remote surgery (by means of new supervisory control techniques), very few heart operations are being conducted remotely at ranges over 500 miles.

In addition to spending time on her basic surgical training, Jennifer is participating in a research project being conducted at the center that is concerned with the use of microtelerobots for the diagnosis and treatment of circulatory disorders. Microtelerobots with dimensions less than 1 nm are being designed to enter the circulatory system, make measurements of various circulatory parameters at various locations within the system, and then perform local microsurgery under remote supervisory control.

Shopping

Samantha Roberts, the 12-year-old daughter, is spending the early hours of the evening shopping for a dress via her SE station.

A week ago she underwent the periodic body scan essential for individuals whose measurements are rapidly changing. This scan was performed using the video camera associated with the SE station, a standard body-scan program provided by the shopping network to which the family had subscribed, and a special body stocking with grid lines to facilitate the automatic measurement process.

Dress material (design and fabric) is selected through the use of an interactive program in which a sequence of samples is displayed to Samantha (visually and tactually); she responds to each element in the sequence by rating the material on a scale of 1 to 10. The presentation sequence is adaptive in the sense that the choice of material to be presented on step n *in the sequence is based on the presentations and responses for steps 1 through* n – 1. *All such sequences are stored in the Samantha file of the shopping system to provide background for future marketing efforts. The universe of patterns and fabrics considered is defined by the currently available manufacturing techniques. The notion of inventory is no longer relevant because all clothing is manufactured to design specifications of individual clients after the specifications are determined. The subjective ratings supplied by Samantha to guide the presentation sequence are based on feeling the fabric by means of a special tactile display as well as seeing it by means of the standard visual display. The tactile display, an experimental component of the special SE clothing package sold by the shopping network, consists of a rectangular array of microactuators that allow one to feel different fabrics and textures by stroking the array with the hand.*

After a tentative decision is made on the material to be used, a similar process is employed to pick out a style. In this case, the ratings are based solely on visual displays—tactile images are irrelevant.

Given a tentative choice of fabric and style, the next step involves virtual modeling of the dress by Samantha herself. Given the record of Samantha's physical measurements and images in the shopping network's file, the system now synthesizes visual images of Samantha wearing the dress she picked out by her sequence of rating responses. Moreover, the actions of the synthesized Samantha model are controllable by Samantha herself by means of the shopping network's clothing-model interface (again supplied by the shopping network as part of the special shopping package sold to the family). After a modest amount of practice with the interface, Samantha is able to cause her image to perform routines similar to those she has seen professional models perform in conventional fashion shows. Initially, the shopping network's synthesis program was intentionally distorted to make the client's image appear more like his or her ideal image (derived from a "get-to-know-you" program included in the initial package) than it actually does; however, a special regulatory rule was introduced to control such distortions. In the future, Samantha may also be able to "feel" the fit of each dress. The industry realizes the importance of the sense of fit and has initiated an intensive, long-range research effort to develop the complex tactile displays required.

The cost of each of the virtual dresses considered by Samantha is presented to her as soon as the fabric and style are selected. Occasionally, Samantha scans through the fabric/style/cost matrix of the dresses she is considering in order to refresh her memory about these dresses.

By paying a special fee, it is possible for the shopper to inspect the files of other shoppers on the system. In particular, when considering the purchase of a specific dress, it is possible to call up a file that provides information on all the other shoppers in the network who purchased a similar dress. (A substantial fraction of the fees collected in this manner are paid out to the shoppers on the network in order to entice them to grant permission for such file inspection by other shoppers.)

Once a final decision is made as to the dress to be purchased, the shopper's decision is communicated to the manufacturing component of the shopping network; the selected "marketing design" is mapped into the appropriate "manufacturing design" and then manufactured in the shopping network's programmable factory—a demand-activated, computer-controlled, manufacturing system in which marketing and sales are fully integrated with design and production. Simultaneously, the funds for purchasing the dress are deducted from the shopper's account in the shopping network bank. In most cases, the dress is delivered within three days of its selection.

No returns of merchandise are permitted until the subscriber has spent a threshold amount of funds via the network; thereafter, returns are permitted, but the cost of such returns to the shopping network is factored into the cost of the shopping service.

High School Education

Peter Roberts, the 16-year-old son, is doing what previously was called school homework. The distinction between doing school work at school and doing homework for school at home has become very muddy; in both cases, much of the time is spent interacting with teachers, other students, and virtual worlds via networks of SE stations. In addition, along with the deemphasis of school as a geographic location, the distinction between school and not-school has diminished. The defining characteristics of students' experiences are the network to which they belong and the network courses or projects in which they become involved. Among the major consequences of these fundamental changes are adaptive time sched-

ules to accommodate collaboration among individuals who live in different time zones across the world and the inclusion of children and teachers who are homebound because of severe physical disabilities.

Currently, Peter is participating in four network courses: mathematics, environmental science, empathy, and dance-music-dance.

Mathematics

In the mathematics course being taken by Peter, the SE facilities are used to provide course participants with an intuitive understanding of various non-Euclidean geometries. Participants in the course enter into virtual worlds in which the properties of the space are determined by the axioms of the particular geometry being studied. These properties are explored not only by virtually traveling through the space, but also by building virtual towers, bridges, and houses within the space. The effects of changing the axioms of the geometry, which the students are encouraged to explore, are immediately realized in terms of the virtual world structure. When the changes lead to axiomatic systems that are internally inconsistent, the space "blows up" and a tombstone appears with an inscription describing the inconsistency.

In the special term project Peter has selected (each student is required to teach some aspect of mathematics to younger children), he is designing a method for showing younger children how speed is represented by the slope of the distance-versus-time graph (all in Euclidean space). Peter's basic idea is to construct a virtual train that will move across a virtual horizontal track at variable speed and to present associated graphs in virtual graph space. The first graph will plot the distance the train goes from its initial position in the train station as a function of time, and the second the instantaneous speed as a function of time. In addition to augmenting the graphs with a conventional clock face icon, Peter plans to display the tangent to the first course at the current time on the first graph continuously as the curve evolves over time, using the same color for plotting this tangent line on the first graph and for plotting the speed curve on the second graph.

Environmental Science

In this course, Peter is participating in three projects, each of which is led by a professional meteorologist in Indiana who is donating four hours per week to the course via the network. The first project focuses on the gathering of information on atmospheric conditions around the world by means of atmospheric measurement kits located in the homes of all the students taking the network course around the world, entering this information into the network computer assigned to the course, and then studying the atmospheric condition displays generated by the system. Inasmuch as each measurement kit records not only temperature, pressure, and humidity, but also certain molecular constituents of the sampled air, there are many parameters that have to be represented in the display. Some of the course participants are comparing current conditions with those predicted by the model developed by the meteorologist in Indiana. Others, including Peter, are working on improved methods for displaying and interacting with the empirical information, the model's predictions, and the deviations between the two.

In the second project, which the meteorologist introduced by explaining the concept of a microclimate, Peter and his student collaborators are studying a hypothetical environmental accident in Birmingham, Alabama. The specific question being addressed by the students on this day is the following: If a hypothetical accident released 8,000 kg of chlorine gas into the air at the Vulcan Tower in Birmingham at 9:00 a.m. today, what portions of the city should be evacuated? A geographic profile of Birmingham's topography was made available to the network course, and virtual sensors indicating current temperature, precipitation, barometric pressure, and wind velocity were distributed throughout virtual Birmingham. Based on the readings from these sensors and information provided by the meteorologist on how chlorine gas dissipates by bonding with various materials, estimates are being made of the chlorine-content contours associated with the chlorine gas cloud as a function of time. These estimates, combined with information on human tolerance to chlorine gas and on the capacity of various transportation facilities in Birmingham, are being used to construct evacuation plans.

In the third project, the students are learning about the formation and behavior of tornadoes. The meteorologist provided a computational model of tornadoes to the network school and the students are learning about tornadoes by virtually locating themselves at different positions in the tornado and also observing how the velocity vector at different points of the tornado varies as a function of the values assigned to the parameters of the computational model. Production of the visual images seen by the students results from a blending of visual images generated by the computational model and images derived from video recordings of real tornadoes. Appreciation of the forces associated with the tornado is facilitated by providing the students with an ability to place different virtual objects (people, cars, houses, etc.) in the path of the tornado and observing the effects of the tornado on these objects.

Empathy

This course was developed by a multidisciplinary team of physical scientists, biologists, anthropologists, sociologists, and psychologists.

The goal of the course is to familiarize the network students not only with the behavior of different kinds of people, animals, and physical system, but also with how it feels to participate in these other worlds. Many of the techniques used in this course are refinements of techniques previously developed in connection with interactive virtual environment theater, and many of the supporting personnel for the course are college students participating in various types of internship programs.

Each student visiting a virtual world is assigned a virtual actor in this world and must learn to control this actor in a manner that satisfies the constraints designed into the specific scenario considered. In general, these constraints are used to give the participants experience in living in different physical environments (e.g., in the desert, in the arctic, on the moon); in different social or anthropological settings (e.g., as a member of an ancient culture, a highly discriminated-against minority, a person with severe physical disabilities; or even as a member of a different animal species (e.g., as a member of an insect society or as a sea-dwelling creature low down on the food chain). The role assignments are typically a month in duration and students are expected to refine their understanding of the assigned role to the point at which other observers cannot distinguish the behavior of the characters controlled by the students from the "natives" (played by trained personnel or by highly developed computer-controlled autonomous agents).

In order to make the simulations employed in this course practical for real-time use by the students, they are greatly simplified. In most cases, such as those concerned with other physical environments and other animal species, these simplifications are readily accepted. However, in some cases, particularly those that focus on human social issues, such simplifications have occasionally been seen as offensive stereotypes and strongly resented. Thus, these portions of the courses have become very controversial and have led to considerable turmoil.

Dance-Music-Dance!

Dance-music-dance is a new course introduced by a research associate in the arts department of the network. She is a professional dancer who has become highly skilled in the use of the human body as a musical instrument and in composing musical compositions by means of dance routines. Her main work during the past year had been concerned with evaluating different mappings from the outputs of the body tracker she has been using (an optical system that does not encumber her body or interfere with her dance movements) into the control parameters used for generating sounds via the computer-music system at her disposal. Recently, she has become interested in the relations between the dance routines used to generate the music and the dance routines the resultant music inspires in other dancers listening to the music. Similarly, she is studying the relations between the initial music generated by her own dance routines and the secondary music generated by the other dancers. In the network course she has constructed, students select dance-to-music mappings to be used, choreograph their own dance routines and thereby compose their own music, arrange for the other students to dance to this music, and then do an analytic study of the above-described relationships. Peter, who believes himself to be rather clumsy and is rather inhibited about performing a full free-body dance, choreographed his initial dance piece using only the index finger of his right hand.

Network Telemeetings

Later in the evening, Jennifer, Samantha, and Peter participate in the weekly network telemeeting. The main focus for this particular meeting is a discussion with the network candidate for Congress. Except for a few minor functions, the members of Congress now represent networks rather than geographic regions such as states. Whereas the last portion of the meeting is intended to accommodate free-ranging discussion, the first portion is structured to cover four specific sets of issues.

The first set concerns the cost structure of network participation, the extent to which members of different networks are becoming isolated from each other and incapable of communicating across network boundaries, and the increasing problem of "ghetto networks."

The second set of issues concerns the rapid rise of network gambling. The amounts of money involved in this activity now exceed that involved in medical care and education combined, and gambling taxes now exceed income taxes. The principal issue of concern is how the tax money collected should be split between the federal government and the network.

The third set of issues focuses on the creation of appropriate laws for governing behavior within virtual environments. The number of cases in which VE crimes and misdemeanors are occurring is increasing, and no significant body of law is available to handle these cases. Also, problems are arising that involve crossing the VE boundary, i.e., unlawful acts being committed in the real world in response to injuries suffered in VEs. Thus, for example, when a virtual pet salamander who was being guided by a man's son was deliberately stepped on by a virtual actor being guided by a stranger on the network, the man located the home of the stranger and went over to his home and shot him.

The fourth set of issues concerns the use of the "information highway" and SE stations for purposes of lovemaking at a distance. Although the use of these facilities for this purpose was clearly predictable (e.g., based on the use of the mail for transmitting love letters, the use of the telephone for vocal lovemaking, and the use of interactive video for including visual images), it nevertheless is the center of considerable controversy. Of particular concern is the commercial introduction of special devices that contain sensors and actuators to facilitate tactual lovemaking at a distance ("telesex kits"). The inclusion of the tactual channel, and the associated increased blurring between "direct sex" and "telesex," has caused a number of attitudinal tolerance thresholds to be exceeded. For example, in addition to the usual strong feelings evidenced by different groups in society about who should have what kinds of sexual relations with whom under what circumstances, and about which kinds of commercial exploitation of sex should be allowed and which kinds disallowed, the legal community is now concerned about having to handle cases in which sexual relations are conducted across the boundaries of states with different laws governing sexual behavior. Similarly, on the basis of a study in France that documented a decrease in sexually transmitted diseases associated with the increased use of telesex, the Center for the Control of Communicable Diseases is now considering adding the use of telesex kits to the use of condoms as an important means for controlling the still-exploding incidence of the AIDS virus.

Two kinds of facilities are available for participating in these network telemeetings. With the first kind, each individual in the family wears a head-mounted display and works through his or her own SE station; in the second, rather than using the traditional head-mounted displays, the family sits together in a special room outfitted with a wall-sized visual display, a set of acoustical loudspeakers, and a set of video cameras. Each individual is assigned one of the video cameras, which then tracks that individual as he or she moves around the room. The Roberts family purchased both kinds of facilities because they believed that neither one alone is adequate for all purposes.

SE Overdose

Henry Roberts, the father, is not participating in the telemeeting because he is suffering from SE overdose. Although problems associated with conventional simulator sickness were brought under control years ago, a set of deeper problems emerged as individuals began to spend substantial portions of their waking hours living in synthetic environments.

One aspect of these problems is evidenced in the choices Henry makes about how to spend his leisure time. Initially, he spent much of his leisure time playing SE games, taking SE trips to foreign lands and planets (both real and imaginary), participating in interactive SE theater, and exploring the world from the viewpoint of different types of creatures (again, real and imaginary). One of his favorite activities had been to interact with real bees inside a real beehive using a telerobotic model bee system. (A whole variety of such telerobotic model animal systems was developed in connection with scientific study of animal behavior at the Tinbergen-Lorenz Institute in Munich.) Now, however, he wants to get away from all this "electromechanically mediated stuff" and interact with the world directly. Accordingly, he spent his last three vacations in the surrounding mountains camping out with some friends who are having similar problems. Air quality has improved substantially with the introduction of SE systems because of the associated reduction in auto travel. The fact that Henry's desire to get back to nature is rather common is evidenced by the enormous growth taking place in the camping equipment industry.

Henry has also increased the amount of time he spent exercising in the real world. When he first acquired his SE station, he made extensive use of SE jogging (which involved the use of a six-degrees-of-freedom treadmill and synthetic scenery) and SE golf; however, as part of his reaction to too much SE, he has switched back to the "real thing." Also, he refuses to join the political movement concerned with the large amounts of energy being wasted by the exercise mania that has swept the country and with finding a practical scheme for capturing, storing, and making use of this energy.

Henry is also undergoing therapy in connection with his SE experiences. Although many aspects of these experiences seem disturbing to Henry and his therapist, one aspect of central importance concerns Henry's body image, his sense of presence, and his underlying identity. Apparently, the ease with which Henry is able to transform himself into other creatures in other environments, and become realistically immersed in these other roles and other worlds, is becoming a real psychological problem for him.

Ordinarily, the therapist administers treatment to his patients via an SE network that incorporates a biofeedback mode. However, in cases such as Henry's (loss of presence and loss of identity due to SE overdose), the use of an SE system in the therapeutic procedure seems unwise.

Another factor of critical importance to Henry's mental state concerns the articles he has been reading about research on human-machine interfaces that are designed to tap directly into the human's neural system. Although he fully understands the advantages of such interfaces for individuals with severe sensorimotor disabilities, the idea of them makes him uneasy.

Henry's job as an architect adds an additional important dimension to his mental state. Initially, SE played only a supporting role in his work; it was used merely as a tool for design or as a tool for marketing to the client. Recently, however, the company for which Henry works received some large contracts to design virtual spaces for use in virtual worlds. Apparently, the large amounts of time now being spent in virtual spaces, combined with the limitations of computer scientists in their abilities to design virtual spaces that are not only functional but also aesthetically pleasing, are leading to a new market for architectural firms. However, this new market is of no interest to Henry; in fact, it increases his desire to switch fields. Unfortunately, when he scans the job opening information available to him on his SE network, he finds that the most common type of opening involving interaction with the real world concerns the installation and maintenance of SE systems.

CURRENT STATE OF THE SE FIELD

Although some of the technologies assumed in our visions of the future are already available and others are the subject of current research, these visions are without doubt visions of the future, not the present. In this section, we briefly depict the current state of the SE field. We begin by describing the application areas that are currently receiving the most attention. We then discuss a number of topics in the field of psychology relevant to the design, use, and evaluation of SE systems and the human component of these systems. Next we summarize the status of the associated technologies that make SEs possible: the interfaces used to link the machine and the human operator, the computer hardware and software used to generate VEs, the telerobots used in teleoperator systems, and the communication networks used to integrate multiple SE systems. The section ends with a brief assessment of the SE evaluation efforts to date. More detailed information on most of these topics can be found in the chapters of the report.

Application Domains of SE Systems

The range of potential applications for SE systems is extremely large. Application domains currently receiving considerable attention include: (1) entertainment, (2) national defense, (3) design, manufacturing, and marketing, (4) medicine and health care, (5) hazardous operations, (6) training, (7) education, (8) information visualization, and (9) telecommunication and teletravel.

The entertainment domain is serving both as a massive informal test bed and as a major economic driving force for the development of new VE technology. Although some of this technology can be expensive (particularly that associated with the entertainment of large groups), on the whole the VE technology associated with the entertainment industry is "low end." For example, the head-mounted displays being used for entertainment purposes are—as they would have to be to make the enterprise commercially viable—orders of magnitude less expensive (and correspondingly less capable) than those being developed for military purposes. Even though applications in the entertainment domain are still in their infancy, they are by far the most widely implemented of all VE applications. In essentially all of the other domains, the activities are in the stage of research and development rather than commercial application or practical use. Not only is there much to be learned about how best to utilize SE technology, but also the cost-effectiveness of most current SE technology (i.e., the bang for the buck) is inadequate for any application domain other than that of entertainment.

The national defense domain, like the entertainment domain, constitutes both a major test bed and a major driving force for VE technology. It differs in that (with the exception of the use of traditional simulator systems) research and development activities still dominate, the associated technology tends to be high end rather than low end, the systems of interest include teleoperators as well as VEs, and the networking of large numbers of active participants is emphasized.

This report discusses neither the entertainment domain nor the national defense domain in detail. The former domain is omitted because it is already receiving extensive commercial support, many of the scientific and technical research issues that arise in this domain also arise in other domains, and improved entertainment technology does not appear to us as one of society's most pressing needs. The latter domain is omitted because it is currently receiving substantial attention within the government (e.g., Thorpe, 1993), significant information may be classified for security reasons and therefore inaccessible, and, again, many of the research issues that arise in this domain also arise in others. This last reason is especially relevant to the national defense domain because so many of the other domains considered, such as training, information visualization, and telecommunication and teletravel, are directly relevant to national defense.

Finally, although not included here as a formal application domain, VE systems are beginning to be envisioned as highly desirable facilities for research groups concerned with experimental psychology. Clearly, not only is knowledge and understanding of psychological phenomena essential for efficient design and productive use of VE systems, but also a high-quality VE system that makes available a wide variety of precisely controlled stimuli, response measurements, and adaptive testing procedures constitutes an ideal tool for conducting research in experimental psychology.

In the following subsections, we discuss briefly the other application domains listed above. As indicated in these subsections, significant research and development is taking place in a wide variety of applications and, in a few cases, the results of these efforts are beginning to be applied on at least an experimental basis. It is not yet clear, however, how to choose the tasks that will eventually prove most appropriate for the application of SE technology. Not only are the results obtained in the various application domains still too meager to allow one to specify the nature of such tasks from empirical data, but also there is no evidence that much effort has been given to answering the question "What is SE technology good for?" theoretically.

Individuals with computer graphics backgrounds usually point to tasks involving three-dimensional spatial information and to immersion in three-dimensional space; those focused on multimodal interactive interfaces often point to tasks that depend strongly on sensorimotor involvement. In any case, in order to fully specify what SE is good for, one must estimate the cost-effectiveness of the envisioned SE system both compared with the way in which the same task is now being performed and compared with alternative new systems that could be developed (e.g., that might achieve equivalent task performance at substantially reduced cost). Eventually, of course, in addition to comparative cost-effective estimates to help select tasks for which SE systems are likely to be appropriate, one must evaluate such systems once they are developed. The important and often neglected topic of SE evaluation is considered further in Chapter 11.

Design, Manufacturing, and Marketing

Design, manufacturing, and marketing are generally recognized as a major application domain for SE technology, and it is currently receiving substantial attention. Although much of the activity in this domain is still in the development phase, it is clearly in the process of moving to actual usage. The procedures and technologies used for design are progressing from those associated with conventional computer-assisted operations to those involving VE and augmented-reality systems. Similarly, we are beginning to see at least experimental use of VE in the marketing of products and services. It appears that it will not be many years before design, manufacturing, and marketing will all take place within a unified system that makes substantial use of SE technology.

Independent of whether the item to be sold is a haircut, a kitchen, or an office building, the ability of the client to see and interact with realistic representations of a variety of possible versions or realizations of the item can positively influence both the evolution of the design and the attitude of the client. Furthermore, when very complex and expensive systems, such as an aircraft or submarine, are being designed, the potential for cost savings by using virtual mock-ups and prototypes rather than real physical ones is enormous.

Medicine and Health Care

Medicine and health care, like design, manufacturing, and marketing, are considered to be a major SE application domain. Although much of the work is still at the experimental stage, applications of both VE technology and teleoperator technology are being pursued very actively.

In addition to developing improved communication networks for providing the right medical information to the right place at the right time, much of the current research is directed toward improved methods for diagnosis; planning of treatment; provision of information to the patient; provision of treatment; and training of medical personnel. VE systems are being developed and studied experimentally to extend conventional consultations and telediagnosis performed over the telephone to include interactive visual displays of both participants and medical information. Such systems are also being studied for use in planning surgical procedures and in helping to increase patients' awareness and understanding of these procedures and of the possible outcomes. Augmented-reality systems are being studied to present visual displays in which information previously obtained from special imaging techniques is overlaid on the normal direct view of the patient; integrated VE and teleoperator systems are being developed for use in telediagnosis and telesurgery and for the training of surgeons. In general, the potential benefits of telemedicine that are being considered include not only the ability to obtain medical information and perform medical actions at a distance, but also the ability, as in any other application of teleoperation, to effectively transform the sensorimotor system of the operator to better match the task at hand. The rapidly increasing use of laparoscopic surgical procedures illustrates the importance of these other benefits.

Aside from the efforts required to realize technology that is adequate for the various medical applications, substantial research is being initiated to realize adequate physically based models of the human body (e.g., for VE training of surgeons). However, current success in creating virtual human skeletons, organs, and physiological subsystems constitutes only a tiny fraction of what needs to be achieved over the long term.

Additional health-related research and development activities in the SE area are taking place in connection with physical rehabilitation. Individuals with sensory or motor disabilities constitute a uniquely challenging domain for application of SE systems with specially designed human-machine interfaces (e.g., gestural tracking and recognition devices for individuals who have lost both the ability to articulate speech and the manual dexterity required to operate a keyboard). The application of SE technology to psychological rehabilitation (for example, to reduce phobic reactions) is also beginning to receive attention.

Hazardous Operations

One of the driving forces for the creation and development of teleoperator systems has been the need to perform operations that are hazardous, and the application of SE systems to this domain is certainly one of the older applications areas considered. Thus, unlike the situation in some of the other domains, current activities in this domain include actual use as well as research and development. Among the specific applications in this domain that are receiving attention are the handling of dangerous materials, operating heavy machinery, firefighting and policing, conducting military operations, and exploring the ocean floor and outer space.

Despite the potential benefits that can be obtained by using teleoperator systems in many of these areas, and despite the benefits that have already been demonstrated in some of them (e.g., handling nuclear materials, undersea exploration) neither the government nor the public has evidenced great enthusiasm about this domain. Aside from the general lack of excitement engendered by visions of teleoperation compared with visions of virtual reality, perhaps interest in the use of teleoperator systems for hazardous operations is limited by the lack of personal experience most people have with hazardous operations (e.g., defusing a bomb or locating and carrying to safety a child from a burning building). It is even conceivable that the use of teleoperation for hazardous operations may lack support from potential operators because it is inconsistant with a macho self-image.

Although for many applications in this area further research and development is required to achieve teleoperator systems that are both reliable and cost-effective, there is no evidence that such goals cannot be achieved. Also, and quite apart from the use of teleoperator systems in conducting hazardous operations, substantial opportunity for the application of SE systems in this area arises in connection with the training of individuals to conduct hazardous operations (with or without the use of teleoperation). As discussed in the next subsection, the use of VE systems for training constitutes a major application domain for such systems.

Training

Because most activities require at least some training, it is not surprising that the use of VE technology for training is a major application area in almost all domains considered. Thus, for example, it is of major interest for national defense, medicine and health care, and hazardous operations, among others.

On one hand, the use of simulators for training is quite extensive. Simulations of various types have been used for a long time and, judging by their continued use, are relatively cost-effective (although appropriate analyses have rarely been performed). The apparently successful results obtained with simulators in training various tasks (e.g., flying an airplane) constitute a major motivation for interest in the exploitation of VE technology for training.

On the other hand, the extent to which current VE technology is actually being used in the training area is very limited: essentially all current work on VE training is at the stage of research and development. Given the existing background in the use of simulation for training, it is clear that one of the factors responsible for this situation is the inadequacy of the currently available VE technology. However, that is not the only problem; others relate to our inadequate understanding of basic psychological issues related to training and training transfer. The flexibility inherent in the use of VE systems for training and, in particular, the opportunity to create learning situations that are superior to those that are realizable without such systems (e.g., by the use of special multisensory instructional cues, by purposefully distorting the real situation, by providing multiple viewpoints and various levels of abstraction, and by adapting the system automatically to the individual and the individual's state of training) seriously challenge our basic understanding of the learning process. Of particular concern is the issue of training transfer. Much remains to be known about which of the possible differences between the real task and the task as realized in the envisioned VE training systems are likely to be important, either positively or negatively, and which are insignificant.

Education

Although the term *education* can be used very broadly to cover almost any situation in which learning takes place, in this report we use the term to refer to the goals and activities normally associated with K-12 education in schools.

One major set of applications currently being explored in this domain focuses on the communication component of SE networks. Examples include communication between students, between teachers, and between students and teachers at different sites; televisits to places of interest that would normally involve costly travel (to explore another culture, to learn a foreign language, to visit a site in outer space or under the ocean); and even teleoperation of remote telerobots. Other applications focus more on the use of VEs as immersive, interactive, experimental, and play facilities. At one extreme, a VE can be used to present a well-defined situation in a highly structured course. At another extreme, a VE system can be used to encourage free play and various types of model building, or even the construction of virtual environment tools.

As in the training domain, much of the current work in the education domain is being directed toward research to determine the ways in which technology can be usefully applied.

Despite the potential of SE technology to provide cost-effective improvements in K-12 education, many people judge societal infrastructure problems in education to be so overwhelming that attempts to exploit SE technology within the current education system would have only marginal benefits. The history of attempts to introduce computers into the classroom is cited as an example of useful technology being available but not well used. In general it is believed that, unless the infrastructure surrounding the education system is radically changed, the best opportunity for using SE technology to help educate children is likely to occur through the entertainment industry and the entertainment facilities that will be available in many homes (leading to a new meaning for the phrase *dual use* that is now being so frequently used in government circles). It is also conceivable, of course, that SE technology, together with associated networking features, can play an important role in helping to change the infrastructure.

Information Visualization

The dependence of our culture on information and the amount of information that one needs to perceive, digest, understand, and act on are steadily increasing. In attempts to prevent information overload or, alternatively, to prevent ignoring information that is vital to action, research is being conducted to determine methods of information visualization that are superior to those now used. (The term *visualization* is used here in its most general sense and only for historical purposes; we do not mean to imply that the information is necessarily presented only through the visual channel.) This application domain, like the training domain, cuts across the other application domains considered: effective visualization of information is important in essentially all domains.

In general, the problem of information visualization is an extremely old one. Inventive pictorial representations of important events go back to cave paintings. Descartes' invention of analytic geometry and the associated use of graphs to represent tables of numbers constitutes a truly major advance in this area. Less dramatic, but more technologically relevant advances have taken place in the area of computer graphics. The extension of two-dimensional to three-dimensional graphics and controlled manipulation of scale and viewpoint are illustrative examples of these advances. Unfortunately, there is relatively little guiding theory available, and relatively little systematic evaluation has been performed to determine the benefits of these new graphics techniques (although they are clearly commercially successful). Furthermore, the use of modalities other than vision and exploration of the benefits of different kinds of sensorimotor involvement in understanding information are only now beginning to be seriously considered. Perhaps the most advanced application area in this domain concerns the visualization of scientific information. Scientific visualization is generally recognized as a major and growing application of VE and is starting to receive substantial attention. Specialists in various fields of science are beginning to make use of advanced computer graphics techniques for improved visualization, and preliminary research and development are being conducted on the use of the auditory and haptic channels for this purpose.

Telecommunication and Teletravel

The domain of telecommunication and teletravel, like training and information visualization, cuts across essentially all of the other domains considered. Telecommunication, which is intended to include teleconferencing as a special case, enables two or more people at disparate locations to interact in any manner permitted by the technology and chosen by the participants. Although sophisticated telecommunication systems involving multiple participants and real-time video as well as audio have been envisioned for many years, their use is still primarily experimental. Currently, only phone, facsimile transmission, and electronic mail are being extensively used by large groups of people. The incorporation of a real-time haptic channel into the telecommunications network by means of which individuals can communicate tactually is only beginning to be considered. Clearly, the ability to hold conferences with people at different locations in a manner that closely approximates real conferences (i.e., that permits one to see all the participants interacting together in a room, to focus one's attention arbitrarily on any member of the group, to direct remarks to specific individuals in the group by directing one's gaze toward that individual, etc.) will require substantial advances in the application of SE ideas to the telecommunications domain.

Teletravel, which would allow individuals to effectively visit remote locations for purposes of work or pleasure, also has not advanced very far; only the phone and noninteractive video are commonly available. The use of teleoperation to facilitate active exploration and work at distant locations still appears to be confined to the domain of hazardous operations and to experimental work in the domain of health care. Little consideration is being given to the use of teleoperation to enable individuals who are homebound (e.g., because of physical disabilities or because they are incarcerated) to work at jobs that are located elsewhere and that require substantial interactive sensing and manipulation.

Finally, advances in the technology for telecommunications and teletravel are being accompanied by new ventures in the sex-for-profit business. Unless appropriate societal strictures are imposed, it appears that interactive live audio-video will be heavily exploited for this purpose within the next 2 to 3 years, with equivalent exploitation involving the tactual channel occurring within the next 5 to 10 years.

Some Psychological Considerations

By definition, human operators constitute a major component of all SE systems. Furthermore, the range of experiences to which the operator is subjected in these systems can be extremely broad. Thus, there are very few topics concerning human behavior (sensorimotor performance, perception, cognition, etc.) that are not relevant to the design, use, and evaluation of SE systems. A number of the modality-specific topics in this general area are discussed in the section below in connection with human-machine interfaces; some of the more general ones are considered here.

One set of such topics focuses on human performance characteristics and includes, for example, sensorimotor resolution, perceptual illusions, information transfer rates, and manual tracking. Knowledge about all of these topics is essential to cost-effective design of SE systems. For example, the limits on human sensory resolution place an upper bound on the resolution required in sensory displays. Similarly, unintentional variability (noise) in motor responses puts an upper bound on the resolution required in control devices. Both sensory input and motor output characteristics are included in human operator models used to interpret performance in various types of manual tracking tasks. And information transfer rates help characterize the operator's ability to receive information via displays, process information centrally, and transmit information via controls. Perceptual illusions can be used to simplify (and thereby reduce the cost of) stimulus generation procedures. If not thought about in advance, however, they can also lead to unexpected failures in performance.

A second set of such factors arises in connection with the alterations in sensorimotor loops that occur when a human operator "drives" an SE system, and the extent to which and the manner in which the human operator adapts to such alterations as a function of his or her experience with the alteration. Consider, for example, the case in which the operator is viewing a virtual image of his or her hand as his or her real hand moves through space. The alteration may be caused by defects in the technology and can involve a spatial discrepancy between the position of the seen hand and the position of the kinesthetically sensed (felt) hand, a time delay between the felt motion and the seen motion, or a statistical decorrelation between the seen and felt hand positions (e.g., due to noise in the visual display channel). Similar examples are often encountered in the sensorimotor loop involving the visual (or auditory) scene presented by a head-mounted display and the kinesthetically felt orientation of the viewer's (listener's) head being tracked by an appropriate sensor mounted on the head to control the displayed scene.

Although alterations involving time delays or noise generally have negative effects and are therefore to be avoided in designing and constructing SE systems, alterations involving fixed transformations or distortions may be introduced intentionally to enhance performance. For example, such conditions arise automatically in any teleoperator system that employs a nonanthropomorphic telerobot (e.g., a telerobot that has four eyes, six arms, and moves about on wheels). Because the telerobot and human operator in such cases are nonisomorphic, a special (unnatural) mapping must be employed to relate the sensing and motor actions performed by the telerobot to the sensing and motor actions performed by the human operator. Similar issues arise when sensory substitution is used (e.g., auditory signals are used to represent force feedback) or when VE interfaces are designed to achieve supernormal resolution by magnification (e.g, by simulating increased distance between the eyes to achieve improved visual depth perception or increased distance between the ears to achieve improved auditory localization).

Additional significant issues that arise in connection with such alterations concern the role these alterations play in eliciting the sopite syndrome (chronic fatigue, lethargy, drowsiness, nausea, etc.) and reducing the subjective sense of telepresence, and the extent to which subjects can adapt to such alterations. Unfortunately, there are as yet no adequate theoretical models for enabling SE system designers to predict how subjects adapt to such alterations as a function of the alteration and the kinds and magnitudes of exposure to the alteration. Issues related to interactive effects among multiple alterations (e.g., a distortion plus a time delay plus some jitter in time or space) have hardly begun to be considered.

A third set of topics of major importance in this general area concerns the development of appropriate cognitive models. In the design of systems, like VE systems, in which the goal is to alter the human operator, understanding the human mental processes involved in knowledge acquisition and knowledge organization, and the application of knowledge to tasks, such as concept understanding, problem solving, decision making, and skill mastery, is critical. Of particular importance is the organization, sequence, amount, and pace of information presented. If the information characteristics of the system correspond to the cognitive processing features of the individual, then the system will be more effective in facilitating training or education and enhancing task performance.

Cognitive scientists have been working for many years to describe the processes used by humans to acquire and build knowledge structures. This work has led to a variety of hypotheses and some research results suggesting that knowledge acquisition strategies and features of effective information presentation depend on a person's level of knowledge, the characteristics of the content area, and the type of task performance required.

Although significant progress has been made and strong research efforts by cognitive scientists will undoubtedly continue, a large number of questions remain about the compatibility between types of tasks and preferred features of information presentation. For example, should novices be presented with different visualizations of scientific data than experts? If so, what features should be different? What is the relationship between image fidelity, amount of information presented, and knowledge or skill acquisition for different types of tasks? And how do we facilitate transfer from a training task to an operational task? These questions are not new, but efforts to date have only begun to develop preliminary answers. In general, we do not yet understand the relationship between information presentation in an immersive SE environment and learning and performance by the SE user.

A fourth set of topics in this area concern what might be called cognitive side effects. Quite apart from how various features of the SE influence the learning and performance exhibited by the user with respect to the specific task of interest, it is important to know how these features influence other aspects of the user's cognitive structure and behavior. For example, might not extensive use of certain kinds of SE systems significantly decrease the user's sense of presence in his or her usual environment or alter the mental model held of his or her own body? At present, experience with truly immersive SE systems is too limited to provide reliable answers to such questions. Another set of questions involves how experiences in SEs might affect an individual's attitudes toward such social behaviors as violence, sex, and fantasy role playing. There is at least some anecdotal evidence of a connection between aggressive behavior in children and playing violent video games. In addition, there have been several cases of individuals who were reportedly so completely drawn into computer role-playing games that they devoted all their time to them. It is possible that experience in immersive SEs could have even greater impact; however, the current experience with SE systems is too limited to allow one to draw any conclusions in this area.

Current State of SE Technology

Generally speaking, at the time of this writing, a substantial gap exists between the SE technology that is commercially available and the SE technology that is needed to realize the potential envisioned in the various application domains. Even the demonstrations of what are considered advanced SE research systems that can be seen at various universities, military installations, and industrial laboratories sometimes leave technically sophisticated observers who have no vested interest in the technology unimpressed.

There are, of course, important exceptions to this relatively negative assessment. Current VE technology is certainly adequate to be used for some applications in the entertainment domain. Similarly, current teleoperator technology is adequate to be used for some applications in the hazardous operations domain. Also, newly developed SE technology of various types (including that associated with augmented reality) is beginning to be applied in the domains of design, manufacturing, and marketing; medicine and health care; and information visualization. Nevertheless, it is clear that the development of significantly improved technology is a major requirement for most SE applications to be truly successful.

In the following sections, we summarize the current state of technology in the areas of human-machine interfaces, computer generation of VEs, telerobotics, and networks. As in the body of the report, the material on human-machine interfaces and networks has been separated from the material on computer generation of VEs and telerobotics because it is generally applicable to both VE and teleoperator systems. Furthermore, the section on human-machine interfaces covers the visual, auditory, and haptic channels, whereas the section on the computer generation of VEs covers the visual channel only. This asymmetry arises because the overwhelming majority of previous work on computer generation of VEs has been restricted to the visual channel. As the field matures and the computer science community becomes more involved in the generation of auditory and haptic images as well as visual images and the community concerned with the auditory and haptic channels becomes more involved with computer synthesis of environmental signals and objects, this imbalance will become less severe.

Human-Machine Interfaces

The *human-machine interface* in SE systems consists of all devices used to present information to the human operator and to sense the actions and responses of the human operator that control the machine in question. Although the problems associated with human-machine interfaces differ to some extent according to whether the system is a VE system or a teleoperator system (i.e., whether the given machine is a computer or telerobot), there is clearly substantial overlap between these two sets of problems. In the following sections, we briefly discuss interface issues for the visual channel, the auditory channel, and the haptic channel. In addition, we consider position tracking and mapping, motion interfaces, speech communication, physiological responses, and a display system that presents information by means of odors and radiant heat.

Visual Channel Of all the devices associated with the human-machine interface component of VE systems, visual displays have received the greatest attention. In addition to continuing efforts directed toward the general use of displays (home television, scientific research, etc.), substantial efforts have been directed toward the development of visual displays specifically for SE systems.

The visual displays currently available for SE use include both head-mounted displays (HMDs) and off-head displays (OHDs). HMDs, which also include devices for presenting auditory signals (earphones) and for measuring the position and orientation of the head (head trackers), would be ideally suited to the SE field; however, they still suffer from information loss (poor resolution, limited field of view) as well as a variety of ergonomic problems, including excessive weight and poor fit (both mechanically and optically). They may also cause the user to experience the sopite syndrome.

In addition to HMDs in which all of the visual images are computer-generated, HMDs are being developed in which the computer-generated images are combined with directly sensed environmental images (see-through displays) or with environmental images sensed and possibly transformed by a telerobotic optical sensing system. To date such augmented-reality systems have not been much in demand by the entertainment industry and therefore remain largely in the research domain.

Low-end HMDs, the development of which has been driven mainly by the entertainment industry, can be obtained for less than $10,000; high-end HMDs, the development of which has been supported mainly by the military, can cost as much as $1 million. Although the HMD area has been and continues to be extremely active, and although there is a wide range of HMDs now available and a substantial number of research projects exploring new technologies for use in HMDs, all of the HMDs now available have major drawbacks. In fact, given the current drawbacks, we think it is extremely unlikely that any individual would choose to wear an HMD on a regular basis (e.g., 40 hr/week) without special incentives. At present, and for some years to come, OHDs will provide significantly better performance than HMDs for most tasks in most application domains. One intermediate technology currently available is lightweight stereographic glasses and desktop stereo display screens. Another includes a display mounted on a boom that can be moved about (its position and orientation appropriately tracked) manually. Many other types of OHDs, some of which involve tracking and some of which do not, are discussed in the chapter on visual displays.

To a large extent, the design of the visual displays being used in the SE field takes little account of the dual structures found in the visual system (foveal versus peripheral vision, focal versus ambient systems, etc.). Some research has been done on the use of a high-resolution inset and the tracking of gaze direction to locate the inset within the field so that it is always at the right place for stimulating the fovea. However, the design and use of such systems, which reduce the computational requirements associated with presenting continuously varying complex scenes at the cost of the complexities associated with continuous eye tracking, are still mainly in the experimentation phase.

An important set of issues concerning perceptual effects in the visual channel that are only now beginning to be addressed concern the abovementioned augmented-reality displays. Not only is relatively little known about the detailed perceptual effects of *misregistration* (misalignment) of visual images, but there are few guidelines available to help designers choose how to merge real and synthesized information (a problem with a strong cognitive component) even when there is no detailed registration problem.

Many of the perceptual issues that have important implications for the design of technology for the visual channel, and about which current knowledge is inadequate, concern how humans respond to various types of sensorimotor alterations associated with the visual display. As mentioned earlier, such alterations can result from the intentional introduction of distortions to achieve superior performance (e.g., simulating greater interocular distance to achieve improved depth perception, using a telerobot with a nonanthropomorphic optical sensory system), from unintentional optical distortions in an HMD, or from time delays and noise generated somewhere within the visual channel. Also, we still know relatively little about how various characteristics of the visual display influence performance on various types of tasks. Although some deficiencies, like those that are likely to induce the sopite syndrome, may be more or less task independent, other deficiencies are likely to depend strongly on the task.

Auditory Channel Unlike the situation for the visual channel, currently available hardware for the auditory component of HMDs is adequate for essentially all SE applications. Earphones to present desired signals and "ear defenders" (active or passive) to attenuate unwanted signals from the immediate real environment are effective, inexpensive, and ergonomically reasonable. (The main ergonomic problem occurs when earphones and visual displays have to be used together and the design of the helmet that includes the visual display does not take proper account of the need to also stimulate the auditory channel.) Although considered much less frequently in the context of SE systems, roughly the same conclusions apply to off-head displays and hear-through displays. Loudspeaker technology is sufficiently advanced to provide effective off-head displays in various types of spaces, and hear-through displays can easily be achieved by the use of earphones with controlled acoustic leakage or by placing microphones in the environment and adding the synthetic signals and the environmental signals electronically. Although earphones are preferred to loudspeakers for most applications, in some areas, such as those in which the system is intended to simulate battlefield sounds with sufficient energy to shake portions of the body other than the eardrums, loudspeakers are clearly preferred.

The main current inadequacies associated with the use of the auditory channel in VE systems concern the synthesis of the signals to be presented via the interface. One component of this problem concerns the spatialization of sounds. Despite extensive recent work on spatialization using earphones, the results are far from perfect. Particularly noteworthy is the inability of most current HMD spatialization systems to cause sound sources to be perceived as located in front of the listener (as opposed to behind the head, above the head, or inside the head). Similarly, although the audio industry has devoted a great deal of attention to issues of spatialization for loudspeaker systems, the results are still overly sensitive to the precise location of the listener's head; relatively small movements away from the designated "sweet spot" can cause serious degradation of spatialization. The character of the spatialization response depends on both technology factors and perceptual factors.

A second set of inadequacies involves the generation of acoustic signals. Record and playback (sampling) methods suffer from the need to store an enormous number of sounds or be satisfied with crude approximations of the desired sounds. Reasonably satisfactory sound synthesis methods have been developed only for speech and music; they do not yet exist for the generation of environmental sounds. Also, if the sounds are generated in real time rather than ahead of time and then stored, the process may consume substantial portions of the system's computing power.

Relevant perceptual issues that are being studied include those related to the spatialization of sounds and issues similar to those already discussed in connection with the visual channel and in the section on human responses to alterations in sensorimotor loops (e.g., concerning human responses to distortions, time delays, and noise within the channel). A further set of issues that are beginning to receive attention involves the use of the auditory system for sensory substitution. Such substitution is being considered both when the visual channel is overloaded and when appropriate haptic feedback is unavailable. Another set of issues being studied, which is highly relevant to synthetic auditory displays (independent of whether they are being used for sensory substitution), concerns the manner in which the auditory system organizes temporal sequences of acoustic signals into a coherent perception of the auditory scene (auditory scene analysis). Understanding scene analysis in the auditory system is quite different from understanding that of the visual system, because in the auditory system there is no peripheral representation of source location (i.e., most information on source location is derived by comparing the signals received at the two ears, a process that involves central processing).

Position Tracking and Mapping By *position tracking* is meant the real-time measurement of the pose (defined as the three-dimensional position and three-dimensional orientation) of a moving object. Position tracking is required in VEs to control computer-generated stimuli and in teleoperators to control the behavior of the telerobot. In many applications, position tracking of the head, the hand, or the fingers is crucial. In other applications, position tracking of the eyes, the torso, and the arms and legs may also be required. In such applications, partial pose measurement may be sufficient—for example, three-dimensional orientation just for head tracking or three-dimensional position just for hand tracking.

By *position mapping* is meant the determination of a surface, such as that of the body or of the environment, by measuring a dense set of three-dimensional positions on that surface. Position mapping is required for determining bodily dimensions, for recognizing facial expressions, and for environment mapping to create a geometrical model for simulation. In applications such as environment mapping, real-time requirements may be absent. When position mapping is used for body tracking, however, real-time constraints must be met. Constraints on position mapping are likely to be exceptionally severe when augmented-reality applications are considered and registration issues arise.

Knowledge of the values that can be assumed by the various motion parameters (e.g., velocity, acceleration, bandwidth) for different kinds of bodily actions is reasonably adequate for purposes of tracker design. Less adequate is our knowledge of how various deficiencies in tracker technology degrade performance or contribute to the sopite syndrome in various types of tasks.

Currently, there are four basic technologies for position tracking and mapping in SE work: mechanical linkages, magnetic sensors, optical sensors, and acoustic sensors. SE systems are likely to include a mix of such systems, because each type of system has particular strengths and weaknesses and the requirements depend on the particular application. Although none of the inertial trackers currently available is adequate for SE applications, research is now under way to develop such trackers. As for many other kinds of devices, commercial specifications of position trackers and mappers are not reliable or consistent.

Mechanical trackers are relatively inexpensive, have very small intrinsic latencies, and can be reasonably accurate. Yet body-based linkage devices (called goniometers) may be cumbersome, whereas ground-based linkage devices (e.g., hand controllers) suffer from workspace limitations. The use of goniometers involves problems of fit and measurement related to alignment with joints, rigidity of attachments, calibration of linkages mounted on human limbs, and variations among individuals. The use of ground-based linkage devices involves the difficulty of tracking multiple limb segments and limb redundancies. Hybrid systems, in which body-based and ground-based devices are combined, are also likely to be required for some applications (e.g., both to track finger motion and to provide force feedback to the hand without causing forces to be applied to other portions of the body).

Magnetic trackers are commonly used because of their convenience, low cost, reasonable accuracy, and lack of obscuration problems. Significant current disadvantages that limit their usefulness include modest accuracy, short range, high latency (20-30 ms), and susceptibility to magnetic interference.

Optical sensing is one of the most convenient methods to use for certain kinds of tracking and is capable of providing accuracies and sampling rates that meet many VE requirements. The main drawbacks include visibility constraints and especially high costs.

Acoustic trackers are very attractive for VE because the costs are relatively modest and the accuracies and sampling rates are often sufficient. Efforts are being made to improve accuracies by taking into account atmospheric effects and by using echo rejection.

Inertial trackers, despite having played a distinguished role in the field of long-range navigation, have received little attention in the SE area. Their unique advantage is that they are unconstrained by range limitations, interference, and obscuration; also, latencies are low. Further reductions in sensor size and cost are needed to make inertial trackers a convenient and economical alternative to other trackers.

An ideal eye tracker would satisfy three requirements: linear response over a large range (roughly 50 deg), high bandwidth (1 kHz), and tolerance to relative motion of the head. Although many eye-tracking devices are available, none of them satisfies all three requirements.

Haptic Channel The haptic channel differs from the visual and auditory channels in two major ways. First, it involves manipulation as well as sensing. Second, it has received less attention than the other two channels with respect to both basic science and device development. One reason for the relatively backward state of the haptics field appears to be the intrinsic difficulties in studying haptics associated with the complexity of combining sensory functions with manipulative functions and with the use of electromechanical systems (for example, the control and measurement of the effective stimulus in haptics has always been difficult). Another reason is the lack of a recognized societal need to develop the haptics field. Whereas research and development related to the visual and auditory channels has been strongly driven by both medical needs and entertainment considerations, haptics-related research and development has had no such support. Until very recently, the main support for research and development on haptic interfaces has come from the field of telerobotics and, as indicated previously, this field has had limited support. It should be noted, however, that the introduction of the relatively simple haptic interfaces known as mice is having a major effect on the interest in such interfaces by computer scientists as well as those concerned primarily with human-machine interface issues.

The most useful currently available haptic interfaces fall into one of two categories: (1) body-based gloves or exoskeletons that track the position and posture of the hand (as discussed in position tracking) and (2) ground-based devices, such as "joysticks," that both sense certain actions of the hand and provide force feedback. Whereas many of the latter devices have been developed in connection with human-machine interfaces for teleoperators and have a relatively long history, many of the former devices have been developed recently with VE systems in mind. Although exoskeletons that provide force feedback have been developed for research and development purposes, they tend to be both cumbersome and expensive and are not in widespread use even experimentally. At present, one of the main experimental thrusts is directed toward the development of *tool-handle systems*, in which the human user manipulates a real tool handle, the actions of the tool handle are used to control some feature of a telerobot or a VE, and force feedback is displayed through the tool handle according to interactions of the telerobot with the real environment or of the virtual tool with virtual objects in the VE.

Relatively little work, even at the research and development level, has been directed toward haptic interface devices that provide feedback (related to perception of texture or temperature) of kinds other than simple force feedback, and no such device is yet available commercially. To convey more detailed information through the skin, tactile displays of the type used to convey visual or auditory information to individuals who cannot see or hear need to be considered. Although a variety of such displays has been developed (e.g., involving vibratory or electrocutaneous arrays), none has been successfully incorporated into SE systems.

Current research associated with the development of haptic interfaces involves investigation of human haptics, development of technology, and optimizing the interactions between the two. Basic research on human haptics includes biomechanical studies of the hand and psychophysical studies of the sensorimotor and cognitive systems associated with the hand. Illustrative issues of particular concern in this research include determination of the mechanical properties of the soft tissues in contact with haptic interfaces; quantification of limits on human sensing and control of contact forces and hand displacements; identification of stimulus cues in the perception of contact conditions and object properties; and characterization of human sensorimotor performance in the presence of time delays, distortions, and noise. Basic research on technology includes development of novel technologies for sensor and actuator hardware; design of computer architectures for fast computation of physical models; and development of algorithms for real-time control of devices that render tactual images.

Among the more applied topics that are beginning to be studied are those that are related to the inclusion of tactual images in multimodal VEs. These include the design of high-performance haptic interfaces with appropriate sensors, actuators, linkages, and control as well as their evaluation in uni- or multimodal VEs. The interfaces may be ground- or body-based and may or may not include force reflection or tactile displays. Avoidance of mechanical instabilities and false cues in contact tasks requires capabilities that are at the limits of current technology in the areas of range, resolution, and bandwidth of forces and displacements. Evaluation of the effectiveness of the interfaces is critical to the design of improved versions. Systematic studies to evaluate the human user's comfort in operating the devices and to investigate multimodal display methods for achieving optimal task performance and telepresence (immersion) are barely at the planning stages at this point.

Motion Interfaces In the real world, many kinds of motion occur, including whole-body passive motion (passive transport), whole-body active motion (locomotion), and part-body active and passive motion (e.g., when an arm is moved passively or actively). Also, in many cases such motion is accompanied by a wide variety of stimuli in a wide variety of sensory channels: motion cues may be contained in signals from the vestibular system, the motor system, the visual and auditory systems, and the proprioceptive/kinesthetic and tactile systems. It is no great surprise, therefore, that, in addition to the existence of many types of motion and many types of motion cues, there are many ways in which motion can be simulated and many types of motion interfaces.

Currently, motion interfaces for passive transport are being used primarily in flight simulation for flight training, in the entertainment industry for "thrill rides," and in research projects directed toward improved understanding of human perception and performance (including motion sickness) in a wide variety of contexts involving real or simulated passive transport.

Motion interfaces for passive transport can be divided into two categories: inertial displays, in which body mass is actually moved, and noninertial displays, in which motion is simulated without moving body mass but by stimulating various sensory channels in appropriate ways. Often, inertial and noninertial techniques are used in combination.

With inertial displays, patterns of force vectors are applied to the body that approximate to varying degrees of completeness and accuracy the patterns that would be present in the real situation being simulated. Such displays are generated by the use of centrifuges, rotating rooms, motion bases, tilting platforms, spinning chairs, etc. In some sense, G-seats, in which the seat and back can be inflated or deflated as well as vibrated, also fall into this category.

With noninertial displays, the body remains stationary, but patterns of stimuli are presented that are usually associated with movement of the body through the environment. The most obvious example in this category, one that has been studied in some detail and frequently applied, involves altering the visual scene in a manner that corresponds to the changes that would occur if the body moved through the given environment (e.g., in a car or plane). Similar results can be achieved using the auditory channel; however, auditory-induced self-movement has been studied less and applied less and, as one would expect because of the poor resolution, the results appear to be less dramatic. Other techniques for inducing the perception of self-motion include stimulating the vestibular system by changing temperature (caloric irrigation) or applying electrical currents (galvanic stimulation); stimulating the cutaneous system by sliding surfaces over the skin (e.g., beneath the soles of the feet) or by stimulating muscle spindles by vibrating muscle tendons. However, none of these techniques, with the possible exception of the one involving cutaneous stimulation, appears to be practical for use in SE applications.

Motion interfaces for active transport (i.e., locomotion) permit the user to experience active sensations of walking, running, climbing, etc., while remaining within a constrained volume of space. The best known and most widely used interfaces of this type are the common linear, one-dimensional treadmill and the stair-climbing exercise machine. The ideal system would incorporate a six-degrees-of-freedom platform (shoe) for each foot with position and force sensors and force feedback. Although a number of systems that are considered more advanced than the common treadmills and exercise machines are beginning to be developed and evaluated by research groups, no such systems are yet available commercially.

Current research in this general area, apart from work associated with attempts to develop improved technology, is focused on evaluating the different methods of movement simulation, with respect to both achieving the desired sensation of movement and minimizing the extent to which the simulation results in motion sickness (or, more generally, the sopite syndrome).

Other Types of Interfaces The above discussion by no means covers all the interface communication channels of interest. For example, nothing has been said about the olfactory (smell) or gustatory (taste) channels, or about interfaces related to the sensing of heat, wind, and humidity. Apart from influencing the general sense of presence and immersion in various types of environments, such sensations might be of specific use for conveying core information in some major application areas—for example, olfactory information in training firefighters or medical personnel concerned with low-tech diagnostic procedures. Furthermore, the technological problems in creating an olfactory interface do not appear overwhelming. Not only have odor-releasing systems for use in theaters and odor (scratch) records for novelty home use been available for many years, but also significant current research is now being conducted in this area.

Perhaps the two most important methods for interfacing that have not yet been discussed concern speech communication (automatic speech recognition and speech synthesis) and direct physiological sensing and control.

Although the discrete nature of speech makes it less appropriate for conveying information that is represented by continuous variables than by discrete variables, it clearly is one of the most natural methods for humans to use in communicating with another entity, human or machine. Previous research on automatic speech recognition and speech synthesis, which has been driven by needs outside the SE area, has produced a variety of systems that are now available commercially and that can be usefully applied in the SE field. Current speech recognition systems differ from the ideal in that they have limited capacity to handle large vocabularies, different voices, continuous speech (as opposed to isolated words or phrases), interference produced by background noise, and degraded speech production. Nevertheless, reasonably high accuracies (e.g., 95 percent correct word identification) can be obtained for task-specific applications in which modest demands are made along the just-mentioned dimensions of difficulty. Current speech synthesis systems appear relatively adequate in their ability to produce speech that is highly intelligible (and in comparison to synthesis systems for environmental sounds); however, they are only now beginning to produce speech that sounds reasonably natural or that mimics the idiosyncratic speech patterns of individual talkers. In general, however, there is no question that speech communication interfaces are now becoming available that can be usefully integrated into SE systems for a variety of practical applications, and that the overall quality of such interfaces will continue to improve over the next few years (with or without help from the SE field).

Physiological interfaces (direct stimulation of neural systems or sensing of physiological responses or states of the human organism) have received very little attention in the SE field. Direct stimulation of neural systems seems relatively inappropriate except in cases in which the subject is disabled by loss of sensory function. Even in these cases, however, the appropriate transducers are likely to become part of the subject, (i.e., to be implanted on a more or less permanent basis, as in the case of cochlear implants to mitigate loss of hearing) so that no special requirements are imposed on the SE interface. The use of internally generated physiological signals associated with activities of the brain, muscles, circulatory system, respiratory system, etc. can be used, at least in principle, to indicate general emotional and cognitive human states and to control specific variables in the SE system. Although significant research is being conducted in this area in an effort to determine and improve the reliability of such signals for purposes of control (by the military as well as by those concerned with aiding individuals who have severe motor disabilities), extensive practical use of such signals for SE control purposes appears several years away.

Computer Generation of VEs

To many people, and certainly to most computer scientists, computer generation of VEs is the core of the SE field; to them, human-machine interfaces, telerobots, and even the human operators, are of secondary importance. Furthermore, most past and current work in this area has focused on the generation of visual images. Apart from individuals who are themselves involved in the development or use of teleoperators, when people think of SE systems they tend to think of interactive, computer-generated visual images. Except for speech synthesis, which has been developed primarily by speech scientists rather than computer scientists, the generation of auditory and haptic images has been ignored. Accordingly, our discussion in this section is focused primarily on the visual channel. Information related to the computer generation of the auditory or haptic components of VEs is found primarily in discussions of human-machine interfaces.

It is possible to imagine a VE system that can create photorealistic images, that can be fully interactive in real time, and that has graphics, computation, and communication capabilities to handle all possible environments of interest with equal ease—that is, a general-purpose system that can generate environments relevant to manufacturing, health care, military training, etc. Such a system is beyond the current technology, and it is anticipated that for a long time to come, trade-offs between realism and interactive capacity will be required. Furthermore, due to these limitations, effective VE implementation will depend on targeted application domains. Some applications, such as architectural visualization, may require photorealistic rendering, whereas others, such as training, may not. Many manufacturing and medical applications may require a much higher level of real-time interaction than an architectural walkthrough. Although there are many applications in which a realistic visual environment is unnecessary (and maybe even undesirable), the ability to generate such an environment is clearly an important target for the development of VE technology.

One requirement for creating a realistic visual environment concerns the *frame rate*, that is, the number of still images that must be presented per second to provide the illusion of continuous motion. It has been demonstrated that frame rates must be greater than 8 to 10 per s to maintain this illusion. A second requirement concerns the *response time* a VE system must exhibit to preserve an illusion of instantaneous interactive control. Research shows that such delays must be less than 0.1 s. A third requirement concerns the *picture resolution* needed for realism. According to some VE technologists, a scene can be rendered in all of the detail resolvable to the human eye with 80 million polygons (Catmull et al., 1984). However, using today's hardware, a system that used 80 million polygons per picture would be far too slow to be truly interactive—thus the current major trade-off between realistic images and realistic interactivity. These requirements, of course, can be highly application dependent. Applications with rapidly moving objects may require significantly higher frame rates and shorter delays, whereas highly abstract or stylized applications may require fewer polygons or lower resolution.

Hardware Maintaining an adequate graphic frame rate is so computationally demanding that special-purpose hardware is often necessary. The main purpose of this graphics hardware is to provide rapid geometric transformations, clipping, hidden surface elimination, polygon fill, and surface texturing.

Several of today's leading graphic workstations are RealityEngine2 (produced by Silicon Graphics), Pixel Planes 5 and PixelFlow (developed at the University of North Carolina), and the Evans & Sutherland Freedom systems. All of these systems run on parallel architectures (in which graphic rendering operations occur on parallel paths); however, they differ on a variety of characteristics, including frame rate, processing speed, and anti-aliasing capabilities. Both RealityEngine2 and Pixel Planes 5 can process approximately 2 million texture-mapped polygons per s. PixelFlow promises significantly higher polygon processing rates than are available in current designs. None of these machines is able to render photorealistic time-varying visual scenes at high frame rates.

Vectorized or massively parallel super computers are also in use in VE applications to improve computational throughput. However, the use of parallel super computers may not significantly reduce the run times of VE applications that cannot be performed in parallel or that require large amounts of data movement. One approach suggested for maximizing versatility is to base computations in VE systems on a few parallel high-powered scalar processors with large shared memories. In such systems, different processors would handle different parts of the VE. Alternatively, different processors might be dedicated to different types of data (e.g., one might handle all computations related to the density field of a fluid, whereas another might handle all those related to its velocity field).

Other limiting factors relate to speed of data access. These include the time required to find the data in a mass storage device (seek time) and the time required to read the data (bandwidth). In circumstances in which very large data sets are needed for a single computation or picture, the bandwidth is critical and semiconductor memory may be the only viable storage medium in the near future.

Software In order to provide a fully interactive, real-time, natural-appearing environment, software development is required in a wide variety of areas. The real-time generation of VEs requires consideration of interaction, navigation, modeling, the creation of augmented reality, hypermedia integration, and operating system software.

INTERACTION Interaction software makes use of the outputs of human-machine control devices to modify the VE. The control devices now in use include position trackers, mice, keyboards, joysticks, and speech recognition systems. As a rule, tasks in VEs are performed using a number of control techniques in combination because none is adequate by itself. The interaction software takes all such control signals, scans them for obvious errors resulting from equipment or user malfunction, and then transfers the resulting information to those portions of the system involved in generation of the appropriate VE.

NAVIGATION Visual navigation software controls what the user sees as he or she moves through the VE and turns his or her head. More specifically, what a user sees is determined by two parameters: the user's location in the VE and the gaze direction. Typically, a head tracker is used to sense the position and orientation of the user's head. Changes in the user's location within the VE can be effected by a virtual vehicle (e.g., a human-operated treadmill, bicycle, or joystick that moves the viewpoint of the human user), by the specification of a new vantage point within the environment and the execution of a logical command to fly to that point, or by simulated teleportals in the distance at which the user can suddenly appear without moving through the intervening space.

Two important sets of software issues that arise when such user movements are taking place concern the mapping of the user's control actions into specifications of how the visual scene should be changed (an interaction issue) and the navigator aids provided to the user to prevent the user from getting lost in the VE.

A further very important software issue concerns the need to minimize the load on the graphics processors. Even if the contents of a VE remain static (as in the case of a virtual building depicting an architectural design), the display to the user changes as his or her point of view changes. A number of techniques have been developed to reduce the polygon flow to the graphics processor, but no general solution is yet available. Current solutions are generally application-specific, and some work well only if the underlying environment does not change dynamically. Two general techniques for minimizing polygon flow are the partitioning of the polygon-defined world into volumes that can be readily checked for visibility by the viewer and the low-resolution rendering of objects that are small in the user's visual field (e.g., as the result of being very far away).

MODELING Models that define the form, appearance, and behavior of virtual objects are the core of any VE. Today, geometric models constructed for VEs are developed, for the most part, using commercially available computer-aided design (CAD) systems. The tools provided by these systems aid designers in specifying object shapes and sizes; however, these objects are often difficult to use in situations that were not considered by the original CAD designer. As a result, a substantial amount of manual manipulation may be required to use an object specified with a CAD system for a VE. When a VE application requires a replica of a real environment, it is generally considered preferable to map the real environment rather than build a model of it. Active mapping techniques, such as scanning laser range finders and light stripes, are used to make three-dimensional measurements directly. The drawback of these methods is that they capture only information visible from a particular viewpoint. To achieve a complete map can require taking multiple views and combining them into a coherent picture. Some passive techniques, such as stereoscopic methods, are also in use. However, none of the stereo algorithms is robust enough to compete with active methods. For many purposes, far more is required of an environmental model than just a map of object surface geometry. If, in applications that attempt to model real-world behaviors, the objects are to be manipulated, the physics of the objects is needed—how they behave, their composition, etc. Even when the relevant physics is well understood, simulations based on this understanding can be tedious and time-consuming (both to construct and to run). In a VE, these simulations must run reliably and automatically; any situation that might arise must be adequately anticipated and handled correctly in real time.

The need for *autonomous agents* may arise in many VE applications, such as entertainment, training, manufacturing, and education. Although the ability to create fully credible simulated humans is well beyond our grasp in the foreseeable future, we do have the capability to develop simple agents. The agent's body in a VE is a physical object to be controlled to achieve coordinated motion. A computer model of a human figure that can move and function in a VE is called a *virtual actor*. A guided virtual actor is one whose movement is directed and controlled by the motions of a real human being. An autonomous virtual actor operates under program control and is capable of independent behavior that is responsive to the VE, including both human participants and simulated objects and events. An autonomous actor may touch and manipulate objects, make contact with various surfaces, or make contact with other humans directly (e.g., shaking hands) or indirectly (e.g., two people lifting a heavy object). Autonomous agents need not be literal representations of human beings but may be represented by various abstractions.

AUGMENTED REALITY In an augmented-reality system, virtual and real objects appear in the user's view simultaneously; the artificial or virtual image is overlaid on the real-world image. Creating adequate software for augmented-reality applications is a difficult task that requires a complete model of the real environment as well as of the synthesized environment. Automatic generation of effective augmented reality is still at the research stage. A major issue is the ability to create and maintain accurate registration between the real and synthetic environments, particularly when they are both rapidly varying.

HYPERMEDIA INTEGRATION Hypermedia is multimedia data composed of audio, compressed video, and text that is linked together in a nonlinear manner. A hypermedia system provides an individual with the opportunity to explore a topic by moving through a series of logically linked information nodes. When a node is reached, the individual may obtain all of the information available at that node in a variety of forms. One example is a virtual museum containing hypermedia nodes that provide significantly expanded information about particular artifacts. Once at a node, an individual can pursue a particular object in depth or explore its relationship to other objects in the exhibit. Hypermedia integration software (which involves the blending of computer graphics, video, sound, and, in the future, haptic images) is used to combine hypermedia software with VE. Embedding hypermedia nodes into a VE system allows a participant in a VE to go to a node and gain additional information about a particular experience. Work in this area is currently at the test bed stage.

OPERATING SYSTEM Current operating systems (UNIX, Windows NT) are not geared to supporting the real-time multimodal requirements of VEs, and significant modifications are required if the development of VEs is to proceed efficiently. There is a major need for systems that ensure that high-priority processes (such as user tracking) receive service at short, regular intervals and to provide time-critical computing and rendering with negotiated graceful degradation algorithms that meet frame rate and lag-time guarantees—this is a new computing paradigm.

Telerobotics

As indicated previously, a number of important SE applications involve sensing, navigating through, and manipulating objects in real-world environments. Such applications frequently arise in the domains of hazardous operations and medicine. In all such applications, telerobotics plays an essential role; human-machine interfaces and computer-generated VEs are not sufficient.

In many ways, current research activities concerned with teleoperation are similar to those concerned with VEs. Independent of the purpose for which the system is being designed (e.g., to train someone to fly an aircraft or to actually remove some hazardous waste), and independent of whether the relevant environment is real or virtual, in both cases these activities are concerned with the design, construction, and application of multimodal immersive systems that enable the operator to interact usefully with some structured environment. Furthermore, although concern with complex electromechanical systems was previously confined to individuals working on telerobotics, now, as haptic interfaces are becoming recognized as important components of VE systems as well as of teleoperator systems, the VE community is also beginning to show interest in this area. Independent of whether the haptic master in the interface controls the behavior of a telerobot or of some computer-generated virtual entity, design and construction of the master requires consideration of electromechanical phenomena and devices.

The principal differences that currently exist relate to (1) the design and performance of the telerobots and (2) communication time delays, and (3) the demands of real-time input/output operations. Unavoidable time delays (transport delays) arise in communicating between the human-machine interface and the telerobot when these subsystems of teleoperator systems are separated by a large distance (e.g., a delay of 30 ms between Washington, D.C., and Los Angeles, a delay of a little more than 1 s between the earth and the moon). Such transport delays decrease in importance as the distance decreases, as other delays in the overall systems (e.g., resulting from inadequate computational speed) increase, and as the importance of haptic interfaces with force feedback decreases. It should also be noted, however, that transport delays will increase in importance in VE systems not only as such systems increase their use of haptic interfaces with force feedback, but also as VE networks with tightly coupled players at distant locations come into being. The approaches being used to alleviate the time delay problem, which can arise in connection with problems of both physical stability and perception, include supervisory control and predictive modeling. Both of these approaches are being actively pursued by the teleoperator research community and, more recently, by the VE research community as well.

In typical VE systems, only simple haptic interfaces or trackers may be used, thus placing modest real-time input/output demands on the computer. In such circumstances, the real-time performance of current operating systems may be adequate or could become adequate with some modifications. For telerobotic systems and for VE systems with complicated haptic and other human interfaces, the input/output requirements are too massive to be handled by ordinary workstation-based architectures. The approach is to use a separate microprocessor system for real-time operations, connected to a workstation or PC that acts as a front end.

Apart from the research related to time delay problems and control issues, much of the current research activity in the telerobotics area is focused on telerobotic hardware. Although substantial advances have been made in this area and a number of impressive telerobots have been developed and applied to practical problems, the limitations imposed by inadequate hardware are still substantial. For example, sensor technology (to sense object proximity, object surface properties, and applied force) remains inadequate. Similar remarks apply to actuator and transmission technology.

Other current activities in the telerobotic area concern the development and use of new materials and the exploitation of advances in microelectromechanical systems. The availability of extremely small structures (including sensors and actuators) is stimulating exciting new work on microtelerobotic hardware and the interface and control problems associated with scaling down movements and forces from human scale to micro scales.

Also, and partly as a consequence of the advances in microelectromechanics, engineers are beginning to think about the possible benefits and feasibility of creating teleoperator systems that make use of *distributed telerobotics,* that is, a large number of relatively simple, relatively small telerobots with relatively narrow bandwidth communication among these telerobots. The use of multiple patrol telerobots for security purposes is one example of such an application. A major challenge in this area, aside from the development of the telerobots themselves, concerns the nature of the human-machine interface and the design of a display and control system that treats the set of telerobots as an integrated system rather than as a collection of independent entities that require a separate interface for each telerobot (and perhaps a whole set of human operators rather than a single operator).

The general problem of networking telerobotic systems, either in the sense of networking telerobots or networking the human operators, has received relatively little attention.

Networks

Communication networks can transform VEs into a shared environment in which individuals, objects, and processes interact without regard to their location. These networks will allow us to use VE for such purposes as distance learning, group entertainment, distributed training, and distributed design.

Currently, the two application domains in which the most networking activity is occurring are entertainment and national defense. In the entertainment industry, VR companies are in the process of forming cooperative arrangements with cable television companies to develop multiuser games and interactive shopping.

Applications for the military have focused on large-scale simulated network training exercises, such as those offered by SIMNET. In SIMNET, as many as 300 soldiers in tank and aircraft simulators located at different military bases can engage in a realistic battle against an intelligent enemy on a common battlefield. Currently, the Defense Department is using new communication software that upgrades the SIMNET protocol. This upgrade, named Distributed Interactive Simulation, is expected to be used by the military in building future distributed training scenarios as well as in simulating the acquisition process for various pieces of planned equipment. Another example of network use can be found in the experimental program currently being pursued in telemedicine by the state of Georgia. Network applications such as these, all of which are discussed in Chapter 12, constitute only a first step.

To move the communication software development forward to a point at which it can truly support VE applications will require the development of VE-specific, applications-level network protocols. These applications protocols are required for communicating world-state-change information to the various networked participants in the operating VE.

As mentioned in the previous section, the networking of telerobotic systems in diverse locations has also begun to receive some attention. For example, the universities associated with the Space Automation/Robotics Consortium (Texas A&M University, the University of Texas at Austin, the University of Texas at Arlington, and Rice University), in conjunction with the National Aeronautics and Space Administration (NASA), have developed software and protocols to use the Internet to control telerobots in their different locations. This type of effort is under consideration elsewhere and is likely to grow in scope. Advances in networking will strongly impact such developments.

Wide-area network (WAN) hardware is being developed on a variety of fronts. The national telecommunications infrastructure is being radically altered by the installation of fiber optic cabling that is capable of operating at gigabit speeds across the country. As a result, major commercial carriers are installing special switches that handle both synchronous and asynchronous signals at very high speeds. Government-supported networks are also in the process of upgrading. For example, NSFnet, the backbone of the Internet, is currently operating at T-3 speeds (45 Mbit/s) and plans to move to OC-3 (155 Mbit/s) in 1994 and to OC-12 (622 Mbit/s) by 1996. The National Research and Education Network is one of four components in the U.S. High Performance Computing and Communication program, which is supporting the installation of OC-12 networks at five regional test beds for research purposes.

Workstations with three-dimensional graphics will be connected to the WANs discussed above through local-area networks (LANs). Most of these LANs, which currently use Ethernet technologies (10 Mbit/s), do not have the capabilities to support the high-performance demands of VE and multimedia. However, work is proceeding on larger and faster local networks, such as the Fiber Distributed Data Interface. This system currently operates at 100 Mbit/s, but the follow-on, expected by the middle to late 1990s, will operate at speeds up to 1.25 Gbit/s. Moreover, the Institute of Electrical and Electronics Engineers (IEEE) has issued a draft standard, Integrated Services Local Area Network Interface, which defines a LAN that carries voice, data, and video traffic.

Although networks are becoming fast enough to support the development of distributed VE, we need greater bandwidth to support the very large-scale, multimodal, and multiple-user applications that we foresee in the near future. Another problem is network host-interface slowdowns caused by the multiple layers of the operating system software. A further issue is the high cost associated with buying time on a high-speed WAN. The current estimated cost for one year exceeds the entire budget of most research groups. Finally, it should also be noted that, despite the beginning efforts mentioned above to network teleoperator systems, it is important to focus on teleoperator systems as well as VE systems as work in this area progresses.

Evaluation of SE Systems

SE both draws on and provides research and development challenges to several well-established disciplines, including computer science, electrical and mechanical engineering, sensory physiology, cognitive psychology, and human factors. In each discipline, the requirements associated with creating SE technology raise new questions that call for research and evaluation. Some examples include: (1) identifying the capabilities and limitations of human beings as criteria for system design, (2) developing the hardware and software that can deliver SEs in a cost-effective manner, and (3) determining areas in which SE can make an important difference in human experience or performance.

As with the introduction of many new technologies, SE technology has not been adequately evaluated. Moreover, evaluation with regard to ultimate effectiveness is difficult because the technology is at an early stage in its evolution and, as a result, does not provide the high-fidelity environments and the natural interfaces that are planned for the future. In addition, SE offers some particularly complex evaluation issues because of its interdisciplinary nature and the requirement to integrate several technologies to create a full SE system. Human studies are needed to generate requirements for both component and full system design. Moreover, it is desirable to have cost-effectiveness evaluations for each component as well as for prototype systems in each application area. Some evaluation questions concern the engineering reliability and efficiency of components or full systems. Other questions focus on how well the design accounts for human perceptual and cognitive features or for human responses to alterations in sensorimotor loops. In the cognitive area, research indicates that it is difficult to make generalizations about the relationship between types of tasks, task presentation features, and human performance. This indicates a need to conduct studies to explore these relationships as the technology improves and ideas for new SE applications are proposed.

Another set of questions for evaluation includes the possible medical and psychological effects of SE technology on human beings. For example, studies will be needed to ensure that the technology will not have any adverse effects over time on human visual, auditory, or haptic systems. Furthermore, there are the important concerns about induced motion sickness in SEs and the potential after-effects of adaptation to SEs when an individual moves back into the real environment.

There are many types and levels of evaluation that can be used to provide direction, understanding, and a general picture of performance effectiveness. Standard evaluation methodology offers a range of options, including: (1) empirical studies using observation techniques or experimental designs to collect data in laboratory or field settings and (2) analytic studies involving theoretical modeling, heuristic evaluations, and simulations of varying system functions. Each of these methods can be used at various points in the system's evolution.

At present, relatively little evaluation of SE systems is taking place. The extent to which the novel aspects of SE will require new evaluation tools remains to be seen.

RECOMMENDATIONS

The committee's overarching conclusion is that SE systems have great potential both for helping to satisfy various societal needs and for stimulating advances in some important areas of science and technology. In our deliberations, when reservations about the value of investment in some particular SE area were expressed, they often reflected the judgment that the importance of advances in the particular area were dwarfed by the need to modify related social and political factors rather than that the area was unimportant or inappropriate from a scientific or technical point of view.

In recommending topics and areas for concentrated research and development efforts, the committee rejected the approach of developing a small number of high-priority ("star") recommendations. The possible applications cover such a broad range of societal activities, and the advances required to realize truly cost-effective systems for these applications cover such a broad range of research and development activities, that the star approach seemed totally inappropriate.

In constructing this research agenda, the committee used three kinds of criteria. The first is concerned with advancing the state of the art.

Specifically, it is concerned with the extent to which a project under consideration can be expected to lead to improved understanding of important phenomena and/or to improved technology. We refer to criteria of this type as *science and technology criteria.*

The second is concerned with the likelihood that the project in question will have important practical—and positive—consequences within the not-too-distant future (i.e., the next five years). We refer to criteria of this type as *practical applications criteria.*

The third is concerned with factors such as leverage, cost-effectiveness, and ratio of payoff to effort. In addition to technical matters, evaluation of these factors must involve consideration of current conditions and forces in society, which go beyond those the committee was appointed to examine. In general, we refer to criteria in this category as *leverage criteria.*

Finally, it should be noted that the recommendations have not been prioritized in any detailed manner. This is due primarily to our judgment that successful development and application of SE systems depends on an entire matrix of interrelated factors, not on one or two isolated factors. We nevertheless feel that it is important to stress the crucial need for improved hardware technologies to enable development of improved interface devices and improved computer generation of multimodal images. Unlike the situation in the area of teleoperation, in the area of VEs there are relatively few individuals who have primary interests or backgrounds in hardware; most individuals in the VE area are involved primarily in the software end of computer science, in communication or entertainment media, and in human perception and performance. Thus, the importance of adequate hardware, without which the VE field will never come close to realizing its potential, tends to be underplayed by the VE community. A somewhat similar comment concerns the issue of user comfort. To date, a very large fraction of VE usage has occurred in the context of short demonstrations, a context in which the degree of comfort is relatively unimportant. However, if the comfort of VE systems (particularly head-mounted displays) cannot be radically improved, the practical usage of these systems will be limited to emergency situations or to very short time periods. In other words, adequate comfort, as well as technically adequate hardware, are essential to realizing the potential of the SE field.

The research agenda we propose covers four main categories:

(1) Application domains,
(2) Some psychological considerations,
(3) Development of improved SE technology, and
(4) Evaluation of SE systems.

In this section we present recommendations for research and development in the near future in terms of these categories. We also indicate to the extent possible the role played by the various types of criteria in making these recommendations. Further details are provided in the chapters of this report. In addition, we make comments and suggestions for government policy and infrastructure based on the experience and judgment of committee members. They are suggestive of the kinds of tools and mechanisms that federal agencies might use to encourage coherence, integration, and overall development of the field.

Application Domains of SE Systems

RECOMMENDATION: The committee concludes that four application domains show the most promise for SE: (1) design, manufacturing, and marketing; (2) medicine and health care; (3) hazardous operations; and (4) training. We recommend that the research needs in these domains be used as one of the principal means to focus SE technology development and testing.

Our review shows that each of these domains includes tasks that are particularly compatible with the projected capabilities of SE. Each of these domains received high scores with respect to the science and technology and practical applications criteria. The domain of hazardous operations also received a high score with respect to the leverage criterion because of the relative lack of attention and funding given to this domain.

The committee has not assigned priority to the application domains of education; information visualization; and telecommunication and teletravel. Although committee members agreed that the education domain is exceedingly important—perhaps the most important of all the domains considered—it was not assigned priority because of our judgment that the development of improved education technology will have only a minor effect on the quality of education actually received. In other words, the main current obstacles to achieving substantial improvements in education are social, political, and economic, not technological. Thus, even though the education domain can be viewed as a high-leverage domain with respect to funding considerations, it is regarded as a low-leverage domain overall. Also, the committee did not rate this domain highly with respect to the science and technology criterion. Although further scientific research is required to determine how SE technology can best be utilized in K-12 education, it is believed that other application domains are likely to play a more important role in driving SE technology. If the relevant infrastructure undergoes changes that greatly facilitate widespread and in-depth use of technology within this area, then priority for the education domain would be indicated.

For the present, however, it is believed that efforts with the highest payoff relevant to the education domain will be those that are directed toward the general development of improved technology, toward alteration of the relevant social, political, and economic infrastructure, and toward influencing the entertainment industry to include programs, activities, systems, and facilities in the soon-to-be upgraded interactive home entertainment centers that have increased educational value. To the extent that the introduction of improved communication technology that links students and teachers (and parents) at disparate locations can help improve the current educational infrastructure as well as directly improve the quality of education, the committee would strongly support such an effort.

The information visualization and the telecommunication and teletravel domains are not assigned priority for a number of reasons. Paramount among them, and consistent with their cross-cutting characteristics, is that work in the specified priority domains will necessarily include work in these domains as well. Thus, for example, work in design, manufacturing, and marketing as well as medicine and health care will necessarily include work on information visualization. Similarly, work on hazardous operations and training will necessarily involve work on telecommunication and teletravel. An additional reason for denying the information visualization domain priority is the fact that work in the area of scientific visualization using VE is already quite active. Thus we did not feel that this domain deserved a high score on the leverage criteria.

Although equivalent remarks could be made about the cross-cutting nature of the training domain—that is, applications of SE technology to training will occur in connection with each of the other high-priority domains—training was judged to be so important and the potential for achieving substantially improved cost-effectiveness in training so great, that it retained a priority rating.

Special Projects

RECOMMENDATION: The committee recommends two projects for special attention: (1) modeling the human body for purposes of medical education, surgical planning, and providing explanations of procedures and outcomes to patients and (2) studying the transfer of knowledge and skill gained in training in a VE to performance in a real-world task environment.

Modeling the human body is required for many of the VE applications considered within the general domain of medicine and health care. For example, within the subdomain of surgery, physical models of various bodily organs, skeletal structures, physiological subsystems, etc., are needed for the planning of surgery, the training of surgeons, and explaining possible procedures and outcomes to patients. These models must be sufficiently accurate to serve the purpose at hand yet sufficiently simple to satisfy the computational constraints imposed by the limited power of the available VE system. In many cases, the fidelity of the virtual body parts, processes, and systems will be limited not only by inadequate VE facilities, but also by inadequate scientific knowledge (i.e., inadequate empirical data about the phenomena to be modeled or inadequate theoretical techniques for the quantitative modeling of these phenomena). Nevertheless, we recommend this project because it satisfies all three criteria and because even very crude models are likely to provide results that are substantially superior to those now achieved. Thus, for example, it should not be overwhelmingly difficult to develop physical models with enough fidelity to create VE educational experiences (for medical personnel and patients) that usefully supplement those now being realized by means of static two-dimensional illustrations, conventional videos, and spoken or written words. Not only should it be possible to develop models in which the visual components are superior to those now achieved with conventional techniques, but the addition of the auditory and haptic channels in such efforts should also greatly facilitate the education process.

Studies of training transfer, the second special project area cited, are essential to the successful development of VE training procedures for a wide variety of training tasks in a wide range of application domains. Although it is generally recognized that VE systems have great potential for cost-effective training, in order to realize this potential it is necessary to determine and understand how various differences between the situation faced by the trainee in the VE training system and the situation faced by the trainee in the real-world situation for which the individual is being trained influence the effectiveness of the training as measured by task performance in the real-world situation. Research on this transfer-of-training issue has, of course, been conducted for many years in connection with other types of training. However, the class of differences between the training and real-world performance situations that will arise when VE training is used are likely to include numerous elements that have not been studied before (e.g., the differences associated with unique points of view or instructional cues that can be generated by VE systems but not by conventional training systems). Furthermore, even when no such new elements exist, understanding of training transfer issues remains rather limited. As with the human-body modeling project, the training transfer project has high ratings on all three criteria. It is very challenging scientifically, advances will have important and immediate applications, and modest amounts of additional funding are likely to significantly increase the probability of success in VE training applications.

Psychological Considerations

RECOMMENDATION: The committee recommends that support for psychological studies be organized around the following objectives:

(1) Development of a comprehensive, coherently organized review of theory and data on human performance characteristics (including consideration of basic sensorimotor resolution, perceptual illusions, information transfer rate, and manual tracking) from the viewpoint of SE systems.

(2) Development of a theory that facilitates quantitative predictions of human responses to alterations in sensorimotor loops for all channels, with special emphasis on: (a) degradations in performance resulting from deficiencies in SE technology (e.g., in the form of distortions, time delays, and system noise), (b) supernormal performance achievable through introduction of purposeful enhancing distortions, (c) radical sensorimotor transformations that arise, for example, in connection with the use of sensory substitution or strongly nonanthropomorphic telerobots, (d) methods of accelerating both adaptation to various types of alterations and readaptation to normal conditions, (e) channel-interaction effects that occur with multimodal interfaces, (f) factors governing the occurrence, kind, and magnitude of sopite sickness from SE exposure, and (g) factors governing the strength of subjective telepresence and its relationship to objective performance.

(3) Development of cognitive models that will facilitate effective design of VE systems for purposes of education, training, and information visualization.

(4) Development of improved understanding of the possible deleterious effects of spending substantial portions of time in SE systems.

Although a wide variety of research on human performance characteristics has already been performed, the results have not yet been coherently organized to reflect the viewpoint of SE systems. Such a review would not be too difficult or expensive to prepare, and would be extremely useful to a large segment of the SE research community. In addition to providing important relevant information to this community, it would help delineate the further research that is required in this area to guide design of improved SE systems.

The task of characterizing and modeling human responses to alterations in sensorimotor loops constitutes a major challenge. Knowledge in this area, which is fundamental to the design of essentially all SE systems for all types of applications, is seriously inadequate. Currently, it is difficult to predict how such alterations will influence either objective performance or subjective state, and in particular, how performance and state will change over time as the user gains experience with the alteration.

Some of the subareas in which we urge special emphasis here may be strongly application-dependent. We also recognize that past efforts in some of these subareas have not always been as fruitful as one would like. Nevertheless, because of the importance of these subareas to the progress of the SE field, the committee's recommendation to pursue this research received high ratings according to all three classes of criteria.

The development of improved cognitive models for characterizing the manner in which experience gets organized, problems get solved, world views are formed, and learning takes place is essential to a wide variety of VE uses; however, it is particularly important in the application domains of education, training, and information visualization. Without such models, and without the proper integration of such models into a general framework that includes the effects of multimodal sensorimotor experience on perception and cognition, the design of VE systems for these application areas will be seriously handicapped.

A final set of topics for research concerns the study of the possible deleterious effects of spending substantial portions of time in SE systems. To some extent, problems of this type will be addressed automatically as a consequence of the government-mandated human-use monitoring of experimental research involving human subjects; however, the precautions associated with this monitoring are not likely to adequately address possible negative effects of continued usage in everyday life. In any case, it is obviously essential to cover issues related to the effects of SE on relatively deep psychological factors such as self-image, cognitive style, affective state, and motivation.

Development of Improved SE Technology

Human-Machine Interfaces

Human-machine interfaces include all the devices used in an SE system to present information to the human or to sense the actions or responses by the human that control the machine in question.

> **RECOMMENDATION: The committee recommends support of research on visual displays, haptic interfaces, and locomotion interfaces, with emphases as outlined below.**

These recommendations were highly rated on all three criteria: science and technology, applications, and leverage.

Visual Displays The development of adequate head-mounted displays is very important to the SE field. The main deficiencies in current HMDs involve the quality of the visual display and the ergonomics of the helmet used to mount the visual display on the user's head. (Both the quality and the ergonomics of the auditory component of HMDs are currently adequate.) Despite the substantial effort that is already being directed toward reducing these problems by the entertainment industry (whose focus is on low-end systems) and by the military, because of the large payoff that would result if substantial improvements were made, the committee recommends that strong support be provided to research in this area. Of particular importance for the visual display is substantially improved combinations of resolution and field of view. Also to be considered in this area is the inclusion of see-through options for use in augmented-reality applications.

Work on the ergonomics associated with the HMD should include not only consideration of mass and center of gravity (as well as fit), but also consideration of wireless, broadband communication to eliminate the customary tether used. Integration of visual, auditory, and position tracking in one device (possibly helmet, but much preferably, sunglass-sized) should take account of detailed data and understanding of norms and variations in human heads. Of particular importance is exploring alternative materials and configurations and evaluating them in a VE context.

Because there is no guarantee that HMDs that are fully adequate for all important tasks will be available even within the next 10 years, attention must also be given to improving off-head displays. In particular, the committee recommends that OHD research be carefully monitored and support be provided to those display projects that appear promising from the performance point of view (e.g., some of those concerned with autostereoscopic displays) but are too risky from the business point of view to be supported by industry. The projects supported by the military in this area have not led to displays that are affordable, and the projects conducted by private hobbyists and inventors in this area are likely to die out because of inadequate funding.

Special attention should also be given to the study of perceptual effects associated with the merging of displays from different display sources (as in augmented-reality applications). Independent of whether the images that are combined with computer-generated VE images are derived directly from the real environment by means of see-through displays or indirectly through the artificial eyes of a telerobot, the psychophysics of such merged displays must be carefully studied. Not only is there relatively little past work in this area, but also the use of such merged displays is likely to become increasingly common. Equivalent psychophysical studies in the auditory and haptic channels are regarded as worthwhile but less urgent.

Haptic Interfaces Many important potential SE applications cannot be realized adequately without the development of substantially improved haptic interfaces. Although the haptic devices that can be developed over the next 5 to 10 years will undoubtedly fall short of the ideal device, they can vastly improve the range and quality of haptic interactions that are now available in SE systems. Particular attention should be given to the development of tool-handle interfaces in which the possible haptic interactions are constrained by the nature of the tool.

Special support also should be given to basic haptic science relevant to the development of improved haptic interfaces. Empirical data and theoretical models describing phenomena for this channel are much less adequate than for the visual and the auditory channels, and considerable scientific work is needed to support technology development for this channel. This work should include studies confined entirely to the haptic channel (e.g., exploration of haptic illusions), as well as studies concerned with the interaction effects of this and other channels (especially vision).

Locomotion Interfaces Another technology area for which the committee recommends special support is that of locomotion interfaces. The range of possible applications would be substantially increased by the availability of improved locomotion interfaces, and there seems to be no fundamental obstacle to the creation of such interfaces.

Position Tracking and Mapping The development of improved devices for position tracking and mapping is extremely important.

> **RECOMMENDATION: In this area the committee recommends a multiphased approach: (1) conduct research and development on mechanical trackers and inertial trackers, (2) explore the possibility of obtaining improved cost-effectiveness in tracking by using hybrid systems, and (3) carefully monitor commercial developments in magnetic, acoustical, and optical trackers, in eye trackers, and in trackers directed toward registration problems in augmented reality. If market forces do not drive the development of these trackers, federal research support is urged.**

The committee's recommendations in this area are complex because of factors relating to the leverage criteria.

Market forces in the development of mechanical trackers, inertial trackers, and hybrid systems currently appear minimal, so research in these areas requires support. In the committee's judgment, the other devices mentioned in the recommendation will continue to improve without special support. For example, it is believed that commercial interest in the manufacturing and health care domains is sufficient to drive development of improved optical systems for registration of images from different sources in augmented-reality applications. If this should turn out to be incorrect, however, the committee would assign high ratings to research in this area.

Testing and Evaluation A further important recommendation concerning interface technology concerns the physical testing and evaluation of interface devices.

> **RECOMMENDATION: The committee recommends the establishment of a set of standards or an independent laboratory to evaluate SE interface devices.**

Because of the lack of reliability in manufacturer's specifications, and because the techniques and standards employed by different laboratories in such efforts vary widely (even if the laboratory performing the work has no vested interest in the device), it is recommended that an independent laboratory be established to evaluate SE interface devices or, alternatively, strict standards be set for such evaluations.

Other Interface Issues We have not included recommendations concerning support for auditory displays, speech communication interfaces, physiological interfaces, or other types of displays previously mentioned (related to olfaction, temperature, wind, etc.). For auditory, speech, and physiological interfaces, this determination results from our judgment that any required advances in these areas will be driven by forces outside the SE field and thus no special support is required. Although work on the other types of interfaces may not be driven by such outside forces, we do not regard the need for these other types of interfaces to be as important for most application domains. It should be noted, however, that their inclusion is likely to provide a major increase in the sense of presence or immersion in VEs and, to the extent that such subjective effects are likely to have a strong positive influence on performance, they may be important.

Computer Generation of VEs

Hardware Advances in computational and communication hardware are essential to the full realization of VEs. The hardware capabilities available today have given researchers, entrepreneurs, and consumers just a taste of virtual worlds and a promise of possible applications. Because of the potentially wide appeal of VEs and the large variety of applications with differing performance requirements, it is important to continue hardware development at several levels, from high-end supercomputer workstations to low-end personal workstations with modest capabilities.

Extrapolating current trends, we expect that VE applications will continue to saturate available computing power and data management capabilities for some time to come (dataset size will be the dominant problem for such applications as physical modeling and scientific visualization). In the future, high-end VE platforms will require the following features: very large physical memories (> 15 Gbyte), multiple high-performance scalar processors, high bandwidth (> 500 Mbyte/s), low latency (< 1 millionth s) mass storage devices, and high-speed interface ports for various input and output peripherals.

The research and development required to achieve the hardware capabilities stated above gets high ratings in terms of both the science and technology criteria and the practical applications criteria. However, we are much less clear about the leverage criteria. On one hand, current market forces do not appear to be driving adequate development of coherent and integrated architectures for multimodal modeling, representation, and rendering. Even within the confines of the visual channel, it appears that relatively little attention is being given to time-deterministic generation of images (i.e., systems that guarantee compliance with appropriate bounds on graphics update rate and lag, possibly at the expense of resolution). On the other hand, if the overall commercial market for computer hardware continues to grow at the same rate as it has in the past, or if the SE field looks very promising to the industry, then such advances will probably require no special support. Therefore:

> **RECOMMENDATION: The committee recommends no aggressive federal involvement in computer hardware development in the SE area at this time. Rather we conclude that hardware development remain largely a private-sector activity. Should serious lags in development occur, the government might then consider strategies for leveraging private-sector development efforts.**

Software In the past, research on and development of software for VE has been conducted through small, independent research programs.

> **RECOMMENDATION: The committee recommends that a major unified research program be created that focuses on those areas of development directly related to the generation, implementation, and application of VEs. The basic topics that need to be considered in such a program include: (1) multimodal human-computer interactions, (2) rapid specification and rendering of visual, auditory,**

and haptic images, (3) models and tools for representing and interacting with physical objects under multimodal conditions (including automated model acquisition from real data), (4) simulation frameworks, (5) a new time-critical, real-time operating system suitable for VEs, (6) registration of real and virtual images in augmented-reality applications, (7) navigational cues in virtual space, (8) behavior of autonomous agents, and (9) computer generation of auditory and haptic images.

There is a need to develop methods and software to interpret and respond to multimodal inputs from a wide variety of devices, including those associated with position tracking, haptic manipulation, and speech commands, often occurring in concert. Improved software must be developed to determine and filter out errors in these control signals and to provide the appropriate resulting information to the software generating the VE.

Fast and realistic specification and rendering of visual, auditory, and haptic images is a fundamental topic for VE research. The combined requirements of realism and interactive performance are extraordinarily severe. For example, in the visual modality, it has been estimated that the creation of a totally realistic interactive environment would require 80 million polygons per picture and a minimum of 10 pictures/s—a total of 800 million polygons/s, far beyond the capacity of current graphics workstations. Across modalities, there is a need to develop representations and rendering algorithms that dramatically increase effective throughput without sacrificing realism. A key issue will be to generalize static rendering methods to effectively handle dynamic scenes, for example by exploiting temporal coherence and by automatically adjusting the level of detail to match what the user will be able to perceive. Research and development in parallel rendering will also become important.

Real-time rendering for interactive systems poses additional problems when the auditory and haptic modalities are included. For each modality, the software for rendering images receives the output of the physical model and generates the commands needed to drive and control the interface devices. The major issues are: (1) the accuracy of rendering in relation to computed output of the model, the capabilities of the display device, and the sensory resolution of the human user in each of the modalities; (2) minimization of time delays in the rendering for each modality; and (3) synchronization of displays among multiple modalities. Efficient rendering requires that the capabilities of the rendering software should be commensurate with those of the physical model, the display device, and the human user. The higher the display resolution, the more time-consuming is the rendering, leading to time delays. Such delays, if excessive, generally lead to perceivable lags and distortions. In the case of force-reflecting haptic interfaces, they may also cause mechanical instabilities. Therefore, rapid rendering software that minimizes time delay while retaining optimal display resolution is critical for each of the modalities. Under multimodal conditions, the additional condition of synchronization of the modality-specific displays needs to be satisfied. We recommend support for development of software tools for rapidly driving visual and auditory display devices, together with fast, real-time control of haptic interfaces. Such software can have both device-dependent and device-independent components.

There is a major need to develop more powerful methods for acquiring and representing realistic models of physical objects and for realistic simulation of the physical behavior of these objects. To construct realistic models of truly complex environments is all but impossible with current computer-assisted design tools. Creation of such complex models will require a combination of automated model acquisition from real data and automated model synthesis based on concise descriptions. Such models will ultimately need to capture not only the object characteristics relevant to the visual channel, but also all of the physical properties that must be specified to realistically simulate objects' appearance and behavior in the broad multimodal sense. Simulating the mechanics of the everyday world will be of central importance in giving virtual environments a sense of solidity and allowing users to effectively manipulate virtual objects through haptic interfaces. The problems that arise in generalizing standard batch-simulation methods to handle interactive VEs are analogous to those that arise in the extension of static rendering techniques.

Research into the development of environments in which object behavior as well as object appearance can rapidly be specified is an area that needs further work. We call this area *simulation frameworks*. Such a framework makes no assumptions about the actual behavior (just as graphics systems currently make no assumptions about the appearance of graphical objects). A good term for what a simulation framework is trying to accomplish is *meta-modeling*. Such frameworks would facilitate the sharing of objects between environments and allow the establishment of object libraries. Issues to be researched include the representation of object behavior and how different behaviors are to be integrated into a single system.

Because most current operating systems are built on commercial versions of UNIX, which is not designed to meet real-time performance requirements, the committee recommends that approaches to a new operating system suitable for VEs be studied. In principle this could be achieved by creating a new operating system architecture or providing upgrades or enhancements to existing operating systems (e.g., UNIX, Windows NT).

The operating system capabilities required for VE include support of very large numbers of lightweight processors communicating by means of shared memory, support of automatic or transparent distribution of tasks to multiple computing resources, support of time-critical computation and rendering, and very high resolution time-slicing and guaranteed execution for high-priority processes (to within 0.001 s resolution). Although not specifically addressing all of these concerns, the efforts of the IEEE Posix standards committee are starting to bring real-time capabilities to the open-system workstation environment.[1] Supporting these capabilities in the operating system will significantly facilitate the development of many VE applications, especially larger, more ambitious efforts). The commercial sector cannot be expected to perform the necessary research and development in this area without incentives from the federal government. Specifically, we recommend that the government participate with industry in funding the upgrades and enhancements needed to provide an operating system that will meet the performance requirements for implementing VEs. Moreover, these joint funding efforts should be accompanied by a plan to move the new or upgraded systems to commercial adoption. To ensure that VE systems are written using an appropriate operating system, a financially sound transition plan must be formulated, funded, and executed.

Another important area for development is registration of real and synthetic images for augmented-reality applications. To create the illusion that synthetic and real objects exist in the same world requires highly accurate registration. For example, to make a synthetic object appear to rest on a real table, the object must move with the table as the observer moves, and accurate registration requires both a good geometric model of the scene and good measurements of observer motion. In addition to the purely geometric aspects of registration, illumination effects (casting synthetic shadows onto real objects) must be handled. Note also that significant misregistration may be disastrous in certain applications (e.g., surgery).

In addition to the general areas discussed above, the committee recommends that research and development on the following topics be supported: navigational cues in virtual space, the behavior of autonomous agents, and the computer generation of both auditory and haptic images for VE. Navigational cues are important because there is a great tendency in current VE systems for users to lose their way during virtual travel (or even simply during rotations of the head). Work on autonomous agents

[1] For VEs with relatively simple input/output, the real-time requirements are different from those of the telerobotics community. The fundamental difference is that the massive input/output requirements for complicated haptic and other human interfaces associated with telerobotics cannot be handled by an ordinary workstation-based architecture.

is important because many future applications are likely to require such agents, and the task of designing appropriate psychological and physical models for "driving" these agents is an extremely difficult one. With regard to computer generation of auditory images, spatialization, synthesis of environmental sounds, and auditory scene analysis are judged to be the most critical; in the haptic channel, because so few results are currently available, a wide array of research projects should be supported. Although certain components of some of these problems relate primarily to design of human-machine interface devices, others relate primarily to software.

Telerobotics

Recommendations in the area of telerobotics that are not already included elsewhere concern: (1) the effects of communication time delays on teleoperator performance, (2) telerobotics hardware (structures, actuators, and sensors), (3) microtelerobotics, (4) distributed telerobotics, and (5) real-time computational architectures.

> **RECOMMENDATION: The committee recommends that support be given to improving control algorithms, improving methods for constructing and using predictive displays, and improving methods for realizing effective supervisory control strategies.**

Unless communication delays are properly handled, teleoperator performance will be severely degraded and may, under certain circumstances, become unstable. In order to combat the effects of such delays, continued efforts should be directed toward the development of improved control algorithms that ensure stability and yet, to the extent possible, provide reasonable gains. At the same time, continued effort should be directed toward the development of improved methods for constructing and using predictive displays and for realizing effective supervisory control strategies. Advances in combatting the delay problem are required not only in connection with hazardous operations, but also in connection with certain components of telemedicine (particularly telesurgery).

> **RECOMMENDATION: The committee recommends work in four areas of hardware development: (1) multiaxis, high-resolution tactile sensors, (2) robot proximity sensors for local guidance prior to grasping, (3) multiaxis force sensors, and (4) improved actuator and transmission designs.**

Multiaxis high-resolution tactile sensors are needed to provide the telerobot with an effective sense of touch. Robot proximity sensors are required to provide local guidance prior to grasping. Such guidance would greatly facilitate the development of adequate supervisory control. Multiaxis force sensors are needed to measure the net force and torque exerted on end effectors. For example, miniature force sensors of this type could be mounted on finger segments to accurately control fingertip force. Improved actuator and transmission designs are required to provide high-performance joints and improved performance of telerobotics limbs.

RECOMMENDATION: The committee recommends that research be conducted on issues that arise when microtelerobots are used in teleoperation.

As the field of microelectromechanics evolves, and smaller and smaller telerobots can be constructed, the need for both basic and applied research in this area will steadily increase. For example, it will be necessary to address problems associated with the scaling of movements and forces. Because the mechanical behavior of objects in the micro domain are radically different than in the macro domain, such scaling will require the development of new types of telerobotic controllers.

RECOMMENDATION: The committee recommends that consideration be given to the development and application of distributed telerobotic systems.

Relatively little attention has been given to teleoperator systems in which the human operator is interfaced to a distributed set of telerobots. Because many functions require sensing or acting over a region that is large relative to the size of an individual telerobot (e.g., patrolling land or structures for security reasons), such systems, if appropriately designed and developed, would have many important applications. Issues that need to be addressed in this area include the careful selection of specific applications, the design of the communication system for transmitting information among the telerobots and between the set of telerobots and the human operator, and the design of human-machine interfaces that are well matched to human sensory and control capabilities in situations involving multiple telerobots.

RECOMMENDATION: The committee recommends the establishment of intercommunication standards for point-to-point connections in coarse-grained parallel computational architectures. However, for applications with demanding input/output operations, the committee does not recommend new real-time development systems or operating systems.

The most demanding VE system will require powerful real-time input/output capabilities to handle haptic interfaces, trackers, visual displays, and auditory displays. In robotics and telerobotics, in which the requirements are similar, there is a general movement toward coarse-grained systems based on point-to-point communications, such as transputers and C40 systems. Commercial development environments and real-time operating systems are adequate for such systems. However, there has not yet emerged a high- speed intercommunication standard for point-to-point computational architectures, which would offer users great flexibility in mixing and matching components across vendors and different processor types.

Networks

The committee anticipates that in the future most VE applications will rely heavily on network hardware and software. Although networks are now becoming fast enough for distributed VE applications, development is needed to provide the enormous bandwidth required to support multiple users, video, audio, haptics, and possibly the exchange of three-dimensional interaction primitives and models in real time. Moreover, handling the mix of data over network links will require new applications level protocols and techniques. Because of the central nature of network technology to the implementation of VEs, the committee sees network hardware and software development as critical to advancing the science of VE and its applications. However, we believe that the hardware necessary to support VE applications will be developed without intervention from the VE research community. In other words, there are forces in both the federal government and the private sector that are driving major advances in hardware. As a result, we do not recommend additional investment in network hardware development at this time.

Nevertheless, it is important to acknowledge the existence of significant infrastructure problems that could impede the use of networks for VE applications. For these problems, specific effort should be provided in support of VE requirements. One infrastructure issue is the high cost of research on large-scale networked VEs. A very limited number of universities can afford to have dedicated T-1 lines (with installation expenses of $40,000 and operating costs of $140,000 per year, as for the Defense Simulation Internet currently) needed to support these activities. Various approaches, such as an open VE network and the necessary VE applications protocol, should be considered for providing research universities with access to the needed facilities. Unless costs are significantly reduced, it will not be possible to initiate a concerted effort to develop software solutions for networked VE.

Perhaps our greatest infrastructure concern is the need for the development of network standards that will be compatible with the long-range needs of distributed VEs. One danger is that the entertainment industry, with its interest in interactive games for the home, will set the networking protocol standards at the low end, and the military community will set the standards at the high end. Therefore:

RECOMMENDATION: The committee recommends that the federal government provide funding for a program (to be conducted with industry and academia in collaboration) aimed at developing network standards that support the requirements for implementing distributed VEs on a large scale. Furthermore, we recommend funding of an open VE network that can be used by researchers, at a reasonable cost, to experiment with various VE network software developments and applications.

Evaluation of SE Systems

RECOMMENDATION: The committee recommends that the federal government encourage the SE system developers it supports to include a comprehensive evaluation plan in the early design stages of their research projects. It also recommends that the federal government help coordinate the development of standardized testing procedures for use across studies, systems, and laboratories, particularly in those areas in which the private sector has not acted.

SE technology is in the early stages of development, is growing rapidly, and is the subject of highly optimistic projections about its usefulness. In contrast, the extent to which its usefulness has actually been seriously evaluated is vanishingly small.

In general, evaluations are required not only to compare overall cost effectiveness of SE approaches with other approaches addressed to the same goals, but also to provide insights to guide modifications and new design directions. To be optimally effective, such evaluations must take place both at the overall system level, at the component level, and at all stages of the development process. Although many of the specific questions to be addressed in an evaluation effort are likely to depend to some degree on the structure and purpose of the system or component in question, it should be possible to determine a common framework for a substantial portion of the evaluation needs.

In order to help ensure adequate SE evaluation, the federal government should encourage individuals involved in federal supported research and development to include serious evaluation plans in the design of their projects. Such plans should address questions about engineering performance, user needs and acceptance, dependence of human performance and safety on various system or component features, and costs of development, implementation, and marketing. Furthermore, in order to facilitate consistency across SE projects, the federal government should help coordinate the development of standardized testing procedures for use across studies, systems, and laboratories, particularly in those areas in which the private sector has not acted. These procedures should include methods for identifying key system dimensions that affect task performance, developing special metrics uniquely suited to evaluating SEs, and comparing SE system performance to performance of other systems intended to meet the same or similar goals.

Suggestions for Government Policy and Infrastructure

The magnitude, quality, and effect of the SE-oriented research and development that is accomplished will clearly depend on the role played by the federal government. The current status of the SE field is sufficiently embryonic, compared with what is likely to develop over the next 10 years, that the federal government now has a rare opportunity to foster coherent planning in this area. Furthermore, the recently established National Science and Technology Council at the White House would appear to be an appropriate organization to provide oversight for such a planning effort. Also, in conducting such a planning effort, substantial benefits would be gained by attending carefully to the developments that are already taking place in the other areas of the administration's planning effort—for example, the Advanced Technology Program of the National Institute of Standards and Technology, the High Performance Computing and Communications program, and the programs associated with defense conversion.

In this section, we discuss a number of mechanisms that illustrate the kind of leadership role that the government could play. We see that role as both informing and complementing the federal agencies' strategic planning for their support of research and development programs.

Establish an Effective Information Infrastructure

A national information system that provides comprehensive coverage of research activities and results in the SE field in a user-friendly way to a wide variety of users could be a useful tool for promoting cross-fertilization and integration of the research and development efforts. The free flow of ideas and information among researchers, users, and individuals in government, academia, and industry who require information for SE planning and decision making is crucial to the development of this new field. Also, in order to diminish the increasing threat of a major societal division between the technologically advantaged and the technologically deprived (as well as to counter the current hype about virtual reality), the public should have information of the appropriate type in an easily available form. Although information by itself cannot prevent such a division, it is a necessary ingredient of any program that could.

We suggest that the federal government consider establishing a national information system in order to promote these vital communication goals. To reduce costs and to realize potential benefits as soon as possible, consideration could be given to integrating the SE information system with other public information systems currently being developed. For example, such a system might be an ideal component of the national digital library based on high-speed networking envisioned in the National Information Infrastructure (NII) initiative of the Clinton administration. Issues of ownership and control, as well as technological issues, will be important to consider in the design of an SE information system.

To some extent, the technology, procedures, and ideas being developed within the SE field itself could be usefully exploited in the design of the SE information system. Such a system might eventually have uses well beyond those initially envisioned; for example, it might include a library of computational models. Although for many years there has been a tendency for scientists to express their understanding of various systems and phenomena in terms of computational models, this tendency is clearly being accelerated by the role such models play in the generation of VEs. Indeed, it seems possible that, in the near future, computational models will constitute one of the society's primary forms of knowledge representation. Thus, for example, reading a book about Newtonian mechanics is likely to be augmented by interacting with a virtual world based on a computational model that includes Newtonian mechanics and then, perhaps, "reading" the computational model. The same kind of evolution is, of course, occurring with fiction and imaginary worlds; independent of whether a structure or a series of events is real or imaginary, much of the relevant information can be stored in the form of a computational model. In order to make such computational models available to society, the federal government might consider establishing a national system for standardizing, collecting, storing, and disseminating such models. In view of society's current concerns with health care, initial efforts in this area might be focused on computational models related to the structure and function of the human body and modifications of the human body associated with injury, trauma, disease, aging, and medical and surgical treatments.

Encourage Appropriate Organizational Structures and Behaviors

Two major factors that could inhibit advances in SE involve the ability of researchers to communicate and cooperate across disciplines and across organizations. Because the creation of effective SE systems requires contributions from many different disciplines (with many different associated cultures), special efforts are required to ensure adequate communication and cooperation across disciplines. Similarly, because of the high value placed on competition within our society, special efforts are needed to ensure adequate communication and cooperation across government agencies, military branches, industrial firms, and academic institutions. At present, the organizational barrier appears to be more debilitating than the disciplinary (or cultural) barrier. In fact, the lack of cooperation among competing organizational entities (for example, competing companies) probably constitutes the main obstacle to achieving a truly satisfactory solution to the information infrastructure problem discussed above. Consideration of explicit incentives for cooperative behavior might be very useful.

In order to reduce these problems, the committee suggests that the federal government consider establishing a small number of national research and development teams, each of which would focus on a specific application area. These teams could involve government, industry, and academia, as well as the various disciplines relevant to the given application area. Funding could be provided jointly by the federal government and the private sector.

The work to be performed by each national research and development team would include basic research, technology development, functional prototypes, technology evaluation, and technology transfer to industry. Despite the emphasis on applications that is implied by how the teams are defined, the work could be directed toward long-term as well as short-term goals, and the basic research needed to achieve these goals would then be a priority for support. Also, to the extent feasible, it might make sense to connect these collaborative teams or applications consortia not only to already existing federal activities (as has been the case, for example, with the textile partnership AMTEX that is being managed by the Department of Energy), but also to already existing professional societies. In setting up these teams, the choice of leadership for the activity will be crucial. In some cases, federal leadership may be appropriate; in others, industrial; in still others, academic. Finally, each of the envisioned teams might well find it appropriate to develop a powerful networked communication system among its members to ensure true collaboration at the working level.

Use SE Systems Within the Government

It might be useful for some federal agencies and offices to explore the use of SE to meet their own administrative and program needs. In addition to the application of SEs to the defense and space programs already under way, other application domains, such as training, telecommunication and teletravel, and information visualization, are relevant to the activities of many agencies. One way for the government to facilitate development of the SE field would be to select a few agencies to serve as test beds for synthetic environment technology in these general domains.

There are a number of reasons for suggesting that the government make use of SE systems in conducting its own activities. Government agencies (local as well as federal) are natural early users of new technology: they could help spearhead development efforts and provide feedback to the developers. Also, such use could increase the cost-effectiveness of government activities. In addition, such use could create a market for SE systems and thus stimulate private industry to become involved in the design and production of SE systems.

At present, uses within the government that are receiving the most attention are those associated with the Department of Defense; however, other entities, such as NASA and the Department of Energy, are also involved. Although military applications trail behind those associated with the entertainment industry as an economic driving force, they nevertheless constitute a force that is significant. This significance is derived not only from the overall magnitude of the associated economic activity, but also from the special role played by defense agencies in stimulating the development of relatively high-quality systems for military applications. Also of interest in this connection are the current efforts to explore the use of SE systems in the Department of Defense for education and training. If the results of these studies are positive, they could play a significant role in stimulating the use of SE systems for education and training throughout the nation as a whole.

The use of SE systems in NASA appears to hold great potential not only with respect to training people for operations in hostile environments, but also with respect to performing the operations themselves. The enormous expense associated with manned space flights and space stations may well serve as a strong stimulus to the use of teleoperation in space activities.

Developing National Standards and Regulations

Although it is probably too early in the development of SE systems to establish national standards and regulations, it is not too early to begin to evaluate the work already under way in connection with the formulation of standards and regulations for the telecommunications and entertainment industries. Problems that are already of concern but are likely to become of even greater concern as the SE field develops relate to technological compatibility and interoperability issues, enforcement and control issues, and social and ethical issues. For example, in the technological area, problems related to the timing of information flow in SE networks merit special consideration. Similarly, in the social and ethical area, the potential of SE for providing participants with powerful emotional experiences (including those related to sex and violence) needs to be addressed.

In general, it appears that SE, because of its mass entertainment potential, is likely to become one of the largest uses of high-speed communication networks, and its use should have an early and continuing part in the development of standards, regulatory principles, and tariff-setting models for such networks. The recent congressional attention that has been given to the kinds of material that are appropriate for the media to present is but a mild precursor to the public debate that is likely to arise when advanced VE technology becomes widely available.

It will be critical for the federal government to consider VEs in the formulation of national standards and regulations. Studies could be undertaken to illuminate issues related to technological compatibility and interoperability, enforcement and control, and social and ethical problems raised by the use of VEs in society.

Analyze and Evaluate Market Forces and Societal Impact

The extent to which government funds will be directed toward specific SE research depends, at least in part, on the likelihood that such projects will be funded independently, i.e., by industry. Estimating this likelihood requires not only an analysis of current market forces, but also predictions of how market forces will evolve in the future. Although such predictions are notoriously difficult to make with accuracy, and market forces are as likely to be shaped by the results of the research and development as they are to shape the research and development that is performed, failure to consider market forces in making funding decisions is likely to seriously reduce the extent to which the funding is effective in advancing the field. For these reasons, it would be prudent for the federal government to monitor market forces as part of developing its strategic plan for the allocation of scarce resources.

As with most other technologies, the effects of the advances in SE are likely to be mixed; some effects will be positive and others negative. And as with the predictions of market forces, although accurate predictions of societal impact are difficult to derive, serious attempts to consider such factors would be decidedly worthwhile. It cannot be assumed that all technological advances, even those that are likely to have substantial practical applications, will necessarily be beneficial.

II

RESEARCH AND TECHNOLOGY

The research and technology that is relevant to the SE field covers an enormous range because of the many disciplines involved, the multimodal aspects of most SE systems, and the wide variety of potential SE applications. Thus, the topics covered in this part are not all-inclusive; they have been selected because they were judged to be relatively crucial by the committee.

The chapters in this part, the heart of the report, mirror the topics in the overview. Specifically, we present detailed information on psychological considerations (Chapter 1) and on the available research and technology involved in the creation of synthetic environments (SE) including human-machine interfaces (Chapters 2-7), computer generation of virtual environments (VEs) (Chapter 8), telerobotics (Chapter 9), networks (Chapter 10), and evaluation (Chapter 11). Much of the material in Chapters 2-7 applies to all kinds of systems (including augmented-reality systems). Chapter 8 is directed specifically toward VE systems, and Chapter 9 covers teleoperator systems.

It should also be noted that the visual channel is treated differently from the other channels and appears in more than one place. The material on the visual channel in Chapter 2 is restricted rather rigorously to human-machine interface issues. However, because most previous work by computer scientists on the computer generation of VEs has focused on the visual channel (i.e., graphics), Chapter 8, which deals with these hardware and software issues, is necessarily focused mainly on the visual channel. In order to obtain a comprehensive overview of the visual channel, it is necessary to read both Chapters 2 and 8.

In contrast, for the auditory, haptic, and other channels, for which the majority of past work has been performed by individuals from other disciplines and has been directly concerned with interface issues (or issues traditionally lumped together with such issues by these individuals), essentially all of the relevant material is contained within each chapter.

It should also be noted that our descriptions of the various channels differ in the extent to which previous knowledge on the part of the reader is assumed. For example, because it is expected that most readers are less familiar with issues related to haptic interfaces than those related to auditory interfaces, more general information is provided.

In principle, the chapter on networks (Chapter 10) is relevant to both VEs and teleoperators; however, most current activities in this area are directed toward the networking of VE systems rather than teleoperator systems. Furthermore, even within the domain of VEs, relatively little attention is being given to the communication of signals required for haptic interactions. These factors too, like those mentioned above, are reflected in the way in which the material is presented.

1

Some Psychological Considerations

Because human beings are an essential component of all synthetic environment (SE) systems, there are very few areas of psychology that are not relevant to the design, use, and evaluation of SE systems. For example, if the system under consideration is a virtual environment (VE) system that is intended to provide realistic simulations, then all the issues relevant to the identification of the effective stimulus in real environments, as well as the issues that focus on how equivalent perceptions or responses can be achieved with more simply synthesized artificial stimuli, must be examined. If it is a VE system that is intended to maximize information transfer to the user and incorporates special distortions for this purpose, or if it is a teleoperator system that incorporates a non-anthropomorphic telerobot, then all the issues relevant to the perception of, adaptation to, and learning about altered perceptual cue systems must be considered. In addition, to the extent that the system can be thought of as an extension of traditional manual control systems, many of the concepts and findings relevant to such systems are likely to be applicable. Further issues arise in connection with higher-level processes related to learning and the formation of problem-solving strategies and cognitive models, as well as with the effects of SE experience on affect, motivation, personality, etc. A similarly broad range of issues is generated when one scans across the various application areas of SE systems.

The topics covered in this chapter—which represent only a minute sample of all relevant topics—were chosen to illustrate some of the types of issues that need to be considered. Although the topic of discomfort obviously contains elements outside the domain of psychology, it is included here for convenience.

RESOLUTION, ILLUSIONS, AND INFORMATION TRANSFER

Perhaps the most obvious kinds of knowledge about human perception and performance that are needed to design cost-effective SE systems concern the resolution of the human's input and output systems and the way in which effective resolving power is changed as these systems are integrated with SE interface systems having various kinds of displays and controls. (The term *resolution* refers here to the ability to separate out and independently sense different signals as well as to detect small changes in isolated signals.) Given such knowledge, one can then examine implications for task performance for various types of tasks, and the cost-performance trade-offs for these tasks.

Knowledge of normal human resolving power on the input side, i.e., the sensory side, allows one to predict the display resolution beyond which finer resolution could not be perceived and would therefore be wasted. A similar statement holds for the output, i.e., control, side. Although knowledge of human resolving power in vision and audition is incomplete, it is sufficiently advanced to provide designers of SE systems with solid background for design choices. Areas in which current knowledge is considerably less adequate include both the input (sensory) side and the output (motor) side of the haptic system, as well as the ways in which performance is degraded when displays and controls (in any of the modalities) with less-than-human resolution are used. Information on resolution for specific modalities (e.g., vision) is provided in the chapters concerned with these modalities.

A further and related set of issues that is important to consider in the design of SE systems concerns perceptual illusions. Generally speaking, a given perception is thought of as illusory to the extent that it appears to be generated by a stimulus configuration that is different from the actual one. VEs themselves can be regarded as integrated sets of illusions. Detailed study of both intrasensory and intersensory illusions is important because, in many cases, the existence of illusions enables SE system design to be simplified and therefore to increase its cost-effectiveness. At the opposite end of the spectrum, the occurrence of unexpected illusions can seriously interfere with the expected performance of the system. Elicitation of motion sickness often involves the occurrence of illusions concerning the position, orientation, and movements of various portions of the body.

It is possible to regard certain types of illusions, such as the illusion of continuous motion that can be generated by sequences of static images at rates of 30 Hz, merely as reflections of imperfect sensory resolution and therefore to assume that studies of resolution will automatically include studies of illusions. However, other types of illusions, such as the Muller-Lyer illusion, are more appropriately characterized in terms of response bias and therefore cannot be regarded in this manner. Thus, it is necessary to consider illusions as a separate issue from resolution.

Much of the past work on illusions has focused on the visual channel and on the implications of these illusions for theories of visual perception and cognition. However, some results, such as the continuous motion illusion just cited, clearly have direct implications for SE design. Other illustrative results relevant to SE design include those on the dominance of vision over audition and haptics in cases of intermodality conflict (e.g., as evidenced in the ventriloquist effect) and on the use of auditory stimuli to improve the perception of events that are represented primarily in the visual or haptic domains (as in the use of sound effects). Material on illusory effects for vision can be found in Howard and Templeton (1966); for audition in Bregman (1990); and for haptics in Loomis and Lederman (1986), Hogan et al. (1990), and Fasse et al. (1990).

It should also be noted that relatively little work has been done on sensorimotor illusions associated with whole-body movements. The factors involved in these illusions, which usually involve the perception of body movement, support surface stability, and visual field stability, are likely to be of considerable importance in SE designs that include voluntary locomotion through virtual space. Further material on these kinds of illusions can be found in Chapter 6.

Finally, it should be noted that the merging of data from different sources in augmented-reality systems is likely to lead to a whole new set of illusory effects that will require study. Relatively little is known about the effects of different merging techniques (even if one restricts one's attention solely to the merging of visual images).

Issues related to information transfer rates tend to be very complex because such rates depend not only on basic resolving power, but also on factors related to learning, memory, and perceptual organization. With respect to information transfer rate, an SE system can be thought of as consisting of a human operator, an artificial machine (a computer or telerobot), and a two-way communication link consisting of displays that send information from the machine to the human operator and controls that send information from the human operator to the machine. One of the main goals in such systems is to optimize the efficiency of communication in both directions. For many purposes, it is useful to characterize the imperfections in the communication channels in terms of statistical variability (noise), to include in this noise both channel noise and noise internal to the human and/or the machine, and to measure the efficiency of the communication by the information transfer rate. Crudely speaking, the information transfer is defined as the information gain resulting from the communication, which in turn can be defined as "how much more the receiving system knows about the state of the transmitting system after the communication signal is received than before it is received." The information transfer rate is then defined simply as the rate at which information is transferred. Within this context, a good human-machine interaction technique is one in which the information transfer rate is high and the amount of training required to achieve this high rate is low. Extensive background on the use of information theory concepts in characterizing human performance is available in Quastler (1955), Garner (1962, 1974), Sheridan and Ferrell (1974), Stelmach (1978), and Rasmussen (1986).

In general, in order to get high information rates into the human operator via displays or out of the human operator via controls, the human operator must be very familiar with the information coding scheme employed. Perhaps some of the coding schemes with which individuals are most familiar are those related to language. Estimates of maximum information transfer rates involving language reception and transmission in various modalities have been presented by Reed and Durlach (1994). The results indicate that maximum rates for reading English (vision), listening to spoken English (hearing), and observing the signs in American Sign Language are all roughly the same and lie in the range 60-70 bits/s. Reception of language in the haptic domain (by means of Grade 2 Braille, the feeling of signs in sign language, or by the Tadoma method, in which certain deaf and blind individuals receive speech tactually by placing one's hand on the face of the talker and monitoring the mechanical actions of the speech articulation system) shows maximum rates of roughly one-half those obtained via the visual and auditory channels.

The maximum output rate for the motor actions of the speech articulators in speech production is estimated to be roughly the same as the maximum rate for listening to speech (60 bits/s) and for the motor actions of the hands in typing to be roughly 20 bits/s. To the extent that (1) the assumptions that underlie these results are valid (in particular, that the estimate of 1.4 bits of information per letter for sentence length segments of speech is reasonable) and (2) these results provide an upper bound on the information transfer rates that can be achieved in communicating between human operator and machine in SE systems, the amount by which the results achieved with a given SE system fall below these rates provides a measure of the room for improvement. It should also be noted that the figure of 20 bits/s appears to provide an upper bound on the rate that can be achieved in simple discrete spatial tracking tasks (e.g., one in which the transmitted signal consists of lighting up a randomly selected square in a checkerboard array presented on a visual display and the correct response consists of touching the lit square with a finger or directing one's gaze at the lit square).

Unfortunately, although there are a number of general statements that can be made about the properties of a good coding system (e.g., it should be well matched to the properties of the sensory or motor system involved, it should make use of perceptually high-dimensional stimuli or responses, it should have cognitive properties that make it easy to learn, etc.), there is no theory available that enables one to make reliable predictions of performance as a function of the detailed coding scheme and detailed training procedures employed.

These problems become even more challenging when one replaces the individual human operator by a collaborative team. In such cases, the system designer must also consider how to best break up the input information and output control among a number of individuals.

MANUAL CONTROL, TRACKING, AND HUMAN OPERATOR MODELS

In SE systems, motor outputs of the operator are used to control the behavior of a simulation or a telerobot. Furthermore, these outputs usually take account of feedback and occur in real time. In an important sense, therefore, such systems can be regarded as descendants of traditional manual control systems, with their emphasis on tracking paradigms and human operator models. The term *manual control* is taken to mean the receipt of sensory information about the desired state of a system and its current state by a human operator and the use of that information by the operator to command inputs to the system through mechanical devices (hand controllers, pedals, etc.) so as to minimize some function of the error between those two states (Sheridan and Ferrell, 1974). The principal differences between current SE systems and traditional manual control systems consist of the increased variety and complexity of the displays and controls, as well as of the constraints imposed on the relevant transfer functions.

Research in manual control was most active in the 1950s, 1960s, and 1970s due to the interest in operator control of aircraft, automobiles, and other vehicles (a comprehensive overview is found in Sheridan and Ferrell, 1974). Additional useful references include Jex (1971), Kleinman et al. (1974), and McRuer et al. (1965). In manual control systems, there is a closed loop that includes the behavior of the human operator (human operator dynamics), the system or plant being controlled, and feedback to the operator regarding the performance goals and plant state. Classification of manual control systems according to the type of input to the human operator is as follows.

• *Compensator* The operator controls the system to reduce the error between the state of the system and a fixed reference state.

• *Pursuit* The operator controls the system to reduce the error between the state of the system and a changing reference state.

• *Preview* The operator, having knowledge of future values of the reference state, controls the system to reduce the error between the state of the system and a changing reference state.

• *Precognitive* The operator, having foreknowledge of input in terms other than a direct view (for example, statistics on the input), controls the system to reduce the error between the state of the system and a changing reference state.

Human operator dynamics influence the closed-loop performance in manual control systems as much as do the plant dynamics. For example, the control of aircraft cannot be predicted unless a model of the pilot is factored in as well (Sheridan, 1992c). Initially it was thought that a human operator model independent of the manual control task could be formulated; however, research shows that, depending on the controlled process, the behavior of the human operator is modified to achieve satisfactory performance. A number of models of the human operator of varying complexity have been proposed; the more well-known of these are listed below.

• *Linear, quasilinear models* The simplest is a linear model of the operator with adjustable gain and a remnant (noise). These models have been further extended to handle reaction-time delay (Sheridan and Ferrell, 1974).

• *Crossover model* Experiments by Elkind (1956) and McRuer et al. (1965) show that operator behavior was dependent on the error signal; the operator was found to change his dynamics so that the combined system had good servo behavior at crossover frequency. These results were valid for compensatory tracking tasks but were not valid for high-bandwidth tasks or control of high-order linear and highly nonlinear systems.

• *Optimal control model* These are another class of models of the human operator in manual control tasks based on results obtained from compensatory tracking experiments in which the operator is modeled as an optimal controller within limits of internal constraints and knowledge of task objectives (Bryson and Ho, 1975).

As human manual control behavior became better characterized and the role of the human operator in teleoperation changed to performing complex manual and decision-making tasks in unstructured environments, research shifted to modeling the higher-level functions performed by the operator (Baron et al., 1982; Kok and Van Wijk, 1977; Rasmussen,

1983; Sheridan, 1976, 1992c). Current related work has focused on the incorporation of manual control models of the operator in the characterization of the human operator as a supervisor and decision maker.

Subsystem Decomposition of the Human Operator

The human operator models mentioned above are black box models based on control theory formulations. The original ideal of developing human operator models independent of the task is still a good one; however, this requires a detailed understanding of cognitive, perceptual, and motor functions that is still far from complete. A subsystem decomposition of the human operator for pursuit tracking based on physiology would include the following elements (Jones and Hunter, 1990):

- Sensory processing, in which the central nervous system (CNS) measures tracking error (actual minus desired target position) either visually or kinaesthetically;
- Cognitive processing, in which the CNS generates appropriate central commands to the motor neuron pool in the spinal cord;
- Excitation/contraction coupling (muscle), which reflects the propagation of action potentials through muscle as a result of efferent nerve input and the initiation of contractile mechanisms (cross-bridges), which produce muscle force;
- Limb mechanics, in which muscle force generates limb displacement; and
- Reflexes and nerve delays in which displacements of the limb excite muscle spindles that feed back to the motor neuron pool in the spinal cord over afferent nerves and sum with the CNS-generated motor neuron input to produce muscle activation over efferent nerves.

Each subsystem contributes its own dynamics to the overall human operator dynamics. Delays associated with each subsystem are different. Whereas the transformations are linear and time-invariant for some subsystems, they are nonlinear and time-varying for others. For the neuromuscular subsystems, recent advances in stochastic system identification techniques now make it possible to determine the dynamics of each subsystem for every individual operator (Kearney and Hunter, 1987).

Methodologically, the dynamics of the sensory and cognitive subsystems are what remain when the dynamics of all the other subsystems have been subtracted from overall human operator performance. Of course, modeling the cognitive aspect is the most difficult and incomplete. Certain aspects of the modeling of the sensory attributes are discussed in Chapter 1 for vision, Chapter 2 for audition, and Chapter 3 for mechanical interface variables. In particular, a knowledge of the sensory resolution limits for length, force, stiffness, viscosity, and mass is important for understanding human operator performance and for design of haptic interfaces (Jones and Hunter, 1992; see also Durlach et al., 1989; Pang et al., 1991; Tan et al., 1992, 1993). Increasing the stiffness of a manipulandum decreases the response time but also the accuracy (Jones and Hunter, 1990) while increasing the viscosity decreases the delay and the natural frequency of the human operator (Jones and Hunter, 1993). This enhancement of performance with the addition of stiffness and viscosity to the manipulandum may be due to an increase in proprioceptive feedback from the periphery.

Sensory Substitution

An ongoing area of research is the substitution of feedback to alternative sensory channels of the operator, that is, presentation of sensor information to the operator in a sensory channel other than that in which it was sensed by the teleoperation system. There are a number of reasons for choosing sensory substitution:

- Shielding the operator from hazards but still conveying information on the conditions of the environment, for example, in chemical spill, high temperature, and radioactive environments.
- Presenting sensor information in visual form, for example, dials, gauges, etc., for lack of other suitable choices.
- Using the higher sensitivity available in the alternative operator sensory channel, for example, the representation of temperature may be to tenths of degrees on a visual display; this is far above the ability of the operator to discriminate when sensing actual temperature.
- Reducing the cognitive load on the operator.
- Overcoming the drawbacks of force reflection due to instabilities when operating with time delay. Additional reasons are to reduce the size of hand controllers and eliminate reaction forces on the operator.

Reports of research in this field have concentrated on sensory substitution of force reflection. Some early work in this area was that of Bertsche et al. (1977) and Bejczy (1982) on the use of visual displays of force feedback. More recently, Massimino and Sheridan (1993) documented the use of tactile and auditory displays for force feedback. There are many more applications yet to be tested, particularly in situations in which data from different types of sensors are to be fused into a single measure of interest. Further results on sensory substitution are available from studies concerned with aiding individuals who are deaf or blind or deaf and blind (e.g., Bach-y-Rita, 1972, 1992; Reed et al., 1982; Warren and Strelow, 1985; Reed et al., 1989).

In addition to sensory substitution, the provision of visual aids and haptic aids may enhance operator performance (see Chapters 8 and 11 for further discussion).

THE SUBJECTIVE SENSE OF TELEPRESENCE

One major feature of a user's experience when operating an SE system concerns the extent to which the user is immersed in, and actually feels present in, the remote or synthesized environment, i.e., the extent to which subjective telepresence occurs. Whereas objective telepresence refers to the use of teleoperator technology for sensing and manipulating remote entities, subjective telepresence refers to the sensations and perceptions experienced by the user. For simplicity, in the following remarks on subjective telepresence, the modifier *subjective* is dropped and we use simply the term *telepresence*. Also, because essentially all the issues with which we are concerned here are independent of the distinction between VE and teleoperation, in the following remarks we ignore this distinction; the term *telepresence* is used in connection with all types of SE systems. (Those wishing to make such a distinction might use the term *virtual presence* for VEs, as suggested by Sheridan, 1992a.)

At present, there appear to be three main questions of interest in relation to the concept of telepresence: How should telepresence be defined operationally? How can one create telepresence? What is telepresence good for?

Although there have been a number of discussions of telepresence in the literature (e.g., Akin et al., 1983; Fontaine, 1992; Heeter, 1992; Held and Durlach, 1991, 1992; Loomis, 1992; Pepper and Hightower, 1984; Schloerb, 1994; Sheridan, 1992b; Steuer, 1992; Zeltzer, 1992), there is still no generally agreed-on operational definition of telepresence. It should be noted, however, that a serious effort in this direction has recently been initiated by Schloerb (1994) and that a number of individuals are now attempting to conduct empirical research on telepresence.

Almost all of the articles just cited attempt not only to define telepresence, but also to identify the factors that play a role in the creation of telepresence. Although the particular set of factors considered or emphasized varies with the author's viewpoint, there are a number of factors that are relatively obvious. One set of such factors concerns the exclusion of stimuli originating in the immediate environment. In other words, the sense of telepresence is likely to be reduced if the operator is constantly reminded of his or her presence in the real environment by stimulation originating in this environment. Such stimulation can arise from sources outside the system (e.g., the auditory component of the human-machine interface provides inadequate attenuation to prevent the operator of the

SE system from hearing a door slam in the room that houses the SE system) or from the system itself (e.g., the helmet used for the visual display is too intrusive to be ignored).

A second set of such factors concerns the existence of user-predictable interactivity. Telepresence is likely to increase when the user's actions, and the consequences of these actions as represented by the subsequent stimuli sensed by the user (i.e., the feedback), constitute a rich and easily perceived and influenced interaction pattern. When the synthetic world is highly realistic, such conditions will be satisfied automatically. When it is unrealistic, the extent to which they are satisfied will depend on the extent to which the user can adapt to the new world. As the user adapts, the degree of telepresence (and the transparency of the interface) will generally tend to increase. The extent to which the user can adapt to the new world, however, will depend strongly on both the nature of this world and the nature of the user's exposure to it. If the world is incomprehensible (either because the relations between the user's actions and the effects of these actions are random or simply because they are so complicated that they appear random), adaptation will not occur. Further discussion of adaptation is presented in the next subsection.

A third set of such factors relates to higher-level, more cognitive features of telepresence and to similar experiences that occur outside the domain of SE technology. In general, the ability of humans to be transported into unreal worlds is the basis of most art, literature, theater, and entertainment, not to mention hypnosis and the use of hallucinogenic drugs. The extent to which consideration of these other forms of transportation will prove useful in the study of SE telepresence is not yet clear. It does appear likely, however, that the variation among individuals in their susceptibility to transportation by these other methods is likely to also occur with SE telepresence.

The question "What is telepresence good for?" has not yet been adequately answered. The interest in the concept of telepresence is due in part to intrinsic philosophical and scientific interest in issues concerned with reality and illusion. It is also due in part, however, to an implicit assumption that a high degree of telepresence is positively correlated with good performance. That this is not generally the case, however, can be easily demonstrated merely by noting that one of the primary motivations for the use of teleoperator systems in hazardous environments is to prevent the operator from experiencing noxious stimuli present in the real environment (i.e., reducing the sense of presence in the real environment). In general, the relationship between telepresence and performance has not yet been determined. Furthermore, even if telepresence were adequately defined in an operational sense, and even if it were determined using this operational definition that in most situations telepresence and performance were highly correlated, it still would not be clear that the concept of telepresence has practical significance. In particular, it is not clear that it would enable one to design better SE systems. In order for this to be the case, it would be necessary to show that models and measurements of telepresence can be usefully substituted for models and measurements of performance, or at the very least, that models and measurements of telepresence provide significant added value to the results that can be achieved solely through the use of models and measurements of performance.

ALTERATIONS OF SENSORIMOTOR LOOPS

In practically all SE systems, the human operator's normal sensorimotor loops will be altered by the presence of distortions, time delays, and noise (statistical variability) in the system. In many cases, such alterations will be introduced unintentionally and will degrade performance. For example, time delays may result from the need to communicate over long distances, and time delays, noise, and unwanted distortions may result from the inclusion of imperfect system components. In some cases, however, these alterations (specifically distortions) may be introduced intentionally in an attempt to achieve performance that is better than normal—for example, in a teleoperator system that incorporates a telerobot that is intentionally nonanthropomorphic. Because of the lack of isomorphism (i.e., structural and functional similarity) between the operator and the telerobot in such systems, the mapping between the human operator and the telerobot will necessarily result in altered sensorimotor loops for the operator. Similar conditions will exist in all virtual environment systems in which special features of the environment are artificially emphasized and unrealistic methods for interacting with this environment are employed in an effort to achieve superior task performance. Attempts to achieve improved resolution by magnification of perceptual cues represent only one line of investigation in this area.

Independent of the nature and origin of the alteration, in order for a system designer to predict the performance of a candidate system, theoretical models must be available for characterizing human responses to the alterations associated with the use of the system. Such models should be able to predict the effect of the alterations on such variables as simulator sickness (and also, perhaps, telepresence), as well as on objective task performance, and to describe how the various response components change over time due to sensorimotor adaptation and learning.

Although considerable work has been performed in this area (e.g., see the extensive review by Welch, 1978), there are as yet no adequate models available for predicting performance. For example, no adequate models are available for specifying the amount of sensorimotor adaptation that is achievable with different kinds of distortions using different types of training procedures. Similarly, no attention has been given in the research on adaptation to changes in *resolution;* attention has been focused almost exclusively on changes in *response bias* (i.e., the deviation between the mean response and the correct response). Furthermore, with only minor exceptions, interactions among different kinds of alterations (distortions, time delays, and noise), many of which are likely to be present simultaneously in SE systems, have been ignored. A few modality-specific comments on sensorimotor adaptation are available in Chapters 2, 3, and 4. Extensive further discussion of sensorimotor adaptation in the context of whole-body motion is available in Chapter 6.

DISCOMFORT

An important prerequisite for widespread use of SE systems is that they be comfortable for people to use. Independent of whether the discomfort caused by the system is most appropriately considered under the heading of "motion sickness," "poor ergonomics," or the "sopite syndrome" (see Chapter 6 for a definition of this syndrome), such discomfort must be reduced sufficiently to permit individuals to make effective use of the system over extended periods of time. Despite significant previous research on some components of this problem, substantial further research in this area is warranted for a number of reasons.

First, as the situation now stands, discomfort is a real threat to the effective use of SEs. For example, quite apart from the deficiencies in currently available helmet-mounted displays with respect to the visual information provided, they tend to cause such a high degree of discomfort that daily long-term use seems almost out of the question. In fact, the combination of relatively limited visual information and relatively high discomfort is leading some individuals to seriously consider using off-head displays in their SE systems (discussion of both helmet-mounted displays and off-head displays is presented in Chapter 2).

Second, past work on the sources and effects of discomfort has not yet resulted in adequate understanding of the phenomena involved. More specifically, we do not yet know how the magnitude of each discomfort component depends on the characteristics of the system (the properties of the visual and auditory displays, the weight of the devices mounted on the head, the method by which movement through space is simulated, etc.) or on the characteristics of the individual user (including the user's prior experience with the system). Although some progress has been made in related areas (e.g., studies of motion sickness conducted in connection with flight simulators, sophisticated use of anthropometric measurements for cockpit design), this progress has not yet led to adequate comfort for SE users.

Third, many of the situations created by evolving SE technology are new; the stimulus-response configurations to which individuals are exposed in SE systems are often not covered by those previously studied. In other words, the increased flexibility associated with the new technology provides us with opportunities not only to perform certain tasks with increased cost-effectiveness, but also to expose ourselves to situations that exhibit new kinds or magnitudes of discomfort.

Fourth—and illustrative of the point made elsewhere in this report about the use of SE systems as basic laboratory facilities for psychological research—evolving SE technology provides us with new tools to study some of these issues. Not only is work in this area essential to the realization of practical applications, but it should also advance our understanding of the human organism.

Further material related to the discomfort issue is available in later chapters. In particular, extensive discussion of motion sickness and the sopite syndrome appears in Chapter 6.

LEARNING AND PROBLEM SOLVING

Understanding how humans learn and solve problems is critical to the development of educational and training systems regardless of the nature of the instructional tools employed. VE is one such tool, and the more we know about how the mind works, the better able we will be to create experiences that facilitate learning by its use. Thus, in general, an appreciation of human cognition is an important element in using synthetic environment technology to alter human behavior.

Work in the development of cognitive models has a long history. As early as 1932, Bartlett developed the notion of a schemata as a large knowledge structure—the basic unit of memory and thought. Schemata are conceived to exist at all levels of abstraction and to be hierarchically organized and interrelated; they are used to comprehend new and complex situations and are, in turn, modified by experience. This concept was revived in the late 1960s by cognitive psychologists and computer scientists who used it as a paradigm to test hypothesis about human mental processes.

A similar concept was brought into use by computer scientists who were also concerned with modeling cognitive processes. They used the term *frames* to describe the organization, structure, and developmental process of human memory (Minsky, 1975; Kuipers, 1975). Much of their work involved creating computer programs that acquired knowledge, followed procedures as described in scripts, and solved problems in ways that were hypothesized to mimic human problem-solving strategies (Newell and Simon, 1972). The results of these studies led to new hypotheses and more fully elaborated theories.

More recently, researchers have used cognitive models to develop intelligent tutors. One of the more long-standing research efforts in this area is the work of John Anderson and his colleagues at Carnegie Mellon University in which the ACT (Adoptive Control of Thought) theory of learning and problem solving was used to build intelligent tutors in algebra, geometry, and LISP programming language. The ACT theory makes a distinction between factual or declarative knowledge and procedural knowledge. Declarative knowledge involves the acquisition of facts (the content of a theorem); procedural knowledge in the form of production rules relates to the development of cognitive skill (the ability to apply a theorem). The early stages of learning are dominated by declarative knowledge; the later stages by procedural knowledge. According to a recent review by Anderson et al. (1993), the 10-year effort has led to further understanding of human cognition as well as to an appreciation for how to implement the system in the classroom. One important finding was that the original conception of tutoring as a process of human emulation changed to the notion of a tutor as a learning environment in which helpful information can be provided and useful problems can be selected.

Another important line of research in cognitive science is modeling the knowledge structures and judgments of experts and novices and comparing the two as a basis for understanding the nature of expertise and for training novices to become experts. For example, Chi et al. (1981) have examined the differences in the knowledge structures and problem approaches of expert and novice physicists as a way to better understand how the acquisition of knowledge and rules changes problem-solving strategies. The schema, algorithms, and heuristics used by experts were explicated by using such methods as cognitive task analysis or think-aloud protocols (Newell and Simon, 1972). According to Glaser et al. (1991), several knowledge models representing various stages in moving from a novice to an expert would be useful in guiding the learning process.

Using cognitive task analysis, Lesgold and his colleagues have described various stages of expertise in electronic troubleshooting as a basis for developing dynamic assessments of learner competence. The resulting computer system, known as Sherlock, uses information on the stages of expertise to track a learner's performance, to diagnose strengths and weaknesses in both knowledge and process, and to provide corrective feedback (Lesgold et al., 1990; Lajoie and Lesgold, 1992).

Although cognitive researchers have made considerable progress in understanding how the mind works, there are still many issues to be examined. Specifically, we need to know much more about what aspects of synthetic environments will facilitate learning. Research is needed to further understand the relationship between content, types of tasks, the individual's knowledge state, and preferred information presentation features. Moreover, it will be important to examine the special contributions of immersive environments that are feasible through the use of VE compared with other formats for enhancing education and training. Of particular interest here is the opportunity for extensive sensorimotor involvement provided by VE systems.

MOTIVATION

A question that cuts across the objective-subjective boundary is whether immersive environments contain intrinsic advantages with respect to motivation or incentives to participate in entertainment and educational experiences. Again, it is too early for definitive answers. However, anecdotal reports suggest that young people can become so immersed in computer-based group games that they neglect such basic activities as eating and sleeping. In education, Neuman (1989) has shown that young people with learning difficulties have taken to computer-based instruction with enthusiasm that lasted longer than could be attributed to a novelty effect. Observations suggested that the children were transforming the otherwise ordinary lessons into competitive games. For young people, computers and game playing seem to be conceptually linked. Perhaps this link can be constructively exploited by giving VE instructional experiences a suitable gamelike quality to strengthen student engagement.

In more recent studies, another explanation for student enthusiasm has surfaced (Coombs, 1993; Morrison et al., 1993; Coldevin et al., 1993; Dubriel, 1993). According to this explanation, the computer appears to empower students who are otherwise educationally disadvantaged. The promise of VE is that its effects should be stronger than those of other computer-based instructional facilities. For example, students with physical handicaps could be given experiences via VE that they otherwise could not have. Likewise, students in remote settings could participate in experiences otherwise impossible to obtain—such as a stroll through a big city. Finally, VE might be an ideal means for simulating certain types of experiences for which there are emotional, political, or economic barriers. For example, animal rights advocates have made a political issue of the dissection of animals for biology instruction. VE could provide a means for learning anatomical details that would be free of negative emotional overtones and might be more economical.

Although some skeptics have argued that most of the observed advantages attributed to all forms of computer-based instruction derive from novelty effects, there is some additional evidence to the contrary. Introspective reports from adult learners confronting difficult material (Vasu and Vasu, 1993) and from teachers for whom the novelty had certainly worn off after repeated use of the same courseware (Novelli, 1993) indicate a well-sustained sense of continued interest in using computer-based resources.

ATTITUDE, OPINION FORMATION, AND PERSONALITY EFFECTS

From both scientific and social perspectives, problems exist with respect to the possibility that VE might change individuals' attitudes or self-perceptions. The most credible hypotheses are that such powers are limited by both technical and psychological factors. With regard to the psychological aspect, changes in attitudes and opinions are seen to be most effectively brought about by person-to person interactions. For example, it has long been known that personal contact plays a major role in expediting changes in the attitudes that control acceptance or rejection of innovations. Likewise, of the development of a sense of self appears to depend strongly on social transactions (e.g., Mead, 1934). In other words, one's idea of self—including self-regard or self-esteem—depends on what messages are sent day after day by a set of significant others. We do not yet know the extent to which and manner in which such social transactions change as the world in which the person is operating becomes more imaginary.

This emphasis, in both attitude formation and self-concept development, on social processes highlights specific technical properties of the VE situation. That is, how cost effectively can VE generate images of humans who can interact realistically with the subject? In particular, can VE generate images of specific significant others, such as family members? And how rigidly can or must such images be programmed in advance? Although social interaction might not be absolutely necessary to engender attitude change, such change may be greatly facilitated by social persuasion.

Finally, in the domain of personality development, there are possible changes that might be situationally but not socially induced. A good example is risk tolerance. A subject could be exposed via VE to situations in which normal physical laws do not operate, as in the mode of Wiley Coyote in a Roadrunner cartoon. After experiencing in a VE situation any number of events whereby the action of gravity was significantly delayed and the ultimate consequence of being hammered down by two tons of rock was only a slight dizziness, would one tend to act less cautiously in the real world? The answer, based again on arguments by analogy, appears to be that some such shifts in perception are possible, but only if the conditions are tightly controlled. One crucial parameter to be controlled is task ambiguity. Research on effects of social influence (Kidd, 1958) shows that some effects can be induced very quickly if the situation is ambiguous and there are no serious consequences for making the wrong choice.

Another potentially crucial parameter is the initial attitudes of participants. If they are young, inexperienced, and uncertain about underlying event probabilities or vulnerable to certain forms of peer pressures, change can be readily induced. Such subjects could possibly be more influenced by elaborately contrived experiences in a VE situation—again, particularly if the VE system fabricated specific images of other people, such as important peers (Sjoberg and Torell, 1993; Benthin et al., 1993).

Finally, there is some anecdotal evidence that computer role-playing games and video games that portray violence may have some influence on individual attitudes and behavior (much of this preliminary research grows out of the efforts to demonstrate the influences of television on attitudes and behavior). For example, there are cases of individuals becoming so involved in playing Multiple User Dungeons (a computer fantasy game) that they leave little or no time for other activities. A recent article in the *Washington Post* (Schwartz, 1994) described a college student who flunked out of college and stayed up almost all night every night to play a fantasy character and interact with other fantasy characters in the Multiple User Dungeons game. Basically, this young man lost his real identity to a character in a game.

With regard to violent video games, there are some recent studies that suggest that children who play these games are more aggressive as a result. Some preliminary studies in this area include Fling et al. (1992), Funk (1992), and Cesarone (1994). These studies are based on questionnaire and survey results rather than empirical evidence of changes in performance. Nevertheless, they suggest that further attention should be given to the potential effects of such games, particularly as these games become more realistic and more interactive.

In summary, VE could probably be used to engender substantial changes in the psychic structure of participants. The magnitude of such changes will depend on the quality and types of images that the VE system can generate, the congruence of the inclusion of some form of controlled message content in the VE programming, the initial susceptibility of participants, and the ongoing willingness of such participants to accept messages or situations that appear to contradict or deviate from their other, non-VE experiences.

RESEARCH NEEDS

Many of the needs in the area of psychological research, implicitly outlined in the above discussion, will be automatically satisfied as the field of experimental psychology follows its normal evolutionary course of development. Without special effort, however, much of the information and understanding required to guide evolution of the SE field will be available too late to be useful. In order to significantly increase the cost-effectiveness of the SE research and development work, as well as to determine the likely psychological effects of heavy SE usage before these effects are prevalent throughout the society, substantial work must be done within the next few years.

As indicated previously, the goals of this work should be to develop (1) a comprehensive, coherently organized review of human performance characteristics from the viewpoint of SE systems; (2) a theory that facilitates quantitative predictions of human responses to alterations in sensorimotor loops that are likely to occur in SE systems; (3) cognitive models that will facilitate effective design of VE systems for purposes of education, training, and information visualization; and (4) increased understanding of the possible deleterious effects of spending substantial portions of time in SE systems. The important issue of user comfort is partially addressed by item 2 in that feelings of discomfort such as those associated with simulation sickness constitute a particular type of response to alterations of sensorimotor loops. Many other aspects of discomfort, such as those related to poorly fitting helmets for visual displays, are best thought of purely in terms of physical effects.

2

The Visual Channel

The visual interface will, in many cases, provide the human user with the most salient and detailed information regarding the synthetic environment (SE). Given current technical limitations on the real-time display of detailed, continuous-motion imagery, making optimal use of normal visual sensory capabilities constitutes a formidable challenge for system designers. Of the many display options that currently exist, none is completely adequate across all applications.

STATUS OF THE RELEVANT HUMAN RESEARCH

It is clear that the limits imposed by technological constraints on stimulation interact with the information-processing capabilities of the human observer. Consequently, the effectiveness of current technology must be assessed against knowledge of the full range of human sensory capabilities. A proper understanding of both the technological and sensory aspects of visual displays defines the challenge for subsequent research and development (Boff et al., 1986; Haber and Hershenson, 1980; Sekuler and Blake, 1990).

An effective visual interface provides an appropriate match of the parameters of stimulation to the characteristics of the sense organ. The production of visual stimuli in such a system depends in part on technological capabilities for monitoring movements of the observer (to generate both movement-contingent and noncontingent sensory information), graphics-processing power, and display characteristics. For a visual interface to be effective in an SE context, the sensory information need not exceed, and may often be less than, the sensory-processing capabilities of the average human observer. The latter observation is borne out by the experience of people with deficient vision (or hearing): people with many kinds of sensory disabilities can still experience both real and virtual environments (VEs). Indeed, perturbations of normal visual stimulation may be disturbing but not necessarily destructive of perception in either real or synthetic environments. When sensory conditions are maintained with minimal variation, various forms of adaptation can occur and can be used to advantage in overcoming technical limitations (Held and Durlach, 1991).

SE-Relevant Aspects of Visual System Organization

The superficial similarity between the human visual system and created imaging systems (such as a television camera) has the potential to misguide efforts to advance the state of the art of the visual channel of SE systems. One example of a fundamental difference between the design goals of the two systems lies in the manner in which images are collected. Created systems generally strive to collect a uniformly resolved image of a scene. For applications in which any area of the scene might be attended to (by the ultimate viewer), this is an appropriate goal. However, the ideal proximal stimulus in an SE system is one in which the information is presented in a manner that is complementary to the normal operation of the sense organ. In the case of the eye, an optimal system would not ignore the role of eye motions, the uneven distribution of photoreceptors in the retina, and the limitations of the eye's optics (Westheimer, 1986). Because the eyes are used to actively probe a scene (Gregory, 1991) in a way somewhat similar to an insect probing the world with antennae, the concentration of image quality at the center of the fixation point may be a highly appropriate use of resources. In a sense, the fovea is like the sensitive tip of a blind person's cane; the rest is just context.

Although these concepts have begun to be addressed by designers of eye-slaved displays, the design of the visual channel of SE systems is likely to benefit from more widespread conceptualization of the eye as an output device as well as an input device. There are a number of potential benefits of utilizing such fixation information. It can guide the allocation of resources by matching the information transfer rate to the receiver (e.g., high resolution can be limited to the fovea). It can reduce the computational burden of image generation in VE systems. The fixation point information collected can be used to adapt the information presented (in the case of teleoperator systems) or generated (in the case of VE systems) to more seamlessly match the interests of the viewer.

The effects of spatiotemporal blur can be managed using fixation point information. In this case, the analogous phenomena of display persistence (due to the time constants of the imaging device) and retinal persistence or smear (due to the time constants of the receptors and light-adaptation mechanisms) have different perceived outcomes based on the mode in which they are used. Direct view of a panoramic display requires that moving images present in the display be carefully managed so that attention is not directed to motion blur. This usually involves creative application of cinematographic techniques (Spottiswoode, 1967). The human observer, however, remains largely unaware of motion blur because of the links between eye motion, retinal and higher-level "suppression," and the perceptual processing involved in active viewing of a scene (Matin, 1975; Matin et al., 1972). Thus, rapid target motions that elicit saccadic (ballistic) eye movements are not perceived as blurred because visual sensitivity to the target is reduced during the period beginning about 50 ms before the saccade is initiated to 50 ms after the new fixation point is reached (Latour, 1962). Slower target motions in the range of 5 to 40 deg of visual angle per second resulting in pursuit eye movements are much less subject to suppression effects, thus presenting a challenge for the display designer.

In the visual system there are many such opportunities to achieve the dual goals of more closely accommodating the sense organ while allocating technological resources economically. The development of color head-mounted displays (HMDs), for example, has presented difficulties due to the need to present three channels of chromatic information. Examination of the photoreceptor population of the retina reveals first that the color-sensing receptors (cones) are not evenly distributed throughout the retina. The greatest concentration of cones is found where the visual axis intercepts the retina, but beyond 10 deg of visual angle the cone density is uniform at about 5 percent of the central value. Moreover, the noncolor-sensing rod population at this intercept point is zero but increases with eccentricity to a maximum at 18 deg. Thus rods are present in far greater density throughout the periphery than are cones, with the result that the resolution of color information is limited in the periphery (Hood and Finkelstein, 1986).

Cones come in three general types as defined by the photopigment present. By measuring the differential response of the three cone types, the visual system is able to determine (within a metameric equivalent) the wavelength of stimulation. But unlike the case of created imaging systems, the quantity and distribution of the different cone types is not uniform. Overall, the ratio of short (S), medium (M), and long (L) wavelength peak cones is about 3:7:14. Short cone representation in the retina is relatively sparse, drastically limiting the resolving power of the eye on this color dimension. (It also follows that the foveal vision system is particularly dependent on the medium and long cones and relatively independent of the short cones and rods.)

These nonhomogeneities provide another opportunity to match the needs of the visual system while achieving technological savings. Since the bandwidth for color information is not uniform throughout the retina, the image-generating component of the display system is relieved of a technological constraint.

There are about 120×10^6 rod and 6.5×10^6 cone photoreceptors per eye. This represents an extraordinary amount of information to be processed by the visual system. In order to accommodate this flow of information, initial image processing occurs at the retinal level. The mosaic of receptors is not connected one-to-one with the optic nerve fibers. In fact, there are only about 1×10^6 fibers leaving the eye that contain the codified information from the over 126×10^6 photoreceptors. Intermediate cells (horizontal, bipolar, and amacrine) make excitatory and inhibitory connections to groups of receptors in various spatial configurations. These connections result in receptive fields that preferentially respond to retinal excitation of specific spatial configurations and temporal qualities. (The duplex nature of the retina alluded to earlier also becomes evident in the connectivity pattern: rods are part of one network and the various classes of cones are part of another, with minimal synaptic communication between the two.) Once again we find that the visual system allocates the smaller, higher spatial resolution channels to the fovea.

The rather long axons of the retinal ganglion cells constituting the optic nerve regroup into left and right optic tracts containing left field or right field information from both eyes. About 20 percent of these fibers split off from here to make synaptic connections at the superior colliculus (SC), while the remaining 80 percent terminate in the lateral geniculate nucleus (LGN). (Of the ganglion cells destined for the LGN, there are two classes: P ganglion cells make connections to parvocellular areas and M ganglion cells make connections to the magnocellular areas of the LGN.) These connectivity patterns yield two important insights to the functioning of the visual system that have relevance for the display designer.

The primary flow of information through the lateral geniculate nucleus is from the retina to the visual cortex. But, in addition to the initial processing of motion, color, etc., described above, the lateral geniculate nucleus also serves as a modulator of visual stimulation. The thalamus, of which the lateral geniculate nucleus is a part, generally serves as a sort of volume control for sensory stimulation. (There are corresponding nuclei of the thalamus that provide analogous modulatory control for most other senses.) In this capacity, feedback signals from the reticular activating system provide control that is based on the general level of arousal, whereas signals from the visual cortex help direct visual attention.

The first insight based on the postretinal connectivity patterns concerns the emergence of separate temporally acute and spatially acute processing channels. These channels roughly correspond to the periphery and fovea, respectively. There are six layers observable in the structure of the LGN: four outer layers made up of small cells (parvocellular layers) and two inner layers containing large cells (magnocellular layers). The parvo cells, although slower, process finer details in the image and support color perception (opponent color connection patterns, etc.). The color-blind magno cells respond quickly and are involved with motion perception.

The disproportionately large representation of the fovea, along with the "receptive field" neural coding of higher visual features (including more abstract visual characteristics such as orientation and contour) repeats throughout the cortex and is a basic mechanism of information extraction in the visual system (Graham, 1989). These representations may have implications for creating appropriate image compression and generation algorithms in SE systems.

A second insight (based on connectivity patterns) that should be more carefully studied and exploited involves the ambient visual system. As mentioned above, 20 percent of fibers from the optic tracts converge on the superior colliculus. This midbrain structure has close links to lower autonomic centers responsible for emotion, arousal, and motor coordination. The control of eye movement depends in large part on these signals (Hallett, 1986). One important neighboring area, the pretectal region, controls the pupillary reflex. Also, vestibular and tactile inputs, as well as signals coming from muscle tension and joint position sensors, converge in various ways with these signals from the superior colliculus (and other brainstem nuclei) yielding a concrete feeling of bodily position and configuration that ultimately guides motion.

We have seen many dualities present in the visual system: rod versus cone and their retinal distribution with respect to the fovea; P ganglion versus M, with polysynaptic and the analogous parvo connections supporting the perception of detail and color; superior colliculus versus cortical processing, with cortical signals underlying conscious perception of visual images. This latter distinction supports the notion that there are two visual systems: the focal system and the ambient system. One example of the separateness of the focal system is the ability (often called blindsight) of cortically blind people to point to objects they cannot "see."

This focal-ambient duality lends support to Bridgeman's (1991) demonstrations of separate cognitive and motor-oriented maps. The implications for display designers include customizing the display based on its intended function: spatial displays, such as representations of instruments and dials, are subject to motion-induced illusions and cognitive biases; displays designed to serve as a means of facilitating visually guided behavior (such as tracking or piloting tasks) are not subject to these illusions, but are subject to constraints based on limited memory for this information and need to be continuously displayed.

The integration of vestibular and visual information leads to interactions of form and orientation that can be exploited by the display designer (Mittelstaedt, 1991). Given the technical difficulties involved in artificially communicating inertial and gravitational cues in an SE, and considering the potential for enhancing the effectiveness of environments emphasizing orientation and motor control (as almost any application of SEs does), the area of ambient visual system effects should be explored. In addition, further knowledge of the autonomic effects of the ambient visual system are likely to be instrumental in mitigating the sopite syndrome.

The implications of the technical factors presented above for immersion and performance are not well known. Given the present limitations in generating veridical displays, much of what is known about vision will not be accommodated by the technology in the near term. However, if we are to realize the full potential of the visual interface, we must not let display development be driven solely by current trends in electronic design.

Sensory Constraints for Displays

Spatial distortion and image degradation may be caused by either inadequate channel capacity in the display unit or by inadequate optical arrangements between the display and the user's eyes. Channel capacity sets a fundamental limit on all attributes of the displayed imagery, including spatial resolution, temporal update rates and lags, contrast, luminance, and color resolution. Distortion of optical factors can arise in connection with (among other things) the choice of focal distance, the stereoscopic arrangement, the alignment of the optic axis, and the relation between convergence and accommodation.

Special optical arrangements are required to transfer the image on the display screen to the image on the retina of the human observer. Undesirable distortions can result from this transfer, especially when the display screens are positioned just a few centimeters from the eyes (as in most virtual environment displays). To project the image on a tiny display screen into a real-world-size image on the retina, a high degree of optical magnification is required. Furthermore, at a fixed close viewing distance, the larger the magnification, the greater the geometric distortion (particularly at the greater eccentricities that can arise in displays with large fields of view).

In general, the optics must allow for clear focusing, minimization of geometric distortion, and off-axis eye movements without significant image vignetting. It must also be lightweight and easily adjustable for a wide variety of human users. In the case of stereoscopic displays, great care must be taken to ensure adequate alignment of the images provided to the two eyes and minimization of the inherent conflict between ocular convergence and accommodation (e.g., see Miyashita and Uchida, 1990). For example, in the case of near-field imagery, the image disparity between the two eyes will necessarily be large to elicit the proper stereo depth effect. In natural viewing this disparity in the near field is accompanied by an appropriate level of ocular convergence and lens accommodation. In a collimated-optics HMD, these additional depth cues are not present, so the percept is degraded.

Temporal resolution limitations center on three separable factors: frame rate, update rate, and delays (Piantanida et al., 1993). Frame rate is related to the perception of flicker and is hardware-determined. Because the human temporal contrast sensitivity function (CSF) shifts toward higher frequencies in response to increases in the brightness level of the display, higher light levels result in greater flicker sensitivity. Although models of the CSF can predict the human response to these stimuli (Wiegand et al., 1994), precise recommendations of frame rates must take into account the emission characteristics of the display as well as the time course of the stimuli. The typical luminances produced in current non-see-through displays are sufficient to cause flicker for rates of 30 Hz and below. The higher light levels required for see-through displays, coupled with the limited control of real-world lighting in many augmented-reality situations, may result in the need for higher frame rates for such applications. The CSF is also dependent on retinal eccentricity: peripheral sensitivity to flicker is greater than foveal (Pirenne and Mariott, 1962). As field-of-view limitations are removed, peripheral flicker will become objectionable.

Update rate is determined primarily by computational speed. When the view in an HMD is updated at a rate above 12 Hz, motion is perceived as smooth, and motion parallax cues support depth perception. Below this rate illusory motion or even simulator sickness may be induced (Piantanida et al., 1993).

At the current time, orientation and position tracker system delays are the major contributing factor to viewpoint update delays in HMD images. These "phase lag" delays can be as long as 250 ms for commonly available electromagnetic tracking systems (Meyer et al., 1992). Because all of the above delays are additive, motion artifacts and desynchronization of the visual channel with respect to the other channels of a multimodal SE are likely to dwarf the more subtle characteristics of human temporal response based on the CSF.

The Third Dimension

Distance in the third dimension (from the eyes) and its corollary, depth perception, are ubiquitous properties of realistic vision. A variety of monocular stimulus conditions are capable of eliciting the appearance of relative depth. These include relative motion, perspective, occlusion, shadows, texture coarseness, aerial perspective (changes of color and luminance with distance), and others. Beyond approximately 1 m, these monocular depth cues predominate in creating the percept of depth, whereas in the near field (arm's length), the combined use of motion cues with stereopsis becomes crucial for eliciting three-dimensional perception that is convincing.

Binocular vision provides all the stimulus conditions of monocular vision plus the advantage of stereopsis with its high resolution (a few seconds of arc) as well as the disadvantage of requiring alignment of the two eyes to preclude the appearance of double images and binocular rivalry.

Our ability to simulate full binocular vision requires better technology than is now available for the following reasons. In stereoptic vision not only is resolution of each image good to about 1 minute of arc, but also relative displacements between corresponding image features on the two retinas (disparities) can be discriminated to at least an order of magnitude better. No currently available display can meet such specifications. As a consequence, resolution of detail in all three dimensions is considerably less than optimal—it corresponds to the vision of a visually disabled observer. However, inasmuch as depth perception is also elicited by the relative motion of retinal images and the use of triangulation (parallax), some degree of distance discrimination can be achieved without stereopsis.

Augmented-Reality Systems

Many of the factors discussed above become especially critical when augmented-reality systems are considered. The difficulties stem largely from the immediate comparison between the real-world stimulus and the synthetic elements displayed as an overlay. When the subtle shifts in light qualities, zero delay viewpoint changes, and three-dimensional qualities of objects present in the real world are combined with the less-responsive synthetic stimuli, separation of the real and the synthetic is obvious. Similarly, the perception of the surface qualities of reflective objects, the distinction between self-luminant and reflective objects, the perceptual invariance of color and brightness of objects based on the perceptual integration of light source and other information present in the entire scene, are all likely to constrain the type of information that can be effectively and unambiguously displayed (Roufs and Goossens, 1988; Westerink and Roufs, 1989).

The visual system has evolved to resolve ambiguities in the proximal stimulus through knowledge about how visual scenes behave. In using see-through HMDs, it will be hard not to violate this knowledge. Further research is needed to evaluate these effects and to develop robust presentation techniques that are effective within the technological limitations.

Sensory Adaptaton to Visual Distortion

Many simple alterations involving the visual channel are effortlessly adapted to. For example, most users have little difficulty learning to position a cursor on a vertical computer screen through the use of a mouse constrained to a horizontal tabletop. Stable homogeneous alterations can be adapted to, even if they are extreme (Kohler, 1964 [1951]). Nonhomogeneous alterations that are stable with respect to the visual field (such as those introduced by prism goggles) can also be adapted to (Held and Freedman, 1963). Because HMD optics (or any lens system placed in front of the eyes) exhibits these prismatic distortions, the design of such systems can be informed by what is known about the process of adaptation to prismatic distortions.

Often, distortions are purposely introduced to enhance perception (e.g., magnification to increase visual acuity). By incorporating appropriate distortions, displays can be better matched to specific environments or tasks, thus enhancing the visual capabilities of the observer. These enhancements would typically involve local changes in magnification (similar to the use of bifocal lenses), but they can also be based on higher-order properties of the visual stimulus such as alterations of optical flow fields in motion-oriented tasks.

Another example of an enhancing visual distortion involves changes in the effective interocular distance. Depth through stereopsis is limited to a maximum retinal disparity that can be accommodated by ocular geometry and the size of the retinal fusional areas. The result is that stereopsis starts to substantially degrade at approximately 10 m and is not effective beyond about 135 m from the viewer (Haber and Hershenson, 1980). Increased interocular distance has been used in binoculars and other optical instruments to overcome this limitation. Enhancement of the near field is not so straightforward, however. The ability to fuse the disparate retinal images into a perception of depth is limited by fairly unchangeable neural connectivity patterns. The limit of maximum disparity (Parnum's limit) suggests an upper bound to the degree of enhancement of stereopsis that is possible through eye position manipulation. Adaptation to extreme interocular distance changes, if they occur at all, are likely to have extremely long time courses and may never become complete.

It is clear that the purposeful introduction of alterations has the capacity to both overcome equipment limitations and enhance the sensory capabilities of the observer. Development of visual displays that capitalize on these effects must combine existing knowledge of perception with findings from our growing experience with SEs.

STATUS OF THE TECHNOLOGY

An ideal visual display would be capable of providing imagery that is indistinguishable from that encountered in everyday experience in terms of quality, update rate, and extent. Clearly, however, current technology will not support such highly veridical visual displays, nor is it clear that such a high level of verisimilitude is required for most applications. The relative importance of various display features including visual properties (e.g., field of view, resolution, lumination, contrast, and color), ergonomics, safety, reliability, and cost must therefore be carefully evaluated for any given application. McKenna and Zeltzer (1992) provide a thorough comparison among several of these features for five major three-dimensional display types.

A complete visual interface consists of four primary components: (1) a visual display surface and an attendant optical system that paints a detailed, time-varying pattern of light onto the retina of the user; (2) a system for positioning the visual display surface relative to the eye of the operator; (3) a system for generating the optical stimuli (either from camera sources of real-world scenes, stored video imagery, or computer synthesis or some combination of the three); and (4) a system for sensing the positions and motions of the head and/or eyeballs. Items (3) and (4) are treated in Chapters 5 and 8; they are not treated here except insofar as they have direct implications for the first two topics.

Two major classes of visual display systems are available for SE applications: head-mounted and off-head displays. The imagery displayed by either form of device may be coupled to the motions of the user's head either directly (using sensors that measure head position and orientation) or indirectly (typically using manual control devices such as joysticks or speech input).

At present, the majority of visual displays for synthetic environments are physically coupled to the head of the operator by mounting display hardware on a helmet or headband worn by the user. A significant advantage of head-mounted displays is that the display positioning servomechanism is provided by the human torso and neck. This allows the generation of a completely encompassing viewing volume with no additional hardware and eliminates lag introduced by display surface positioning systems required by some off-head strategies. In many HMDs, all the imagery is synthetic and generated by computer. For certain augmented-reality displays, however, a semitransparent display surface is used and the synthetic imagery is overlaid on imagery resulting directly from objects in the environment. In other augmented-reality displays, the synthetic imagery is combined with imagery derived from optical sensors on a telerobot.

Among the disadvantages of head-mounted displays are the weight and inertial burden that interferes with the natural, unencumbered motions of the user, the fatigue associated with these factors, and the increased likelihood of symptomatic motion sickness with increased head inertia (DiZio and Lackner, 1992). In general, it is difficult to build head-mounted display devices that exhibit good spatial resolution, field of view, and color yet are lightweight, comfortable, and cost-effective.

Although often not associated with synthetic environment applications, a variety of off-head displays are available for SE tasks. These displays range from relatively inexpensive monoscopic and stereoscopic panel and projected displays to experimental autostereoscopic displays. High-resolution color panel and projected display systems are available at relatively low cost due to common use in the computer graphics and entertainment fields. These systems, at most, require a lightweight pair of active or passive glasses to generate high-quality stereo and therefore impart a minimal inertial burden on the user and are comfortable to use. Within the limits of comfortable viewing range, the static field of view and spatial resolution of panel and projected displays is determined by user distance from the display surface. Relatively large field of view (e.g., > 100 deg horizontal field of view) can be readily achieved with larger display surfaces without requiring correcting optics. Panel and projection displays are typically larger and heavier than HMDs, a disadvantage in volume- and weight-limited applications. In addition, servo-controlled or multiple static display surfaces must be used to provide a completely encompassing visual environment. Another approach to increasing viewing volume is to use head position and orientation sensors of the same type and form used on head-mounted displays. Although the relationship between the user's head position and orientation and the remote or virtual viewpoint is not so straightforward as that encountered when using head-mounted displays, this form of viewpoint control can nonetheless be used to advantage to both control the viewpoint into the synthetic environment (Spain, 1992) and to correct for off-axis viewing perspective (McKenna and Zeltzer, 1992).

Hybrid panel display systems exist that use smaller display surfaces servo-controlled or manually steered to the user's head position and orientation (Oyama et al., 1993; MacDowall et al., 1990). These systems remove the inertial burden of head-mounted displays yet allow the use of more readily available, nonminiature, high-resolution color panel display devices to generate the imagery. In addition, user head position and orientation information is readily available through instrumentation of the display support structure, and consequently a large effective viewing volume is achieved without additional complexity. In servo-controlled variants, delay and distortions introduced by the electromechanical control mechanisms must be dealt with, in addition to the safety issues associated with coupling a powered device to the human head. Manually steered devices do not have the same safety concerns but do limit the applications to which these devices may be applied, since one or both hands must be used to move the visual display and therefore are not available for manual interaction tasks.

Autostereoscopic off-head display techniques, such as slice-stacking and holographic video, do not require special viewing aids (e.g., field-sequential or polarized glasses). Displays of this type, currently in the research and development phase, are discussed further in the next section.

Components and Technologies

Display Surfaces and Optics

The most frequently used and technically mature display types for synthetic environment applications are cathode ray tubes (CRTs) and backlighted liquid crystal displays (LCDs). Although these technologies have proven to be very useful for near-term HMD applications, several shortcomings compromise their long-term promise. CRT technology has been able to deliver small, high-resolution, high-luminance monochromatic displays. These CRTs, however, are relatively heavy and bulky and place very high voltages on the human head. In addition, the development of miniature, high-resolution, high-luminance, color CRTs has been difficult. In contrast, current LCD technology can produce color images at low operating voltages, but only at marginal picture-element densities. Both technological approaches require bulky optics to form high-quality images with sufficiently large optical exit pupils. The following paragraphs discuss emerging technological approaches that have the potential to produce high-resolution color (Tektronix or custom built devices) imagery while reducing some of the weight, bulk, and cost of current displays.

A near-term approach that is bringing high-quality, color, CRT-based HMDs to the commercial marketplace is the use of mechanical or electronic color filtering techniques applied to monochrome CRTs. In this approach, the CRT is scanned at three times the normal rate, and red, green, and blue filters are sequentially applied. Some of the integrated systems discussed in subsequent sections of this report use this technique.

Commercially available SE displays have almost exclusively relied on TV-quality liquid crystal display technology. This technology was developed and chosen because of the demand for large-area displays for computer and television screens; it is limited in its maximum pixel density and size by the thin-film manufacturing techniques used for its fabrication. The quality of transistors, lithographic resolution, and parasitic resistance of wires in the thin-film technologies is considerably inferior to that capable of being realized in silicon very large scale integrated (VLSI) technologies. In the VE and teleoperator fields, large-area displays are not desirable. In these applications, very high resolution in compact, lightweight form is needed.

Thomas Knight of the MIT Artificial Intelligence Laboratory is pursuing such display characteristics using silicon VLSI chip technology. The potential for density and performance can be seen by comparing the available resolution of liquid crystal (LC) displays with the density of commercially available image sensors based on silicon technology. The LCDs have maximum resolutions of about 640 × 400, with a pixel pitch of about 330 μ. Silicon sensors have been fabricated with resolutions of 4,000 × 4,000 pixels, with dot pitches of 11 μ. Smaller pixel dimensions are lithographically possible, but light sensitivity demands the use of larger pixel areas. Knight believes that a feasible prototype in today's technology is a 2,000 × 2,000 image display with pixels on an 8 μ pitch, producing a display with an active area of 16 mm × 16 mm.

There are several approaches to coating such a high-resolution substrate with optically active materials. One simple approach is to use conventional field-sensitive liquid crystal material as a reflective display. An advantage of SE applications is the ability to control the illumination environment, making many problems in LC display technology, such as angular field of view, less critical. Some indication of the possible resolution available with LC techniques on silicon is the application of liquid crystals to diagnosis of failing integrated circuit parts, in which lines of under 3 μ are routinely imaged (Picart et al., 1989).

Kopin, Inc., in Taunton, Massachusetts, is also investigating high-resolution LCDs on silicon, with the use of very thin (1,000 A) silicon substrates, which are optically transmissive. The Kopin approach relies on fabricating very thin layers of recrystallized silicon on a silicon dioxide substrate, forming circuitry on the substrate, and finally releasing the recrystallized silicon by etching. The released thin silicon sheet is then floated onto a glass substrate and used as a transparent display. Devices with a resolution of 1,100 lines per square inch over a 0.5 × 0.5 inch area have been manufactured.

Another optical interface approach is the use of electrochromic polymers, such as polyaniline (Kobayashi et al., 1984). This material changes color reversibly from a pale yellow to a dark green when reduced by current flow in solution. By coating an area array of noble metal electrodes fabricated on an active silicon structure, reduction in pixel-sized regions is achievable. With either of these display technologies, the major advantages of cost-effective, mass producible, lithographically defined devices remain, leading to potentially dramatic system-level economic benefits.

The optical portion of a SE visual display is potentially amenable to significant cost savings with the use of holographic optical elements (HOEs) as a replacement for bulky and heavy lens assemblies. The combination of a silicon display and HOEs could lead to a light, cost-effective SE interface suitable for mass production. Work in this area is ongoing at the University of Alabama, where the focus is primarily on computer-generated holograms using LC materials on silicon substrates as a high-resolution binary display.

A unique alternative to the classic HMD based on CRT or LCD technology is being pursued at the University of Washington's Human Interface Technology (HIT) Laboratory (LaLonde, 1990). The HIT laboratory is developing a display based on laser microscanner technology; it will use tiny solid state lasers to scan color images directly onto the retina. The advantage of this approach is that it does not require the use of the heavy, bulky optics typically used in CRT/LCD-based collimated aerial image systems and has the potential for developing high-resolution, lightweight, and low-cost display systems. The laser microscanner display, however, still faces substantial technical obstacles. A thoughtful analysis of this form of display may be found in Holmgren and Robinett (1994).

Autostereoscopic Displays

Autostereoscopic displays are interesting in that no viewing aids (e.g., field-sequential or polarized glasses) are required and no inertial burden is added to the human user. Depending on the size of their viewing zone or viewing volume, autostereoscopic displays can be seen by multiple viewers. Some displays, such as *lenticular* or *parallax barrier* displays, can present a small number of precomputed perspective views, such that limited motion parallax and lookaround—usually restricted to the horizontal plane—can be generated without tracking viewer head position. *Holographic* and *slice-stacking* displays can present continuously varying perspective views, although in a restricted viewing volume. Since autostereoscopic displays entail either restricted viewing volumes or large, resolution-limited display surfaces, however, they may not be appropriate for SE applications requiring a completely encompassing viewing volume (McKenna and Zeltzer, 1992).

Lenticular Displays A *lenticular sheet* is an array of cylindrical lenses that can be used to generate an autostereoscopic three-dimensional image by directing different two-dimensional images into viewing subzones. The subzones are imaged out at different angles in front of the lenticular sheet. When an observer's head is positioned correctly, each eye will be in a different viewing zone and will see a different image, thus allowing for binocular disparity.

Lenticular imaging requires very high resolution to image a large number of views. With CRT technology, the pixel size limits the upper resolution and thus the number of views. The bandwidth requirements can also become very large, since *N* views are displayed. Furthermore, *N* views must be rendered in real time, with the imagery "sliced" and placed into the vertical strips behind the lenticules. The number of displayable views is limited by the imperfect focusing ability of the cylindrical lenses. Lens aberrations and diffraction of the light reduce the directivity of the lenses, so that the focused imagery from the back screen does not emerge with parallel rays, but rather spreads with some angle. This spread limits the number of subzones that can be differentiated from each other. Another key issue with lenticular sheet displays is that the back screen imagery must be closely aligned with the slits or lenticules, otherwise the subzone imagery will not be directed into the appropriate subzone.

Parallax Barrier Displays A *parallax barrier* is a vertical slit plate that is placed in front of a display, simply to block part of the screen from each eye. A parallax barrier acts much like a lenticular screen, except that it uses barriers to obstruct part of the display rather than lenticules to direct the screen imagery. The screen displays two images, each of which is divided into vertical strips. The strips displayed on the screen alternate between the left and right eye images. Each eye can see only the strips intended for it, because of the slit plate. More than two images can be displayed on the screen to create multiple views from side to side. When a CRT monitor is used with a parallax barrier, the horizontal resolution is divided by the number of two-dimensional views provided. Multiple projecting monitors can be used to maintain a higher horizontal resolution with a large number of views. Each projector images a different viewpoint, and the barrier and diffusion screen direct the light back to the viewing zones. Parallax barriers are not commonly used because they suffer from several drawbacks. First, the displayed imagery is often dim, because the barrier blocks most of the light to each eye. Also, with small slit widths, the diffraction of light from the slit gap can become problematic, as the light rays spread. As discussed above, the CRT imagery must be segmented into strips, as with a lenticular sheet display.

Slice-Stacking Displays *Slice-stacking displays*, also referred to as *multiplanar* displays, build up a three-dimensional volume by layering two-dimensional images (slices). Just as a spinning line of light emitting diodes (LEDs) can perceptually create a planar image, a rotating plane of LEDs can create a volumetric image. A similar volume can be scanned using CRT displays and moving mirrors. Rather than using a planar mirror, which would have to move over a large displacement at a high frequency, a variable-focus, or varifocal, mirror can be used. A 30 Hz acoustic signal is commonly used to vibrate a reflective membrane. As the mirror vibrates, its focal length changes, and a reflected monitor is imaged, over time, in a truncated-pyramid viewing volume. The mirror continuously changes its magnification, so that imagery scanned over time (as a CRT operates) will be continually changing in depth (not in discrete slices). A variant on this approach is under development by Texas Instruments. In this technique, micromechanical mirrors, 17 μ square, are supported by silicon beams on diagonal corners. The two unsupported corners are metalized and used as electrodes in an electrostatic actuator, which allows the mirrors to be pulled to one side or the other. Actuation rates of about 10 microseconds and angular deflections of about 10 deg allow these microscopic mirrors to deflect incoming light to form high-resolution displays. A version containing approximately 700 × 500 pixels using frame sequential color derived from a color filter wheel and six bit pulse width control of the intensity of each pixel has been demonstrated.

Another method for generating a volumetric image is to illuminate a rotating surface with a random access light source. Some experimental systems have employed a spinning double helix, illuminated by lasers and controlled by acousto-optic scanners (Soltan et al., 1992). In order to illuminate a specific location in the volume, the laser is timed to strike the helical surface as it passes through that location.

Slice-stacking methods trace out a luminous volume, such that objects are transparent, and normally obscured objects, further in depth, cannot be hidden. This can be ideal for volumetric data sets and solid modeling problems, but it is poorly suited to photographic or realistic images with hidden surfaces. The addition of head-tracking would allow hidden surfaces to be approximately removed in the rendering step for one viewer. Not all surfaces can be correctly rendered, however, because the two eyes view from differing locations; each eye should see some surfaces that are obscured to the other.

Holographic Displays Computer-generated *holograms* fall under two main categories, computer-generated stereograms and computer-generated diffraction patterns. Computer-generated stereograms are recorded optically, from a set of two-dimensional views of a three-dimensional scene. The final hologram projects each two-dimensional image into a viewing zone, and stereo views can be seen with horizontal parallax (Benton, 1982). Full-color, high-resolution images have been generated, as well as large, wide field-of-view holograms. This is a non-real-time imaging technique, however; it requires off-line recording. A large amount of information is needed to generate the hologram as well, since every view (typically 100 to 300) must be generated.

Rather than record a set of two-dimensional views holographically, a true diffraction pattern can be generated by computer. When illuminated, the hologram will create a three-dimensional wavefront, imaging three-dimensional objects and light sources in space (Dallas, 1980; Tricoles, 1987). The methods used to compute the diffraction patterns are complex and computationally expensive. Until recently, computer-generated holograms had to be recorded using plotter or printing techniques, as an off-line process. A new method, however, allows a holographic image to be displayed in real time, from a fast-frame buffer storage (St. Hilaire et al., 1990; Benton, 1991). Because the holographic signal can be scanned in real time and potentially broadcast, this system is referred to as holographic video by its creators at the Massachusetts Institute of Technology.

The information contained in a hologram with dimensions of 100 × 100 mm with a viewing angle of 30 deg corresponds to approximately 25 Gbyte (25 billion samples), well beyond the range of current technology to update at frame rates of 30 Hz (St. Hilaire et al., 1990). The MIT system addresses this problem by reducing the information rate in three ways—by eliminating vertical parallax (saving several orders of magnitude), by limiting the viewing zone to approximately 14 deg (wider angles require higher spatial frequency diffraction patterns), and by limiting the image size. The diffraction patterns for a frame are computed on a supercomputer (Connection Machine II) in under 5 s for fairly simple objects composed of luminous points. The hologram is stored in a high-resolution frame buffer (approximately 6 Mbyte per frame) and is transmitted to a high-bandwidth acousto-optical modulator (AOM). The AOM modulates a coherent light source to create the three-dimensional image. Both monochrome and tricolor displays have been demonstrated. Recently, a larger version of the system with a modified scanning architecture has been developed and can display a holographic image at 140 mm × 70 mm and a 30 deg angle of view, each frame of which requres 36 Mbyte of computed diffraction pattern.

Integrated Systems

In the United States, following the pioneering work of Ivan Sutherland and his students in the late 1960s, research and development of display systems for synthetic environments have been carried out at MIT's Media Lab, the Ames and Langley Research Centers of the National Aeronautics and Space Administration (NASA), Wright-Patterson Air Force Base, the Naval Ocean Systems Center, the University of North Carolina, LEEP Optics, the University of Washington, the Japanese government's Mechanical Engineering Laboratory at Tsukuba, CAE Electronics, VPL Research, Virtual Research, Technology Innovation Group, Kaiser Electronics Electro-Optics Division, Hughes Electro-Optical & Data Systems Group, Stereographics Corporation, and Fake Space Labs, to name a few of the more prominent research and development centers in academia, government, and private industry. Applications of the various research systems include scientific visualization, vehicle guidance, and remote manipulation. A variety of relevant references are available in Aukstakalnis and Blatner (1993), Earnshaw et al. (1993), and Kalawsky (1993).

Head-Mounted Displays

Over the years, advanced, high-performance HMDs have been developed for various military research applications (Merritt, 1991). These systems, although quite capable, often cost in excess of $.5 million. The highest-resolution HMD technology used to date has been developed by CAE Electronics of Canada for a variety of military applications. High-resolution color images produced by a pair of General Electric liquid crystal light valve (LCLV) projectors are condensed onto fiber optic bundles that convey image information to an HMD incorporating wide-angle eyepiece optics. The Phase V version of this fiber optic HMD provides a 160 deg horizontal (H) by 80 deg vertical (V) field of view with a 38 deg (H) stereoscopic overlap centered in the operator's field of view. In contrast, commercial HMDs until recently consisted mainly of low-resolution, wide-field-of-view LCD-based systems. Historically, the most widely used commercial head-mounted systems were the Virtual Research Flight Helmet, the VPL EyePhone, and LEEP Systems CyberFace

II. All three provide a large total horizontal field of view of roughly 100 to 110 deg using LCD screens with effective spatial resolutions of around 300 × 200 pixels in the standard NTSC (National Television Standard Code) aspect (width/height) ratio. All of these devices are or were sold for under $10,000. VPL Research introduced but then withdrew a higher-resolution HMD titled the HRX, due to maintainability problems. W Industries (in the United Kingdom) markets a rugged HMD titled the Visette, with properties similar to that of the VR Flight Helmet. Neither the VR Flight Helmet nor the VPL EyePhone is currently being sold, and few CyberFace devices remain in use.

The past few years have witnessed the introduction of commercially available systems that can serve as relatively simple, robust, and inexpensive developmental tools for those interested in developing SEs. These systems are often commercial variants of military systems and are introducing color CRT-based and advanced solid-state-based display devices to the commercial marketplace.

Virtual Research, upon discontinuing its successful Flight Helmet, introduced a lightweight helmet-mounted display using miniature monochrome CRTs with color wheels known as the EYEGEN3. This 28-ounce system takes two NTSC inputs, has a resolution of 250 (H) × 493 (V), and a field of view that ranges from 32 deg (H) at 100 percent binocular overlap to 48 deg (H) at 50 percent binocular overlap. It has outstanding brightness and contrast and costs under $8,000. In a similar price range, Liquid Image Corporation's Mirage HMD uses a 5.7-inch diagonal, thin-film transistor, active-matrix LCD display to provide a 240 (H) × 720 (V) bioccular (i.e., nonstereoscopic) color display over a 110 deg (H) field of view. Although the system accepts an NTSC input, its response time is 80 ms (typical for LCD displays). At the higher end of the commercial performance spectrum (approximately $150,000), Kaiser Electro-Optics has introduced the Color Sim Eye. A direct result of Kaiser's military work, the Color Sim Eye provides 1,280 (H) × 1024 (V) interlaced lines over a 40 deg diameter field of view. Variants with 60-80 deg field of view are under development; a monochrome version is also available. CRTs are mounted low alongside the head to minimize their inertial burden during turns of the head. Simple relay optics are used to convey images into the operator's eyes. They also provide a 44 percent see-through capability. Surface-stabilized ferroelectric liquid crystal shutters are used to sequentially display the three primary colors at a refresh rate of 60 Hz (i.e., 180 Hz scan rate); it weighs 3.5 lb (1 lb less than the monochrome version). Technology Innovation Group's recently developed HMD systems provide (at somewhat higher cost) greater adjustability, more sophisticated relay optics that yield a wider field of view, and improved helmet features. Using an overall configuration similar to that of the Color Sim Eye (i.e., CRTs mounted close to the turning axes of the head), the system is capable of providing an expanded field of view while minimizing geometric distortions.

Off-Head Displays

Classic CRT-based panel and projection displays with sufficient scan rates and bandwidths can generate high-resolution (i.e., 1,000+ horizontal line) stereoscopic displays using field sequential techniques. High-resolution, color, high-bandwidth displays of this form start at approximately $2,000 for panel versions and $13,000 for wide-screen projection versions. This form of stereographic display uses a temporal sequencing approach to provide alternating stereo display pairs to the right and left eyes. The premier provider of field sequential devices in the computer graphics market place is Stereographics Corporation. Stereographics manufactures a series of LCD-shuttered glasses that use infrared technology for field synchronization. These glasses and the infrared transmitter can be added to appropriate computer consoles and projection systems for well under $2,000. A variety of even lower-cost, wire-synchronized, field sequential systems are available from 3D TV. Total system costs are approximately one order of magnitude less than that of comparable head-mounted displays.

Several interesting hybrid systems are available. Fake Space Labs has developed a class of displays known as the BOOM. The user holds the display head by two handles, pointing and moving the view direction by turning the display head as if it were a pair of binoculars. Thumb buttons can be used to control forward and backward motion within the SE. Several configurations of this rugged, floor-mounted device are available. The low-end system uses twin CRT displays to provide 640 (H) × 480 (V) resolution in monochrome. The high-end three-color version of the BOOM provides 1280 (H) × 960 (V) interlaced resolution using color filtering techniques. BOOM viewers have integral tracking sensors in the six-degrees-of-freedom support structure. This tracking system is very fast, returning orientation and position information to the computer in 5 ms or less after a change of viewing direction. LEEP Optics has introduced CyberFace III, a higher-resolution monochrome system with a single image source and a countersprung support arm to neutralize the weight on the wearer's head. CyberFace III is aimed at the PC-based computer-aided design (CAD) market and is a lightweight, low-priced alternative to the Fake Space BOOM. It can be used with a face-fitting head mount or in a hand-guided mode. Costs for these hybrid systems fall between those of off-head panel displays and head-mounted displays.

Advantages and Disadvantages of Head-Mounted and Off-Head Displays

In summary, head-mounted displays provide a straightforward approach to the generation of a seamless, all-encompassing viewing volume. High-performance HMDs, however, are costly, and current technological challenges result in less than ideal performance. In particular: (1) high-resolution, miniature, lightweight, low-cost display surfaces are yet to be realized; (2) weight and inertial burdens imposed by most HMDs affect the incidence of symptomatic motion sickness, the ability of users to make proper judgments concerning orientation, and their long-term habitability; and (3) due to size, performance, and cost constraints, fixation/focus compensation is utilized in most HMDs and conflicting visual depth cues are provided to the user. Furthermore, the proper operation of HMDs is intimately tied to the performance of head-tracking systems (i.e., update rate and lag), which is currently less than ideal.

Off-head display approaches can help alleviate some of these shortcomings. Since the display is not worn on the head: (1) relatively inexpensive, widely available display surfaces (i.e., panel and projection) can provide high-resolution color imagery; (2) the effects of placing additional weight and inertial burdens on the user's head are minimized; (3) some of the volume and weight constraints on fixation/focus compensation systems are relaxed; and (4) some off-head configurations (i.e., those that supply an instantaneous encompassing viewing volume) do not necessarily require head-orientation tracking, and therefore, lags and distortions due to the tracking system are minimized. Off-head displays, however, are by no means perfect. If an instantaneous encompassing viewing volume is not provided, some form of head tracking is required. Not only must the computer- generated imagery be slaved to the user (as in HMDs), but also the display surfaces must be servo-controlled. Finally, the overall volume and weight of the display system is typically much greater for off-head systems than for head-mounted systems.

RESEARCH NEEDS

Within the past few years, the design and development of visual display systems appropriate for use in synthetic environments have engaged unprecedented commercial interest and involvement. High-end HMD technology, originally developed for military applications, has been transferred to the commercial sector and new commercial markets have resulted in innovative lower-end designs. Lightweight, intermediate-resolution (approximately 1,000 horizontal lines), color, see-through capable HMDs are now available for under $200,000; lower-resolution (i.e.,

NTSC level) systems are available for under $10,000. Intermediate-resolution, off-head, field-sequential, panel and projection systems are available for well under $20,000. Equivalent BOOM-type systems cost under $100,000. No special government-directed research effort seems to be required as an impetus to drive the efforts to refine system design and reduce system costs in the intermediate resolution display technologies under discussion. That is not to say that there remains no significant research and development to be done, but rather that commercial pressures seem adequate to motivate and fund these efforts.

Current commercial and technological trends, however, are still likely to result in displays for synthetic environments that have several shortcomings. The best of the current commercial systems, with resolutions roughly equivalent to high-definition TV, cannot provide eye-limited resolution except across a relatively small field of view (i.e., approximately 38 deg [H] field of view). It may be some time before commercial market pressures by themselves will drive the development of the higher-resolution display devices required for wider field-of-view, eye-resolution-limited display systems. Eye-tracking approaches (i.e., in which a high-resolution display area is kept within the foveal region of the eye and a lower-resolution image is presented elsewhere), do not alleviate the required absolute display surface resolution but rather mitigate the computational requirements for generating visual images in virtual environments or the camera resolution requirements for teleoperators. Some government involvement therefore may be appropriate for encouraging high-resolution display design that will enable wide-field-of-view (i.e., perceptually seamless), eye-resolution-limited display systems to be built. In order to provide a 180 deg (H) by 120 deg (V) field of view (i.e., the instantaneous field of view of the human visual system), display devices providing approximately 4,800 (H) × 3,800 (V) lines of resolution are required (McKenna and Zeltzer, 1992). It seems prudent for both CRT-based and solid-state device-based approaches for display devices to be pursued. A particular area for emphasis should be the development of miniature, high-resolution CRT and solid-state display surfaces, since commercial pressures tend toward larger versions of these devices and HMDs require smaller versions. Autostereoscopic off-head displays (e.g., video holographic displays), which are especially attractive for situation assessment applications, should also be considered for continued government-funded research and development.

In designing a visual display system for synthetic environments, it is important to consider the specific tasks to be undertaken and, in particular, the visual requirements inherent in these tasks. None of the available technologies is capable of providing the operator with imagery that is in all important respects indistinguishable from a direct view of a complex, real-world scene. In other words, significant compromises must be made. The compromises made by designers should be based on the best available evidence regarding the relationships between vision system features and objective performance metrics—not on wholly subjective criteria. Although many important lessons can be learned from the work conducted over many years on large-scale simulators concerning these issues, further research is clearly required.

Unanswered questions include: For a defined task domain, how should one compromise between spatial resolution and field of view? Is color worth the added expense and loss of spatial resolution? Is a stereoscopic display worth the trouble? If so, what are the appropriate parameters (baselines, convergence angles, convergence/accommodation mismatches) for various types of tasks? Can supernormal cues (e.g., depth, contrast, etc.) be used to advantage? Can the simulator-induced queasiness that accompanies the use of wide-field-of-view HMDs for some users be minimized or eliminated entirely? Which forms of delays and distortions induced by the visual display system can be adapted to? How are the required visual display system parameters affected within multimodal systems? Can visual display system requirements be relaxed in multimodal display environments? What are the perceptual effects associated with the merging of displays from different display sources? How do we design a comfortable HMD that integrates visual, auditory, and position tracking capabilities? These are but a few of the many research issues that impact the practical design of any visual display system for synthetic environments.

3

The Auditory Channel

As indicated previously, the accomplishments and needs associated with the auditory channel differ radically from those associated with the visual channel. Specifically, in the auditory channel, the interface devices (earphones and loudspeakers) are essentially adequate right now. In other words, from the viewpoint of synthetic environment (SE) systems, there is no need for research and development on these devices and no need to consider the characteristics of the peripheral auditory system to which such devices must be matched. What is needed, however, is better understanding of what sounds should be presented using these devices and how these sounds should be generated. Accordingly, most of the material presented in this section is concerned not so much with auditory interfaces as with other aspects of signal presentation in the auditory channel.

STATUS OF THE RELEVANT HUMAN RESEARCH

There is no topic in the area of auditory perception that is not relevant to the use of the auditory channel in some kind of SE system. The illustrative topics included here have been chosen for discussion because we believe they have exceptional relevance or are receiving considerable attention by investigators in the field of audition who are concerned with the use of the auditory channel in SE systems. General background on audition can be found in Carterette and Friedman (1978); Pickles (1988); Moore (1989); Yost (1994); and Fay and Popper (1994).

Resolution and Information Transfer Rates

General comments on resolution and information transfer rates, most of which are not modality specific, are presented in the Overview, in the discussion of the current state of the SE field. Here we supplement those previous, more general remarks with information that is specific to the auditory channel.

Most work on auditory resolving power has focused on artificially simple stimuli (in particular, tone pulses) or speech sounds. Except for a small amount of work directed toward aiding individuals with hearing impairments, relatively little attention has been given to the resolution of environmental sounds. Nevertheless, knowledge of the normal auditory system's ability to resolve arbitrary sounds is quite advanced. Thus, for example, there is an extensive literature, both experimental and theoretical, on the ability to discriminate between two similar sounds, or to perceive a target sound in the presence of a background masking sound (presented simultaneously or temporally separated). On the whole, knowledge in this area appears to be adequate for most SE design purposes. Useful information on auditory resolution can be found in the general texts on audition cited above, in the *Handbook of Human Perception and Human Performance* (Boff et al., 1986), in the *Engineering Compendium* (Boff and Lincoln, 1988), and, most importantly, in the many articles published each year by the *Journal of the Acoustical Society of America.*

Issues related to information transfer rates are much more complex because such rates depend not only on basic resolving power, but also on factors related to learning, memory, and perceptual and cognitive organization. It appears that the upper limits on information transfer rates for spoken speech and Morse code, two methods of encoding information acoustically for which there exist subjects who have had extensive training in deciphering the code, are roughly 60 bits/s and 20 bits/s, respectively. Unfortunately, we are unaware of any estimates of the information transfer rate for the perception of music. We would guess, however, that the rate lies between the above two, with a value closer to that of speech than of Morse code (because of the much higher dimensionality of the stimulus set in music than in Morse code). Although we cannot prove it, we suspect that the rate achieved with spoken speech is close to the maximum that can be achieved through the auditory channel. We say this not because we believe that speech is special, in the sense argued by various speech experts (e.g., Liberman et al., 1968), but rather because of the high perceptual dimensionality of speech sounds and because of the enormous amount of learning associated with the development of speech communication. Furthermore, except possibly for the case of music, we believe that it would be extremely difficult to achieve a comparatively high rate with any other coding system. Certainly, none of the individuals we know would be willing to spend an equivalent amount of time attempting to learn an arbitrary nonspeech code developed for purposes of general research or for use in SE systems. Unfortunately, there is no theory yet available that enables one to reliably predict the dependence of information transfer rates on the coding scheme or training procedures employed.

Auditory Displays

Much of the current work concerned with communicating information to individuals through the auditory channel falls under the heading of "auditory displays." A comprehensive view of work in this area, together with hundreds of references, can be found in the book edited by Kramer (1994).

The focus of work in this area has been the creation of different kinds of displays. Relatively little attention has been given to questions of evaluation, to an analysis of the information transfer achieved, or to training issues and how performance changes with practice. The kind of displays include *audification*, in which the acoustic stimulus involves direct playback of data samples, using frequency shifting, if necessary, to bring the signals into auditory frequency range; and *sonification*, in which the data are used to control various parameters of a sound generator in a manner designed to provide the listener with information about the controlling data. In general, such displays are being used both for monitoring tasks (e.g., to monitor the condition of a hospital patient or the state of a computer) and for data exploration (e.g., in physics, economics).

In the attempt to create effective displays, investigators are using perceptually high-dimensional stimulus sets and display sounds and codes that capitalize on the special sensitivities and strengths of the auditory system. Generally accepted positive characteristics of the auditory system include the ability to detect, localize, and identify signals arriving from any direction (despite the presence of objects that would cause visual occlusion), the tendency for sounds to cause an alerting and orienting response, and the exceptional sensitivity to temporal factors and to changes in the signal over time. These characteristics make auditory displays particularly useful for warning and monitoring functions.

A further set of issues in the design of acoustic displays concerns the extent to which the display makes use of everyday sounds and natural codes so that learning time is minimized. Although such codes are always preferred, some applications, such as the sonification of financial data, require codes that are highly abstract. Furthermore, when an abstract code is required, attention must be given to the way in which the resulting complex of auditory signals will be analyzed into distinct perceptual streams (see the detailed discussion on auditory scene analysis below).

Spatial Perception

The topic of auditory spatial perception is important because the perception of the spatial properties of a sound field is an important component of the overall perception of real sound fields, because the location of a sound source is a variable that can be used to increase the information transfer rate achieved with an auditory display, and because it has been a central research focus in the simulation of acoustic environments for virtual environment (VE) systems.

Auditory localization of isolated sound sources in anechoic (i.e., echo-free) environments is relatively well understood. The direction of a source is determined primarily by comparing the signals received at the two ears to determine the interaural amplitude ratio and the interaural time delay or phase difference as a function of frequency. Whereas interaural phase provides the dominant cue for narrowband low-frequency (< 1,500 Hz) signals, interaural amplitude ratio provides the dominant cue for narrowband high-frequency signals. For signals of appreciable bandwidth, interaural time delay also plays a significant role at high frequencies. These empirical psychophysical results are consistent with what one would expect based on the relevent physical acoustics: at the low frequencies, the effects of head shadow are relatively small because of diffraction; at the higher frequencies, measurement of time delay suffers from phase ambiguities unless there is sufficient bandwidth to eliminate these ambiguities. The results are also consistent with the known physiology: the neural firings in the auditory nerve are able to follow the individual cycles of a narrowband signal only at the lower frequencies; they can, however, follow modulations in the envelope of high-frequency signals with significant bandwidth at the output of the auditory "critical band" filters.

Ambiguities in the determination of direction achieved using interaural amplitude ratio and interaural time delay are well predicted by assuming that the head is a sphere with sensors located at the end of a diameter, and then determining the three-dimensional surfaces over which both the interaural amplitude ratio and the interaural time delay remain constant. All such surfaces are surfaces of revolution about the interaural axis, which can be approximated by cones for sources far removed from the head, and the median plane (in which the signals to the two ears are identical if one assumes the head is symmetrical) constitutes a limiting case of this family of surfaces.

The methods used by the auditory system to resolve these ambiguities involve (1) head movements and (2) monaural processing. Thus, for example, front-back ambiguities are easily eliminated by rotating the head in azimuth. Directional information from monaural processing is achieved by estimating properties of the direction-dependent filtering of the transmitted signal that occurs when the transmitted signal propagates from the signal source to the eardrum. The directional information achieved in this manner, however, depends strongly on (1) the listener having adequate a priori information on the spectrum of the transmitted signal (so the spectrum of the received signal can be factored into the spectrum of the transmitted signal and the transfer function of the filter) and (2) the existence of high-frequency (> 5,000 Hz) energy in the transmitted signal (so that the listener's pinnae can have a strong directional effect on the spectrum of the received signal).

For isolated sound sources in an anechoic environment, the just-noticeable difference (JND) in azimuth for sources straight ahead of the listener is roughly 1 deg. However, the JND in azimuth for sources off to the side, the JND in elevation within the median plane, and the JND in distance are relatively large. In all these cases, the JND constitutes a substantial fraction (e.g., one-fifth) of the total meaningful range of the variable in question.

When reflections (echoes and reverberations) are present, direction finding tends to be degraded below that which occurs in anechoic environments. However, this degradation is limited by the tendency of the auditory system to enhance perception of the direct acoustic wave and suppress the late-arriving echoes (the precedence effect), at least with respect to direction finding.

Perception of sound source distance is generally very poor. Whatever ability there is seems to be based on the following three changes in the received signal, as source distance is increased: a decrease in intensity, an increase in high-frequency attenuation, and an increase in the ratio of reflected to direct energy. Distance perception is poor because none of these cues is reliable: they all can be influenced by factors other than distance. The influence of reflections on both direction perception and distance perception, together with an inadequate understanding of how listeners separate out the characteristics of the transmitted signal from the characteristics of the acoustic environment that modify the characteristics of the transmitted signal as it approaches the ear, make the study of reflection effects in audition a high-priority item.

Even though distance perception is poor, under normal circumstances, sound sources are perceived to be located outside the head, that is, they are perceptually externalized. However, under special circumstances, the sources can be made to appear inside the head. Although in-head localization seldom occurs under normal conditions, it is a major problem when sounds are presented through earphones as they are in head-mounted displays for VE systems.

Further important issues in the area of auditory spatial perception concern the identification of source position (as opposed to discrimination) and the phenomena that occur when multiple sources are simultaneously active at different locations. Identification of source position, like identification of many other variables, is limited by inadequate short-term memory and evidences constraints on information transfer that are much stronger than those that would be implied solely by the JND data. In the area of multiple sources, a great deal of research has been conducted on the ability to detect a given source in a background of interference emanating from sources located at other locations; however, very little is known about the ability to localize a given source in such a background of interference. Furthermore, most of the work in this area is limited to the case in which the interference arises from a single location in space and the space is anechoic.

Detailed information on the topics discussed above is available in the following articles, chapters, and books, and in the many references cited in these publications: Blauert (1983); Colburn and Durlach (1978); Durlach and Colburn (1978); Durlach et al. (1992c); Durlach et al. (1993); Gilkey and Anderson (1994); Mills (1972); Searle et al. (1976); Wenzel (1992); and Yost and Gourevitch (1987).

The parameters of the human auditory system pertinent to real-time system dynamics include spatial resolution, as estimated by the JND in angle or, as it is sometimes called, the minimum audible angle or MAA (Mills, 1958); the velocity-dependent minimum audible movement angle or MAMA (Perrott, 1982); and the corresponding minimum perceivable latencies. As noted above, the auditory system is most sensitive to changes in the azimuthal position of sources located in front of the listener, so that the necessary angular resolution and accuracy of a head tracker are determined by the JND in azimuth for sources in front of the listener. This localization blur (Blauert, 1983) is dependent to some extent on the nature of the signal, but is never less than 1 deg. The angular accuracy of the commonly used magnetic trackers now available is on the order of 0.5 deg, and thus meets this requirement. The necessary translational accuracy of trackers depends on the distances of the sources to be simulated. However, since perception of source distance itself is very poor, permissible tracker translational error is not bound by error in simulated source distance but rather by error in simulated source angle. Again considering sources in front of the listener, source distance must be large enough that the maximum angular positional error (equal to the angular tracker error plus the error in angular position due to translational tracker error) is smaller than the JND in angular position. If the angular error for the tracker is assumed to be 0.5 deg, then, as long as the translational error causes no greater than about 0.5 deg of angular error, it will be perceptually insignificant. For a given translational accuracy, sources above some minimum distance (dmin) from the listener can be simulated without perceptual error. The value of dmin is given by dmin = e / 2tan ($a/2$), where a is the allowable angular error due to translational error (0.5 deg) and e is the translational accuracy of the tracker. For an accuracy of $e = 1$ mm, dmin is about 12 cm, a value that seems sufficient for most practical cases. It should also be noted that the limited accuracy of the tracking device will, in most cases, cause no audible effects even if the sources are somewhat closer to the subject, because the localization blur (even in azimuth) is larger than 1 deg for many kinds of signals and is considerably larger for directions other than straight ahead.

The questions of the latency constraints and update rates required to create natural virtual auditory environments are still largely unanswered. Some relevant perceptual studies include the work of Perrott et al. (1989) and Grantham and his colleagues (e.g., Grantham, 1986, 1989, 1992; Chandler and Grantham, 1992). These studies have begun to examine the perception of moving auditory sources. For example, for source speeds ranging from 8 to 360 deg/s, minimum audible movement angles were found to range from about 4 to 21 deg, respectively, for a 500 Hz tone burst (Perrott, 1982; Perrott and Tucker, 1988). The recent work on the perception of auditory motion by Perrott and others using real sound sources (moving loudspeakers) suggests that the computational latencies currently found in fairly simple auditory VEs are acceptable for moderate velocities. For example, latencies of 50 ms (associated with positional update rates of 20 Hz as found in the Convolvotron system from Crystal River Engineering) correspond to an angular change of 18 deg when the relative source-listener speed is 360 deg/s, 9 deg when the speed is 180 deg/s, etc.

As yet, no relevant studies have been performed that investigate the delays that may be tolerable in rendering dynamic cues when subjects actively move their heads. Such psychophysical questions are currently being investigated in the United States by the National Aeronautics and Space Administration (NASA) and in Europe by the SCATIS (Spatially Coordinated Auditory/Tactile Interaction Scenario) project. In any event, the overall update rate in a complex auditory VE will depend not only on tracker delays (nominally 10 ms for the Polhemus Fastrak), but also on communication delays and delays in event processing or handling as well. The tracker delays in such systems will probably contribute only a small fraction of the overall delay, particularly when multiple sources or larger filters (e.g., required to simulate reverberant rooms) are used.

Auditory Scene Analysis

Unlike the visual system, in which there is a direct mapping of source location to image on the retina, in the auditory system, there is no peripheral representation of source location. There are only two sensors (the ears), each sensor receives energy from all sources in the environment, and the spatial analysis takes place centrally after the total signal in each ear is analyzed by the cochlea into frequency bands. Thus, the auditory picture of the acoustic scene must be constructed by first analyzing the filter outputs into components, then combining these components both across frequency bands and across the two ears. This signal-processing architecture differs not only from that found in the visual system, but also from that which would be used in the design of an artificial acoustic sensing system. Ideally, in an artificial system, the spatial processing would be performed prior to the frequency processing: one would use an array of microphones and parallel processing of the set of microphones outputs to achieve separate channels corresponding to different regions of space, and then separately analyze the frequency of the signal arising from each spatial cell (Colburn et al., 1987; Durlach et al., 1993).

Independent of the evolutionary developments underlying the way in which the auditory system is designed, the existing design presents a unique problem for creating a coherent auditory scene. The lack of a peripheral representation of source location complicates not only the task of determining source location, but also that of determining the number and character of the sources. Somehow, the higher centers in the auditory system must decompose the output of each auditory filter in each ear into elements that, after the decomposition, can be recombined into representations of individual sources.

In general, understanding auditory scene analysis is important to the design of SE systems because properties of this analysis play an important role in determining the effectiveness of auditory displays. Background information on this topic can be found in Hartman (1988), Handel (1989), Bregman (1990), and Yost (1991). Briefly, auditory scene analysis refers to the fact that, analogous to the visual domain, one can conceive of the audible world as a collection of acoustic objects. The auditory system's job is to organize this world by identifying and parsing the frequency components belonging to individual objects from the mixture of components reaching the two ears that could have resulted from any number of "real" acoustic sources.

Historically, research in this area has been concerned with the phenomena of auditory stream segregation, in which an auditory stream corresponds to a single perceptual unit or object. Studies have shown that, in addition to spatial location, various acoustic features such as synchrony of temporal onsets and offsets, timbre, pitch, intensity, frequency and amplitude modulation, and rhythm, can serve to specify the identities of objects and segregate them from one another. A single stream or object, such as a male human voice, will tend to be composed of frequencies that start and stop at the same time, have similar low-frequency pitches (formants), similar rhythmic changes or prosody, and so on. A female voice is parsed as a separate object because it has higher frequency formants, a different prosody, etc. Nonspeech sounds can possess the same sorts of distinguishing characteristics. Furthermore, Gestalt principles of perceptual organization normally applied to visual stimuli, such as principles of closure, occlusion, perceptual continuity, and figure-ground phenomena, have their counterparts in audition as well. For example, these principles operate in a drawing when one object, say the letter *B* is occluded by another, say an irregular blob of ink. The visual system has no trouble seeing the discontinuous visible fragments of the occluded *B* as a unitary and continuous object. However, if the blob is removed, it is much more difficult to recognize the same fragments as belonging to a *B*. A similar effect can be heard when a series of rising and falling tonal glides are interspersed with bursts of white noise (Dannenbring, 1976). When the noise is not present, one hears bursts of rising and falling tones that seem like isolated events. However, when the noise is present the auditory system now puts the tonal glides together into one continuous stream that seems to pass through the noise, even though the glides are not physically present during the noise.

While such illusions may seem merely like perceptual curiosities, these kinds of effects can have a profound impact on whether information is perceived and organized correctly in a display in which acoustic signals are used to convey meaning about discrete events or ongoing actions in the world and their relationships to one another. For example, if two sounds are assigned to the same perceptual stream, simultaneous acoustic masking may be greater but temporal comparisons might be easier if relative temporal changes cause a change in the Gestalt or overall percept or cause the single stream to perceptually split into two streams. Suppose two harmonically related tonal pulses with similar rhythms are being used to represent two different friendly ship signatures in a sonar display. The information content of this nicely harmonic, unitary percept might be that all is well. However, if the signal recognition system of the sonar system now detects a change in one ship's acoustic signature, this might be represented as an inharmonicity of one component relative to the other. A further change in their relative temporal patterns might break the sound into two separate, inharmonic objects signaling "potential enemy in the vicinity." Conversely, if the inharmonic and temporal relationships are not made sufficiently distinct, the warning signal might not be recognized. Whereas the above discussion is rather speculative, it illustrates how acoustic signals could be used, or misused, for monitoring important events in situations like a sonar display in which the operator's visual channel is already overloaded.

In the context of display design, the notion of auditory scene analysis has been most influential in the recent interest in using abstract sounds, environmental sounds, and sonification for information display (Kramer, 1994). The idea is that one can systematically manipulate various features of auditory streams, effectively creating an auditory symbology that operates on a continuum from literal everyday sounds, such as the rattling of bottles being processed in a bottling plant (Gaver et al., 1991), to a completely abstract mapping of statistical data into sound parameters (Smith et al., 1990). Principles for design and synthesis can be gleaned from the fields of music (Blattner et al., 1989), psychoacoustics (Patterson, 1982), and higher-level cognitive studies of the acoustical determinants of perceptual organization (Bregman, 1990; Buxton et al., 1989). Recently, a few studies have also been concerned with methods for directly characterizing and modeling such environmental sounds as propeller cavitation, breaking or bouncing objects, and walking sounds (Howard, 1983; Warren and Verbrugge, 1984; Li et al., 1991). Other relevant research includes physically based acoustic models of sound source characteristics, such as radiation patterns (Morse and Ingard, 1968). Further discussion of some of these issues is presented in the section below on computer generation of nonspeech audio.

Adaptation to Unnatural Perceptual Cues

There are many SE situations in which a sensory display, or the manner in which such a display depends on the behavior of the operator of the SE system, deviates strongly from normal. Such situations can arise because of inadequacies in the design or construction of an SE system or because intentional deviations are introduced to improve performance. Whenever a situation of this type exists, it is important to be able to characterize the operator's ability to adapt.

There have been a number of instances in which acoustic signals have been used as substitutes for signals that would naturally be presented via other modalities (e.g., Massimino and Sheridan, 1993). Unfortunately, however, much of this work has not included careful experimentation on how subjects adapt to such sensory substitutions over time. A major notable exception to this can be found in the work on sonar systems for persons who are blind (Kay, 1974; Warren and Strelow, 1984, 1985; Strelow and Warren, 1985).

Most of the work on adaptation to unnatural perceptual cues has focused on spatial perception and on transformations of cues within the auditory system, i.e., no sensory substitution has been involved. Extensive reviews of this work, with long lists of references, can be found in Welch (1978) and Shinn-Cunningham (1994).

Current research in this area directly relevant to SE systems continues to be concerned with spatial perception and is of two main types. First, studies are being conducted to determine the extent to which the signal-processing and measurement procedures required to achieve realistic simulations of real acoustic environments using earphone displays can be simplified and made more cost-effective. An important part of this effort concerns the extent to which listeners can adapt to the alterations in cues associated with these procedures. Thus, for example, a number of investigators have begun to explore the extent to which listeners can learn to make use of the direction-dependent filtering associated with someone else's ears (Wenzel et al., 1993a). Second, research is under way to determine the extent to which listeners can adapt to distortions that are purposefully introduced in order to improve spatial resolution. In particular, a number of transformations are being studied that present the listener with magnified perceptual cues that approximate in various ways the cues that would be present if the listener had a much larger head (Durlach and Pang, 1986; Van Veen and Jenison, 1991; Durlach et al., 1993). Also being studied now are transformations that enable the listener to perceive the distance of sound sources much more accurately (see the review by Durlach et al., 1993). Although it is fairly obvious that effective spatial resolution can be improved by such transformations, the degree to which the response bias introduced by these transformations can be eliminated by adaptation is not obvious.

In general, although adaptation is a topic that has been of interest to psychologists for a long time, there is still no theory available that will enable one to predict the rate and asymptote of adaptation as a function of the transformation and training procedure employed.

STATUS OF THE TECHNOLOGY

A recent comprehensive review of technology for the auditory channel in VE systems is available in Durlach et al. (1992b), and much of the material presented in this section is taken from that report.

An auditory interface for virtual environments should be capable of providing any specified pair of acoustic waveforms to the two ears. More specifically, it should (1) have high fidelity, (2) be capable of altering those waveforms in a prescribed manner as a function of various behaviors or outputs of the listener (including changes in the position and orientation of the listener's head), and (3) exclude all sounds not specifically generated by the VE system (i.e., real environmental background sounds). Requirement (3) must be relaxed, of course, for an augmented-reality system in which the intention is to combine synthetically generated sounds and real environmental sounds (see the section on hear-through displays below).

Generally speaking, such results can be most easily achieved with the use of earphones; when loudspeakers located away from the head are employed, each ear receives sound from each loudspeaker and the control problem becomes substantial. Although commercial high-fidelity firms often claim substantial imaging ability with loudspeakers, the user is restricted to a single listening position within the room, only azimuthal imaging is achieved (with no compensation for head rotation), and the acoustic characteristics of the listening room cannot be easily manipulated. In addition, since the ears are completely open, extraneous (undesired) sounds within the environment cannot be excluded. Finally, although the tactual cues associated with the use of earphones may initially limit the degree of auditory telepresence, since the user will be required to transit back and forth between the virtual and the real environments, such tactual cueing may actually prove useful. In any case, such cuing is likely to be present because of the visual interface. One set of situations for which loudspeakers might be needed are those in which high-energy, low-frequency acoustic bursts (e.g., associated with explosions) occur. In such cases, loudspeakers, but not earphones, can be used to vibrate portions of the body other than the eardrums.

Headphone Displays

Earphones vary in their electroacoustic characteristics, size and weight, and mode of coupling to the ear (see, for example, Shaw, 1966; Killion, 1981; Killion and Tillman, 1982). At one extreme are transducers that are relatively large and massive and are coupled to the ear with circumaural cushions (i.e., that completely enclose the pinnae). At the other extreme are insert earphones through which the sound is delivered to some point within the ear canal. The earphone may be very small and enclosed within a compressible plug (or custom-fitted ear mold) that fits into the ear canal, or the earphone may be remote from the ear and its output coupled via plastic tubing (typically 2 mm inner diameter) that terminates within a similar plug. Intermediate devices include those with cushions that rest on the pinnae (e.g., Walkman-type supraaural earphones) and those with smaller cushions (about 1.5 cm diameter) that rest within the concha (so-called earbuds).

All earphone types can deliver sounds of broad bandwidth (up to 15 kHz) with adequate linearity and output levels (up to about 110 dB sound pressure level). Precise control of the sound pressure at the eardrum requires knowledge of the transfer function from the earphone drive voltage to the eardrum sound pressure. Probe-microphone sound-level measurements proximal to the eardrum can be used to obtain this information. These measurements, particularly at frequencies above a few kHz, are nontrivial to perform. In general, this transfer function is expected to be more complex (versus frequency) and more different across individuals as the size of the space enclosed between the eardrum and the earphone increases. Thus, circumaural earphones that enclose the pinnae are expected to have more resonances than insert earphones. One might similarly expect greater variability with repeated placement of the earphones on a given individual, but recent measurements of hearing thresholds at high frequencies indicate that, even with relatively large supraaural earphones, test-retest variability measures are small (a few dB up to 14 kHz—Stelmachowicz et al., 1988). Thus, repeated measures of this transfer function on a given listener may be unnecessary. For some applications, even interindividual differences may be unimportant and calibration on an average (mannequin) ear may be sufficient. Insert earphones (or earplugs), by virtue of having a high ratio of contact area (with the ear canal wall) to exposed sound transmission area (the earplug cross-section), afford relatively good attenuation of external sounds (about 35 dB above 4 kHz, decreasing to about 25 dB below 250 Hz). Circumaural earphones can achieve similar high-frequency attenuation but less low-frequency attenuation. Recently developed active-noise-canceling circumaural headsets (e.g., Bose Corp.) provide up to 15 dB additional low-frequency attenuation, thereby making their overall attenuation characteristics similar to that of insert earphones. Supraaural earphones, which rest lightly on (or in) the external ear, provide almost no attenuation (consistent with their commonly being referred to as open-air design). The greatest attenuation can be achieved by combining an insert earphone with an active-noise-canceling circumaural hearing protector. Even including such a protector, it is unlikely that the cost of such a sound delivery system will exceed $1,000.

Most of the past work on auditory interfaces for virtual environments has been directed toward the provision of spatial attributes for the sounds. Moreover, within this domain, work has focused primarily on the simulation of normal spatial attributes (e.g., Blauert, 1983; Wenzel et al., 1988; McKinley and Ericson, 1988; Wightman and Kistler, 1989a,b; Wenzel, 1992). Relatively little attention has been given to the provision of supernormal attributes (Durlach and Pang, 1986; Durlach, 1991; Van Veen and Jenison, 1991; Durlach et al., 1992a).

Normal spatial attributes are provided for an arbitrary sound by multiplying the complex spectrum of the transmitted sound by the transfer function of the space filter associated with the transformation that occurs as the acoustic waveform travels from the source to the eardrum. (In the time domain, the same transformation is achieved by convolving the transmitted time signal with the impulse response of the filter.) For binaural presentation, one such filter is applied for each of the two ears. Inasmuch as most of the work on virtual environments has focused on anechoic space, aside from the time delay corresponding to the distance between source and ear, the filter is determined solely by the reflection, refraction, and absorption associated with the body, head, and ears of the listener. Thus, the transfer functions have been referred to as head-related transfer functions (HRTFs). Of course, when realistic reverberant environments are considered, the transfer functions are influenced by the acoustic structure of the environment as well as that of the human body.

Estimates of HRTFs for different source locations are obtained by direct measurements using probe microphones in the listener's ear canals, by roughly the same procedure using mannequins, or by the use of theoretical models (Wightman and Kistler, 1989a,b; Wenzel, 1992; Gierlich and Genuit, 1989). Once HRTFs are obtained, the simulation is achieved by monitoring head position and orientation and providing, on a more or less continuous basis, the appropriate HRTFs for the given source location and head position/orientation.

The process of measuring HRTFs for a set of listeners is nontrivial in terms of time, skill, and equipment. Although restriction of HRTF measurements to the lower frequencies is adequate for localization in azimuth, it is not adequate for vertical localization, for externalization, or for elimination of front-back ambiguities (particularly if no head movement is involved). Thus, ideally, a sampling frequency of roughly 40 kHz is required (corresponding to an upper limit for hearing of 20 kHz). Similarly, if the HRTFs are measured in an acoustic environment with reverberation, the associated impulse responses can be very long (e.g., more than a second). These two facts, combined with the desire to measure HRTFs at many source locations, in many environments, and for many different listeners, can result in a monstrously time-consuming measurement program.

Preliminary investigations have demonstrated that, without training, nonindividualized HRTFs cause greater localization error, particularly in elevation and in front-back discrimination, than do HRTFs measured from each subject (Wenzel et al., 1993a). These results have been interpreted as showing that HRTFs must be measured for each individual subject in order to achieve maximum localization performance in an auditory SE. However, essentially all of the work done to date has been done without regard for the effects that could be achieved by means of sensorimotor adaptation (e.g., see Welch, 1978). We suspect that if subjects were given adequate opportunity to adapt in a closed-loop SE, much of the current concern with detailed characteristics of the HRTFs and with the importance of intersubject differences would disappear. In addition, significant advances are now being made in the modeling of HRTFs and the development of parametric expressions for HRTFs based on abstractions of the head, torso, and pinnae. In fact, in the not-too-distant future, it may be possible to obtain reasonably good estimates of an individual's HRTFs merely by making a few geometric measurements of the outer ear structures. These two factors, combined with the use of models for describing the effects of reverberation, should greatly simplify the HRTF estimation problem.

Given an adequate store of estimated HRTFs, it is then necessary to select the appropriate ones (as a function of source and head position/ orientation) and filter the source signal in real time. Although some ability to perform such processing has been achieved with relatively simple analog electronics (e.g., Loomis et al., 1990), the devices available for achieving the most accurate simulations employ digital signal processing. The earliest commercially available systems (e.g. the FocalPoint system from Gehring Research and the Convolvotron) employed simple time-domain processing schemes to "spatialize" input sound sources. The Acoustetron (successor to the Convolvotron) quadrupled the computational capabilities of these earlier systems. It stores 72 pairs of spatial filters sampled at 50 kHz for spatial positions sampled at 30 degrees in azimuth and 18 deg in elevation. The spatial filters that most closely correspond to the instantaneous position of the source relative to the listener's head are retrieved in real time, interpolated to simulate positions between the spatial sampling points, and then convolved with the input sound source to generate appropriate binaural signals. The system can spatialize up to 32 sources in parallel, enabling simulation of simple acoustic room models (with first- and second-order reflections) in real time.

Another time-domain spatialization system was developed by McKinley and his associates in the Bioacoustics and Biocommunications Branch of the Armstrong Laboratory at Wright-Patterson Air Force Base in order to present three-dimensional audio cues to pilots (R.L. McKinley, personal communication, 1992). The HRTFs incorporated into this system, derived from measurements using the mannequin KEMAR, are sufficiently dense in azimuth (HRTFs are measured at every degree in azimuth) to eliminate the need for interpolation in azimuth. In elevation, the measurements are much less dense and linear interpolation is employed. The researchers at Wright-Patterson, in conjunction with Tucker Davis Technologies of Gainesville, Florida, have recently developed a new time-domain processor based on their earlier efforts. This machine, which is now commercially available, contains more memory than has been available in previous spatialization systems. In addition, the product has been designed to maximize the flexibility of the system, allowing researchers to allocate the available processing power as necessary for the individual application.

Frequency-domain filtering is being employed in the Convolvotron II, a next-generation spatialization device under development at Crystal River Engineering. By performing frequency-domain filtering, the Convolvotron II will be capable of greater throughput for relatively low cost. In addition, because the Convolvotron II is being built from general-purpose, mass-produced digital signal processing (DSP) boards, the system will be completely modular and easily upgraded.

Room-Acoustics Modeling

The state of the art in spatial auditory displays is not yet quite adequate for high-quality VE applications. For example, there are serious questions about the adequacy of the current techniques used for constructing interpolated HRTFs (Wightman et al., 1992; Wenzel and Foster, 1993). Similarly, with the current-generation processors, limitations in memory and filter size, as well as in processing speed and algorithm architecture, have limited the ability to simulate nonisotropic sources or reverberant environments (measured or synthesized). To date, the only real-time auditory spatialization system that included even simplified real-time room modeling was that developed by Foster and his associates (Foster et al., 1991) for the Acoustetron. Even though the acoustic model used was rather simple and provided only a small number of first- and second-order reflections, this system provides increased realism (e.g., it improves externalization). The role played by reverberation (as well as other factors) in the externalization of sound images produced by earphone stimulation is discussed in Durlach et al. (1992c).

More sophisticated and realistic sound-field models have been developed for architectural applications (e.g., Lehnert and Blauert, 1991, 1992; Vian and Martin, 1992), but they cannot be simulated in real time by any of the spatialization systems currently available. An overview of the current state of the art for sound-field modeling and a representative collection of contemporary papers may be found in special issues in *Applied Acoustics* (1992, 1993). As the computational power of real-time systems increases, the use of these detailed models will become feasible for the simulation of realistic environments.

The most common approach to modeling the sound field is to generate a spatial map of secondary sound sources (Lehnert and Blauert, 1989). In this method, the sound field due to a source in echoic space is modeled as a single primary source and a cloud of discrete secondary sound sources (which correspond to reflections) in an anechoic environment. The secondary sources can be described by three major properties (Lehnert, 1993a): (1) distance (delay), (2) spectral modification with respect to the primary source (air absorption, surface reflectivity, source directivity, propagation attenuation), and (3) direction of incidence (azimuth and elevation).

In contrast to the digital generation of reverberation (which has a long history—e.g., Schroeder, 1962), very few people have experience with real-time sound-field modeling. In order to achieve a real-time realization of sophisticated sound models, it has been suggested that early and late sources (Lehnert and Blauert, 1991; Kendall and Martens, 1984) be treated differently. Early reflections would be computed in real time, whereas late reflections would be modeled as reverberation.

Two methods are commonly used to find secondary sound sources: the mirror-image method (Allen and Berkley, 1979; Borish, 1984) and variants of ray tracing (Krokstadt et al., 1968; Lehnert, 1993a). Lehnert and colleagues (Shinn-Cunningham et al., 1994) have compared the computational efficiency of the two methods with respect to their achievable frame rate and their real-time performance for a room of moderate complexity with 24 reflective surfaces. A total of 8 first-order and 19 second-order reflections were calculated for a specific sender-receiver configuration.

For this test scenario, the mirror-image method was more efficient than the ray-tracing method. In addition, the mirror-image method is guaranteed to find all the geometrically correct sound paths. For the ray-tracing method, it is difficult to predict the required number of rays to find all desired reflections. The ray-tracing method does have the advantage that it produces reasonable results even when very little processing time is available, and it can easily be adapted to work at a given frame rate by adjusting the number of rays used, whereas the mirror-image method cannot be scaled back easily since the algorithm is recursive. Ray tracing will yield better results in more complex environments since the dependence of the processing time on the number of surfaces is linear not exponential, as is the case for the mirror-image method. Thus, although for the given test case the mirror-image method was more efficient, there will probably be some scenarios in which the mirror-image method is superior and others in which the ray-tracing method offers better performance.

Since calculation of the sound-field model will be the most time-consuming part of the auditory pipeline, optimization of these calculations is worthwhile (e.g., see the discussion in Shinn-Cunningham et al., 1994). Computational resources can be assigned as necessary to achieve the necessary accuracy of the simulation. If, for example, a primary source is to be presented, no reflection filters are required and more resources (i.e., more filter coefficients) can be assigned to the HRTFs to obtain more precise spatial imaging. For a second-order reflection, spectral filtering must take place, but since the directivity of the delayed reflection is less salient psychoacoustically, less accuracy may need to be used for the HRTF filtering. The structure of the auralization unit should allow for task adequate assignment of resources. Efficient algorithms and signal-processing techniques for real-time synthesis of complex sound fields are currently being investigated in a collaborative project by Crystal River Engineering and NASA.

Off-Head, Hear-Through, and Augmented-Reality Displays

Apart from work being conducted with entertainment applications in mind, most of the research and development concerned with auditory displays in the SE area has been focused on stimulation by means of earphones. However, as indicated previously, such stimulation has two drawbacks: (1) it encumbers the listener by requiring that equipment be mounted on the user's head and (2) it stimulates only the listener's eardrums.

In considering item (1), it should be noted that for many of the head-tracking techniques now in use (see Chapter 5), head-tracking devices as well as earphones must be mounted on the head. If the visual display in the system is also head mounted, then the ergonomic difficulties associated with adding the earphones are minor: not only would a head-tracker presumably be required for the visual display even if none were required for the auditory display, but also the incremental burden of adding earphones to the head-mounted visual display would be relatively trivial (provided, of course, that the use of earphones was envisioned when the mounting system was designed and not added later as an afterthought).

In considering item (2), it should be noted that, even though earphones can generate sufficient power to deafen the user, stimulation via earphones is totally inadequate for delivering acoustic power to the user in a manner that affects body parts other than the ear. Although for most applications in the SE field, stimulation of the auditory system via the normal acoustic channel (outer ear, eardrum, middle ear, cochlea, etc.) is precisely what is desired, if one wants to provide realistic simulations of high-energy acoustic events in the environment, such as explosions or low-altitude flyovers by fast-moving aircraft, then acoustic stimulation of the rest of the body (e.g., shaking the user's belly) may also be important.

The design of off-head auditory displays, i.e., loudspeakers, has been a central focus of the audio industry for many years. Many of the loudspeaker systems now available are, like earphones, more than adequate for all SE applications with respect to such characteristics as dynamic range, frequency response, and distortion. They are also adequate with respect to cost, although they tend to be more expensive than earphones, particularly if the application requires the production of very high intensity levels throughout a very large volume (e.g., loud rock music in a big theater). For SE applications, the main problem with loudspeaker systems, as it is with earphones, is that of achieving the desired spatialization of the sounds (including both the perceived localization of the sound sources and the perceived acoustic properties of the space in which the sources are located).

The major problem that arises in spatialization for loudspeaker systems that does not arise when earphones are used concerns the difficulty of controlling the signals received at the two eardrums (and the differences between these two signals). Unlike the situation with earphones, in which the signal received at a given eardrum is determined simply by the signal transmitted by the earphone at that ear (and the fixed filtering associated with acoustic transmission from the earphone transducer to the eardrum), when a loudspeaker system is used, the signal received at the given eardrum is influenced by all the signals transmitted by all the loudspeakers in the room, together with all the transformations that each of the signals undergoes as it propagates through the room from the loudspeaker to the eardrum. Among the names used to designate various approaches to one or more components of this problem are *binaural stereo, transaural stereo, spectral stereo, multichannel stereo, quadrophony, ambisonics,* and *surround sound.* Even when a given system is tuned to provide an adequate perception for a given position of the listener's head, this perception is likely to rapidly degrade as the listener's head is moved out of the "sweet spot." To date, no loudspeaker system has been developed that incorporates head-tracking information and uses this information to appropriately adjust the inputs to the loudspeakers as the position or orientation of the listener's head is altered.

Background material on the spatialization problem for signals presented by means of loudspeakers is available in Eargle, 1986; Trahiotis and Bernstein, 1987; Cooper and Bauck, 1989; Griesinger, 1989; and MacCabe and Furlong, 1994.

Within the VE area, one of the best-known systems that makes use of off-head displays is the CAVE, a VE system developed at the Electronic Visualization Laboratory, University of Illinois at Chicago (Cruz-Neira et al., 1993). The current CAVE system employs four identical speakers located at the corners of the ceiling and amplitude variation (fading) to simulate direction and distance effects. In the system now under development, speakers will be located at all eight corners of the cube and reverberation and high-frequency attenuation will be added to the parameters that can be used for spatialization purposes. In a new off-head system being developed by Crystal River Engineering, attempts are being made to utilize protocols similar to those used in the earphone spatialization systems (e.g., the Convolvotron), so that the user will be able to change from one type of system to the other with minimal waste of time and effort.

Finally, it should be noted that relatively little attention has been given to augmented reality in the auditory channel. As in the visual channel, there are likely to be many applications in which it is necessary to combine computer-synthesized or sampled audio signals with signals obtained (1) directly from the immediate environment or (2) indirectly from a remote environment by means of a teleoperator system. In principle, the signals from the immediate environment can be captured from controlled acoustic leakage around the earphones (hear-through displays) or by positioning microphones in the immediate environment (perhaps on the head-mounted display, HMD) and adding signals in the electronic domain rather than the acoustic domain. However, because one may want to manipulate the environmental signals before adding it, or because one may want to use the same system for the case in which the sources of the environmental signals are remote and sensed by a telerobot, the latter approach seems preferable. Ideally, an acoustic augmented-reality system should be capable of receiving signals sensed by microphones in any environment (immediate or remote), transforming these signals in a manner that is appropriate for the given situation, and then adding them to the signals presented by the VE system. Currently, the most obvious use of acoustic augmented-reality systems is to enable an individual who is deeply immersed in some VE task to simultaneously monitor important events in the real world (e.g., the occurrence of warning signals in the real world).

Computer Generation of Nonspeech Audio

While much work remains to be done in the areas of sound spatialization and modeling of environmental effects, even less has been accomplished in the area of physical modeling and the computer generation of nonspeech sounds. How can we build a real-time system that is general enough to produce the quasi-musical sounds usually used for auditory icons, as well as purely environmental sounds, like doors opening or glass breaking? The ideal synthesis device would be able to flexibly generate the entire continuum of nonspeech sounds described above as well as be able to continuously modulate various acoustic parameters associated with these sounds in real time. Such a device or devices would act as the generator of acoustic source characteristics that would then serve as the inputs to a sound spatialization system. Thus, initially at least, source generation and spatial synthesis would remain as functionally separate components of an integrated acoustic display system. While there would necessarily be some overhead cost in controlling separate devices, the advantage is that each component can be developed, upgraded, and utilized as stand-alone components so that systems are not locked into an outmoded technology.

Synthesis Techniques

Many of the algorithms likely to be useful for generating nonspeech sounds will be based on techniques originally developed for the synthesis of music as well as speech. The main goal in speech synthesis is the production of intelligible (and natural sounding) speech waveforms. To do this, one must accurately synthesize the output of a specific type of instrument, the human voice. Also, in speech synthesis, the final acoustic output can be rated according to the measurable amount of information conveyed and the naturalness of the speech. In synthesizing music, typically the goals are not as specific or restricted: they are defined in terms of some subjective criteria of the composer. Usually, the goal is to produce an acoustic waveform with specific perceptual qualities: either to simulate some traditional, physical acoustic source or to produce some new, unique sound with appropriate attributes.

Because the aims of synthesized music are more diverse than those of speech synthesis, there are a number of different, acceptable methods for its synthesis. Choice of the method depends on the specific goals, knowledge, and resources of the composer. In any synthesis method, the composer controls some set of time-varying parameters to produce an acoustic waveform whose perceptual qualities vary accordingly. These computer-controlled parameters may be related to a physical parameter in a modeled instrument, to the shape or spectrum of the acoustic waveform, or to the perceptual quality desired for the sound. Often, these varying techniques are combined to get a specific effect. Some of the most common techniques are described below.

One method used for computer-controlled synthesis is known as *additive synthesis*. In this method, a synthesized voice (or instrumental line) is generated by the addition of simple sine waves using short-time Fourier methods. One of the problems with this method is that the number of parameters needed to generate an acoustic signal with specific qualities is large, reflecting the fact that many important music percepts are far removed from the short-time Fourier parameters. Thus, synthesizing a particular sound quality can be cumbersome. Often, additive synthesis is used to simulate known sounds by first analyzing the desired sound and measuring these parameters directly (Grey and Moorer, 1977; Moorer, 1978; Portnoff, 1980; Dolson, 1986). Small alterations in these known parameters can then fine-tune the acoustic waveform.

Another common technique, which Cook et al. (1991) have described as an example of an abstract algorithm, is *frequency modulation* or FM synthesis, in which the composer specifies the carrier frequency, modulation index, and modulation frequency of a signal (Chowning, 1973; Chowning and Bristow, 1986; Schottstaedt, 1977). By varying the relation of the carrier to modulation frequency, the resultant sound can be harmonic or inharmonic. If harmonic, the relation defines what overtones exist in the sound (important in timbre perception). Changing the value of the modulation index controls the spread of the spectrum of the resultant sound, and therefore its perceptual brightness. Since the relationship between the values of the frequency modulation parameter and the corresponding perceptual aspects of the synthesized output is reasonably straightforward, the frequency modulation technique has proven both powerful and versatile. This technique is employed by many commercially available synthesizers, such as the Yamaha DX-7.

Subtractive synthesis is a term in music synthesis that refers to the shaping of a desired acoustic spectrum by one or more filtering operations and is a precursor to the more modern approaches that have come to be known as physical modeling. Subtractive synthesis is often used when a simple physical model of an instrument can be described by an excitatory source that drives some filter(s). Both the source waveform and the filters, which may or may not bear any relation to a real instrument, are specified by the composer.

Most traditional instruments can also be mathematically modeled in this way: they are excited by some vibratory noise source, and their physical properties acoustically filter that source. For brass-like instruments, vibrations of the player's lips are the acoustic source; in a stringed instrument, it is the string vibration; for a singer, it is the actions of the vocal chords. The physical properties that affect the acoustic output include the shape of the instrument, its effective length, the reflectivity of the material out of which the instrument is constructed, etc. By modeling the vibratory mode of the excitatory source and the acoustic effects of the physical properties of the instrument, subtractive synthesis has been used to synthesize known instruments (Jaffe and Smith, 1983; Risset and Mathews, 1969). As an instrument is played, the effective dimensions of the instrument are altered by opening or closing keys or valves. These changes cause the modeled filters to change in time. Because such modeling is closely tied to physical parameters, it may be less cumbersome (and intuitively easier) to adjust parameters to achieve a desired effect.

Another alternative is to develop methods of generating sounds by modeling the motions of the physical sound events, i.e., by numerical integration of the wave equation. Generating sound by solving the equations of motion of a musical instrument captures a natural parameterization of the instrument and includes many of the important physical characteristics of the sound. The conventional and perhaps most general way of representing an acoustics system is to use a set of partial differential equations (PDEs) in the temporal and three-dimensional spatial domain. However, it is not practical considering the intensive numerical computation and the constraint of real-time computing. To reduce the computational complexity of the sound generation tasks without giving up the physical essence of the representation, one approach is to use the aggregate properties of the physical model instead of solving the problem at the microscopic level, i.e., solving the PDEs.

As noted above, subtractive synthesis was the first attempt at this type of modeling of aggregate properties. Several more recent physical modeling techniques based on aggregate properties have been developed, including the digital waveguide technique, transfer function techniques, modal synthesis, and mass spring models to synthesize sounds ranging from musical instruments to the singing voice of the human vocal tract (Borin et al., 1992; Cadoz et al., 1993; Cook, 1993; Djoharian, 1993; Keefe, 1992; Morrison and Adrien, 1993; Smith, 1992; Wawrzynek, 1991; Woodhouse, 1992).

Collision sounds are an example of a simple auditory event in a virtual environment. We may predict the potential collision of two objects by observing their paths with our eyes but, short of actual breaking or deformation of the objects, it is the sound of the collision that best reveals how the structure of the objects has been affected by the collision.

To illustrate how to generate collision sounds by using physically based models, we choose a uniform beam as an example structure. The vibrational mechanics theory of beam structures has been examined carefully and can provide a solid groundwork for collision sound synthesis. The collision can be decomposed into two parts, the *excitation* component, which initiates the impact event and the *resonator* component, in which the interesting vibration phenomena take place.

The excitation module is affected by the force, position, and duration of the impact; the resonator module is determined by the structure boundary conditions, material density, modulus of elasticity, and object geometry (e.g., length, width, and height). Because the uniform beam has a simple structure, one can derive the equations to depict its major vibrational modes and calculate the natural resonant frequency associated with each mode based on the aggregate physical properties. The natural resonant frequency reveals the strong linkage between material types and object shapes and can show objects' attributes. For complex free-form objects, finite element analysis can be used to calculate the associated resonant frequency for the major vibrational modes.

Additional topics that are often grouped into the broad field of music synthesis and may be relevant to the use of sound in SEs include the use of special computational structures for the composition of music, specific hardware and software developed for music synthesis, and notational or computational conventions that are specialized for music synthesis. Of particular interest is the research by Cadoz and associates that makes use of human-machine interfaces with tactual feedback and focuses on the production of music by gestural control of simulated instruments (Cadoz et al., 1984, 1990; Cadoz and Ramstein, 1991). Further information on music synthesis is available in Moore (1990), Mathews and Pierce (1989), Roads and Strawn (1988), and Richards (1988).

Synthesis Technology

Current devices available for generating nonspeech sounds tend to fall into two general categories: *samplers*, which digitally store sounds for later real-time playback, and *synthesizers*, which rely on analytical or algorithmically based sound generation techniques originally developed for imitating musical instruments (see Cook et al., 1991; Scaletti, 1994; and the discussion above). With samplers, many different sounds can be reproduced (nearly) exactly, but substantial effort and storage media are required for accurately prerecording sounds, and there is usually limited real-time control of acoustic parameters. Synthesizers, in contrast, afford a fair degree of real-time, computer-driven control.

Most widely available synthesizers and samplers are based on MIDI (musical instrument digital interface) technology. The baud-rate of such devices (31.25 kbar), especially when connected to standard serial computer lines (19.2 kbar), is still low enough that continuous real-time control of multiple sources/voices will frequently "choke" the system. In general, synthesis-based MIDI devices, such as those that use frequency modulation, are more flexible than samplers in the type of real-time control available but less general in terms of the variety of sound qualities that can be reproduced. For example, it is difficult to generate environmental sounds, such as breaking or bouncing objects, from an FM synthesizer (see Gaver, 1993).

Large-scale systems designed for sound production and control in the entertainment industry or in music composition incorporate both sampling and digital synthesis techniques and are much more powerful. However, they are also expensive, require specialized knowledge for their use, and are primarily designed for off-line sound design and post-production. A potential disadvantage of both types of devices is that they are primarily designed with musical performance and/or sound effects in mind. This design emphasis is not necessarily well suited to the generation and control of sounds for the display of information and, again, tends to require that the designers and users have specialized knowledge of musical or production techniques.

The most general systems would be based on synthesis via physical or acoustical models of sound source characteristics. A simpler but less versatile approach would use playback of sampled sounds or conventional MIDI devices, as in most current systems. Since very general physical models are both difficult (perhaps impossible) to develop and computationally intensive, a more practical and immediately achievable system might be a hybrid approach that uses techniques like real-time manipulation of simple parameters, such as the pitch, filter bandwidth, or intensity, of sampled sounds, and real-time interpolation between sampled sounds, analogous to "morphing" in computer graphics; the E-mu Morpheus synthesizer is an example of this kind of approach. Recently, several commercial synthesizer companies have announced new products based on physical modeling techniques. A sound card being developed by MediaVision is based on digital waveguides (Smith, 1992); the Yamaha VL1 keyboard synthesizer is based on an unspecified physical modeling approach; and the Macintosh-based Korg SynthKit allows construction of sounds via interconnection of a visual programming language composed of modular units representing hammer-strikes, bows, reeds, etc.

A few nonspeech-sound-generation systems have been integrated specifically for virtual environment applications (e.g., Wenzel et al., 1990; VPL Research's Audiosphere; see also Wenzel, 1994); some designers are beginning to develop systems intended for data "sonification" (e.g., Smith et al., 1990; Scaletti and Craig, 1991; Wenzel et al., 1993b). Related developments in auditory interfaces include the work on audio "windowing" systems for applications like teleconferencing (Ludwig et al., 1990; Cohen and Ludwig, 1991). However, far more effort needs to be devoted to the development of sound generation technology specifically aimed at information display (see Kramer, 1994). Even more critical, perhaps, is the need for further research on lower-level sensory and higher-level cognitive determinants of acoustic perceptual organization, since these results should serve to guide technology development. Furthermore, relatively little research has been concerned with how various acoustic parameters interact to determine the identification, segregation, and localization of multiple, simultaneous sound streams. Understanding such interaction effects is likely to be critical in any acoustic display developed for SE systems.

With the advances in algorithms and hardware to produce such simulations, there is also a need to develop an extensible protocol for virtual audio. Such a protocol will need to encompass all the acoustic models in use today and those expected to be developed in the near future. This protocol should allow developers and designers of sonification systems to utilize SE technology, even as that technology makes dramatic improvements in its capabilities.

RESEARCH NEEDS

Most of the needs associated with the auditory channel lie in the domain of perceptual studies. With one major exception, discussed below, the technology for the auditory channel is either adequate now or will be adequate in the near future.

Perceptual Issues

Many of the perceptual issues in the auditory channel that require attention fall under the general heading of adaptation to alterations in sensorimotor loops. Some specific examples in this category, all of which relate to the spatialization problem, concern the extent to which (and rate at which) listeners adapt to various types of distortions, including those associated with (1) the use of simplified HRTFs or transformations designed to achieve supernormal localization performance in VE systems and (2) various mappings between telerobotic sensor arrays and the human auditory system in teleoperator systems that make use of nonanthropomorphic telerobots. Knowledge of how well and how fast individuals can adapt to distortions or transformations of these types under various training conditions is essential to the design of effective systems. Other closely related examples focus on the use of the auditory channel for sensory substitution purposes, e.g., the presentation of auditory signals to substitute for haptic feedback that is difficult to provide because of current equipment limitations.

Another major area in the perceptual domain that requires substantial work falls under the heading of auditory information displays. Current knowledge of how to encode information for the auditory channel in ways that produce relatively high information transfer rates with relatively small amounts of training is still rather meager. It is obvious that encoding all information into speech signals is inappropriate and the statement that the encoding should be natural is simply not adequate to guide specific design choices. This general encoding and display problem is judged to be both important and difficult to solve. Also, it is believed that progress in this area will depend, at least in part, on an improved understanding of auditory scene analysis.

Technology

The main technology area that requires attention concerns the computer generation of sounds: software and hardware are required to enable VE system users to specify and generate the desired acoustic behavior of the objects and agents in the VE under consideration. Physical modeling of environmental sounds, together with the development of appropriate mathematical approximations and software and hardware that enables such sounds to be computed and rendered in real time, constitutes a major task in this area. Subsequently, it will be necessary to develop a user-friendly system for providing speech, music, and environmental sounds all in one integrated package. And eventually, of course, the whole sound generation system will have to be appropriately integrated with the software and hardware system for generating visual and haptic images. Compared with these sound generation problems, most of the remaining technological problems in the spatialization area for both HMDs and off-head displays appear relatively minor. It should be noted, however, that current technology is still unable to provide interactive systems with real-time rendering of acoustic environments with complex, realistic room reflections.

4

Haptic Interfaces

Haptic interfaces are devices that enable manual interaction with virtual environments (VEs) or teleoperated remote systems. They are employed for tasks that are usually performed using hands in the real world, such as manual exploration and manipulation of objects. In general, they receive motor action commands from the human user and display appropriate tactual images to the human. Such haptic interactions may or may not be accompanied by the stimulation of other sensory modalities, such as vision and audition. Computer keyboards, mice, and trackballs constitute relatively simple haptic interfaces. Other examples of haptic interfaces available in the market are gloves and exoskeletons that track hand postures and joysticks that can reflect forces back to the user. Even more sophisticated devices have been built and implemented successfully in research laboratories. To realize the full promise of VEs and teleoperation, further development of haptic interfaces is critical. In pursuing this goal, many of the issues and technologies described in the sections on position tracking (Chapter 5) and telerobotics (Chapter 9) are relevant. To achieve success, a comprehensive research program is needed in human haptics, technology development, and interactions between the two.

In contrast to the purely sensory nature of vision and audition, only the haptic system can both sense and act on the environment. The human hand is a versatile organ that is able to press, grasp, squeeze, and stroke objects; it can explore object properties such as surface texture, shape, and softness; it can manipulate tools for repairing a watch and breaking concrete. In the words of Paul Valéry (1938), the hand is "a device which can, in turn, strike, receive and give, feed, take an oath, beat a musical rhythm, read for the blind, speak for the mute, reach to a friend, stop a foe." Being able to touch, feel, and manipulate objects in an environment, in addition to seeing (and hearing) them, provides a sense of immersion in the environment that is otherwise not possible. It is quite likely that much greater immersion in a VE can be achieved by the synchronous operation of even a simple haptic interface with a visual and auditory display, than by large improvements in, say, the fidelity of the visual display alone. Real environments or VEs in which one is deprived of the touch and feel of objects seem impoverished, seriously handicap human interaction capabilities, and, at worst, can be disorienting.

Although haptic interfaces are typically designed to be operated by the user's hands, alternative designs suitable for the tactual and motor systems of other body segments are conceivable. However, not all interfaces that interact with the human mechano-sensorimotor systems are haptic interfaces. The distinction is based on the nature of the tasks for which the interface is used. For example, whole body motion interfaces (Chapter 6) concerned with conveying a sense of mobility to the user are not haptic interfaces in a strict sense.

STATUS OF THE RELEVANT HUMAN RESEARCH

The Human Haptic System

In order to develop cost-effective haptic interfaces, it is necessary to understand the roles played by the mechanical, sensory, motor, and cognitive subsystems of the human haptic system. The mechanical structure of the human hand consists of an intricate arrangement of 19 bones, connected by almost as many frictionless joints and covered by soft tissue and skin. Altogether, the bones are attached to about 20 intrinsic and extrinsic muscles through numerous tendons, which serve to activate 22 degrees of freedom (DOF) of the hand. The sensory system includes large numbers of various classes of receptors and nerve endings in the skin, joints, tendons, and muscles. Appropriate mechanical, thermal, and chemical stimuli activate these receptors, causing them to transmit electrical impulses via the afferent neural network to the central nervous system (of which the brain forms a part), which in turn sends commands through the efferent neurons to the muscles for desired motor action.

Haptic *exploration* and *manipulation* of solid objects covers a wide range of haptic functions yet provides a task framework within which the roles of the biomechanical, sensory, motor, and cognitive subsystems can be understood. Exploration is concerned mainly with the extraction of object properties, and it is therefore a sensory dominant task, although well-controlled motor actions are necessary to obtain reliable information about the object. It consists primarily of discrimination or identification of surface properties (for example, shape and surface texture) and volumetric properties (for example, mass and compliance) of objects. Manipulation is concerned mainly with modification of the environment and thus it is a motor dominant task, although sensory feedback is essential for successful performance. Manipulation tasks can be grossly subdivided into precision tasks (for example, watch repair) and power tasks (for example, using a hammer).

In any task involving physical contact with an object, be it for exploration or manipulation, the surface and volumetric physical properties of the skin and subcutaneous tissues play important roles in its successful performance. For example, the finger pad, which is used by primates in almost all precision tasks, consists of hairless ridged skin (about 1 mm thick) that encloses soft tissues composed of mostly fat in a semiliquid state. As a block of material, the finger pad exhibits complex mechanical behavior—inhomogeneity, anisotropy, and rate and time dependence. The compliance and frictional properties of the skin, together with the sensory and motor capabilities of the hand, enable one to both glide over a surface without losing contact, to explore the shape of the surface, and to stably grasp a smooth object to manipulate it. The mechanical loading on the skin, the transmission of the mechanical signals through the skin, and their transduction by the cutaneous mechanoreceptors are all strongly dependent on the mechanical properties of the skin and subcutaneous tissues (Phillips and Johnson, 1981b; Srinivasan, 1989; Srinivasan and Dandekar, 1992).

Tactual sensory information from the hand in contact with an object can be divided into two classes: (1) *tactile* information, referring to the sense of contact with the object, mediated by the responses of low-threshold mechanoreceptors innervating the skin (say, the finger pad) within and around the contact region and (2) *kinesthetic* information, referring to the sense of position and motion of limbs along with the associated forces, conveyed by the sensory receptors in the skin around the joints, joint capsules, tendons, and muscles, together with neural signals derived from motor commands. (The term *proprioceptive* is used almost equivalently to *kinesthetic* by many authors.) For discussion of terminology see Darian-Smith (1984); Loomis and Lederman (1986). Only tactile information is conveyed when objects contact a passive, stationary hand, except for the ever-present kinesthetic information about the limb posture. Only kinesthetic information is conveyed during active, free (i.e., no contact with any object or other regions of skin) motion of the hand, although the absence of tactile information by itself conveys that the motion is free. Even when the two extreme cases just mentioned are included, it is clear that *all* sensory and manipulatory tasks performed actively with the normal hand involve both classes of information. In addition, free nerve endings and specialized receptors that signal skin temperature, mechanical and thermal pain, and chemogenic pain and itch are also present (Sherrick and Cholewiak, 1986).

The control of contact conditions is often as important as sensing those conditions for successful task performance. In humans, such control action can range from a fast spinal reflex to a relatively slow conscious deliberate action. In experiments involving lifting of objects held in a pinch grasp, it has been shown that motor actions such as increasing grip force are initiated as rapidly as within 70 ms after an object begins to slip relative to the finger pad, and that the sensory signals from the cutaneous afferents are critical for task performance (Johansson and Westling, 1984; Johansson and Cole, 1992). Clearly, the mechanical properties of the skin and subcutaneous tissues, the rich sensory information provided by a wide variety of sensors that monitor the tasks continuously, and the coupling of this information with the actions of the motor system are responsible for the human abilities of grasping and manipulation. In the following three subsections we employ a systems viewpoint to briefly review the results on haptics in the psychophysics and neurophysiology literature.

Input-Output Variables of Haptic Interactions

Haptic interfaces in teleoperation or VE systems receive the intended motor action commands from the human and display tactual images to the human. The primary input-output variables of the interfaces are displacements and forces, including their spatial and temporal distributions. Haptic interfaces can therefore be viewed as generators of mechanical impedances that represent a relationship between forces and displacements (and their derivatives) over different locations and orientations on the skin surface at each instant of time. In contact tasks involving finite impedances, either displacement or force can be viewed as the control variable, and the other is a display variable, depending on the control algorithms employed. However, consistency among free hand motions and contact tasks is best achieved by viewing the time-varying geometrical configuration of the hand (for example, the vector of all joint angles and their derivatives with respect to time) as the control variable, and the resulting net force vector and its distribution within the contact regions as the display variables.

Because the human is sensing and controlling the position and force variables of the haptic interface, the performance specifications of the interface are directly dependent on human abilities. In a substantial number of simple tasks involving active touch, one of the tactile and kinesthetic information classes is *fundamental* for discrimination or identification, whereas the other is *supplementary*. For example, in the discrimination of length of rigid objects held in a pinch grasp between the thumb and the forefinger (Durlach et al., 1989), kinesthetic information is fundamental, whereas tactile information is supplementary. In such tasks, sensing and control of variables such as fingertip displacements are crucial. In contrast, for the detection of surface texture or slip, tactile information is fundamental, whereas kinesthetic information is supplementary (Srinivasan et al., 1990). Here, the sensing of spatiotemporal force distribution within the contact region provides the basis for inferences concerning the contact conditions and object properties. Both classes of information are clearly necessary and equally important in more complex haptic tasks.

We now summarize briefly the psychophysical and neurophysiological results available on human haptic abilities in real environments at two levels: (1) sensing and control of interface variables and (2) perception of contact conditions and object properties. Although humans can feel heat, itch, pain, etc. through sensory nerve endings in the skin, we refrain from discussing these sensations here because the availability of practical interface devices employing them is unlikely in the near future.

Sensing and Control of Interface Variables

Limb Position and Motion

Our awareness of the relative positions and motions of our limbs arises from the kinesthetic sensory system, which consists of sensory receptors in the joint capsules, tendons, muscles, and skin around the joints, as well as the signals derived from motor commands during voluntary motion (see reviews by Clark and Horch, 1986; Matthews, 1982). The joint capsules are innervated by three different types of mechanoreceptive nerve terminals, namely, free nerve endings, Ruffini corpuscles, and Paciniform corpuscles (Darian-Smith, 1984), each of which have distinct response characteristics. In addition, the tendons contain Golgi organs, which seem to respond to tension, and the muscle spindles measure the muscle stretch and its rate of change. The skin around the joints contains four types of sensory endings (discussed below) in the hairless skin, in addition to receptors in the hair follicles in the hairy skin, with each receptor type coding different aspects of the mechanical loading imposed on the skin. Furthermore, the *efferent* copy (also referred to as *corollary discharge*) of the command signals generated to drive the muscles during voluntary movements gives information about the intended motor action to the perceptual portions of the brain. Because of the presence of multiple, simultaneously active subchannels that are not individually accessible to experimentation, even basic questions about the functioning of the kinesthetic sensory system have not been answered unequivocally.

The source of kinesthetic information that enables us to know the relative positions of limb segments or joint angles is still controversial (Clark and Horch, 1986). Initially, it was proposed that the receptors in the joints were the source (Skoglund, 1956; Mountcastle and Powell, 1959). Later it was found from neurophysiological experiments that these receptors were activated only in the extremities of the range of joint rotations (Burgess and Clark, 1969; Grigg and Greenspan, 1977). Also, patients with artificial joints did not seem to lose their joint angle sense significantly (Grigg et al., 1973). It also should be noted that Ferrel (1980) has argued that joint afferent discharge is sufficient to help signal the joint angle over its full range, but does not claim that the afferents are exclusively responsible for position sense (Ferrel et al., 1987). Muscle spindles, which are believed to be muscle length detectors, have also been proposed as candidates that provide position sense (Matthews, 1982). Support for this hypothesis comes from the well-known haptic illusion that, when vibration is imposed on muscles and tendons, the corresponding limbs are perceived to be moving (Goodwin et al., 1972). However, because of cocontractions of agonist and antagonist muscles, the lengths of muscles may change without any change in the joint angle. Thus, computations involving all the muscular forces imposed on the joint are needed to extract the joint angle information from muscle spindles. Nevertheless, Matthews (1988) has proposed that it might be possible to recover angular velocity independent of position by combining the spindle signals with corollary discharges from motor centers. The third possible source of joint angle information is the stress and strain field in the skin surrounding the joint, which is directly related to the angle of rotation of the joint. Although this possibility has been mentioned in the literature, we are not aware of any systematic investigation of this hypothesis. Recently, Edin (1993) has shown that the strains produced in the skin can be large enough to signal the joint angle.

A large variety of psychophysical experiments have been conducted concerning the perception of limb position and motion (Clark and Horch, 1986; Jones and Hunter, 1992). It has been found that humans can detect joint rotations of a fraction of a degree performed over a time interval of the order of a second. The bandwidth of the kinesthetic sensing system has been estimated to be 20-30 Hz (Brooks, 1990). It is generally accepted that our sensitivity to rotations of proximal joints is higher than that of more distal joints. The just noticeable difference (JND) is about 2.5 deg for the finger joints, 2 deg for the wrist and elbow, and about 0.8 deg for the shoulder (Tan et al., 1994). In locating a target position by pointing a finger, the speed, direction, and magnitude of movement, as well as the locus of the target, can all affect accuracy. In the discrimination of length of objects by the finger-span method (Durlach et al., 1989; Tan et al., 1992), the JND is about 1 mm for a reference length of 10 mm, and increases to 2-4 mm for a reference length of 80 mm, thus violating Weber's law (i.e., JND is not proportional to the reference length). In the kinesthetic space, psychophysical phenomena such as anisotropies in the perception of distance and orientation, apparent curvature of straight lines, non-Euclidean distance measures between two points, etc., have been reported (for a review, see Loomis and Lederman, 1986; Hogan et al., 1990; Fasse et al., 1990). Investigations of the human ability in controlling limb motions have typically measured human tracking performance with manipulanda having various mass, spring, and damping characteristics (Brooks, 1990; Poulton, 1974; Sheridan, 1992; Jones and Hunter, 1992). The differential thresholds for position and movement have been measured to be about 8 percent (Jones and Hunter, 1992). Human bandwidth for limb motions is found to be a function of the mode of operation: 1-2 Hz for unexpected signals, 2-5 Hz for periodic signals, up to 5 Hz for internally generated or learned trajectories, and about 10 Hz for reflex actions (reviews by Brooks, 1990). In summary, the sensing and control of limb position and motion are complex at all levels, ranging from psychophysical measures to the inner neurophysiological mechanisms.

Net Forces of Contact

When we contact or press objects through active motion of the hand, the contact forces are sensed by both the tactile and kinesthetic sensory systems. Overall contact force is probably the single most important variable that determines both the neural signals in the sensory system as well as the control of contact conditions through motor action. It appears that the JND for contact force is 5-15 percent of the reference force value over a wide range of conditions involving substantial variation in force magnitude, muscle system, and experimental method, provided that the kinesthetic sense is involved in the discrimination task (Jones, 1989; Pang et al., 1991; Tan et al., 1992). In closely related experiments exploring the human's ability to distinguish among objects of different weights, a slightly higher JND of about 10 percent has been observed (see reviews by Clark and Horch, 1986; Jones, 1986). An interesting illusion first observed in the late nineteenth century by Weber is that cold objects feel heavier than warm ones of equal weight (see review by Sherrick and Cholewiak, 1986). In experiments involving grasping and lifting of objects using a two-finger pinch grasp, Johansson and Westling (1984) have shown that subjects have exquisite control over maintaining the proper ratio between grasping and lifting forces (i.e., the orientation of the contact force vector), so that the objects do not slip. However, when tactile information was blocked using local anesthesia, this ability deteriorated significantly because the subjects could not sense contact conditions such as the occurrence of slip and hence did not apply appropriate compensating grasp forces. Thus, good performance in tasks involving contact requires the sensing of appropriate forces as well as using them to control contact conditions. The maximum controllable force that can be exerted by a finger pad is about 100 N and the resolution in visually tracking constant forces is about 0.04 N or 1 percent, whichever is higher (Srinivasan and Chen, 1993; Tan et al., 1994).

Perception of Contact Conditions and Object Properties

Although humans experience a large variety of tactile sensations, these sensations are really combinations of a few building blocks or *primitives*. For simplicity, normal indentation, lateral skin stretch, relative tangential motion, and vibration are the primitives for *conditions of contact* with the object. Surface microtexture, shape (mm size), and compliance can be thought of as the primitives for the majority of *object properties* perceived by touch. The human perception of many of these primitives is through tactile information conveyed by cutaneous mechanoreceptors. The associated neural codes can be classified as intensive, temporal, spatial, and spatiotemporal (for a review, see Loomis and Lederman, 1986). We first describe neurophysiological findings of receptor response characteristics from experiments involving indentations with rounded probes, and then discuss results from experiments involving pressing and stroking of objects configured to emphasize specific geometric or material properties.

Monkeys are used as experimental models for physiological mechanisms in humans since the types of mechanoreceptors, their spacing in the skin, and the sensory capacities to detect and discriminate vibratory stimuli are similar for the two species. In monkey skin, on the finger pads and palm, mechanoreceptive afferents have been classified on the basis of their response properties to ramp and steady indentations of a probe with or without vibration (Knibestol and Vallbo, 1970; Mountcastle et al., 1972; Pubols, 1980; Pubols and Pubols, 1976, 1983; Talbot et al., 1968). They fall into three distinct classes. (1) Slowly adapting afferents (SAs), believed to originate from Merkel cells, respond both during ramp onset and steady indentation by a probe. When the probe is vibrated sinusoidally at the most sensitive spot on the skin, the SAs are tuned (i.e., one nerve impulse per stimulus cycle) at the lowest amplitudes (about 20 μ) when the frequencies are low (less than 20 Hz). (2) Rapidly adapting afferents (RAs), emanating from Meissner corpuscles, respond to ramp onset but are quiet during steady indentation. Their tuning threshold amplitudes (about 5 μ) are the lowest in the middle frequency range (20-50 Hz). (3) Pacinian corpuscle fibers (PCs), while behaving in a similar manner to RAs for ramp and steady indentations, have very low tuning threshold amplitudes (about 1 μ) at high frequency ranges (100-300 Hz). Microneurographic techniques of recording single-nerve fiber responses from awake humans (Vallbo and Hagbarth, 1968; Knibestol and Vallbo, 1970; Johansson and Vallbo, 1979) have revealed another class of slowly adapting afferents that are primarily sensitive to skin stretch and are associated with Ruffini endings. The response properties, such as thresholds and bandwidths of each of the receptor types obtained through neurophysiological experiments, give some of the design specifications for tactile display part of haptic interfaces.

Considerable research effort has been invested on psychophysics of vibration perception and electrocutaneous stimulation using single or multiple probes (for a review, see Sherrick and Cholewiak, 1986). These studies are mostly directed at issues concerned with tactile communication aids for individuals who are blind, deaf, or deaf and blind, areas that are beyond the scope of this chapter. A comprehensive list of references describing such tactile displays can be found in Kaczmarek and Bach-y-Rita (1993) and Reed et al. (1982). In designing these devices, human perceptual abilities in both temporal and spatial domains are of interest. The human threshold for the detection of vibration of a single probe is about 28 dB (relative to 1 μ peak) for 0.4 to 3 Hz. It decreases at the rate of –5 dB/octave for 3 to 30 Hz, and decreases further at the rate of –12 dB/octave for 30 to about 250 Hz, after which the threshold increases for higher frequencies (Rabinowitz et al., 1987; Bolanowski et al., 1988). Spatial resolution on the finger pad, as measured by the localization threshold of a point stimulus, is about 0.15 mm (Loomis, 1979), whereas the two-point limen is about 1 mm (Johnson and Phillips, 1981).

To answer questions concerning perception and neural coding of roughness or spatial resolution, precisely shaped rigid surfaces consisting of mm-sized bar gratings (Lederman and Taylor, 1972; Morley et al., 1983; Phillips and Johnson, 1981a,b; Sathian et al., 1989), embossed letters (Phillips et al., 1983, 1988), or Braille dots (Lamb, 1983a,b; Darian-Smith et al., 1980) have been used in psychophysical and neurophysiological experiments (see review by Johnson and Hsiao, 1992). The perception of surface roughness of gratings is found to be solely due to the tactile sense and is dependent on the groove width, contact force, and temperature but not the scanning velocity (Loomis and Lederman, 1986). Spatial resolution on the finger pad, as measured by the localization threshold of a point stimulus is about 0.15 mm (Loomis, 1979), whereas the two-point limen is about 1 mm (Johnson and Phillips, 1981).

Some of the salient results on the perception of slip, microtexture, shape, compliance, and viscosity are given below. Humans can detect the presence of a 2 μ high single dot on a smooth glass plate stroked on the skin, based on the responses of Meissner-type rapidly adapting fibers (RAs) (LaMotte and Whitehouse, 1986; Srinivasan et al., 1990). Moreover, humans can detect 0.06 μ high grating on the plate, owing to the response of Pacinian corpuscle fibers (LaMotte and Srinivasan, 1991). Among all the possible representations of the shapes of objects, the surface curvature distribution seems to be the most relevant for tactile sensing (Srinivasan and LaMotte, 1991; LaMotte and Srinivasan, 1993). Slowly adapting fibers respond to both the change and rate of change of curvature of the skin surface at the most sensitive spot in their receptive fields, whereas RAs respond only to the rate of change of curvature. Human discriminability of compliance of objects depends on whether the object has a deformable or rigid surface (Srinivasan and LaMotte, 1994). When the surface is deformable, the spatial pressure distribution within the contact region is dependent on object compliance, and hence information from cutaneous mechanoreceptors is sufficient for discrimination of subtle differences in compliance. When the surface is rigid, kinesthetic information is necessary for discrimination, and the discriminability is much poorer than that for objects with deformable surfaces. For deformable objects with rigid surfaces held in a pinch grasp, the JND for compliance is about 5-15 percent when the displacement range is fixed, increases to 22 percent when it is roved (varied randomly), and can be as high as 99 percent when cues arising out of mechanical work done are eliminated (Tan et al., 1992, 1993). Using a contralateral-limb matching procedure involving the forearm, Jones and Hunter (1992) have found that the differential thresholds for stiffness and viscosity are 23 and 34 percent, respectively. It has been found that a stiffness of at least 25 N/mm is needed for an object to be perceived as rigid by human observers (Tan et al., 1994).

Summary

In this section, we summarize the available quantitative research results on human haptics separately for tactile, kinesthetic, and motor systems, as well as results when all the three systems are involved under active touch conditions.

Tactile Sensory System

Humans can distinguish vibration sequences of up to 1 kHz through the tactile sense. The human threshold for the detection of vibration of a single probe is about 28 dB (relative to 1 μ peak) for 0.4 to 3 Hz; it decreases at the rate of –5 dB/octave for 3 to 30 Hz, and decreases further at the rate of –12 dB/octave for 30 to about 250 Hz, after which the threshold increases for higher frequencies. Spatial resolution on the finger pad, as measured by the localization threshold of a point stimulus is about 0.15 mm, whereas the two point limen is about 1 mm. Human detection thresholds for features on a smooth glass plate are a 2 μ high single dot and a 0.06 μ high grating.

Kinesthetic Sensory System

Humans can detect joint rotations of a fraction of a degree performed over about a second. The bandwidth of the kinesthetic system is estimated to be 20-30 Hz. The JND is about 2.5 deg for the finger joints, 2 deg for the wrist and elbow, and about 0.8 deg for the shoulder.

Motor System

Human bandwidth for limb motions is found to be a function of the mode of operation: 1-2 Hz for unexpected signals, 2-5 Hz for periodic signals, up to 5 Hz for internally generated or learned trajectories, and about 10 Hz for reflex actions. The differential thresholds for position and movement have been measured to be about 8 percent.

Active Touch Involving All Three Systems

The JND for length is about 1 mm for a reference length of 10 mm, and increases monotonically to 2.4 mm for a reference length of 80 mm. The JND for contact force is 5-15 percent of the reference force value. The maximum controllable force that can be exerted by a finger pad is about 100 N and the resolution in visually tracking constant forces is about 0.04 N or 1 percent, whichever is higher. The JND for compliance of deformable objects with rigid surfaces can range from 5 to 99 percent depending on the cues available to the human subject. A stiffness of at least 25 N/mm is needed for an object to be perceived as rigid by human observers. The differential threshold for viscosity sensed by activating the forearm is about 34 percent.

STATUS OF THE TECHNOLOGY

Terminology and Variables of Haptic Interfaces

Since haptic interfaces are devices composed of mechanical components in physical contact with the human body for exchange of information with the human nervous system, it is natural to borrow the terms used in mechanics, human physiology, and robotics to describe the subsystems of the interfaces. In performing tasks with a haptic interface, the human user conveys desired motor actions by physically manipulating the interface, which, in turn, displays tactual sensory information to the user by appropriately stimulating his or her tactile and kinesthetic sensory systems. Thus, in general, haptic interfaces can be viewed as having two basic functions: (1) to measure the positions and contact forces (and time derivatives) of the user's hand (or other body parts) and (2) to display contact forces and positions (or their spatial and temporal distributions) to the user. Among these position (or kinematic) and contact force variables, the choice of which ones are the motor action variables (i.e., inputs to the computer) and which are the sensory display variables (i.e., inputs to the human) depends on the hardware and software design, as well as the tasks for which the interface is employed.

Although a force-reflecting haptic interface needs only to display forces, the sensing of forces by the interface (in addition to position sensing) is likely to be needed for several reasons. First, the presence of noise in the system, as well as the need to compensate for friction and inertia, requires closed-loop force control and hence force sensing. Second, the limitations on available VE technology make it necessary to achieve reconfigurability through changes in hardware as well as software (see below). In other words, a general-purpose VE system might need to augment the exoskeleton with a variety of hardware manipulanda, some of which would include force sensing. Third, in certain applications, it may be desirable to create nonnatural environments. For example, in certain cases it might be appropriate to use a fixed-position, force-sensing joystick together with a visual display of tactile information. Alternatively, one might find it helpful to employ a position-displaying joystick, with or without force sensing, to present certain kinds of spatial information (e.g., for guiding a passive hand through a maze).

Classification of Haptic Interfaces

A primary classification of haptic interactions with real environments or VEs that affects interface design can be summarized as follows: (1) free motion, in which no physical contact is made with objects in the environment; (2) contact involving unbalanced resultant forces, such as pressing an object with a finger pad; (3) contact involving self-equilibrating forces, such as squeezing an object in a pinch grasp. Depending on the tasks for which a haptic interface is designed, some or all of these elements will have to be adequately simulated by the interface. For example, grasping and moving an object from one location to another involves all three elements. The design constraints of a haptic interface are strongly dependent on which of these elements it needs to simulate. Consequently, the interfaces can be classified according to whether they are force-reflecting or not, as well as by what types of motions (e.g., how many degrees of freedom) and contact forces they are capable of simulating.

An alternative but important distinction in our haptic interactions with real environments or VEs is whether we touch, feel, and manipulate the objects directly or with a tool. The complexity in the design of a haptic interface is seriously affected by which of these two types of interactions it is supposed to simulate. Note that an ideal interface, designed to provide realistic simulation of direct haptic exploration and manipulation of objects, would be able to simulate handling with a tool as well. Such an interface would measure the position and posture of the user's hand, display forces to the hand, and make use of a single hardware configuration (e.g., an exoskeleton with force and tactile feedback) that could be adapted to different tasks by changes in software alone. For example, the act of grasping a hammer would be simulated by monitoring the position and posture of the hand and exerting the appropriate forces on the fingers and palm when the fingers and palm were in the appropriate position. However, the large number of degrees of freedom of the hand, extreme sensitivities of cutaneous receptors, together with the presence of mass, friction, and limitations of sensors and actuators in the interface, make such an ideal impossible to achieve with current technology. In contrast, an interface in the form of a tool handle, for which reconfigurability within a limited task domain is achieved through both hardware and software changes, is quite feasible. Thus, one of the basic distinctions among haptic interfaces is whether they attempt to approximate the ideal exoskeleton or employ the tool-handle approach.

Another set of important distinctions concerning haptic interfaces results from a consideration of the force display subsystems in an interface. Broadly speaking, force display systems can be classified as either ground-based, such as joysticks and other hand controllers, or body-based, such as gloves and exoskeletons. Frequently, the distinction between grounding sites is overlooked in the literature. For example, exploration or manipulation of a virtual object requires that force vectors be imposed on the user at multiple regions of contact with the object. Consequently, equal and opposite reaction forces are imposed on the interface. If these forces are self-equilibrating, as in simulating the contact forces that occur when we squeeze an object, then the interface need not be mechanically grounded. However, if the forces are unbalanced, as in pressing a virtual object with a single finger pad, the equilibrium of the interface requires that it be attached somewhere. A force-reflecting joystick attached to the floor would be a ground-based display, whereas a force-reflecting exoskeletal device attached to the user's forearm would be a body-based display. The grounding choice affects whether the user experiences throughout his or her entire body the stresses induced by contact with a virtual object. The consequences of using a body-based display to simulate contact forces that really stem from ground-based sources are not known and warrant investigation. A further example of improperly grounded displays occurs with most tactile stimulators. If a tactile stimulator array is attached to the finger pad via a strap surrounding the finger, then the net applied force by the stimulator is balanced by a reaction force on the back of the finger. Whether this reaction force can be distributed with a low enough pressure distribution to be imperceptible, and whether the absence of stresses throughout the rest of the musculoskeletal system is inconsequential, are not known. Although most devices built to date are either ground-based or body-based, hybrid interfaces that are a combination of the two (such as the Dextrous Teleoperation System Master built by Sarcos, Inc.) are also possible.

Current Technology

Hardware

Haptic interface hardware for synthetic environments (SEs) is in the very early stages of development. Many of the devices available today have been motivated by needs predating those of VE technology. Simple position/motion-measuring systems have long been employed to provide control inputs to the computer. These have taken many forms, such as those that involve contact with the user without controlled force display (e.g., keyboards, computer mice, trackballs, joysticks, passive exoskeletal devices) and those that measure position/motion without contact (e.g., optical and electromagnetic tracking devices). Applications motivating development of these devices have ranged from the control of equipment (e.g., instruments, vehicles) to biomechanical study of human motion (e.g., gait analysis, time and motion studies). The requirements for position trackers and a variety of design approaches and devices are described in Chapter 6 on position tracking and mapping.

The early developments in force-displaying haptic interfaces were driven by the needs of the nuclear energy industry and others for remote manipulation of materials (Sheridan, 1992). The force-reflecting teleoperator master arms in these applications were designed to communicate to the operator information about physically real tasks. The recognition of the need for good-quality force displays by early researchers (Goertz, 1964; Hill, 1979) continues to be relevant to today's VE applications. Although Sutherland's (1965) pioneering description of VEs included force-reflecting interfaces, development of practical devices has proven to be difficult. The current state of kinematics, actuators, sensors, and control of master manipulators described in Chapter 9 on telerobotics is directly relevant to haptic interfaces.

A rough breakdown of major types of haptic interfaces that are currently available or being developed in laboratories and companies around the world is as follows:

(1) Ground-based devices
- joysticks/hand controllers

(2) Body-based devices
- exoskeletal devices
  - flexible (gloves and suits worn by user)
  - rigid links (jointed linkages affixed to user)

(3) Tactile displays
- shape changers
  - shape memory actuators
  - pneumatic actuators
  - microelectromechanical actuators
- vibrotactile
- electrotactile

Joysticks are probably the oldest of these technologies and were originally conceived to control aircraft. Even the earliest of control sticks, connected by mechanical wires to the flight surfaces of the aircraft, unwittingly presented force information about loads on the flight surfaces to the pilot. In general, they may be passive (not force reflecting), as in the joysticks used for cursor positioning, or active (force reflecting), as in many of today's modern flight-control sticks. For example, Measurement Systems Inc. has marketed several 2- and 3-DOF position-sensing joysticks, some of which can sense but not display force. Examples of force-reflecting 2-DOF joysticks designed for relatively high bandwidth are the AT&T mini-joystick (Schmult and Jebens, 1993) and one built in the MIT Newman Laboratory (Adelstein and Rosen, 1992).

Many of the force-reflecting hand controllers available today have been developed for the control of remote manipulators (Jacobus et al., 1992; Meyer et al., 1992). Generally, these devices employ at most 6 DOF (plus grip control) and have a wide range of performance qualities. Particularly good reviews of performance characteristics are found in Brooks (1990) and McAffee and Fiorini (1991), and a broad overview of the devices is available in Honeywell (1989). A great deal of work concerning ergonometrics (shape, switch placement, motion and force characteristics, etc.) has gone into the design of the hand grip of these devices (Brooks and Bejczy, 1985). One of the first applications of force-reflecting hand controllers to VEs was in project GROPE at the University of North Carolina (Brooks et al., 1990). The Argonne Mechanical Arm (ARM) was used successfully for force reflection during interactions with either simulations of molecule docking or with data from a scanning tunnelling microscope. Recently, high-performance devices have been specifically designed for interaction with VEs. The MIT Sandpaper is a 3-DOF joystick that is capable of displaying virtual textures (Minsky et al., 1990). In Japan, desktop master manipulators have been built in Tsukuba (Iwata, 1990; Noma and Iwata, 1993). At the University of British Columbia, high-performance hand controllers have been developed by taking advantage of magnetic levitation technology (Salcudean et al., 1992). PER-Force is a 6-DOF hand controller that delivers high performance (Cybernet Systems, 1992). The PHANToM, built in the MIT Artificial Intelligence Laboratory, is a multilink, low-inertia device that can convey the feel of virtual objects (Massie and Salisbury, 1994).

Sophisticated teleoperation masters have been built that can be used to feel and manipulate virtual objects as well. At Harvard, Howe (1992) has developed a teleoperation system with a two-finger master that can be used to execute precision tasks with a pinch grasp between the thumb and the index finger. One of the most complex force-reflecting devices built to date is the Dextrous Teleoperation System Master designed by Sarcos, Inc., in conjunction with the University of Utah's Center for Engineering Design and the Naval Ocean Systems Center (NOSC). Although it is primarily ground-based, by having attachment points at the forearm and upper arm of the user it has the advantages of an exoskeleton, such as a large workspace comparable to that of the human arm. This device utilizes high-performance hydraulic actuators to provide a wide dynamic range of force exertion at relatively high bandwidth on a joint-by-joint basis for 7 DOF. Another high-performance force-reflecting master is a ground-based system built by Hunter et al. (1990) to enable two-handed teleoperation of a microrobot that can meet the dual requirements of wide bandwidth (exceeding 1 kHz) and high accuracy (as low as a few nanometers). Improved versions of these devices have been built for teleoperated eye surgery and represent the state-of-the-art performance that can be achieved using currently available technology (Hunter et al., 1994).

Exoskeletal devices are characterized by the fact that they are designed to fit over and move with the limbs or fingers of the user. Because they are kinematically similar to the arm and hands that they monitor and stimulate, they have the advantage of the widest range of unrestricted user motion. As position-measuring systems, exoskeletal devices (gloves, suits, etc.) are relatively inexpensive and comfortable to use. The well-known VPL DataGlove and DataSuit use fiberoptic sensors to achieve a joint angle resolution of about a degree. The Virtex CyberGlove achieves a higher resolution of about half a degree by using strain gauges. EXOS and the Utah/MIT Dextrous Hand Master consist of rigid link exoskeletons and use Hall effect sensors to obtain a resolution of about 0.2 to 0.5 deg. Rigid link exoskeletons that provide force reflection in addition to joint angle sensing have also been designed and built. Shimoga (1992) provides an excellent review of these devices and design issues, including both human factors and technology. The Utah hand-wrist master (Jacobsen et al., 1989), the Rutgers Portable Dextrous Master (Burdea et al., 1992), the JPL Glove controller (Jau, 1992), the Tsukuba fingertip force display (Iwata et al., 1992), and the EXOS SAFiRE fall into this category of device. However, providing high-quality force feedback with such devices that is commensurate with human resolution is difficult and places great demands on actuator size minimization and control bandwidth.

While the display of net forces is appropriate for coarse object interaction, investigators have also recognized the need for more detailed displays within the regions of contact. In particular, the display of tactile information (e.g., force distributions for conveying information on texture and slip), though technically difficult, has long been considered desirable for remote manipulation (Bliss and Hill, 1971). Tactile display systems in the last two decades have been mostly used in conveying visual and auditory information to deaf and blind individuals (Bach-y-Rita, 1982; Reed et al., 1982). Display systems that attempt to convey information about contact use a variety of techniques. Shape-changing displays convey the local shape of contact by controlling the deformation or forces distributed on the skin. This has been accomplished by an array of stimulators actuated by DC solenoids (Frisken-Gibson et al., 1987), shape memory alloys (TiNi, 1990), and compressed air. The use of a continuous surface actuated by a electrorheological fluid has been proposed by Monkman (1992). Vibrotactile displays deliver mechanical energy through an array of vibrating pins placed against the skin. The Opticon, marketed by Telesensory Systems, and the Bagej Corporation tactile stimulator belong to this class. The EXOS touch master consists of a single voice coil vibrator. A particularly promising desktop tactile array capable of high performance as both a shape changer and a vibrator over 0 to several hundred Hz is being developed at Johns Hopkins University (Schneider, 1988). Electrotactile displays stimulate the skin through surface electrodes. A review of principles and technical issues in vibrotactile and electrotactile displays can be found in Kaczmarek and Bach-y-Rita (1993). Various types of tactile display devices mentioned above are reviewed by Shimoga (1992).

Software

In general, haptic interfaces receive motor action commands from the human user and display appropriate tactual "images" to the user. Tactual images consist of force and displacement fields to be imposed on the observer in order to simulate the observer's desired mechanical interactions with objects in the VE. In general, these images stimulate both tactile and kinesthetic information channels in the observer and are driven by the actions of the observer. Major components of the information conveyed are the mode of contact with the objects (e.g., indentation, slip), mechanical properties of the objects (e.g., texture, shape, compliance), as well as the motions and forces involved in exploration and manipulation in a VE.

Since haptic interfaces for interacting with VEs are in the early stages of development, there is very little software that has been specifically designed for generating tactual images. Commercially developed codes necessary for using position trackers of various manufacturers are available. However, for force-reflecting devices, as in the case of hardware, most of the software has been developed in the context of teleoperation or controlling autonomous robots. Several research laboratories have developed VE systems with visual and haptic displays achieved through appropriate integration of mechanistic models of virtual objects and control of haptic interfaces for rendering tactual images with software used to drive the visual images. For example, the PHANToM interface developed in the MIT Artificial Intelligence Laboratory has been used to tactually display the forces of contact of a stylus held in the user's hand with a variety of static and dynamic virtual objects in synchrony with visual images of the objects and their motion.

Similar to the software needed to generate visual images (Chapter 8), the software necessary to generate tactual images can be classified into three major groups: haptic interaction software, physical models of virtual objects and environments, and software for rendering tactual images. Haptic interaction software mainly consists of reading the state of the haptic interface device. For example, the signal conditioning and noise reduction software necessary for reading position or force sensors would fall within this category. In the case of exoskeletal devices used for tracking hand posture, a higher-level software based on the human kinematic model of the hand is needed as well for interpreting the sensor signals as corresponding to a hand posture.

Physical models of virtual objects and environments receive user's commands through the sensors in the haptic interface and generate force or displacement outputs corresponding to the physical behavior of a simulated object in the VE. As mentioned in the section on world modeling in

Chapter 8, this can either be accomplished by a unified model for all the modalities (visual, haptic, acoustic) or through separate models for each modality together with correlation algorithms for consistency among the displays corresponding to each of the modalities. However, the computations needed for the former approach (e.g., involving finite element methods) tend to be extremely intensive and are difficult to complete in real time, even when one uses supercomputers. Simplifications in generating multimodal images are necessary, not only because of the computational difficulties, but also because the display devices at present have limited capabilities. Therefore, even though the physics governing the visual, haptic, and acoustic behavior of an object is the same, different approximations might be needed for each of the modalities. For example, visual images are scalar, two-dimensional projections of the objects, whereas tactual images are, in general, three-dimensional vector fields. For realistic visual images, all the objects within the visual field need to be displayed and, typically, each object needs to appear as a continuous two-dimensional projection. In the case of tactual images, often only the display of forces within isolated contact regions is sufficient. Also, lumped-parameter models that approximate a continuum through discrete elements may be good enough to generate inputs to the haptic rendering devices. However, these force fields are tightly coupled to the user's actions as well as the mechanical properties of the soft human tissues in contact with the interface device. The mechanics of interaction between the observer and the environment plays a fundamental role in the generation of tactual images. Models of the human operator's behavior and performance developed in teleoperation literature are applicable to VEs as well (Sheridan, 1992).

The software for rendering the tactual images receives the output of the physical model and generates the commands needed to drive the interface device. In the case of the Sandpaper, a 2-DOF joystick capable of force display (Minsky et al., 1990), the authors report success in conveying the feel of exploring rough surfaces by using a simple rule that contact forces to be displayed are proportional to the local gradient of the textured surface. Even when such simple algorithms generate the tactile images, if the user has in addition visual or auditory inputs that are consistent, it is possible that the interactions with VEs will seem sufficiently realistic to him or her. Therefore, the algorithms for the generation of tactual images depend strongly on the particular application as well as the capabilities of the display device, including the available computational speed. Because force displays are prone to mechanical instabilities and human users are sensitive to even low disturbances unrelated to the task, real-time control of the interface devices needs to be of high quality. In the robotics and teleoperation literature (Chapter 9), considerable effort has been directed at implementing conventional proportional-integral-derivative (PID) controllers for contact tasks. Impedance control techniques (reviewed by Brooks, 1990) and the use of the passivity principle have been reported to be successful in combatting instabilities. Substantial theoretical research is currently being pursued in the areas of multivariable control and advanced nonlinear techniques, such as adaptive and robust control.

Summary of Current Technology and Future Possibilities

Computer keyboards, mice, and trackballs are the simplest haptic interfaces and are being widely used to interact with computers. Position-sensing gloves and exoskeletons without force reflection are also available on the market but are used mainly for research purposes. Among the force-reflecting devices, ground-based devices such as joysticks are being used, and modified versions of such devices for different tool handles are feasible in the near future. Force-reflecting exoskeletons are harder to design for adequate performance, and only a few such have been built for research purposes. Tactile displays offer particularly difficult design challenges because of the high density of receptors in the skin to which they must apply the stimulus. There exist a number of examples of tactile stimulators for the finger, including pneumatic shape changers, electrocutaneous stimulators, and vibrating arrays, but none provides convincing tactile images and all are awkward to use (Durlach et al., 1992).

The emerging field of microelectromechanical systems (MEMS) holds promise for providing very fine arrays of tactile stimulators. Arrays of surface-normal, electrostatic actuators currently being developed for sensors could be adapted for use in high-resolution tactile displays (Trimmer et al., 1987). Although capable of relatively small forces and deflections, arrays of such actuators integrated with addressing electronics would be inexpensive, lightweight, and compact enough to be worn without significantly impeding hand movement or function. In addition, the current technology makes feasible a 20 × 20 array of individually controlled stimulators on a 1 cm × 1 cm chip. Finally, recent work on thin-film, shape-memory alloys would enhance the attractiveness of shape-changing displays by increasing stimulator densities and actuation bandwidths. It should be noted that with synchronized multimodal stimulation, such as for simulating the contact between a tool and a rigid object, more realism can probably be achieved by providing an audible "ping" together with low bandwidth force feedback, than by improving the force bandwidth to the maximum value that is possible with current technology. Because of the difficulties in developing good cutaneous stimulator devices, initial efforts on haptic displays should probably focus on devices that apply net forces on the hand or fingertips (the tool-handle approach discussed above). Even with this simplification, large improvements on existing devices can be achieved only by a proper match between the performance of the device and human haptic abilities.

Due to inherent hardware limitations, haptic interfaces can deliver only stimuli that approximate our interactions with the real environment. It does not, however, follow that synthesized haptic experiences created through the haptic interfaces necessarily feel unreal to the user. Consider an analogy with the synthesized visual experiences obtained while watching television or playing a video game. Whereas visual stimuli in the real world are continuous in space and time, these visual interfaces project images at the rate of about 30 frames/s. Yet we experience a sense of realism and even a sense of telepresence because we are able to exploit the limitations of the human visual apparatus. The hope that the necessary approximations in generating synthesized haptic experiences will be adequate for a particular task is based on the fact that the human haptic system has limitations that can be similarly exploited. To determine the nature of these approximations or, in other words, to find out what we can get away with in creating synthetic haptic experiences, quantitative human studies are essential. Basic understanding of the biomechanical, sensorimotor, and cognitive abilities of the human haptic system is critical for proper design specification of the hardware and software of haptic interfaces. In addition, all mechanical devices will have their own intrinsic properties (such as friction, mass, compliance, viscosity, time delay) that will necessarily be interposed between the user and the desired stimulation. This lack of perfect transparency will always be present to some degree and will thus make all stimulators less than ideal. Given the approximate nature of synthetic haptic stimulation, it is clear that there is a need to assess which types of stimulation provide the most useful and profound haptic cues for the task at hand.

RESEARCH NEEDS

Compared with the visual and auditory domains, the capabilities of haptic devices and our understanding of human haptics are quite limited. A comprehensive program to develop a variety of haptic interfaces for VEs and teleoperation needs to include research in three major areas: (1) human haptics, (2) technology development, and (3) matching the performance of humans and haptic devices. It does not mean, however, that such research has to precede any usage of haptic devices. For applications that are simple from a haptic standpoint, such as those requiring relatively low-resolution hand position information, joysticks and gloves currently available off the shelf can be sufficient. More complex applications involving force and tactile displays might need research in some or all of the areas mentioned above. Since progress in the three areas is interdependent, the desirable course of development for a challenging application is to continually build improved versions of haptic devices based on experimental data obtained from the previous versions on the performance of humans, and devices and the interaction between the two. Due to the availability of powerful computers and high-precision mechanical sensors and actuators, it is now possible to exert control over experimental variables as never before.

Human Haptics

As mentioned above, the biomechanical, sensorimotor, and cognitive abilities of humans set the design specifications for devices. Therefore, multidisciplinary studies involving biomechanical and psychophysical experiments together with computational models for both are needed in order to have a solid scientific basis for device design. Perhaps to a lesser extent, neurophysiological studies concerning peripheral and central neural representations and the processing of information in the human haptic system will also aid in design decisions concerning the kinds of information that need to be generated and how these should be displayed. A major barrier to progress from the perspectives of biomechanics, psychophysics, and neuroscience has been the lack of robotic stimulators capable of delivering a large variety of stimuli under sufficiently precise motion and force control.

Biomechanical Investigations

The tight mechanical coupling between the human skin and haptic interfaces strongly influences the effectiveness of the interface. Therefore, the specifications for the design of sensors and actuators in the interface, as well as the control algorithms that drive the interface, require the determination of surface and bulk properties of, say, the finger pad. The measurement of force distributions within the contact regions with real objects is needed to determine how a display should be driven to simulate such contacts in VEs. In addition, computational models of the mechanical behavior of soft tissues will aid in simulating the dynamics of task performance for testing control algorithms, as well as in determining the required task-specific force distributions for the displays. This requires measurement of the in vivo skin and subcutaneous soft tissue response to time-varying normal and tangential loads. Information on such human factors as the size, shape, degrees of freedom, and ranges of motion of the fingers, hand, and arm are generally available in handbooks.

Psychophysical Investigations

Determination of the basic sensorimotor and cognitive abilities of the human haptic system needed for developing haptic interfaces can be subdivided as follows:

(1) *Sensing and control of contact forces and joint angles or end-point displacements:* Even simple questions concerning our abilities (such as what is the resolution, range, and bandwidth in the sensing and control of interface variables) or mechanisms (such as how we perceive joint angles or contact forces) do not yet have unequivocal answers.

(2) *Perception of contact conditions and object properties:* The important connection between the loads imposed on the skin surface within the regions of contact with objects and the corresponding perception has only begun to be addressed. Psychophysical experiments directed at determining the primary cues that signal various object properties need to be undertaken.

(3) *Integration of local contact information with nonlocal perception of the environment:* Tactual perception typically provides local information about an object. To be effective in training tasks, such as cockpit familiarization, that information must be integrated into nonlocal perception of the space within which the hand and arm move. However, haptic perception of mechanical quantities has been found to be significantly distorted (Fasse, 1992; Hogan et al., 1990). The relationship between these haptic distortions and human internal perceptual models of space and the objects in it is unknown. The influence of these distorted perceptions on production of motor behavior has barely been addressed. The theoretical framework to generate testable hypotheses must be built on a fundamental understanding of the relations between haptic perception of geometric and mechanical quantities, such as magnitudes and orientations of lengths, forces, and stiffnesses. Experimentally verified models of the relationship between haptic perceptions and motor actions are critical for the design of effective synthetic haptic environments. Similar studies need to be performed under multimodal conditions as well.

(4) *Performance in the presence of inherent time delays, distortions, and noise:* These experiments are needed for all modalities individually and in combination. Studies directed at sensorimotor and cognitive adaptation and training effects are needed.

(5) *Theoretical developments concerning information flow:* Theoretical developments concerning the task-specific flow of sensory information and control of motor action are needed to generate testable hypotheses on our haptic interactions with both real environments and VEs. Development of improved models of human operator behavior and performance (available in the teleoperation literature) through tests in realistic tasks would be beneficial in both the design and operation of SE systems.

Technology Development

Hardware

Four areas of hardware development are of interest: (1) finger, hand and arm position/joint angle measurement (trackers); (2) displays of forces and torques; (3) tactile displays; and (4) other stimulus distributions applied as two-dimensional fields to the skin, such as thermal stimuli.

The major problems with the position/angle-measuring devices are the intrusion the user feels while wearing, say, an exoskeleton, and the ever-present need for improvements in ranges, resolutions, and bandwidths. In order to display forces, designs with good actuation and control need to be developed such that they have sufficient force range, resolution, smoothness, and bandwidth. Attention needs to be paid to the friction, backlash, mechanical stiffness, apparent mass, inertia, and natural frequencies of the devices. High position resolution is needed to minimize the effect of quantization errors on stability of contact interaction. Force feedback systems need to have vibration rigorously controlled to prevent false cues to the human user. In order to achieve such high performance without mechanical instabilities, robust and adaptive closed-loop control of the devices is necessary. The mechanics of the devices must be intrinsically correct so that the difficult problems of compensating for the mass and inertia of the control arm are avoided or minimized.

Although many of the design specifications for haptic interfaces are task-dependent, we can estimate some of the interface performance requirements based on human haptic abilities. For example, since the human finger joint angle JND is of the order of a deg, the fingertip position resolution is about 1 mm. For the haptic interface to perform well, its fingertip position display resolution should probably be about 0.1 mm, and the bandwidth should be about 30 Hz to match the estimated human kinesthetic bandwidth. The maximum stiffness of the actuators should be in excess of 25 N/mm to have realistic simulation of contact with rigid stationary objects. To fully match human haptic sensory capabilities, the tactile or force displays should have a bandwidth of about 1 kHz, whereas the signals representing the human motor action need to have a bandwidth of only 10 Hz. In order to prevent false cues to the user, vibrations that are not part of the intended display should have amplitudes less than human detection threshold, which is about 25 m at 0.4 to 3 Hz, 3 m at 30 Hz, 0.3 m at 250 Hz, and is higher for higher frequencies. For tactile displays, the spatial density of actuating elements should be at least 1 mm/taxel to match the human tactile resolution. To realistically simulate continuous surfaces of virtual objects, the actuating arrays need to be even more densely packed, or should have a continuous surface over them, because of the high sensitivity of the tactile sensory system to point loads and sharp edges. It should be noted that when visual and/or auditory senses are also stimulated, haptic interfaces with lower performance capabilities than the above estimates may be adequate.

Exploration of novel technologies is needed for quantum improvements in rotary and linear actuators. Use of shape memory alloys (SMAs) and microelectromechanical sytems (MEMS) for tactile displays also needs to be investigated further. It has been estimated that real-time mechanical interactions with typical finite element models need computational speeds on the order of Gflops (Hunter et al., 1990). Similar to graphics engines used commonly with visual displays, special computational hardware specifically designed to accelerate the computations needed for haptic displays will become necessary in the near future.

Software

Modeling of the haptic environment and control of real-time interactions together with synchronous operation of other sensory modalities is a major need in software development that requires substantial research. What needs to be modeled and how to interact and display is task dependent. Trade-offs in precision and computational speed are critical. Standard methods for easily implementing physical models that range from high fidelity to coarse approximations need to be developed. In addition, models of the human operator, the environment, and interaction dynamics available in teleoperation literature need to be adapted and improved for VE applications.

Simulation of multibody environments will be possible only if we address computational efficency and appropriate architectures for modeling and maintaining a mechanical world. It is likely that this problem is much harder than simple graphic simulations. Some parallels exist, like texture, collision detection, and simulation of object dynamics, but to feel right a world model for haptic display must possibly run substantially faster, at least at the points of contact between the user and the synthetic environment. Real-time control algorithms are available to render the calculated outputs of the models to the human user through tactual displays. However, in order for the displays to be robust and feel right, the control bandwidths need to achieve frequencies of the order of several kHz. Efficient methods of implementing the control software need to be developed, including the use of special hardware, such as transputers connected in parallel. Also, theoretical advances in multivariable control and advanced nonlinear techniques, such as adaptive and robust control, are needed.

Matching Performance of Humans and Haptic Devices

Comfort (Ergonomics)

Making the human user comfortable when wearing or interacting with haptic interfaces is of paramount importance, since pain, or even discomfort, supersedes all other sensations. Appropriate attachment methods for ground-based and body-based haptic interfaces need to be developed. Design principles of achieving kinematics and the dynamics of devices that impose minimal constraints or bias on operator's hand/arm operation need to be explored.

Methods of Stimulation

The right balance of complexity and performance in system capabilities is generally task dependent. In particular, the fidelity with which the tactual images have to be displayed and the motor actions have to be sensed by the interface depends on the task, stimulation of other sensory modalities, and interaction between the modalities. Experimenting with the available haptic interfaces, in conjunction with visual and auditory interfaces, is necessary to identify the needed design improvements. Design compromises and tricks for achieving the required task performance capabilities or telepresence (immersion) need to be investigated. One of the tricks might be the use of illusions (such as visual dominance) to fool the human user into believing a less than perfect multimodal display. Techniques such as filtering the user's normal tremor or the use of sensory substitution within a modality (e.g., the use of tactile display to convey kinesthetic information) or among different modalities (e.g., visual display of a force) need to be developed to overcome the limitations of the devices and the limitations of the human user, perhaps to achieve supernormal performance. To tackle the ever-present time delays, efficient and reliable techniques for running model-based and real-time controls concurrently are needed.

Evaluation of Haptic Interfaces

Evaluation of haptic interfaces is crucial to judge their effectiveness and to isolate aspects that need improvement. However, such evaluations performed in the context of teleoperation have been so task-specific that it has been impossible to derive useful generalizations and to form effective theoretical models based on these generalizations. There is a strong need to specify a set of elementary manual tasks (basis tasks) that can be used to evaluate and compare the manual capabilities of a given system (human, robotic, VE) efficiently. Ideally, this set of basis tasks should be such that (1) knowledge of performance on these tasks enables one to predict performance on all tasks of interest and (2) it is the minimal set of tasks (in terms of time consumed to measure performance on all tasks in the set) that has this predictive power.

Two basic psychophysical questions in evaluation are: (1) With a given set-up, how good is the task performance or realism of the subjective experience? (2) How does a *change* in the set-up improve the performance of a given task, realism of the experience, or both? An example of the former is the investigation of the consequences of using an ungrounded display to simulate contact forces that really stem from grounded sources. In the latter question, the word *change* is to be interpreted in a broad sense and includes modifications of the interface hardware, object models, interaction software, and addition/subtraction of visual or auditory modalities. Theoretical and experimental approaches to quantify information transfer rates to and from the user under various single and multimodal conditions need to be developed.

5

Position Tracking and Mapping

Position tracking and mapping of the human body and environments is a basic requirement that permeates virtual environment (VE) systems: (1) head and eye tracking for visual displays; (2) hand and arm tracking for haptic interfaces; (3) body tracking for locomotion and visual displays; (4) body surface mapping for facial expression recognizers, virtual clothiers, and medical telerobots; and (5) environment mapping to build a digitized geometrical model for simulation.

This review is approached from a standpoint of basic sensory systems for position tracking and mapping: mechanical linkages, magnetic sensors, optical sensors, acoustic sensors, and inertial sensors. These systems vary in such aspects as accuracy, resolution, sampling rate, latency, range, workspace, cost, encumbrance, convenience, susceptibility to obscuration, ease of calibration, the number of simultaneous measurements, and orientation versus position tracking. VE systems are likely to include a mix of these basic sensory systems, because each system has particular strengths and weaknesses and the requirements on position tracking depend on the particular application. We now attempt to state some requirements on position tracking for particular applications.

HAND TRACKING

For normal arm movements during reaching, a fast motion is accomplished in about 0.5 ms, and wrist tangential velocities are about 3 m/s (Atkeson and Hollerbach, 1985) and accelerations are about 5-6 *g*. For the fastest arm motions such as throwing a baseball, good pitchers release the ball at 37 m/s and accelerate their hands at more than 25 *g*. Motion bandwidths of normal arm movements are around 2 Hz (Neilson, 1972); the fastest hand motions including handwriting are at around 5-6 Hz (Brooks, 1990a, 1990b). In teleoperation, it is commonly presumed that 10 Hz is the maximum frequency of position commands from the human operator (Fischer et al., 1990).

If hand motion is being used to drive a telerobot or a dynamic simulation of an arm, then the general rule of thumb is that the sampling rate should be around 20 times the bandwidth (Franklin et al., 1990) in consideration of such factors as sensor noise. Taking 5 Hz as defining the frequency content of normal arm motion, then a sampling rate of roughly 100 Hz is called for. Andersson (1993) has proposed a virtual batting cage, in which a batter swings at virtual pitches shown through a head-mounted display (HMD). The bat must be tracked to simulate the hit (or miss). Andersson proposes that sampling rates of about 1 kHz are required to track such motions.

Latency requirements are determined by the psychophysical requirements of the application and are harder to define. For force feedback applications, the hand-tracking latencies must be very low, because the human arm is part of the control loop. For non-force-feedback applications, the hand-motion-to-visual-feedback lag can probably be much longer.

EYE TRACKING

Eye movements can be as fast as 600 deg/s. The smallest time constant for saccades is around 50 ms; the smallest saccades can be finished in 60 ms. The power spectral densities can have significant power up to 50 Hz for position and 74 Hz for velocity (Bahill et al., 1981). Given again the engineering rule of thumb that the sampling rate should be 20 times the bandwidth for noisy measurements, it has been recommended that eye movements be sampled at 1 kHz (Inchingolo and Spanio, 1985). This should allow sufficiently precise tracking of the eye trajectory to characterize the movement time and end-point. As mentioned in Durlach et al. (1992), the eye sees continuous images when display temporal frequency is 60 Hz and above. With 1 kHz sampling rates for eye movement, display targets can be well chosen every 1/60th of a second.

HEAD TRACKING

Head movements can be as fast as 1000 deg/s in yaw, although more usual peak velocities are about 600 deg/s for yaw and 300 deg/s for pitch and roll (Foxlin, 1993). The frequency content of volitional head motion falls off approximately as $1/f^2$, with most of the energy contained below 8 Hz and nothing detectable above about 15 Hz. Tracker-to-host reporting rates must therefore be at least 30 Hz. Delays of 60 ms or more between head motion and visual feedback are known to impair adaptation and the illusion of presence (Held and Durlach, 1987), and much smaller delays may cause simulator sickness. Head-trackers should therefore contribute as little as possible to system latency, no more than 10 ms for high-performance systems. Accuracy requirements are very application dependent. To maintain apparently perfect registration between the real and virtual worlds with infinite-resolution see-through HMDs, the absolute accuracy must be 0.01 deg for yaw and pitch and roll and about 0.03 mm for translation. For purely virtual opaque HMD applications, large offset errors are tolerable; tilt error must not exceed about 15 deg, because that would cause vestibular conflict. However, the resolution needs to be 0.03 deg for orientation and 0.1 mm for translation to achieve perfectly smooth, jitter-free motion.

BODY MAPPING AND TRACKING

For the body-part tracking applications above, it is sufficient to track a few points or landmarks on limb segments. If the position tracker does not directly yield orientation, then multiple-position measurements on a limb segment can be employed to derive orientation. Body motion can be considered to have similar measurement requirements to hand motion tracking.

By constrast, for body surface mapping or real-environment sensing, the position tracker must be able to scan a volume or surface to yield a dense array of points. The real-time requirements are typically absent when environmental reconstruction is the goal, although the image must be captured sufficiently rapidly if the environment (e.g., a body surface) can move. Accuracy requirements depend on the application; for example, in medical imaging systems, accuracies of 1 mm or better are desirable; for environmental mapping, accuracies of a few mm might be acceptable.

Additional information may be found in Durlach et al. (1992) and Meyer et al. (1992). Various implementations of the basic sensory system types in the context of head tracking are well covered in Meyer et al. (1992), and additional information on eye trackers appears in Durlach et al. (1992).

Mechanical Trackers

Mechanical trackers can be an inexpensive, relatively accurate means of tracking head or body-segment positions. Mechanical trackers can measure up to full body motion and do not have intrinsic latencies. Force reflection is readily incorporated by mounting actuators at the linkage joints. We distinguish two types, depending on whether the mechanical linkages are entirely worn (body-based) or are partly attached to the ground (ground-based).

Goniometers

Body-based linkages—exoskeletons or goniometers—have been frequently used in biomechanics for joint angle measurement. For haptic interfaces, they form the basis for master gloves and force-reflecting exoskeletons. Because they are worn, goniometers are portable and facilitate mobility; however, if there is body motion, some other tracking method is also required. They typically have the same workspace as the natural motion of the attached limbs and hence permit the full range of normal motion to be measured.

We may distinguish two ways to use the goniometric data: to infer joint angles and to infer end-point positions. The latter could be employed, for example, to track hand position relative to the body or fingertip position relative to the hand. Then the goniometer may be viewed like a hand controller that is manipulated by the operator; the output is based on calculations from the goniometer's angles, and there is no concern as to how this maps to limb joint angles.

For inferring joint angles, there are significant difficulties. Attachment is problematic, as it is for other tracking technologies, because of the soft tissue and potential for relative motion between the goniometer and limb. How best to clamp perturbation devices to limbs has been a concern for the biomechanics community; for the arm, tight clamping to the wrist is possible because of its bony features (Jones et al., 1991; Xu et al., 1991).

It is difficult to align goniometers with joints, especially for multiple-degree-of-freedom (DOF) joints such as the shoulder. Since goniometers are exoskeletons outside the limb, the centers of rotation of the goniometer differ from the joint rotation centers. Due to this kinematic mismatch, there has to be slippage between the goniometer attachments and the limb during motion. One solution in the context of hand masters is to employ four-bar linkages to project the measured centers of rotation to the finger joints (Rohling et al., 1993); because distances between joint centers are lost with this method, a finger calibration scheme for each individual is required (Rohling and Hollerbach, 1993).

Accentuating this problem is that human joints are not perfect hinge joints or spherical joints: the axes of rotation move with the joint angles. This problem is shared by any tracking method that seeks to determine joint angles. According to Rohling and Hollerbach (1993), a master glove in conjunction with an Optotrak 3D Motion Tracking System (Northern Digital, Waterloo, Ontario) was employed to calibrate the human hand geometry for fingertip control during teleoperation. It was found that the resulting fingertip accuracy of a few millimeters was primarily limited by assumptions of ideal joint structures (e.g., hinge joints) and by the accuracy of joint angle measurements by the master glove due to relative movement.

More information on goniometers can be found in Chapter 4 on haptic interfaces.

Ground-Based Linkages

Ground-based linkages are primarily used for 6-DOF end-point tracking such as the position and orientation of the head or of the hand. Hence the issues surrounding joint angle measurement do not typically arise. It is assumed that the human operator can grasp the manipulandum rigidly, or that the head-mounted system is rigidly attached to the head; both of these assumptions may be problematic. Although the head is bony, it is neverthess difficult to design helmets for visual displays that are fitted for each individual and that do not experience relative movement during fast head motion. Ground-based linkages are easier to actuate for force reflection than are body-based linkages, because actuators do not have to be placed and carried on the body.

One drawback of ground-based linkages is that the operator is tied to the ground and hence the workspace is limited. Even if one is willing to increase inertia and lower the mechanical resonance frequency in order to have longer arms, the range of a two-segment arm is ultimately limited to about 2 m. Nevertheless, they are a good option when operators are seated or not moving very much. Another drawback is the restricted numbers of degrees of freedom that can be measured, usually six. This is a problem for simultaneously measuring multiple limbs or for measuring redundant linkages. For example, multiple-finger motion is not conveniently measured with a ground-based linkage; a goniometer is better. Moreover, the human arm is a redundant 7-DOF mechanism, and a ground-based linkage cannot by itself resolve the redundancy.

For hand tracking, there are many examples of hand controllers, joysticks, and other haptic interfaces mentioned. For head tracking, a commercial example is the ADL-1 six degree tracking system by Shooting Star Technology. A related example is the BOOM viewer from Fake Space Labs, in which the visual display is not worn but supported on a pedestal through the BOOM linkage. Based on experiences in robotics, accuracies of 0.1 mm and high sampling rates should be achievable with such systems, when properly and routinely calibrated (Mooring et al., 1991).

For linkages attached to a helmet, one possibility is to use counterbalancing to reduce the gravity load of a head-mounted display on the wearer (Maeda and Tachi, 1992). One potential drawback is the inertia of the counterbalancing arms, which might impede head movement. One possible solution (not yet implemented as far as we know) is to use an active linkage servo-controlled to remove all dynamic loads on the head; this solution would obviously be complicated and expensive.

An inexpensive method for position measurement is to use three wire potentiometers. This method has been employed in robot calibration (Fohanno, 1982; Payannet et al., 1985), in which submillimeter accuracy has been reported. A cable connected between a torque motor and the human wrist has been employed for movement perturbations (Soechting and Lacquaniti, 1988). Force-reflecting hand controllers have been built using multiple cables attached to motors with pulleys at the edge of a frame (Agronin, 1987; Atkinson et al., 1977; Kawamura and Ito, 1993; Sato, 1991); however, the handle in the middle of this frame has a restricted workspace. Similarly, actuated strings have been employed by Iwata (1991, 1992) to apply force to the feet for walkthrough simulation.

Magnetic Trackers

The most popular position trackers are magnetic because of low cost, modest but reasonable accuracy, and convenience of use. Magnetic trackers do not suffer obscuration problems, although they are sensitive to environmental magnetic fields and ferromagnetic materials in the workspace. Multiple trackers can be employed to map whole-body motion and to increase the range of tracked motion to a small room (Badler et al., 1993).

The commercial trackers from Polhemus and Ascension Technology are currently the most frequently used. The Polhemus Fastrak is a recent introduction and improvement over their original sensor: the commercial brochure states static accuracies of 0.03 in and 0.15 deg and an update rate of 120 Hz. The update rate decreases with the number of sensors, which need to be multiplexed. Although latencies are stated as 4 ms, an independent source has determined a 20-30 ms latency range. The useful range is 1 m. The Ascension Bird sensor has a stated accuracy of 0.1 in and 0.5 deg, and an update rate of 120 Hz; the latencies are 30 ms.

In general, the advertised performance specifications of commercial magnetic sensors should be treated cautiously, as they do not meet their specs in realistic situations. The accuracies for both sensors depend on how close the transmitter and receiver are to each other. In the case of the Fastrak, which employs AC magnetic fields, neighboring metal surfaces will degrade the accuracy because of induced eddy currents; there is no a priori way to gauge the effect other than individual environmental testing. The Ascension Bird sensor employs DC magnetic fields and hence is less sensitive to surrounding metal, although a reluctance effect might alter the magnetic path. In Hollerbach et al. (1993), the Bird sensor was compared with the Optotrak in the context of robot calibration; the relative accuracy was found to be roughly proportional to price.

Optical Sensing

There are a variety of approaches to optical sensing for position tracking and mapping. Distance may be measured by triangulation (e.g., stereo vision), by time of flight (laser radar or ladar), or by interferometry. The passive light of the environment may be employed (stereo vision systems), structured light may be projected (laser scanning), light may be pulsed (ladar), or active markers (infrared light emitting diodes or IREDs) or passive markers may be placed on a moving body. Cameras or detectors may employ linear or planar charge-coupled device (CCD) arrays, position-sensing detectors (PSDs), or photodiodes. Below we discuss some major approaches to position tracking and mapping according to the following categories: passive stereo vision systems, marker systems, structured light systems, laser radar systems, and laser interferometric systems. A more complete review of active range imaging sensors may be found in Besl (1988).

Passive Stereo Vision Systems

Substantial effort is being expended by the computer vision community on developing artificial humanlike vision capabilities. Passive stereo vision systems employ ambient light and square-array CCD cameras, which have a typical resolution of roughly 600 × 400 pixels. A key issue is to solve the correspondence problem: relating the same points in two different images. Although the vision community is far from solving the general stereo vision problem, substantial progress is being made. Advances in algorithms and hardware are resulting in real-time (30 frames/s) three-dimensional imaging at moderate resolutions (Inoue, 1993; Kanade, 1993). Passive stereo vision systems are unlikely to be useful in VE in the near term, as robustness and accuracy are not yet comparable to active ranging systems. In the long term, as the computer vision community continues to advance, the use of passive vision for mapping and tracking is likely to become quite prevalent in VE.

Marker Systems

The stereo correspondence problem is solved in marker systems because a few, easily identifiable fiducial points are tracked on a moving body. The simplest and most accurate approach is to use a number of IREDs, which create very bright spots in the image. The IREDs may be pulsed in sequence with camera detection to uniquely identify each marker. For detection, one frequently employed approach is PSDs (also called lateral effect photodiodes), which are available as roughly 1 cm^2 squares. Incident light induces a current that is measured at each edge of the square to yield the XY location of the centroid of the incident light.

Commercial examples include the Selspot II system developed by Selcom and the Watsmart system developed by Northern Digital. IREDs are multiplexed at high rates (3,000 Hz) and sensed by two cameras, which triangulate the markers to an accuracy of about 5 mm at a distance of 2.5 m. The cameras are mounted on tripods; a calibration cube mounted with precisely known IREDs is employed to calibrate camera poses for triangulation calculation. Multiple markers can be tracked to yield orientation and to follow multiple bodies simultaneously. Workspace is distance-dependent: at 2.5 m, it is about 1 m^3. Researchers at the University of North Carolina have demonstrated a high-performance tracker using four CCD cameras mounted on a helmet pointing up toward a custom ceiling grid with IRED markers mounted at known positions in the tiles (Ward et al., 1992). The inside-out arrangement allows for much better accuracy in orientation and a large, scalable work area.

A fundamental problem with the use of PSDs is reflections of IRED light from environmental surfaces that move the apparent centroid of the sensed light; the amount of reflected light is high, about 25 percent of the total. The result is that it is very difficult to get camera resolutions beyond 1 part in 4,000; for this reason, Northern Digital has abandoned the Watsmart in favor of the Optotrak, which employs multiple cameras with 2,048-element linear CCD arrays. An IRED beam is transformed by a cylindrical lens into a line of light projected onto a linear CCD array. Because of the Gaussian spread of light intensity, an area of the CCD array is illuminated, which allows subpixel localization by area fitting. The result is a camera resolution of about 1 part in 200,000. Reflections are removed by image processing and thresholding: if there is another peak, it is detected and simply removed. This is the substantial advantage over the use of PSDs. At a viewing distance of about 2.5 m, the accuracy is about 0.1 mm and the resolution is about 0.01 mm. According to Marc Rioux (National Research Council of Canada, Ottawa) the limiting factor is air turbulence, whose effect is about 0.01 mm at the 2.5 m distance. The Optotrak comes in two forms. The series 2000 employs two housings, each with two cameras, that can be mounted on tripods; larger viewing distances are possible, at the cost of the use of a calibration cube. The series 3000 is a single housing, with three cameras embedded in an aluminum block; the camera ensemble is calibrated at the factory and hence there is no need for field calibration as for the series 2000 (at the cost of a smaller workspace). The Optotrak is capable of tracking full poses of two moving bodies at 100 Hz through multiple marker placement.

There are several commercial stereo vision systems that employ passive rather than active markers. Accuracies and sampling rates are not as good as those for the commercial active marker systems, although the absence of wires on the body is an attraction. Position and orientation tracking is also possible with a single square-array camera instead of two; Meyer et al. (1992) refer to this method as pattern recognition. When several points on a plane with precisely known relative locations can be recognized, then reasonably accurate estimates can be obtained. This alternative is especially attractive for head tracking because only one camera is used.

Structured Light Systems

Another approach to solving the correspondence problem in stereo vision is to employ structured light, usually a precisely known ray or plane of light. In one common configuration, a plane of light, created by passing a laser beam through a cylindrical lens, is swept across a scene using a galvanometer-driven mirror. At each position of the plane, a light stripe is created, which is sensed by a two-dimensional camera. The intersection of the known plane and the line of sight from the camera determines the three-dimensional coordinates. To reduce the cost of such systems and improve the frame rate, Kanade (1993) has developed a cell-parallel light stripe range sensor, based on custom very large scale integrated (VLSI) design. The camera employs "smart" photosensitive cells, which can detect the time at which peak incident light falls. Resolution is currently 32×32 pixels, acquisition time is 1 ms, and accuracies are about 0.1 percent of the range.

Another common configuration is laser spot scanning, using a variety of different movable mirror arrangements. A common misconception is that the baseline in a laser scanning system (distance between light source and detector) must be large for accurate ranging. Instead, Rioux (1984) developed a synchronized scanner in which the horizontal position detector and beam projector are located nearly collinearly and oppositely. A first scanned mirror between the two directs light on one of its surfaces from the source to the scene via a fixed mirror to a second scanned mirror. Reflected light is redirected by another fixed mirror to the opposite surface of the first scanned mirror to the detector. This reduces the shadow effect over other laser spot scanners and yields a more compact system.

One version of the synchronized scanner is a random access laser scanner, in which the first scanned mirror is a simple two-sided mirror, and the laser spot can be arbitrarily directed to any point in a scene (Beraldin et al., 1993). The advantage of random access is the ability to scan just a portion of the scene, for example to track an object using a Lissajou scan pattern, or to scan a full scene coarsely and then scan a selected scene portion more finely. The working range is from 0.5 to 100 m, the field of view is 40 × 50 deg. In a digitization mode, the sampling rate is 20 kHz (sample points per second) with CCDs and 10 MHz with PSDs. In a tracking mode, single objects can be tracked at 130 Hz; this rate is divided by the number of objects to be tracked. When the range is less than 10 m, the accuracy is typically 0.1 to 0.2 mm. At the far range of 100 m, retroreflectors must be used. Reflectance can also be inferred from light intensity. A VE application of this scanner involved digitizing for simulation the cargo bay and experimental setups of the Space Shuttle Orbiter (Maclean et al., 1990). This simulation was used by the National Aeronautics and Space Administration (NASA) during a mission. Because reflectance is measured, the graphical depiction was reported as being very realistic; during blackout one could hardly tell that a simulation was being used.

A second version is a raster scanner that employs as the first scanned mirror a multifacet pyramidal mirror, which is rotated continuously at a high rate. Linear CCD arrays or PSDs may be used for detection. Frames of 480 lines by 512 pixels are collected at video rates (33 Hz). In an early prototype, the working volume was a 50 mm cube (Beraldin et al., 1992); Hymarc is developing a version with a 1 m cube field of view. Resolution in the field of view is 9 bits (1 part in 512).

Laser Radar

Laser radar, or *ladar* for short, involves calculating the time required for a light beam to travel from the source, reflect off an object, and travel back to a detector; this principle is similar to ultrasonic ranging. The time of flight (in the picosecond range) can be directly measured for laser pulses emitted in rapid succession and scanned. Ladar is more appropriate for long distances than triangulation systems; for example, Maclean et al. (1990) employed a triangulation laser scanner for short range (0.5 to 10 m) and a ladar scanner for long range (up to 2 km).

In one commercial example, IBEO Lasertechnik obtained accuracies of 2 cm at 4,600 point measurements per second. Two other methods for calculating this time are: (1) the phase shift between outgoing and incoming amplitude modulated (AM) light beams and (2) the beat frequency of a frequency modulated (FM) beam (Besl, 1988; Blais et al., 1991).

For the AM method, the diffused reflected beam is typically six orders of magnitude less than the outgoing beam and is detected by a telescope near the outgoing beam (Chen et al., 1993). Mirrors, as in Rioux (1984), are employed to scan the beam in a point-to-point fashion. Because phase is being detected, there is an ambiguity interval of 1/2 wavelength, typically around 1 m. Two notable examples of such ladar systems are the Environmental Research Institute of Michigan (ERIM) system (Hebert and Kanade, 1989) and the Odetics 3D Laser Imaging System. The Odetics system has a resolution of 9 bits, a frame of 128 × 128 pixels, a frame period of 835 ms/frame, and a field of view of 60 × 60 deg. By use of advanced calibration procedures, accuracies of 0.15 in have been reported (Chen et al., 1993).

Further information on ladar sensors, as well as on millimeter-wave radar, may be found in the review of remote vehicles in Chapter 9.

Laser Interferometers

Laser tracking systems employing interferometry have been used in robot calibration and tracking in the past decade; good surveys are presented in Jiang et al. (1988), Kyle (1993), and Mooring et al. (1991). The precision and accuracy of interferometry is very high, although the cost is currently too great for routine use.

A laser beam is steered to a retroreflector on a robot end effector or moving body by a servo-controlled mirror on a two-axis, galvanometer-driven scanner. Either a mirror retroreflector (an open corner) or a solid glass retroreflector (referred to as a cat's eye) is used (Kyle, 1993). The mirror retroreflector has a smaller working range than the cat's eye (±20 deg versus ±60 deg), but can be made smaller. The retroreflector reflects the beam back to the scanner, where a beam splitter directs the beam to a photodetector for interference fringe counting and to a PSD. Based on the PSD output, the outgoing beam is deflected to the center of the retroreflector for tracking. The calculation of the three-dimensional position of the retroreflector is done in one of two ways:

(1) The two mirror angles on the scanner are measured (Lau et al., 1985). Then one scanner is sufficient, as the spherical coordinates are provided. A commercial version is the Smart 310 system by Leica.

(2) The two mirror angles are not employed, but three laser scanners are required for three distances. Chesapeake Laser Systems, Inc., has developed a commercial system.

A problem with laser interferometers is that only incremental displacement is provided. To obtain absolute distance, some calibration procedure must be followed. For the Chesapeake Laser Systems device, it has been shown that by adding a fourth scanner the system can self-calibrate (Zhuang et al., 1992).

Another problem is to provide orientation as well as position. One solution is to use a steerable mirror instead of a retroreflector (Lau et al., 1985), which provides two orientation measurements. More recently, Prenninger et al. (1993a, 1993b) have determined all orientation components by imaging of the diffraction patterns of the edges of a modified mirror retroreflector. Orientation resolutions of 1 arcsec are stated, and motions can be tracked that accelerate at 100 m/s^2.

Acoustic Trackers

Acoustic trackers employ at least three microphones to triangulate an emitter on the moving body. They have been employed in robotics for calibration (Stone, 1987) and in biomechanics for motion tracking (Soechting and Flanders, 1989). Commercial implementations for the VE market include the GP8-3D developed by Science Accessories, the Logitech 3D/6D Mouse, and the Mattel Power Glove. For point tracking at modest accuracies and speeds, ultrasonic trackers are a reasonable and very inexpensive alternative to magnetic sensors: the ranges are larger, and magnetic interference is not a problem. However, a clear line of sight must be maintained and the latency is proportional to the largest distance being measured. Because at least three points are required to infer body orientation, it is difficult to measure full pose at adequate rates.

Most such systems measure the time of flight of ultrasonic pulses. There are a number of technical problems that make it difficult to achieve good accuracy, speed, and range with this technique. The first factor is the frequency of the ultrasonic carrier wave. A shorter wavelength makes it possible to resolve smaller distances, but atmospheric attenuation increases rapidly with frequency starting at about 50-60 kHz. Most current systems use 40 kHz tone pulses with a wavelength of about 7 mm. Metallic sources such as jingling keys produce enormous quantities of energy in this frequency band, making it extremely difficult to achieve immunity to acoustic interference. The use of higher frequencies could avoid some of the interference and increase the resolution, but atmospheric attenuation would limit the range. Furthermore, at high ultrasonic frequencies it is difficult to find an omnidirectional radiator, and microphones are expensive (over $1,000 each) and require high voltage.

Another major problem is echos from hard surfaces, which can have as much as 90 percent reflectivity to ultrasonic waves. At the SIGGRAPH '93 Convention, Bauer of Acoustic Positioning Research, Inc., demonstrated an ultrasonic tracker that uses patented algorithms to achieve robust noise and echo rejection while tracking over a 25 ft range with 1 in resolution. This system, called GAMS, also employs an unusual inverted strategy, with the sound sources mounted on the ceiling and the microphones on the users, so that any number of users may be tracked simultaneously without having to reduce the 30 Hz update rate. By contrast, the GP8-3D has an update rate of 150 Hz divided by the number of emitters being tracked. An advantage of the traditional one-emitter-at-a-time approach is that echo cancellation is easier: the first arrival is always the direct path unless the line of sight is blocked.

Celesco Transducer Products, Inc., is advertising a new wireless 40 kHz time-of-flight system called the V-scope, with 0.1 mm resolution over a 16 ft range.

Instead of time of flight, phase coherence (Meyer et al., 1992) is an incremental motion technique (like interferometry) for which absolute distance must initially be calibrated by some other means. Phase-coherent tracking also has problems with reflections and some environmental noises. In order to overcome the drift problem, Applewhite (1994) has developed a variation called modulated phase-coherence, which can achieve submillimeter accuracy.

Inertial Tracking

The use of accelerometers or of angular rate sensors for motion tracking is becoming increasingly attractive because of advances in sensor design. For example, silicon micromachining has begun to produce very small inertial sensors. IC Sensors markets solid-state piezoresistive accelerometers, which employ a micromachined silicon mass suspended by multiple beams to a silicon frame. The GyroChip developed by Systron Donner employs a pair of micromachined tuning forks, which sense angular velocity through the Coriolis force.[1] The GyroEngine developed by Gyration is a miniaturized spinning wheel gyroscope that is even smaller than the GyroChip.

To derive position or orientation, the output of these sensors must be integrated. The result is sensitive to the drift and bias of the sensors. Another serious problem is that any inertial sensor based on beam bending will suffer inaccuracies due to nonlinear effects. Force-balance accelerometers, such as those from Sundstrand, avoid nonlinear beam-bending problems; they are remarkably stable and have been employed in kinematic calibration after double integration (Canepa et al., 1994).

Nevertheless, some drift is inevitable. Either an inertial package must periodically be returned to some home position for offset correction, or it must be used in conjunction with some other (possibly coarse) position

[1] A deflecting force acting on the body in motion due to the earth's rotation.

sensor and an appropriate method of data fusion. The latter option could be quite attractive. An inertial orientation tracker has been built at MIT using triaxial angular rate sensors with gravimetric tilt sensors and a fluxgate compass for drift compensation (Foxlin and Durlach, 1994). It achieves 1 ms latency, unlimited tracking volume, 0.008 deg resolution, and 0.5 deg absolute accuracy (no drift). The system is now being extended to track translation as well as orientation.

Inertial sensors are also useful in conjunction with other position tracking systems for lead prediction. In the high-end HMD system from CAE Electronics, the outputs of a fast optical head-tracker are combined with angular velocity measurements to predict future head orientation. In this manner, the 100 ms graphics rendering latency is effectively shortened to under 60 ms (Ron Kruk, CAE Electronics, personal communication, 1994). Lead prediction has also been implemented using Kalman filters (Friedmann et al., 1992; Liang et al., 1991). Additional information on the use of inertial sensors is presented in Chapter 9.

Eye Tracking

Eye movement trackers have been generally surveyed by Durlach et al. (1992). The general types are electroocular (measurement of the corneoretinal potential with skin electrodes), electromagnetic (measurement of magnetically induced voltage on a coil attached to a lens on the eye), and optical (reflections from the eye's surface). Only the optical-tracking methods seem particularly suited for general use, because they are less invasive and can be reasonably accurate.

The RK-416 Pupil Tracking System developed by ISCAN Inc. employs a video camera to track the pupil of the eye. The camera coordinates are converted to eye rotation angle through a calibration procedure involving fixation at known targets. Accuracy is stated as 1 deg and bandwidth as 60 Hz. Compensation for minor head movements is accomplished by relative tracking of the first Purkinje image or by head spot tracking.

The Series 1000 Infrared Eye Movement Spectacles developed by Microguide, Inc. employ differential reflections of infrared light from two sides of the iris, detected by photodiodes. Sensitivity is stated as 0.1 deg, bandwidth as 100 Hz, but linearity of only 10 percent. Sources of artifact for such systems due to inaccurate positioning of sensors were considered by Truong and Feldon (1987), due to changes in reflectivity among sclera, iris, and pupil. The sensors are mounted on a post projecting down from a band wrapped around the forehead. The ober2 developed by Permobil Meditech, which operates by a similar principle, employs goggles instead.

RESEARCH NEEDS

In general, position trackers are required that have adequate performance at reasonable costs. For limb segment tracking, some forms of optical sensing are already accurate and fast enough but are too expensive for routine use (in the $50,000-100,000 range), especially when multiple placements to overcome obscuration or to increase workspace are considered. Whatever sensory system is employed for limb tracking, there will be difficulties in identifying reliable fiducial points because of the softness of tissue and clothes. To infer joint angles, some calibration procedure must be applied to set up coordinate systems in limb segments; subsequent joint angle inferences will only approximate the true biomechanical angles because of less than ideal joints and measurements.

For whole-body tracking, the size of the workspace is an important issue. Ideally, one should be able to track a moving person in a sufficiently large space without loss of resolution or worries about obscuration. People should even be able to move from room to room in a building without loss of tracking. If all of the major body segments are to be tracked, then sensory systems that require mounting something on the body (reflectors, markers, goniometers) are less attractive than systems that scan the body as is. For body and environmental surface mapping, some forms of laser trackers are already accurate and fast enough to be generally useful but are still expensive.

The long-term goal for VE purposes is for a high-accuracy, real-time optical mapping system that tracks the human body as required. This could be a stereo vision system with natural light or active laser scanners. One step in this direction has been taken by Mulligan et al. (1989), who employ a model-based vision system to track a moving excavator arm that is controlled in Cartesian coordinates by a hand controller. The excavator joints are not sensed but are calculated from the locations of the boom, stick, and bucket of the excavator. In another project, a human hand is observed in a grasping task, then the grasp is duplicated by a robot hand (Kang and Ikeuchi, 1993). For HMD applications, the user has to wear something anyway, so head-tracking, which has much more demanding requirements than body-tracking, can probably be more accurately accomplished by sensors on the HMD.

Below we comment on particular needs for each basic type of sensory system.

- *Mechanical trackers* For goniometers, research is needed on difficult issues of fit and measurement, such as adjustments to different individuals, alignment with joints, sufficiently rigid attachments, and calibration of the linkage plus human limb. For body-based linkages, the ability to track multiple limb segments and limb redundancies needs to be addressed. Hybrid ground-body based systems are likely to be required; for example, finger motion should be tracked as well as hand motion.

- *Magnetic trackers* Significant current disadvantages that limit usefulness include modest accuracy and high latency (20-30 ms). The high latencies are particularly troubling, as this limits their usefulness in real-time interaction. The accuracies are not competitive with most other tracking technologies. Furthermore, the influence of extraneous magnetic fields in the case of AC sensors makes it difficult to know the accuracy one is getting; there is no simple way of determining and compensating for interfering magnetic fields. It is an open question to what extent the accuracies and latencies can be improved.
- *Optical sensing* Optical sensing is one of the most convenient methods, but has drawbacks due to visibility constraints. These drawbacks can be partially ameliorated by using multiple camera placement or target placement. The ability of passive stereo vision systems to process arbitrary environments is a long-range goal of the computer vision community; when eventually developed, stereo vision would represent an extremely attractive method for position tracking and mapping. In the meantime, developments in laser scanning and laser radar are promising, as sampling rates, fields of view, and accuracies are becoming quite reasonable.

Laser interferometers are capable of the highest accuracy, which does not change with viewing distance. If the cost could be brought down, they would represent an attractive method of end-point or head-tracking. Because of the need for retroreflectors, it will be relatively difficult to track multiple limbs. Visibility constraints are a problem; if beams are ever interrupted, the absolute reference is lost. Relatively robust methods for establishing absolute references are required, perhaps through redundant sensing.

In general, costs will have to be brought down for optical trackers to be more widely used.

- *Acoustic trackers* These trackers have a definite role to play in VEs, because the costs are relatively modest and the accuracies are often sufficient. If in situ calibration of the speed of sound in air could be performed, or if ambient measurements could be taken that feed into a model, then the accuracy of acoustic trackers could be improved. Improvements in detection methods could probably reduce the effect of echos. By using multiple frequencies, it should be possible to track multiple markers simultaneously. Drift problems with phase coherent systems might be resolved by dead reckoning with time-of-flight measurements.
- *Inertial trackers* Further reductions in sensor size and cost are needed to make inertial trackers a convenient and economical alternative to magnetic trackers. Hybrid systems combining inertial sensors with

other technologies need to be developed for high-performance HMD-tracking applications requiring both accurate registration and fast dynamics. A particularly promising combination is an all-inertial orientation tracker combined with a hybrid inertial-acoustic position tracker.

- *Eye-movement trackers* An ideal eye tracker would satisfy three requirements: linear response over a large range (roughly 50 deg), high bandwidth (1 kHz), and tolerance to relative motion of the head. To date, no devices satisfy all three requirements. The CCD imaging systems such as ISCAN have a reasonably linear response, but the sampling rates are too low and range is limited to about 20 deg. The infrared reflection devices have the bandwidth, but are linear only in a small range; calibration to overcome their nonlinear responses is difficult.
- *Characterization methods* One of the most confusing issues when considering which position-tracking solution to adopt is the lack of agreement on the meaning of the performance specifications and how they should be measured. Standards need to be set defining how to measure accuracy, resolution, latency, bandwidth, sensitivity to interference, and jitter. Equipment for making in-house measurements should be made commercially available, and the establishment of an independant testing laboratory would also be beneficial so that consumers would not be forced to rely on manufacturer's specifications, which usually have little relation to actual performance.

6

Whole-Body Motion, Motion Sickness, and Locomotion Interfaces

Many virtual environments (VEs) will give the user a sense of motion either in the form of passive transport in a "vehicle" or through active locomotor movements on a supporting surface. The latter experience couples the voluntary locomotor activity with visual and auditory sensations. In both situations, the support of the user's body by the physical contact surfaces constitutes a haptic interface. The term *haptic interface* is being used here in a more general sense than in Chapter 4. Here we use this term to refer to any place at which mechanical energy interchanges take place between the body and the environment. In the case of passive transport in a vehicle, haptic interfaces with the vehicle are the places at which the vehicle applies accelerative forces to the individual's body. The haptic stimulation associated with these forces contributes to the perception of body motion. Similarly, during voluntary locomotion, such as walking, the energy interchange between the feet and the support surface produces haptic cues that provide information about surface characteristics (e.g., compliance, slope, texture) and body displacement.

Both passive and active displacements of the entire body (and of parts such as the torso during rotary movements of the torso on the hips) reorient sensory receptors vis-à-vis the environment. The resulting changes in receptor activity normally are taken into account, so that perceptual stability of the surroundings is maintained during active or passive self-movement. In VEs, head-trackers and other forms of position monitors are used to update visual and auditory displays to compensate for changes in body position. To the extent that there are time delays beyond 20 or 30 ms or gain changes between the body movement and the stimulus update, performance problems can be anticipated. Similarly, if the optical verticals in a VE do not correspond with the gravitational vertical of the real environment, orientation and movement difficulties may be experienced by the user.

Whole-body movements and locomotion raise a large set of issues concerning the forms of compensation that take place during self-movement, the perception of forces on the surface of the body during movement, the perception of self-displacement through space, and of one's voluntary actions. The way in which adaptive compensations for unusual environments and for maintenance of accurate sensorimotor calibration are achieved is also crucial. All of these issues are critical to understanding how people will adapt to VEs involving locomotory movements, passive transport, and head, arm, and torso movements. However, it is important to recognize that our knowledge of these areas is incomplete. Motion sickness is a factor that is certainly going to affect performance in VEs involving locomotion and experienced self-motion. It is necessary to be aware of the wide range of factors that contribute to motion sickness, the variety of the symptoms, and its sometimes subtle characteristics.

STATUS OF THE RELEVANT HUMAN RESEARCH

Sensorimotor Stability During Self-Motion

Under normal circumstances, an individual accurately perceives his or her voluntary movements and perceives the surroundings to be stable when they are actually stable. This stability that we take for granted is the result of complex sensorimotor adaptations to the 1G force field of earth. The existence of these adaptations becomes apparent during exposure to non-1G force levels or conditions of sensory rearrangement (Lackner and Graybiel, 1981). One aspect of these adaptations includes not perceiving veridically the forces acting on the body surface when they are due to voluntary activity or passive support of the body against the force of gravity. For example, the forces on the bottom of the feet feel roughly comparable when one is standing on one or both feet even though in the former case the force on the stance foot is twice as large (Lackner and Graybiel, 1984c). Similarly, in running, the forces on the feet can vary from 0 to 3G, yet these huge changes are not perceptually registered. When these same force levels are passively applied, the sensation is very strong.

Body movements are accompanied by various sensorimotor consequences. In the case of object manipulation, the pattern of sensory stimulation of the hands in relation to the patterns of motor activity controlling the hands and arms and other parts of the body allows identification of properties of the object such as its shape, texture, and weight. In the case of locomotion, similar types of information along with vestibular cues allow—even in the absence of vision—quite accurate judgments of the distance and direction moved and properties of the support surface, such as its inclination and surface properties.

Manipulation of virtual objects and locomotion in VEs will disrupt many of the sensorimotor regularities that are present in the normal environment. This is a significant issue because we have limited experience with violation of these constraints, and most probably only a relatively small set of the relevant ones have yet been identified. Consequently, movements made in VEs that violate terrestrial sensorimotor constraints may initially cause performance decrements and elicit symptoms of motion sickness (Lackner et al., 1991). In fact, reports of motion sickness in simulated environments are becoming commonplace. As with other unusual sensorimotor environments, continued exposure is likely to lead to adaptive compensations that restore accurate performance and alleviate symptoms of motion sickness. However, on return to the normal environment, there may be persistence of adaptation to the virtual situation leading to performance decrements and motion sickness in the normal environment. It may be possible, however, to create states of dual simultaneous adaptation such that accurate performance is possible in the normal as well as in one or more VEs. The possibility of such dual adaptations has been shown for a number of unusual force situations, such as rotating and nonrotating environments (Lackner, 1990), as well as for other types of situations (e.g., Welch et al., 1993).

Head Movements

Head movements made during exposure to increased or decreased gravitoinertial force background levels tend to bring on symptoms of motion sickness because the normal patterning of sensorimotor control of the head and patterns of sensory feedback are disrupted (Lackner and Graybiel, 1984a, 1985, 1987; Lackner et al., 1991). Virtually any alteration in the normal patterning of control can be provocative (Lackner and DiZio, 1989). For example, wearing a neck brace requires head movements to be achieved by motion of the torso; this alteration in motor control can be quite provocative for many individuals. Passive exposure to various types of motion is provocative because of the labyrinthine stimulation involved (e.g., on shipboard), but such exposure is even more provocative if the head is not passively supported but rather actively controlled (Lackner et al., 1991).

Head movements made during everyday activities are not provocative. By contrast, the same head movements made during passive rotation of the body are extremely stressful and rapidly evoke disorientation and symptoms of motion sickness. Movements of the head during body rotation create unusual patterns of activation of the semicircular canals of the inner ear and generate Coriolis forces on the head (Lackner and Graybiel, 1984b, 1986). These Coriolis forces are proportional to the velocity of rotation and the velocity of the radial motion of the head (or other body part such as the hand) and act in the direction opposite that of the rotation. They are absent prior to and at the end of a movement. Coriolis forces are present when head movements are made during passive and voluntary body rotations. The fact that they are disorienting and produce motion sickness during passive and not active turning movements means that we automatically take them into account in controlling our voluntary movements.

A sense of presence may be a key factor in determining the human's responses to Coriolis stimulation and to expected Coriolis stimulation in VEs. Visual stimulation, especially whole-field visual motion, can elicit sensations of self-motion. If exclusive self-motion is being experienced in an apparently stationary visual field, then head movements can rapidly elicit symptoms of motion sickness and disorientation (Dichgans and Brandt, 1978). By contrast, if visual motion is being experienced and little self-motion, then head movements tend to suppress the sensation of self-motion and less motion sickness will be experienced than from simply looking passively without moving the head (Lackner and Teixeira, 1977; Teixeira and Lackner, 1979). Accordingly, VE displays that induce apparent whole-body motion and that require head movements are likely to elicit greater levels of sickness the greater the fidelity of the experienced self-motion and sense of presence.

Altering the sensorimotor control of the head with a neck brace can be mildly provocative. However, altering the weight of the head and its effective moment of inertia can be extremely provocative, even when the pattern of vestibular input is normal for the actual motion of the head. For example, wearing a helmet that increases the effective weight of the head by 50 percent greatly increases susceptibility to motion sickness during exposure to constant patterns of angular acceleration and also causes natural voluntary head movements to be provocative (Lackner and DiZio, 1989). Head-mounted displays (HMDs) are an integral component of many VE systems and affect the sensorimotor control of the head. This altered inertia of the head may be provocative if the displays are worn for more than several minutes and head movements are made. That is, simply wearing an HMD can be provocative in itself, regardless of the scenes displayed. To the extent that the display induces apparent self-motion, head movements may be extremely provocative. In addition, the time delays in HMDs associated with updating the visual scene to compensate for head movements disrupt the normal patterning of the vestibulo-ocular reflex. This is especially nauseogenic in wide-field HMDs in which large-gaze shifts occur as individuals turn their head and eyes to acquire targets in the peripheral visual field (Lackner and DiZio, unpublished observations). The point to be taken is that there is a spectrum of factors that can be expected to evoke sickness.

Arm Movements

Arm movements made to objects or to control devices are accompanied by patterns of sensory feedback related to the control of the arm and the object manipulated. Object characteristics include inertia, mass, weight, texture, compliance, etc. Expectancies about the properties of objects are exceedingly important. For example, when visually similar objects of different physical sizes but identical masses are hefted in succession, the larger sized one will be perceived as being considerably lighter. The motor plans for hefting the objects probably include compensation for a greater expected mass of the larger object. This effect, known as the size-weight (or Charpentier's) illusion, persists, however, even when the objects are known to be of comparable weight. Insofar as object manipulation in VEs violates cognitive expectancies about object behavior, performance decrements will occur, adaptation will be required, and after-effects can be anticipated on return to the normal environment.

During passive transport of the body in vehicles as well as during voluntary locomotion and turning of the body, arm movements are usually quite accurate. In a familiar vehicle, even one not being self-controlled, compensation can be made for the ongoing and expected motion of the vehicle (e.g., on shipboard, the rolling and scending motions are periodic and can be anticipated). During exposure to linear and angular acceleration, adjustments of the entire body as well as arm movement control may be required. For example, the driver of a vehicle will physically lean into a turn (Fukuda, 1975).

Arm movements made during exposure to passive rotation of the body generate Coriolis forces that deviate the arm from its intended trajectory and target. With repeated reaches, even without visual feedback about movement accuracy, accuracy will be rapidly regained if the hand makes physical contact at the end of the reaching movement. However, in the absence of vision, if terminal contact of the hand is not present at the end of the movement, adaptation does not take place or is greatly slowed (Lackner and DiZio, 1994). On cessation of body rotation, pointing movements made to targets show error patterns that are mirror images of the initial per-rotary reaches. In other words, adaptation to the rotating environment persists and readaptation to the stationary environment is necessary.

These findings are relevant to movement control in VEs for several reasons. First, in everyday behavior, reaching movements made during voluntary body movements of the trunk or whole body generate Coriolis forces far in excess of those used in experimental studies, yet movements are made accurately. This means that motor compensation is made for expected Coriolis forces associated with self-movement. Second, during illusory (e.g., visually induced) self-rotation, compensation may be made during reaching movements for Coriolis forces that would be present during actual body rotation; consequently, reaching errors result. Thus, the perception of self-motion per se may be adequate to induce motor compensations for Coriolis forces expected with actual rotation. VEs that induce apparent motion of the body may lead to compensatory motor adjustments during arm and body movements for expected forces that are actually absent. Third, with prolonged exposure, individuals adapt their movements. Such adaptation can persist on return to the normal environment so that compensation is not initially made during body motion for Coriolis forces that are actually present. Posture and movement control take into account not only Coriolis forces but also a variety of other forces associated with experienced linear and angular accelerations and angular velocity, such as centrifugal forces (Lackner, 1993). To the extent that VEs cause experienced body motion, compensation for expected forces may be generated.

The extent to which an individual feels present in a VE may determine the extent to which compensation for expected inertial forces occurs. Moreover, the significance of these compensations may differ for different VEs. For example, if simulation of a high-performance aircraft is being used for training purposes and a high degree of presence is achieved for dynamic flight conditions, then there potentially could be negative transfer to actual flight situations. For example, compensation might not be made initially for actual inertial forces generated in flight because this compensation has been trained out in the simulator. To deal with such issues, it will be necessary to explore systematically the possibility of adapting to multiple environments simultaneously. It is notable that motion sickness in flight simulators is more common in experienced pilots than in flight trainees (Kennedy et al., 1990). The pilots expect patterns of forces that are absent in the simulator; for example, in a flight simulator, backward tilt is often used to simulate forward acceleration; by contrast, during actual forward acceleration the gravitoinertial resultant force on the individual rotates and changes in magnitude.

Locomotion

During normal walking, the pattern of visual flow is determined by step frequency and stride length. Muscles of the neck and torso are activated cyclically to keep the head and torso from being rotated backward by the reaction forces generated. The walking individual appropriately perceives the visual surroundings as stationary, the walking surface as stationary, and correctly appreciates his or her actual stride frequency and length. These relationships are dependent on dynamic sensorimotor calibrations that are tuned over time to accommodate for changes in body dimensions, for example, in leg length and body weight.

When the relationship between visual flow and locomotory movements is disrupted (e.g., by having a subject walk in a circular treadmill within a large optokinetic drum) so that the visual flow is inappropriate in speed or in direction for the stepping movements being made, a variety of perceptual remappings occur (Bles, 1981; Lackner and DiZio, 1988). These remappings are such as to normalize the experienced body displacement through space and the apparent frequency, direction, and length of stepping movements (Lackner and DiZio, 1988, 1992, 1993). If visual flow is double normal for the stepping movements being made, the individual will perceive either an increase in stepping rate or that the stepping leg lengthens to give the extra displacement to the body. By contrast, if the visual flow is inappropriate in direction, an individual will feel he or she is voluntarily stepping backward when actually making forward-stepping movements, or feel that forward steps paradoxically propel them backward. If the person is holding a stationary bar, the bar will seem to take on a life of its own. With double normal visual flow, the bar will seem to pull the person forward, causing him or her to go faster; with reverse visual flow, the bar seems to move backward, forcing the individual to go backward. This means that the perception of causality and of the forces on the body are being perceptually remapped along with the individual's perception of volitional activity and body configuration. When individuals are exposed to visual velocity increases, such as would be associated with body acceleration, they have difficulty controlling their stepping movements and may have to hold on for support.

Interestingly, when individuals walk at a constant speed on the treadmill during conditions of constant visual flow (regardless of whether the flow is appropriate in direction or magnitude for the actual stepping movements being made), they experience relatively few motion sickness symptoms. By contrast, individuals who are seated and exposed to the same visual flow patterns will report symptoms within a few minutes. It is likely that individuals walking on a treadmill who are exposed to variations in visual flow velocity will experience many more symptoms.

VE systems that embody mismatches between patterns of visual flow and activity associated with locomotion can be expected to distort the perception of body displacement and of voluntary activity and even of body dimensions. However, for VE displays involving locomotion, the degree to which motion sickness is evoked may be relatively minor when visual motion is coupled to voluntary activity.

Body Orientation

Under terrestrial conditions, the force of gravity provides a constant reference for orientation, determining unequivocally the direction of up and down (Howard and Templeton, 1966; Howard, 1982). Our environment, both natural and artificial, also embodies an orientational polarization. The trunks of trees are near the ground (down) and their tops are up. Rooms have floors and ceilings. Within this environment, only certain body orientations and configurations are possible. Locomotion can only take place on the floor of a room, not the walls or ceiling. This means that only certain perspectives of the room are naturally possible. For example, one cannot view the floor of the room or the walls from the perspective of being physically located at the ceiling. Similarly, one cannot (without being artificially suspended) see one's feet spatially separated from the floor when no other part of one's body is in contact with the floor of the room. These considerations raise the possibility that if one explores a VE by means of a locomotory walking interface until one has mastered the geographical layout of the environment, then this same environment may seem unfamiliar if one traverses it in a different way, such as using a control stick allowing one to "fly" through it.

Experience in weightless environments in which terrestrial constraints on orientation can be violated shows that our prior cognitive experience with 1G environments limits the perceptual patterns that are experienced (Lackner, 1992a, 1992b, 1992c, 1993). In particular, patterns of orientation cues that would not be possible on earth are interpreted in such a fashion as to create a perceived terrestrial orientation. For example, an individual floating free in an aircraft that is flying parabolic maneuvers so as to create weightless conditions may not correctly perceive his or her true orientation. The person may experience an orientation that *would* be possible on earth. In other words, a cognitive map of terrestrial possibilities influences the perceived orientation.

Vertigo may also be a problem with VE systems depending on the perspectives generated. Height vertigo—sensations of fright and instability usually accompanied by increased body sway—occur when the viewing perspective is elevated and there are no nearby objects visible (Bles et al., 1980). VE systems will have the clear potential to create such circumstances. On earth, sensations of falling are experienced during actual falling. However, there may be a strong cognitive component to the elicitation of these sensations. During actual free fall (e.g., in parabolic flight or in orbital flight), individuals do not generally experience sensations of falling; they feel stationary (Lackner, 1992a, 1992b). Consequently, visual motion and quick loss of support and knowledge that one is falling, for example, from a ladder, may trigger the sensations under terrestrial conditions. However, VE systems have the possibility of creating situations, for example, of a person going out a window and moving downward, that could well evoke feelings of falling and defensive reactions. One key in designing environments will be to evaluate how different motions through—and positions within—the environment can be expected to affect orientation and movement control.

Illusions of Self-Motion

The perception of ongoing body orientation is influenced by multisensory and motor factors. There are a variety of techniques that can be used in VE systems to provide compelling sensations of motion. However, as we note below, often when individuals experiencing such apparent motion make voluntary head movements, or their heads are passively moved, they begin to experience symptoms of motion sickness or lose their sense of self-motion.

Visual influences on apparent self-motion are often subsumed under the terms *circular vection* and *linear vection* (Dichgans and Brandt, 1978). Individuals exposed to constant-velocity visual motion in a large rotating drum soon feel themselves rotating at constant velocity (in the direction opposite the physical rotation of the drum) and see the drum as stationary. When the illusion of self-motion (vection) is highly compelling, tilting head movements can elicit disorientation (pseudo-Coriolis effects partly analogous to those that occur when head movements are made during actual body rotation). This illusion can also cause motion sickness (Dichgans and Brandt, 1978). By contrast, if vection is not compelling or it is just starting to be experienced, head movements can suppress it (Lackner and Teixeira, 1977). Peripheral visual field stimulation is especially effective in eliciting vection, but even small central fields can have an effect (Dichgans and Brandt, 1978).

Linear vection can be induced by exposure to constant-velocity linear visual motion (Berthoz et al., 1975; Howard, 1982). Its time course and other characteristics are similar to those of circular vection. Depending on the direction of visual flow, horizontal or vertical linear vection can be induced. Rotation of the visual field about the optical axis can elicit sensations of body tilt and rotation, but the exact pattern experienced also depends on head orientation vis-à-vis the body and the gravitational force vector (Held et al., 1975). Whole-field visual motion is commonly used in flight simulators to induce apparent body motion.

Rotating sound fields that have strong spatial volume distribution can elicit apparent motion as well as eye movements compensatory for the apparent motion in subjects in the dark (Lackner, 1977). Head movements strongly suppress the induction of apparent self-motion by sound fields. However, it is likely that moving sound fields in conjunction with visual motion would make the vection illusions more compelling.

Illusions of self-motion can also be induced in seated individuals by having them pedal a platform moving under their feet or asking them to move their hands to turn a circular railing (Lackner and DiZio, 1984). Such illusions can be extraordinarily compelling and, if tilting head movements are made, disorientation and motion sickness can be evoked. The pedaling illusions work for angular motion and undoubtedly will for linear as well. Tactile stimulation of the soles of a seated subject's feet by a moving surface can also induce apparent self-motion, as can stimulation of the palms of the hands. In fact, in situations involving physical motion of the body, somatosensory stimulation can be as important as, and sometimes more important than, vestibular stimulation in determining apparent body orientation (Lackner and DiZio, 1988, 1993).

The spindle receptors within skeletal muscles are important elements in the determination of the overall apparent configuration (i.e., body schema or position sense) of the body. The signals from these receptors are interpreted in conjunction with the motor signals controlling the physical length of the muscles. Mechanically vibrating a muscle excites its spindle receptors, causing it to contract, an effect known as *tonic vibration reflex*. Resisting the motion of the limb controlled by the muscle will cause illusory motion of the stationary limb. For example, vibration of the biceps muscle of the upper arm will cause illusory extension of the restrained forearm (Goodwin et al., 1972). Such illusory motion is in the direction that would be associated with physical lengthening of the vibrated muscle. It is possible by vibrating the appropriate muscles to elicit apparent rotation about a vertical axis in standing subjects, tilt of the body, or tilt of the head (Lackner, 1988; Lackner and Levine, 1979). Such apparent motion is generally accompanied by eye movements compensatory for the experienced motion.

Individuals who are standing and walking in place on a treadmill in darkness or with synergistic patterns of visual flow come to feel that they are locomoting over a stable stationary surface. Such apparent motion through space is totally compelling (Bles and Kapteyn, 1977; Lackner and DiZio, 1988). The precise patterns experienced depend crucially on the relationship between the visual flow patterns and stepping movements (see the discussion of locomotion above). Presenting visual flow that is inappropriate in direction for the stepping movements being made can, for example, make forward stepping individuals feel that they are walking backward over a stable surface.

Intersensory effects provide ways of creating various patterns of compelling apparent self-motion and changes in apparent body orientation in stationary individuals. Such experienced motion is usually accompanied by compensatory eye movements and other compensatory postural changes. It should be recognized that physical motion of the body that elicits eye movements also influences visual and auditory localization. For example, a person in darkness exposed to leftward angular acceleration will experience leftward turning motion, and his eyes will exhibit a compensatory nystagmus with slow phase right. If the person is given a target light to fixate that is stationary in relation to his or her body midline, he or she will perceive it to displace leftward off the midline. This phenomenon is known as the oculogyral illusion (Graybiel and Hupp, 1946): there is also an analogous audiogyral illusion. Similarly, if the gravitoinertial force vector is increased in magnitude and rotated by exposing an individual to centrifugal rotation, various types of oculogravic (Graybiel, 1952) and audiogravic illusions are elicited. Analogous kinds of mislocalizations of auditory and visual signals occur during experienced as opposed to actual body motion.

Motion Sickness

Motion sickness is likely to be a highly significant problem in VEs that involve HMDs and that involve voluntary locomotion or passive displacement of the body. HMDs that significantly influence the effective inertia of the head disrupt the normal sensorimotor control of the head. This in itself can elicit motion sickness and disorientation (Lackner and DiZio, 1989). Likewise locomotion and passive displacement through simulated environments will be unaccompanied by the normal patterns of forces and accelerations associated with such motion through the real environment. The absence of the normally occurring patterns of forces will render many VEs highly provocative in eliciting motion sickness.

Motion sickness is often equated with the nausea and vomiting that can occur during exposure to air and sea travel and, more recently, space travel. In fact, it is a much more complex syndrome and at times can be difficult to identify as such (Graybiel et al., 1968; Lackner, 1989). Motion sickness is rarely experienced during everyday activities except under conditions of passive transport. Individuals without functioning vestibular systems seem to be immune to motion sickness; by contrast, all normal individuals are susceptible to varying extents, although there are tremen-

TABLE 6-1 Diagnostic Categorization of Different Levels of Severity of Acute Motion Sickness

Category	Pathognomonic 16 points	Major 8 points	Minor 4 points	Minimal 2 points	AQS[a] 1 point
Nausea syndrome	Vomiting or retching	Nausea[b] II, III	Nausea I	Epigastric discomfort	Epigastric awareness
Skin		Pallor III	Pallor II	Pallor I	Flushing/subjective warmth > II
Cold sweating		III	II	I	
Increased salivation		III	II	I	
Drowsiness		III	II	I	
Pain					Headache > II
Central nervous system syndrome					Dizziness, eyes closed > II, eyes open III
Levels of Severity Identified by Total Points Scored					
Frank Sickness	Severe Malaise	Moderate Malaise A	Moderate Malaise B	Slight Malaise	
>16 points	8-15 points	5-7 points	3-4 points	1-2 points	

[a]AQS = Additional qualifying symptoms.
[b]III = severe or marked, II = moderate, I = slight.

dous individual differences in susceptibility (Kennedy et al., 1968; Reason and Brand, 1975).

The primary signs and symptoms of motion sickness are summarized in Table 6-1. The pattern of expression of symptoms depends in critical part on individual susceptibility and the type and duration of provocative stimulation (Graybiel et al., 1968; Lackner, 1989). Highly provocative stimulation tends to be more associated with elements of the nausea syndrome, such as stomach discomfort or vomiting. By contrast, less provocative stimulation can evoke more "head" signs and symptoms, such as drowsiness and fatigue. Motion sickness is comparatively easy to identify under laboratory conditions because (1) relatively provocative stimulation is generally used to keep testing periods brief, (2) highly trained observers are present, and (3) subjective and objective measurements of sickness are being taken so that onset and severity can be readily determined. By contrast, under operational and other nonexperimental conditions, motion sickness may be very difficult to recognize. Fatigue, headache, and drowsiness experienced by a worker using a visual motion display could, for example, be due to low-grade motion sickness or from straining too long in front of the display without a rest break. Most low-grade motion sickness probably fails to be recognized as such (Graybiel and Knepton, 1976).

Over the years there have been many attempts to correlate the onset and severity of motion sickness with various physiological parameters, such as heart rate, blood pressure, peripheral blood flow, electrogastrogram activity, etc. (Graybiel and Lackner, 1980; Reason and Brand, 1975; Lawson et al., 1991; Cowings et al., 1990; Stern et al., 1987). Despite many claims of strong correlations, none has so far stood up to systematic experimental scrutiny, nor has any combination of measures been adequate to predict individual sickness and severity (Lawson et al., 1991). At present, training individuals to be aware of the symptoms presented in Table 6-1 is the best that can be done in terms of identifying sickness onset. In fact, this may be adequate under most real-life circumstances because subjective well-being and the ability to perform tasks appropriately is usually all that is necessary.

Progress in studying motion sickness has been hampered by the lack of an adequate theory (Crampton, 1990). Motion sickness is often attributed to sensory conflict because many situations that evoke sickness are associated with various types of conflicts between different receptor system activities, for example, visual versus semicircular canal signals. Conflict theories generally involve neural models of the environment or of the physiological control systems of the body; so, for example, conflict occurs when expectation based on previous experience does not match current inputs during voluntary body movements (Reason and Brand, 1975).

Unfortunately, conflict theories fail to provide a way of understanding which conflicts will be provocative and which will not. Also, such theories do not provide for quantifying the severity of sickness. They only embody the obvious: situations that evoke sickness tend to be ones involving head or body movements under conditions of passive transport.

These limitations make it difficult to develop techniques for predicting how susceptible individuals will be to different forms of stimulation. There have been a number of attempts to relate motion sickness susceptibility to vestibular sensitivity, i.e., thresholds for angular or linear acceleration. However, these attempts have failed. For example, susceptibility to motion sickness during constant velocity, off-vertical rotation of the body (which involves continuous stimulation of the otolith receptors of the inner ear) is not correlated with ocular counterrolling, a measure of otolith sensitivity and gain. Similar lack of correlation has been found between provocative tests involving semicircular canal function and thresholds for perception of angular acceleration, a measure of canal sensitivity (Miller and Graybiel, 1972). Susceptibility to seasickness has been related to the slope of sensation cupolograms (plots of sensation duration versus the log of impulse angular acceleration), but this relationship has proven weak at best. Recently, susceptibility to motion sickness in 0G and 2G force environments has been correlated with the extent to which individuals exhibit velocity storage of vestibular and visual activity (DiZio and Lackner, 1991). This is currently the best predictor for space motion sickness. It is not yet known whether velocity storage[1] correlates with susceptibility to other types of motion sickness, although individuals without functioning labyrinths apparently cannot be made motion sick and they also do not exhibit velocity storage.

A confounding factor in predicting susceptibility to motion sickness is that an individual's susceptibility to one form of provocative motion may not correlate well with susceptibility to another form (Miller and Graybiel, 1972; Calkins et al., 1987). For example, susceptibility in situations primarily involving canal stimulation may not correlate well with those primarily involving otolith stimulation, such as bithermal caloric irrigation (to activate horizontal semicircular canals) and off-vertical rotation. Susceptibilities to similar forms of vestibular stimulation also do not correlate that well for different test situations, such as caloric irrigation

[1] Velocity storage was hypothesized to account for spatio-temporal differences between signals about rotational velocity of the body in space and responses mediated through the vestibular nuclei. For example, the responses of vestibular nucleus cells, slow phase nystagmus velocity, and the sense of self-motion persist temporally and are three-dimensionally reorganized relative to the stimulus velocity input. Such responses are modeled as a sum of the velocity input and a velocity storage signal (Cohen et al., 1977; Robinson, 1977).

versus impulsive angular accelerations in a rotating chair. This means that, to determine reliably an individual's susceptibility in a given situation, it may well be necessary to test him or her in that situation. The one exception to this generalization is that a small percentage of the population, perhaps 10 percent, seems highly susceptible in all motion situations. Such individuals have histories of persistent sickness in cars, boats, and other vehicles and show little change with repeated exposure.

The issue of adaptability, as well as retention and transfer of adaptation, can be as important as basic susceptibility in a given exposure situation. A person with high susceptibility who has rapid adaptation and abatement of symptoms, shows retention of adaptation between widely spaced exposure periods, and shows generalization of adaptation of motion sickness responses to other situations may be better suited for certain situations than an individual of moderate susceptibility who adapts slowly, has poor retention, and shows little transfer (Graybiel and Lackner, 1983). Adaptability and retention of adaptation have not been explored adequately, and only a few studies have even attempted to assess them (Graybiel and Lackner, 1983).

Adaptation to provocative motion can generally be enhanced if exposure is gradual and incremental in intensity. For example, adaptation to rotating environments can be achieved by initially exposing individuals to a very low rate of rotation, one that neither disrupts motor control nor elicits motion sickness, and having them make many head and body movements (Graybiel et al., 1968; Graybiel and Knepton, 1976). By repeating the movements after additional 1 rpm increases in velocity, it is possible to adapt individuals to quite high velocities of rotation without significant performance decrements and without eliciting motion sickness. If the final velocity, say, 10 rpm, were introduced in a single step, most individuals would be incapacitated by motion sickness and unable to adapt regardless of exposure duration. The principle of incremental exposure facilitating adaptation seems to be a general one applicable to all situations so far evaluated (Lackner, 1985; Lackner and Lebovits, 1978) and is likely to apply to VEs.

In thinking about the relevance of motion sickness in VEs, it is critical to remember that it is a complex syndrome with multiple etiological factors, the relative importance of which varies for different individuals and for different intensities of exposure (Kennedy et al., 1992). The presence of more than one eliciting factor (e.g., making head movements as well as looking at a moving visual display) is almost always synergistic in bringing on symptoms. Although nausea and vomiting are often viewed as the most severe manifestations of motion sickness, they can generally be dealt with using antimotion sickness drugs, and in cases of extreme sickness, drug injections (Graybiel and Lackner, 1987). From the standpoint of

VEs, a more severe problem after elements of the nausea syndrome have abated is almost certain to be elements of the sopite syndrome (Graybiel and Knepton, 1976). This refers to the chronic fatigue, lack of initiative, drowsiness, lethargy, apathy, and irritability that may persist even with very prolonged exposure periods. Its elicitation would limit the ability to use virtual reality systems on a regular basis. A further disadvantage of the sopite syndrome is that it can persist for prolonged periods even after exposure to the unusual environment is over.

WHOLE-BODY MOTION AND LOCOMOTION INTERFACES

The control and perception of real whole-body movement and locomotion involves interaction with the world through nearly every possible sensory and motor channel. In the following discussion we sketch out the range of possible levels of technology involvement in whole-body motion and locomotion interfaces relevant to teleoperator and VE systems.

Such interfaces have the potential for extending the capabilities of humans at the expense of producing undesired side effects. Exact specification of the cost-benefit ratio for a given system is not possible because we lack a theory of the sensorimotor regularities to which humans are normally attuned or to which they can become attuned. In the following discussion, we consider a variety of interface systems or system components relevant to whole-body motion for teleoperator and VE systems. The discussion is subdivided into three subsections: inertial displays, locomotion displays, and noninertial displays.

Inertial Displays

These systems induce a sense of *passive* whole-body movement by exposing subjects to a different, highly constrained, body movement through the use of moving-base simulators. In these systems, a constrained motion base generates, in whole or in part, the pattern of force vectors that would be present in the situation being simulated. These systems are extremely expensive and are usually built either for research or for very specific applications.

Full Inertial Displays

The goal of these systems is to simulate continuously the pattern of forces that would be present in the real situation. A high-fidelity example is the Dynamic Flight Simulator at the Naval Air Development Center in Warminster, Pennsylvania. This is a large 16,000 HP gimballed centrifuge that can simulate the angular accelerations and G-loading encountered in

F/4A, F/A-18, and other aircraft. This device can convey the feel of actually flying with no external visual display. These types of devices represent the technology to be replaced by VEs because they are very expensive and are inflexible in terms of simulating multiple aircraft.

One new item in this class that deserves mention is the VEDA Corporation Model 2400 Mark IV Vertifuge. It has two components: one is a short arm (2.5 m) centrifuge with a double gimballed capsule capable of accepting multiple cockpit simulators or a virtual cockpit; the other component is software that maps the pattern of force vectors that would be associated with the manipulation of cockpit controls in an actual aircraft onto the distribution of torques in the degrees of freedom of the simulator that would produce the same patterns of forces. The software attempts to minimize spurious Coriolis affects that are usually associated with short-radius centrifuges by combining the devices' centrifugal accelerations with the primary, secondary, and tertiary Coriolis forces. It also takes into account aspects of vestibular control of perceived orientation in trying to enhance the simulation.

All the systems in this class can reproduce the specified patterns of force vectors imposed on an individual or other object that remains stationary within the simulator. However, all of them fail to simulate both the mechanical dynamics of any voluntary movements made within the simulator and the vestibular stimulation that would be generated during head movements.

Partial Inertial Displays

The goal of these systems is to provide a subset of the forces that would occur in the situation being simulated. There are devices, such as tilting platforms, that can provide the initial cues compatible with inertial body acceleration but cannot simulate sustained acceleration. Complementary devices often provide some sort of cue that sustains the experience of self-motion (e.g., see the discussion of G-seats below).

A six-legged synchronized motion base for a commercial flight simulator is an example of a system in which a subject is seated in position and exposed to a low-fidelity simulation of the forces that would be present in a high-performance aircraft. The base can pitch back and translate forward slightly to partially simulate a forward inertial acceleration. The motion is usually used to enhance the sense of self-motion driven mainly by a visual display. Such systems are widely available commercially, so no specific example is mentioned.

G-seats are used to complement the partial inertial simulations provided by motion bases. A G-seat is itself stationary, but the seat pan and back can be deflated to allow the user to sink into the seat as would be the case during forward and upward accelerations. The cushions can be inflated to simulate the reverse. Some G-seats have an active harness system that controls the pressure applied to the front of the body. Increasing the tension in the harness and simultaneously thrusting the user forward with the seat cushions simulates deceleration. Some G-seats also have the capacity for providing vibration cues. Most vehicles vibrate in a way that is correlated with their acceleration through space. For experienced users, providing vibration cues enhances perceived self-motion, even when the proper accelerative forces are lacking. G-seats may be considered haptic interfaces for the whole body but are listed here because of their close functional association with moving base simulators and the intimate theoretical link between haptic and inertial cues. An example of a G-seat that has all the capabilities just listed is the ALCOGS device at Wright-Patterson Air Force Base.

Variable Gravity Displays

The Graybiel Laboratory slow rotation room at Brandeis University was designed in part to study artificial gravity. This rotating room is an extremely flexible research tool that enables a subject to be exposed to a noninertial, non-1G force background while moving about freely and being monitored by complex on-board equipment.

Examples of the room's flexibility as a research tool is the ability it gives to investigate all of the fundamental problems described in the first section of this chapter, including the nature of automatic load compensation during arm and head movements, the control of eye movements, and sensory localization during body movement. It can also be used to study the control of locomotion in normal and moving environments. Such studies are virtually impossible to conduct in small chambers in which locomotion is limited and there is no room to set up complex equipment for three-dimensional movement analysis.

Another way of simulating non-1G environments is by placing the body in a weighted or counterweighted sling and adjusting the angle of the body in order to increase the effective contact forces between the support surface and the soles of the feet along the body's long axis. The National Aeronautics and Space Administration (NASA) funded a facility at Langley Field in the 1960s that used suspension but still allowed locomotion around a track that was banked such that the component of gravitational force acting along the body's long axis was equivalent to the moon's full gravitational force. Results of studies with this device helped to ascertain what patterns of gait would be most energy efficient and be easiest to control during exploration of the moon's surface.

Locomotion Displays

Ideally, a ground surface interface (which might be called a *haptic interface for the feet*) is a device that permits the user to experience *active* sensations of walking, running, climbing, etc., while remaining within a constrained volume of space. Such a device must allow the user to move his or her limbs in a natural fashion and provide feedback to the user that is matched to the space-time characteristics of the simulated surface (e.g., an inclined plane or a heaving ship deck) and must sense the behavior (positions, movements, forces) of the user in order to both control the actions of the device and to provide appropriate information to the other display systems in the synthetic environment (SE).

Locomotion interfaces can provide the experience of moving about in a large space while actually being confined to a small space. To the extent that such devices reduce the workspace volume of the synthetic environment (SE) system, they reduce the requirements on the other system components. For example, both the subsystem for general monitoring of position and the subsystem for grounding of the haptic interface (for the hands) need not cover large spatial regions. More significant, however, it extends the range of applicability of the SE system to situations in which inclusion of locomotion and/or controlled surface conditions are important, i.e., situations in which conditions underfoot are abnormal, in which visual information is severely reduced, or in which several individuals must coordinate their movements.

Treadmill-Type Displays

The term *treadmill* originated to describe a form of prison punishment: walking on an endless belt driven by rollers has risen socially to a form of voluntary exercise, and the technology has advanced mainly to serve the commercial demands of health clubs. A typical high-performance treadmill (for example, one made by the Quinton Corp.) has some features that make it suitable for conversion to a research or SE tool. The motor and belt have a combined stiffness that allows the horizontal ground contact forces usually applied in walking to be manipulated. It is microprocessor-controlled so that users can preprogram their preferred exercise regimens, but real-time control must be achieved by a custom interface. The entire base can be inclined under the control of a motorized drive system, which could also be controlled in real time by an external custom interface.

A major limitation of commercial treadmills is their one-dimensional nature. A first step in expanding their capabilities would be the development of systems with a belt for each foot. A split-belt system would be useful for simulating turning. The laboratory of Esther Thelen at Indiana University has a custom-built device suitable for such research with infants.

A useful addition to treadmill systems is being developed by the Kistler Corporation. This system has a force-plate under the belt that can sense the vertical ground reaction forces produced during locomotion. This could potentially provide information that could be fed back to the belt through a real-time control algorithm to simulate slippery ground surfaces, for example.

Haptic Interfaces for Individual Feet

Treadmills will be limited to use in SEs that involve locomotion on uniform (but not necessarily horizontal) ground surfaces. Stair climbing exercise machines are just vertical treadmills, in this sense. It may be desirable to generate SEs with arbitrary footing conditions, such as sandy, muddy, and icy conditions, discrete obstacles, stairs, wobbly platforms. A general approach to realizing these conditions would be to develop individual six-degrees-of-freedom platforms or shoes for each foot. Such a system would be analogous to force-feedback haptic devices for the hands. Challenges to development include deciding whether anthropomorphic linkages are necessary, identifying drive systems with sufficient power and bandwidth, and learning more about the role of impedance matching in human locomotion. Programs that have led to the development of improved bicycles and human-powered aircraft have faced some of these problems (especially impedance matching) and may provide an initial strategy and some data for the present application.

Noninertial Displays

Noninertial displays induce a sense of whole-body movement *in stationary individuals* although they can also be used in conjunction with moving bases. Many simulators currently work by presenting stationary individuals with stimuli that are normally associated with body motion so as to enhance a sense of self-motion. This section concentrates on new approaches in this area.

Visual Displays

A variety of devices are being used at present for stimulating individuals with representations of scenes that change in a way that is consistent with body motion through the environment in a vehicle. The major characteristic of visual displays to be used in SEs involving whole-body motion and locomotion is a large stereoscopic field of view. HMDs require hardware and programming to adjust the display to compensate for changes in viewpoint brought about by head movements. A high-resolution display is not critical for visual induction of illusory self-motion. However, if a low-resolution display is used, the visual scene must contain low spatial frequencies and strong pictorial depth cues. Some real visual scenes contain only relatively high spatial frequency textures. If a high-fidelity facsimile of such a scene is moved, an observer experiences self-motion; however, presentation through a low-resolution display unit does not result in perceived self-motion. Obviously, the need for wide (and high) fields of view, the preference for high resolution, and changes in the visual scene to accommodate head movements will generate a high demand for computational and rendering speeds. Although there is no clear psychophysical data on required frame rates, occasional skips in a display that is being generated at an average rate of 30 frames per second can reduce the sense of virtual body motion (assuming there is no moving base or any other supplementary stimulus).

Research that may help guide the integration of visual displays with other SE subsystems for simulating human locomotion is being carried out in several university laboratories (Lackner and DiZio, 1984, 1988; Warren, 1993). One device used for such research includes an independently controllable circular treadmill and visual surround developed by Lackner and DiZio at Graybiel Laboratory, Brandeis University. The circular treadmill is advantageous because combinations of overground and treadmill walking can be combined in the same apparatus. The subject can move through space without walking away from the treadmill and visual display. In this device it is possible to have a subject walk in place at a set pace on the treadmill while the visual display presents a scene whose movement varies in speed and direction. In this situation, the subject perceives whole-body movement with a range of speeds and directions and varying patterns of stepping consistent with it (Lackner and DiZio, 1988). Warren and his colleagues at Brown University have shown that altering the normal ebb and flow of visual feedback (expansion-contraction, vertical oscillation) associated with the walking cycle can alter the pattern of locomotion and the perception of self-motion. This points out the need for developing physical models of the visual feedback from voluntary movement and incorporating it into visual simulations of whole-body movement.

Auditory Displays

Two techniques have been used in this area: one involves presenting a binaural auditory stimulus that exhibits the same spatial and temporal pattern as it would during movement through a natural environment; the other involves cognitive cuing. Induction of illusory self-rotation by simulating a sound field rotating around the head of a stationary subject is an example of the former (Lackner, 1977), and augmentation of visually induced self-motion by virtual wind or engine noise in an aircraft simulator is an example of the latter (Previc et al., 1991). Technology for realizing the first approach is discussed in Chapter 3.

There are no adequate psychophysical data to specify the exact characteristics of auditory displays that are critical for inducing a sense of self-motion. However, neither the number of possible sound sources nor the fidelity with which these sources are simulated is critical for inducing a sense of self-motion. Of more importance is creating the impression of a virtual terrain through which self-motion can occur and, toward this end, the ability to simulate a rich set of sound reflecting surfaces. Technology for simulation of echos and reverberation is also discussed in Chapter 3.

Vestibular Displays

A sense of body motion in a stationary subject can also be elicited artificially by stimulating the semicircular canals of the vestibular system. One method consists of directing streams of cool air or water into one auditory canal and warm into the other. A few seconds of such caloric irrigation can lead to several minutes of perceived self-motion and nystagmus. There are at least two mechanisms governing this response. Altering the temperature of the temporal bone around the external auditory canal (1) locally alters the temperature of the membranous labyrinth of the lateral semicircular canal encased within the bone, and thereby causes a convection current that mechanically stimulates the hair cells of the crista (Barany, 1906) and (2) alters the temperature of the hair cells and primary afferent fibers by direct conduction, leading to modulation of their activity levels (Coats and Smith, 1967). The convection component accounts for about 75 percent of the response (Minor and Goldberg, 1990). This is a standard clinical technique for testing vestibular function that could, in principle, be adapted to SE technology. Its drawbacks for application to SEs are that the latency to onset is about 15 s, the effect is limited to the yaw plane of the head, and it can be nauseogenic. Galvanic stimulation achieved by applying a current between the mastoid bones can also be used to elicit apparent self-motion by directly exciting the vestibular end organs.

Proprioceptive/Kinesthetic Displays

Even in darkness, a sense of moving through the environment, as well as compensatory postural and oculomotor reactions, arises when someone walks in place on a treadmill (Bles and Kapteyn, 1977; Lackner and DiZio, 1988) or manually pushes a revolving railing around (Brandt et al., 1977; Lackner and DiZio, 1984). These results demonstrate that if individuals produce the locomotory movements that normally propel them through space (but without actually displacing), the associated muscular, joint, and tactile feedback, as well as efferent signals, lead to an experience of self-motion.

Even if no locomotor movements are made, patterns of cutaneous or muscular afference that are normally associated with movement through the environment can induce apparent self-motion. One method for achieving this is passing a moving surface against the soles of the feet or the palms of the hands; individuals then report a sense of body motion in the opposite direction. Presenting differential surface speeds to the hands and feet can lead to the feeling that the torso is twisting (Lackner and DiZio, 1984).

Another method of eliciting whole-body motion utilizes muscle vibration. If a standard physiotherapy vibrator oscillating at 120 Hz is applied to the body surface overlying a muscle tendon, such as the Achilles tendon, the spindle receptors of the associated muscle are stretched relative to the extrafusal force-generating fibers because of differential viscoelastic properties. This increases the output of muscle spindle Ia and probably Ib fibers relative to the level appropriate for maintenance of the desired posture. The muscle contracts reflexively to relax the spindle and restore its output to the original level, and the limb controlled by the muscle moves—for example, the ankle extends. This is called a *tonic vibration reflex* (TVR). If a limb is prevented from moving under the influence of a TVR, the spindle activity remains high, and a limb movement will be experienced that is consistent with the muscle's being stretched, for example, flexion of the ankle (Goodwin et al., 1972). When standing on the ground, ankle flexion would ordinarily mean that either the ground is tilted up or the body is tilted forward. Vibrating the Achilles tendons of a subject who is restrained in a standing posture thus elicits an experience of falling forward (Lackner and Levine, 1979). Lackner (1988) has shown that virtually any apparent movement of the body can be elicited by vibration of the proper postural muscles. The movements experienced can be supranormal in the sense that anatomically impossible apparent body configurations are generated—for example, hyperextension of limbs.

RESEARCH NEEDS

The research and development efforts on whole-body motion displays that are needed for development of the SE field, beyond those directed toward achieving more computational power, are summarized briefly in the following paragraphs.

- *Inertial displays* Relatively simple, partial inertial displays need to be developed to replace complex, full-inertial displays because of the cost factor. Currently, however, there is little technology available for reproducing critical subparts of the accelerations that are present in real situations involving active or passive transport through a large volume. Development of such systems should take advantage of the intimate relationship that normally exists between whole-body movement in inertial space and the contact forces that must be applied to the body in order to accelerate it. For example, when a subject is accelerated, the vestibular system senses the inertial motion and cutaneous receptors respond to the contact force propelling the body. Research needs to be focused on this area of haptic-vestibular interactions. An area of technology development that could assist in exploring the roles of these two factors is new padding materials that can allow experimenters to systematically manipulate the distribution of contact forces on the body surface during accelerations in an inertial display. Electrorheological materials hold some promise for research applications and eventually, perhaps, for SE displays.

Another area that could benefit from attention is how active movements affect the perception of whole-body motion induced by noninertial whole-body motion displays. Achieving the same body-referenced limb or head motions requires different muscle forces in stationary and accelerative environments and also leads to different sensory feedback because of noncontact inertial forces on the limb. The lack of expected feedback for the state of body motion being experienced can inhibit the perception of self-motion and lead to a perceptual mapping that better fits all the current sensors. Making head movements during visually induced illusory self-motion can suppress or enhance the sense of self-motion, depending on whether those movements are in the plane of motion and begin when visual stimulation begins or are out of the motion plane and begin after its onset.

Basic research may help determine methods for preventing the inhibition of perceived whole-body motion when the head or arms move or enhancing a weak sense of self-motion. A crucial issue here concerns the extent to which methods can be found for inducing people to perceive contact cues provided by means of haptic VE displays as noncontact inertial perturbations of their limbs.

- *Locomotion displays* Current locomotion displays consist of constant-speed linear treadmills that can provide an individual confined to a small volume with the pattern of visual, auditory, and tactile cues that would be present if they were locomoting through a larger space. However, the only situation that can physically be mimicked by such a device is constant-velocity, linear locomotion.

There are several routes to expanding on these capabilities. First, visual, auditory, and other displays may be used to enhance simple treadmills. To accomplish this, psychophysical work is needed to determine the degree to which perceived acceleration and deceleration (including rotation) can be elicited by such displays, even when it is absent in the mechanical stimulus. Another direction is to improve the treadmills. Linear acceleration can be simulated if hardware and software interfaces are developed that allow control of treadmill acceleration and deceleration when propulsive step forces are generated by the subject. Another advance would be treadmills with a belt for each foot. This would be the simplest version of individual haptic interfaces for each foot and would better allow simulation of changes in direction, i.e., turning. Finally, research should be performed with a view toward providing a system that has separate multidegree-of-freedom platforms for each foot with appropriate sensors and feedback subsystems that can mimic the conditions of walking on level or inclined ground, climbing stairs, and navigating around and over obstacles. The padding materials mentioned above for inertial displays designed for passive whole-body motion might also be useful here for simulating different ground conditions.

• *Visual displays* Requirements on visual displays imposed by consideration of passive whole-body motion or active locomotion are similar to those previously mentioned in other chapters: the best displays would be an HMD that is inexpensive, lightweight, and comfortable; has high resolution and a wide field of view (both horizontally and vertically); and includes both full-color images and refined stereopsis. Of all these characteristics, color is probably the least important.

• *Auditory displays* The most important needs in this area concern those features of the synthesized acoustic field relevant to the illusion of moving through the field. Aside from simulating changes in the direction of sound sources, changes in the apparent distance of the sources and changes in the apparent location of the individual within the reflecting environment are important. Thus, one of the main special needs associated with passive whole-body motion and active locomotion in this area concerns the inclusion of a rich array of reflecting surfaces in the acoustic simulation.

• *Motion sickness* Over the years, motion sickness has arisen as a significant problem with all new modes of passive transport of the body (Guignard and McCauley, 1990). Clearly it will be a problem in SEs as well, especially those involving virtual acceleration and motion of the body (Biocca, 1992). Reports of sickness in SEs are already common

(McCauley, 1984; McCauley and Sharkey, 1992). Research should be directed toward identifying the factors that determine which SEs are especially provocative and how to minimize this while preserving the efficacy of the system. Mechanical factors, such as altered inertial loading of the head by HMDs, as well as sensory factors, need to be considered. Also, attention must be given to elements of the sopite syndrome that are more subtle then those usually associated with motion sickness.

- *Sensorimotor loops* Many SE systems introduce distortions, time delays, gain changes, and statistical variability (noise) between voluntary movements and associated patterns of sensory feedback. Systematic research is necessary to determine the extent to which these factors degrade performance and the subjective state of the user. Acceptable tolerances should be determined for these factors, as well as for the extent to which sensory feedback across different modalities must be in temporal synchrony.
- *Multisensory and motor influences on orientation* This is a critical research area for designing effective VEs that involve locomotion and haptic exploration. Very little is known at present about these influences, except that they are highly complex and pervasive. They are difficult to identify as such because so much of what we take for granted in our everyday activities, such as the perceptual stability of our environment and our bodies during movement, is due to their action.

7

Speech, Physiology, and Other Interface Components

SPEECH RECOGNITION AND SYNTHESIS

Since speech is the most natural form of human intraspecies communication in the real world, it is important to examine the progress and problems associated with research and technology for speech recognition and synthesis by computers for use in communicating with humans in a synthetic environment (SE). Machine recognition and understanding of oral speech has been and continues to be a particularly difficult problem because of the enormous flexibility and variability in speech at both the intersubject and intrasubject levels. Moreover, basic grammar rules are often violated in oral communication. Although humans clearly are able to overcome the problems involved (e.g., by using contextual cues to assist in the interpretation of meaning), it is extremely difficult to duplicate these capabilities with computer software. General background on speech communication, both human and automatic, is available in O'Shaughnessy (1987); information on commercially available automatic speech recognition systems can be found in current newletters and magazines such as *ASR News*, *Voice News*, *Voice Processing*, and *Speech Technology*.

Computer generation of speech also suffers from problems that remain to be solved. In particular, currently available speech synthesis technology does not provide speech that sounds natural or that can be easily matched to the characteristics of an individual speaker. Nevertheless, there are several types of speech synthesis systems available commercially that produce speech with a range of quality and intelligibility. These synthesizers are in use in reading machines for the blind and in certain commercial applications that require information to be provided automatically in response to telephone requests. A comprehensive review of speech synthesis appears in Klatt (1987); information on commercial speech synthesis systems is available in the same newsletters and magazines cited above for speech recognition.

Speech Recognition

There are at least three critical factors contributing to the complexity of speech recognition by machine. The first relates to variation among speakers. For a system to be speaker independent, it must be able to function independently of all the idiosyncratic features associated with a particular talker's speech. In speaker-dependent systems, the computer system functions properly only for the voice or voices it has been trained to recognize. The second critical problem relates to the requirement that the system be able to handle continuous speech input. At the present time, most systems are capable of recognizing only isolated words or commands separated by pauses of 100-250 ms; however, some systems are now becoming available that recognize limited, clearly stressed, continuous speech. The third important factor is the number of words the system is capable of reliably recognizing. Vocabulary sizes in existing systems range from 2 to 50,000 words. Further important factors contributing to the difficulty of speech recognition include intrasubject variability in the production of speech and the presence of interfering background noise and unclear pronunciation. In general, the more predictable the input speech, the better the performance. Thus, for example, a system designed to recognize discretely presented digits spoken by a single person in a sound-isolated room can be made to perform essentially perfectly.

Most of the current successful speech recognition systems rely primarily on an information-theoretic approach in which speech is viewed as a signal with properties that can best be discerned through statistical or stochastic analysis. Recognition systems based on this approach use a simple model to relate text to its acoustic realization. The parameters of this model are then learned by the system during a training phase. Widely accepted practice represents speech as a set of 10 to 30 parameters extracted from the input at a fixed rate (typically every 5 to 20 ms). In this fashion, the input speech is reduced to a stream of representative vectors or numerical indices for each parameter (Davis and Mermelstein, 1980).

Two classes of systems that fall into the information-theoretic category are dynamic time warping (DTW) and hidden Markov modeling

(HMM). In its simplest form, DTW compares the input speech to sets of prestored templates or exemplars corresponding to the prerecorded utterance of each vocabulary item. To add robustness, several templates per word may be recorded by several speakers representing a variety of dialects, speaking styles, and sentence positions. Dynamic programming algorithms are used to evaluate the best match between the input and the templates. DTW refers to the need to time compress (or expand) the input word in order to make it comparable in duration to the stored exemplar. Further discussion of this topic can be found in Dixon and Martin (1979) and Rabiner (1983).

In DTW, both storage and processing times increase, at least linearly, with vocabulary expansion. The system must be trained on every new word in the vocabulary. Furthermore, for speaker-independent recognition, an inordinate number of templates describing the possible variations need to be recorded and stored. As a result, systems with more than a few hundred words are not practical. Because of these limitations, accuracy considerations, and the need to train the system on every word, DTW has been replaced by the HMM approach for the development of high-performance systems.

HMMs represent speech units (words, syllables, phones, etc.) as stochastic machines consisting of several (3 to 20) states. With each state is associated a probability distribution describing the probability of that state's emitting a given observation vector. In addition, there is a set of probabilities of moving from one state to another. Control is transferred between states according to these probabilities each time a new vector is generated (i.e., at the frame rate of the system).

For recognition, a solution to the reverse problem is desired. Given a set of pretrained models and a sequence of observation vectors, the model most likely to have generated the observation must be determined. The utterance associated with this model is then the recognizer's output. The parameters of the model, the probabilities, are determined from labeled training speech data, presumably containing realizations of all the modeled utterances. Efficient algorithms exist for both the training and recognition tasks. The structure of the model (number of states, speech unit, etc.) is chosen judiciously and depends on the language as well as other speech knowledge.

When large-vocabulary recognition is involved, the usual approach is to model speech as a sequence of very basic units such as phones. Coarticulatory phonetic effects are taken into account by having for each phone a different model for every context in which that phone exists (the context is specified by the neighboring phones). As a result, the number of models that must be trained, stored, and evaluated grows as the number of phonetic contexts increases. By taking context into account, the recognition error rate can be cut in half (Chow et al., 1986). In phonetically based HMM systems, it is possible to recognize new words by simply including their phonetic pronunciation in the dictionary; there is no need to train the system specifically on the new words (although the accuracy on those words would be higher if the system is trained on them as well).

Stochastic systems make use of grammar rules that constrain the sequence of speech units that can occur. Grammars are usually either finite-state or statistical. Finite-state grammars describe explicitly the allowable word sequences (a word sequence is either allowed or it is not); statistical grammars allow all word sequences but with different probabilities. A rough measure of a grammar's restrictiveness is its so-called perplexity (Bahl et al., 1983). The lower this number, the more predictable the word sequence. It is easy to see how grammar restrictions can diminish the search that must be performed to determine the identity and order of words in an utterance. Not surprisingly, speech recognition for tasks with low perplexity can be performed much faster and more reliably than tasks with high perplexity.

For some applications, in addition to recognizing the sequence of words spoken, it is important to understand what has been said and give an appropriate response. For this purpose, the output of the recognizer is sent to a language understanding component, which analyzes and interprets the recognized word sequence. To allow for possible errors in recognition, the recognition component sends to the language understanding component not only the top-scoring word sequence, but also the N top-scoring word sequences (where N is typically 10-20). The language understanding component then chooses the word sequence that makes most sense in that context (Schwartz and Austin, 1991).

State of the Art in Speech Recognition

More and more, speech recognition technology is making its way from the laboratory to real-world applications. Recently, a qualitative change in the state of the art has emerged that promises to bring speech recognition capabilities within the reach of anyone with access to a workstation. High-accuracy, real-time, speaker-independent, continuous speech recognition, for medium-sized vocabularies (few thousand words), is now possible in software on off-the-shelf workstations. Users will be able to tailor recognition capabilities to their own applications. Such software-based, real-time solutions usher in a whole new era in the development and utility of speech recognition technology.

As is often the case in technology, a paradigm shift occurs when several developments converge to make a new capability possible. In the case of continuous speech recognition, the following advances have converged to make the new technology possible:

- Higher-accuracy, continuous speech recognition, based on hidden Markov modeling techniques,
- Better recognition search strategies that reduce the time needed for high-accuracy recognition, and
- Increased power of off-the-shelf workstations.

The paradigm shift is taking place in the way we view and use speech recognition. Rather than being mostly a laboratory endeavor, speech recognition is fast becoming a technology that is pervasive and will have a profound influence on the way humans communicate with machines and with each other. For a recent survey of the state of the art in continuous speech recognition, see Makhoul and Schwartz (1994).

Using HMMs, the word error rate for continuous speech recognition has been dropping steadily over the last decade, with a factor of two drop in error rate about every two years. Research systems are now able to tackle problems with large vocabularies. For example, in a test using the ARPA Wall Street Journal continuous speech recognition corpus, word error rates of 11 percent have been achieved for speaker-independent performance on read speech (Pallett et al., 1994). Although this performance level may not be sufficient for a practical system today, continuing improvements in performance are likely to make such systems of practical use in a few years.

Because of the availability of large amounts of training speech data from large numbers of speakers (hundreds of hours of speech), speaker-independent performance has reached such levels that it rarely makes sense to train systems on the speech of specific speakers. However, there will always be outlier speakers for whom, for one reason or another, the system does not perform well. For such speakers, it is possible to collect a relatively small amount of speech (on the order of minutes of speech) and then adapt the system's models to the outlier speaker to improve performance significantly for that speaker.

For information retrieval applications, it is important to understand the user's query and give an appropriate response. Speech understanding systems have reached the stage at which it is possible to develop a practical system for specialized applications. The understanding component must be tuned to the specific application; the work requires significant amounts of data collection from potential users and months of labor-intensive work to develop the language understanding component for that application. In the ARPA Airline Travel Information Service (ATIS) domain, users access flight information using verbal queries. Speech understanding systems in the ATIS domain have achieved understanding error rates of less than 10 percent in speaker-independent mode (Pallett et al., 1994). In these tests, users speak in a normal spontaneous fashion.

Until recently, it was thought that to perform high-accuracy, real-time continuous speech recognition for large vocabularies would require either special-purpose very large system integrated (VLSI) hardware or a multiprocessor. However, new developments in search algorithms have sped up the recognition computation at least two orders of magnitude, with little or no loss in recognition accuracy (Schwartz and Austin, 1991). In addition, computing advances have achieved two orders of magnitude increase in workstation speeds in the last decade. These two advances have made software-based, real-time, continuous speech recognition a reality. The only requirement is that the workstation must have an analog-to-digital converter to digitize the speech. All the signal processing, feature extraction, and recognition search is then performed in software in real time on a single-processor workstation.

For example, it is now possible to perform a 3000-word ATIS recognition task in real time on such workstations as the Silicon Graphics Indigo (SGI) R3000 or the Sun SparcStation 2. Most recently, a 40,000-word continuous dictation task was demonstrated in real time on a Hewlett-Packard 735 workstation, which has about three times the power of an SGI R3000. Thus, the computation grows much slower than linear with the size of the vocabulary (Nguyen et al., 1993).

The real-time feats just described have been achieved at a relatively small cost in word accuracy. Typically, the word error rates are less than twice those of the best research systems.

The most advanced of these real-time demonstrations have not as yet made their way to the marketplace. However, it is possible today to purchase products that perform speaker-independent, continuous speech recognition for vocabularies of a few thousand words. These systems are being used in command and control applications, the routing of telephone calls by speaking the full name of the party being called, the training of air traffic controllers, and the control of workstation applications. Of particular significance to the public will be the development of transactional applications over the telephone, such as home shopping and airline reservations. Practical large-vocabulary, continuous-speech, speaker-independent dictation systems are a few years away.

HMMs have proven to be very good for modeling variability in time and feature space and have resulted in tremendous advances in continuous speech recognition. However, some of the assumptions made by HMMs are known not to be strictly true for speech—for example, the conditional independence assumptions in which the probability of being in a state is dependent only on the previous state, and the output probability at a state is dependent only on that state and not on previous states. There have been attempts at ameliorating the effects of these assumptions by developing alternative speech models, such as the use of stochastic segmental models and neural networks. In all these attempts, however, significant computational limitations have hindered the full exploitation of these methods; in fact, the performance of such alternative models have barely approached that of HMM systems. However, when such models are used in conjunction with HMM systems, the resulting hybrids have achieved word error rate reductions of 10-20 percent (Makhoul and Schwartz, 1994).

Despite all these advances, much remains to be done. Speech recognition performance for very large vocabularies and larger perplexities is not yet adequate for useful applications, even under benign acoustic conditions. Any degradation in the environment or changes between training and test conditions causes a degradation in performance. Therefore, work must continue to improve robustness to varying conditions: new speakers, new dialects, different channels (microphones, telephone), noisy environments, and new domains and vocabularies. What will be especially needed are improved acoustic models and methods for fast adaptation to new conditions. Many of these research areas will require more powerful computing sources. Fortunately, workstation speed and memory will continue to grow in the years to come. The resulting more powerful computing environment will facilitate the exploration of more ambitious modeling techniques and will, no doubt, result in additional significant advances in the state of the art.

In comparison with speech recognition, the field of language understanding, which is a much harder problem, is still in its infancy. One major obstacle for advancement is the lack of a representation of semantics that is general and powerful enough to cover major applications of interest. And even if such a representation were available, there is still a strong need to develop automatic methods for interpreting word sequences, without having to rely on the currently dominant methods of labor-intensive crafting of detailed linguistic rules.

Speech Synthesis

A speech synthesizer is a device that accepts at its input the text of an utterance in orthographic or computer-readable form and transforms that text into spoken form. The synthesizer performs much the same function as a human who reads a printed text aloud. The synthesizer usually contains a component that performs an initial transformation from a written text into a sequence of phonetic units (e.g., transforming *caught* to /kot/), and these phonetic units then control the production of sound in the synthesizer.

The synthesizer can generate the simulated speech in one of two ways. One method, called concatenative synthesis, is to create utterances by stringing together segments of speech that have been excerpted from the utterances of a speaker. These segments consist of pieces of syllables, usually a sequence of consonant and vowel or a vowel and a consonant. For example, a word like *mat* would be synthesized by concatenating two pieces ma + at. Special procedures are used to ensure that the joining of the pieces is done smoothly with minimal artifacts. Rules are used to provide suitable adjustment of timing and other prosodic characteristics. An inventory of several hundred of these prerecorded segments of speech is needed to synthesize most utterances in English. In order to synthesize speech for different speakers, it is necessary to make new recordings for each speaker and to edit these recordings to obtain the necessary inventory of segments.

A second method for generating the sound from a phonetic sequence is to use a synthesis device that models some aspects of the human speech-generating process. This process regards the human speech system as consisting of a set of sound sources that simulate vocal cord vibration or noise due to turbulent airflow, together with a set of resonators that simulate the filtering of the sources by the airways that constitute the vocal tract. The control parameters for this model specify attributes such as the frequency and amplitude of vocal cord vibration and the frequencies of the resonators. Depending on the complexity of the synthesizer, there can be as many as 40 such control parameters or as few as 10. This type of synthesizer has been called a formant or terminal-analog synthesizer.

In this second type of synthesizer, the rules for converting phonetic units into control parameters specify an array of parameter values for each speech sound and describe how these parameters should move smoothly from one speech sound to the next. By proper adjustment of the ranges of the parameters, different male and female voices can be synthesized. The best synthesis based on these formant or articulatory models is somewhat more intelligible than concatenative synthesis.

For both concatenative and formant synthesis, the generation of natural-sounding utterances requires that rules be developed for controlling the temporal aspects of the speech and changes in fundamental frequency that indicate prominent syllables and that delineate groupings of words. The most successful devices for synthesis of speech from text produce speech with reasonably high intelligibility, although not quite as intelligible as human production of speech, and with some lack of naturalness. Continuing research is leading to improvements in naturalness through adjustments of rules for the rhythmic and other prosodic aspects of the speech.

PHYSIOLOGICAL RESPONSES

In all the previous discussion of human-machine interfaces, the inputs to the human (the displays) and the outputs from the human (the controls) have occurred at the limits of the human periphery, i.e., they have made use of the natural input and output mechanisms. In principle, it is possible to construct interfaces that bypass the periphery and display information by stimulating neural structures directly or sense physiological variables for purposes of control. Independent of the extent to which such interfaces provide a useful adjunct to conventional interfaces for normal subjects, they undoubtedly will prove exceedingly important for subjects with certain kinds of sensorimotor disabilities (Warner et al., 1994). In this section, we discuss briefly different kinds of physiological responses that might serve as useful control signals. We have omitted consideration of neural stimulation because, with only minor exceptions, we believe that most devices for providing such stimulation will be employed only for individuals with severe sensory disabilities; for such individuals, these devices will be permanently implanted and become part of the subject, thus eliminating the need to include such devices in human-machine interfaces for SE systems. We have also omitted discussion of the use of physiological response measurements (e.g., associated with muscle actions) to help individuals with severe motor disabilities control computers and telerobots, because this topic is included in our discussion of the medicine and health care application domain in Chapter 12.

Most practical work on physiological responses has been conducted in the laboratory for purposes of establishing the effects of selected experimental conditions on an individual's emotional state or for designing systems and system tasks that take human capabilities and limitations into consideration. In the near term, those measures that have been found useful as indicators of mental, emotional, or physical states in the real world should be equally useful as indicators of the states in an SE. However, apart from use with individuals having severe motor disabilities, it is likely to be several years before physiological-response sensing will be included as a control element in most SE systems. (A possible exception to this statement is suggested by the current research on brain-activated control at Armstrong Medical Research Laboratory by Junker et al., 1988.) The following discussion briefly reviews the current status of work in physiological response measurement.

Measures of physiological responses have been used by researchers to describe the physical, emotional, and mental condition of human subjects, usually in relation to performance of a specific task, involvement in a particular social transaction, or exposure to a particular set of environmental variables. Over the years, ergonomic researchers have conducted an enormous number of studies employing physiological responses as indicators or correlates of fatigue, stress, onset or decline of attention, and level or change in level of mental workload (Wierwille, 1979). The physiological responses traditionally used in such studies focus on involuntary responses controlled by the autonomic nervous system, such as heart rate, blood pressure, stomach activity, muscle tension, palmar sweating, and pupil dilation. The techniques for measuring these responses vary in the ease and intrusiveness of data collection, the statistical properties of the data flow, and the certainty with which analyzed data can be interpreted.

Another area of physiological research activity involves studies of brain organization and cognitive function (Druckman and Lacey, 1989; Kramer, 1993; Zeffiro, 1993). Neurophysiologists have been heavily involved in developing and using physiological measures to map sensory and motor functions in the brain and to identify patterns of brain activity associated with attentional states and cognitive workload.

Autonomic Nervous System Responses

The autonomic nervous system is composed of two subsystems, the sympathetic and the parasympathetic. The sympathetic nervous system is responsible for mobilizing the body to meet emergencies; the parasympathetic nervous system is responsible for maintaining the body's resources. These two systems can either work together or in opposition to one another. Thus, an increase in a physiological response such as heart rate or muscle tension may be interpreted as an increase in sympathetic activity or as a decrease in parasympathetic activity.

Physiological responses that are interpreted as sympathetic nervous system activities are often cited as indicators of emotional response. These include heart rate, systolic and diastolic blood pressure, muscle tension, and skin conductance (Hassett, 1978). In evaluating these measures, it is important to recognize that the sympathetic nervous system does not respond in a unitary way; humans often exhibit an increase in one of these responses such as elevated heart rate, without showing an increase in others. Thus, it is misleading to speak of arousal as if it were a unitary response. Moreover, there is no simple one-to-one correspondence between any single physiological response and a particular emotion or cognitive process (Cacioppo and Tassinary, 1990). Although it may be possible in the future to identify patterns of responses that relate to specific psychological states, the results to date are mixed. For example, it appears to be possible to describe the intensity of an emotion but not necessarily whether it is positive or negative.

Perhaps the most thorough study of the measurement and use of autonomic nervous system responses has been accomplished by researchers interested in identifying physiological indicators of mental workload. Wierwille (1979) provided a complete review of 14 physiological measures and their usefulness. The results of this analysis point to three measures that have some correlation with mental workload: catecholamine excretion as determined by bodily fluid analysis, pupil diameter, and evoked cortical potential. Although each of these was judged to have promise, none is easily measured. In addition, Wierwille suggests that they should be used in combination with behavioral measures when drawing conclusions about levels or types of mental workload.

All physiological measures suffer from some drawback, either in terms of ease of measurement or accuracy of interpretation. One reason it is so difficult to make inferences about possible psychological concomitants of physiological processes is the complexity of the physiological systems and the sensory environments involved. Each physiological process may be internally regulated by multiple control systems and also may be complicated by individual differences (e.g., one person may respond to threat primarily with elevated heart rate, whereas another will respond with elevated blood pressure).

Heart rate can be assessed using electrocardiogram electrodes attached to the chest or by attaching a plethysmograph to the finger or earlobe to detect changes in blood flow. Although heart rate is relatively easy to measure, the results are conflicting. According to a review by Hartman (1980), some studies show increases in heart rate with mental workload; others show a decrease or no change in heart rate with increases in workload. Wierwille's (1979) results suggest that one fairly stable and useful measure is spectrum analysis of heartbeat intervals.

Systolic and diastolic blood pressure are usually monitored by using a self-inflating arm cuff system (Weber and Drayer, 1984). A newer technology from OHMEDA uses a finger cuff to track pulse pressure continuously; it provides a measure of blood pressure and heart rate every two seconds. Systolic pressure is the peak pressure when the blood is being ejected from the left ventricle; diastolic pressure is the pressure between contractions while the ventricle is relaxing (Smith and Kampine, 1984). Heart rate and systolic blood pressure tend to show significant increases when subjects are shown emotionally arousing films, asked to carry out demanding tasks (such as mental arithmetic), or are placed in embarrassing or threatening situations (Krantz and Manuck, 1984). There are many other cardiovascular variables that can be assessed (e.g., pulse transit time, forearm blood flow—Smith and Kampine, 1984). Depending on the precise nature of the research question, additional measures may be needed in order to identify the underlying mechanisms that have produced any observed changes in blood pressure or heart rate. For example, an increase in blood pressure could be due to an increase in cardiac output (either heart rate or stroke volume), or it could be due to an increase in peripheral resistance (constriction of blood vessels or capillaries). Knowing which of these has occurred may in turn point to the neurotransmitters likely to have been active.

Monitoring of muscle tension (electromyography—EMG) involves attaching two or more electrodes over the surface of the target muscle to detect naturally occurring bioelectric potentials. Large skeletal muscles such as the trapezius (located in the back of neck) or frontalis (forehead) may be monitored as a way of assessing muscle tension, or facial expression. Even changes too fleeting or slight to be observed by a human judge can be detected by placing electrodes near key facial muscles (Hassett, 1978; Cacioppo and Petty, 1983). The bioelectric potentials being detected are very small in magnitude and the range of frequencies includes 60 Hz. To ensure clean recording of EMG, isolation from electrical fields is important. This recording procedure is relatively straightforward. Muscle tension may be an indicator of alertness; however, it also changes as a function of fatigue, level, the individual's physical condition, and the physical demands of a task.

EMG signals have been employed for the control of powered prostheses such as the Utah Artificial Arm (Jacobsen et al., 1982). The basis for this control is detailed models of the muscle biomechanics that are used to predict the proper joint torques from the EMGs (Meek et al., 1990). Given the success in prosthetics, it is therefore natural to consider that EMGs could be used to control robotic devices. Hiraiwa et al. (1992) used EMGs to control finger position, torque, and motion of an artificial robot hand. Control was achieved with a neural net, trained by using a VPL DataGlove; accuracies were quite modest. Farry and Walker (1993) are working on EMG control of the Utah/MIT Dextrous Hand Master; various EMG processing methods are being tested using an EXOS Dextrous Hand Master to predict basic grasp types. They also present a concise up-to-date review of EMG signal processing and the use of EMGs in the control of prostheses. The use of EMGs to control telerobots or computers is clearly less appropriate for normal humans than for humans with severe motor disabilities. They are used for persons with disabilities (e.g., to control prosthetics) only because better choices do not exist.

EDA (electrodermal activity, also known as GSR, galvanic skin reflex) is assessed by attaching a pair of gold cup electrodes to the skin (usually on the hand), passing a very small electrical current between them, and measuring the resistance. As palmar sweating increases, the conductance increases (Hassett, 1978). In response to stimuli, people typically show a transient change in EDA; the number and latency of peaks in EDA are often measured as a way of assessing stimulus response. According to Wierwille (1979), stress will cause decreases in skin conductance; however, extensive time averaging is required to demonstrate a change. Moreover, skin response changes as a function of several variables including ambient temperature, humidity, physical exertion, and individual differences in body metabolism.

Research on pupil dilation has shown a change in pupil size as mental workload increases: during high levels of workload the pupil dilates, but when the operator becomes overloaded, pupil size is reduced. Pupil dilation can be recorded by a motion picture or video camera; analysis is accomplished by measuring each frame through manual or automated techniques. The evaluation of pupillary response is complicated by the fact that pupil size changes with the level of ambient illumination, fatigue, and acceleration.

Brain Activity

In recent years, many researchers have made significant contributions to the measurement of brain activity (Kramer, 1993; Zeffiro, 1993; Druckman and Lacey, 1989). Four technologies are of particular interest: event related potentials (ERP), positron emission tomography (PET), magnetic resonance imaging (MRI), and magnetic stimulation mapping.

Cognitive Function

Event-related potentials are found on the electroencephalographic record. They reflect brain activities that occur in response to discrete events or tasks.

Druckman and Lacey (1989) report several important research efforts involving ERPs. For example, a negative voltage with a stimulus-response latency of 100 ms (N100) has been shown to be related to the individual's focus of attention. This finding suggests that N100 might be used to determine whether an individual is following task instructions or has shifted attention to some other aspect of the environment. A negative voltage with a latency of 200 ms (N200) has been found to be related to a mismatch between an individual's expectations and events occurring in the environment. A positive voltage having a 300 ms latency (P300) has been associated with mental workload (Donchin et al., 1986; Kramer, 1993) and analysis of memory mechanisms (Neville et al., 1986; Karis et al., 1984). Kramer (1993) reports the extensive use of P300 in measuring both primary and secondary mental task load. A principal finding across studies is that the P300 elicited by discrete secondary tasks decreases as the difficulty of the primary task increases. Although no formal reliability studies have been conducted, there is a substantial body of evidence that suggests that ERPs are reliable measures of mental workload in the laboratory. The next step is to begin experimentation in field environments (Druckman and Lacey, 1989; Kramer, 1993).

Another important aspect of the ERP is the readiness potential that is manifested as a slow negative wave preceding a voluntary response. This wave has been labeled the contingent negative wave (CNV) by Walter et al. (1964). It is one of a class of events preceding negative waves that have been shown to relate to a subject's mental preparatory processes. There is some evidence that the size of the negative potential and its location by hemisphere can provide evidence regarding the response a subject is thinking about independent of the response that is actually made (Coles et al., 1985).

It is important to note that the time lag associated with obtaining reliable estimates of ERPs for cognitive functions may take several seconds. This time delay limits the potential of ERPs for VE applications requiring real-time, closed-loop control.

Functional and Structural Information

Functional and structural information about the brain can be obtained by a combination of methods, including magnetic stimulation mapping, positron emission tomography, and magnetic resonance imaging. Magnetic stimulation mapping is accomplished by exciting a magnetic coil over the surface of the head. This induces a magnetic field that stimulates the underlying brain. When the coil is placed over one of the motor areas in the brain and a brief burst of pulses is initiated, some of the nerve cells in that area will discharge and the part of the body controlled by those cells will move. Magnetic stimulation mapping is also used to stimulate peripheral nerve cells, such as those in the back or limbs for both clinical and research purposes. The number of cells activated by this technique is unclear.

PET is an imaging technique that portrays the flow of radioactive substances throughout the body. It is based on a combination of principles from computer tomography and radio isotope imaging (Martin et al., 1991). Researchers at the National Institutes of Health use PET to map the motor cortex. They examine blood flow data in the brain from a PET scan as subjects move different parts of their bodies. Areas of high blood flow are matched with the body part being moved; the increase in blood flow has been shown to be proportional to the rate of movement or degree of contraction. Although PET provides accurate information about the function of a particular area of the brain, it does not present particularly useful anatomical information. As a result, MRI, which does provide anatomical information, has been used in conjunction with PET—the images from both techniques can be taken simultaneously for the same subject and then spatially registered with one another. This provides a pattern that links structure and function.

Recently a procedure has been developed at Massachusetts General Hospital to use MRI to collect functional data by imaging oxidation status in response to stimulation and motor responses. This is a completely noninvasive approach to mapping the cortex, and it is much faster than PET. Functional MRI data have been collected that show changes in oxidation status in the premotor cortex and somatosensory cortex when a subject thinks about making a movement. During actual movement, both of these plus the primary motor cortex are activated. Another promising process, magnetic resonance angiography (MRA), images blood vessels and can be used to measure blood flow through the brain (Martin et al., 1991).

Brain-Actuated Control

Finally, there has been some interesting work conducted at Armstrong Medical Research Laboratory (Junker et al., 1988, 1989) on brain-actuated control of a roll axis tracking simulator. The focus of the work has been on providing closed-loop feedback to subjects that allows them to learn to control the resonance portion of the EEG and then to use these resonances to send control signals to a tracking simulator. Essentially, the studies have shown that subjects have been able to accurately roll to the left or right in the simulator by controlling brain resonance signals. Although this research is in early stages, it does appear to hold some promise for training individuals to use physiological responses as simple control signals (e.g., on-off, left-right, etc).

Recently this work on brain-actuated control has extended to its use for rehabilitation (Calhoun et al., 1993). These authors believe that, by employing appropriate evoking stimuli and feedback, people can learn to use EEG responses to control wheelchairs, computers, and prosthetics. Furthermore, they conclude that brain-actuated control may be better than most existing interfaces for people with disabilities.

OTHER INTERFACE COMPONENTS

Other interface components that could be of value concern stimulation of the olfactory (smell) or gustatory (taste) channels, and diffuse sensations related to the sensing of heat, air currents, humidity, and chemical pollutants in the air that affect the eyes or skin surface. In some cases, stimulation can have direct functional significance for the specified task; in others, it may serve merely to increase the general sense of presence or immersion in the environment.

In one research and development project aimed at the training of firefighters, the traditional helmet-mounted display of visual and auditory stimuli is being extended to include the display of odor and radiant heat stimuli (Cater et al., 1994). The current version of this experimental system, which makes use of a 44 lb backpack for housing the special additional equipment required, includes both a computer-controlled odor delivery system and a computer-controlled radiant heat delivery system. The odor system delivers odors by blowing air across microcapsules crushed by pinch rollers; the radiant heat system provides directional as well as overall intensity characteristics by controlling the individual intensity of infrared lamps arranged in a circular array.

The visual display component of the system is driven by video outputs of two Silicon Graphics Indigos (one for each eye); Polhemus Fastrak position information is processed by a PC and communicated to one of the Indigos via an RS-232 serial bus; another Indigo serial bus controls the odor delivery subsystem; a third serial connection on the second Indigo drives a MIDI bus to control the radiant heat subsystem; and an Ethernet connection between the two Indigos is used for synchronization purposes and for sharing peripheral information.

The demonstration program Pyro makes use of two virtual human arms and hands, Polhemus trackers attached to the user's wrists, a virtual torch, and some virtual flammable spheres. The user's task in this demonstration is to light each of the spheres with the torch.

Among the problems encountered in this work are the visualization of the flame, gumming up of the pinch rollers used to crush the odor microcapsules, and the weight of the backpack.

8

Computer Hardware and Software for the Generation of Virtual Environments

The computer technology that allows us to develop three-dimensional virtual environments (VEs) consists of both hardware and software. The current popular, technical, and scientific interest in VEs is inspired, in large part, by the advent and availability of increasingly powerful and affordable visually oriented, interactive, graphical display systems and techniques. Graphical image generation and display capabilities that were not previously widely available are now found on the desktops of many professionals and are finding their way into the home. The greater affordability and availability of these systems, coupled with more capable, single-person-oriented viewing and control devices (e.g., head-mounted displays and hand-controllers) and an increased orientation toward real-time interaction, have made these systems both more capable of being individualized and more appealing to individuals.

Limiting VE technology to primarily visual interactions, however, simply defines the technology as a more personal and affordable variant of classical military and commercial graphical simulation technology. A much more interesting, and potentially useful, way to view VEs is as a significant subset of multimodal user interfaces. Multimodal user interfaces are simply human-machine interfaces that actively or purposefully use interaction and display techniques in multiple sensory modalities (e.g., visual, haptic, and auditory). In this sense, VEs can be viewed as multimodal user interfaces that are interactive and spatially oriented. The human-machine interface hardware that includes visual and auditory displays as well as tracking and haptic interface devices is covered in Chapters 3, 4, and 5. In this chapter, we focus on the computer technology for the generation of VEs.

One possible organization of the computer technology for VEs is to decompose it into functional blocks. In Figure 8-1, three distinct classes of blocks are shown: (1) rendering hardware and software for driving modality-specific display devices; (2) hardware and software for modality-specific aspects of models and the generation of corresponding display representations; (3) the core hardware and software in which modality-independent aspects of models as well as consistency and registration among multimodal models are taken into consideration. Beginning from left to right, human sensorimotor systems, such as eyes, ears, touch, and speech, are connected to the computer through human-machine interface devices. These devices generate output to, or receive input from, the human as a function of sensory modal drivers or renderers. The auditory display driver, for example, generates an appropriate waveform based on an acoustic simulation of the VE. To generate the sensory output, a computer must simulate the VE for that particular sensory mode. For example, a haptic display may require a physical simulation that includes

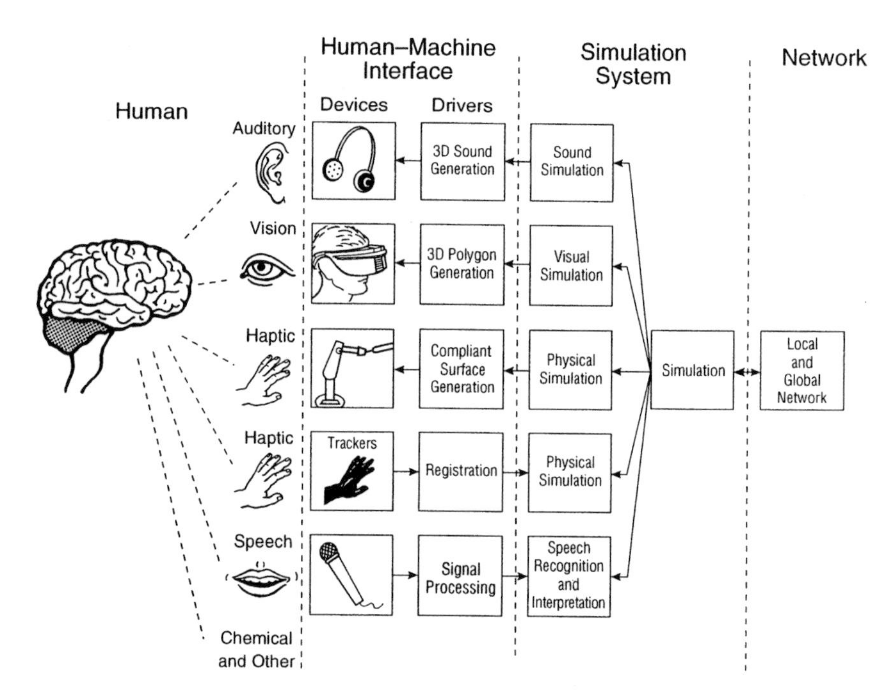

FIGURE 8-1 Organization of the computer technology for virtual reality.

compliance and texture. An acoustic display may require sound models based on impact, vibration, friction, fluid flow, etc. Each sensory modality requires a simulation tailored to its particular output. Next, a unified representation is necessary to coordinate individual sensory models and to synchronize output for each sensory driver. This representation must account for all human participants in the VE, as well as all autonomous internal entities. Finally, gathered and computed information must be summarized and broadcast over the network in order to maintain a consistent distributed simulated environment.

To date much of the design emphasis in VE systems has been dictated by the constraints imposed by generating the visual scene. The nonvisual modalities have been relegated to special-purpose peripheral devices. Similarly, this chapter is primarily concerned with the visual domain, and material on other modalities can be found in Chapters 3-7. However, many of the issues involved in the modeling and generation of acoustic and haptic images are similar to the visual domain; the implementation requirements for interacting, navigating, and communicating in a virtual world are common to all modalities. Such multimodal issues will no doubt tend to be merged into a more unitary computational system as the technology advances over time.

In this section, we focus on the computer technology for the generation of VEs. The computer hardware used to develop three-dimensional VEs includes high-performance workstations with special components for multisensory displays, parallel processors for the rapid computation of world models, and high-speed computer networks for transferring information among participants in the VE. The implementation of the virtual world is accomplished with software for interaction, navigation, modeling (geometric, physical, and behavioral), communication, and hypermedia integration. Control devices and head-mounted displays are covered elsewhere in this report.

VE requires high frame rates and fast response because of its inherently interactive nature. The concept of frame rate comes from motion picture technology. In a motion picture presentation, each frame is really a still photograph. If a new photograph replaces the older images in quick succession, the illusion of motion in engendered. The update rate is defined to be the rate at which display changes are made and shown on the screen. In keeping with the original motion picture technology, the ideal update rate is 20 frames (new pictures) per second or higher. The minimum acceptable rate for VE is lower, reflecting the trade-offs between cost and such tolerances.

With regard to computer hardware, there are several senses of frame rate: they are roughly classified as *graphical, computational,* and *data access.* Graphical frame rates are critical in order to sustain the illusion of presence or immersion in a VE. Note that these frame rates may be independent: the graphical scene may change without a new computation and data access due to the motion of the user's point of view. Experience has shown that, whereas the graphical frame rate should be as high as possible, frame rates of lower than 10 frames per second severely degrade the illusion of presence. If the graphics being displayed relies on computation or data access, then computation and data access frame rates of 8 to 10 frames per second are necessary to sustain the visual illusion that the user is watching the time evolution of the VE.

Fast response times are required if the application allows interactive control. It is well known (Sheridan and Ferrell, 1974) that long response times (also called lag or pure delay) severely degrade user performance. These delays arise in the computer system from such factors as data access time, computation time, and rendering time, as well as from delays in processing data from the input devices. As in the case of frame rates, the sources of delay are classified into data access, computation, and graphical categories. Although delays are clearly related to frame rates, they are not the same: a system may have a high frame rate, but the image being displayed or the computational result being presented may be several frames old. Research has shown that delays of longer than a few milliseconds can measurably impact user performance, whereas delays of longer than a tenth of a second can have a severe impact. The frame rate and delay required to create a measurable impact will in general depend on the nature of the environment. Relatively static environments with slowly moving objects are usable with frame rates as low as 8 to 10 per s and delays of up to 0.1 s. Environments with objects exhibiting high frequencies of motion (such as a virtual handball game) will require very high frame rates (> 60 Hz) and very short delays. In all cases, however, if the frame rate falls below 8 frames per s, the sense of an animated three-dimensional environment begins to fail, and if delays become greater than 0.1 s, manipulation of the environment becomes very difficult. We summarize these results to the following constraints on the performance of a VE system:

Frame rates must be greater than 8 to 10 frames/s.
Total delay must be less than 0.1 s.

Both the graphics animation and the reaction of the environment to user actions require extensive data management, computation, graphics, and network resources. All operations that take place to support the environment must operate within the above time constraints. Although one can imagine a system that would have the graphics, computation, and communications capability to handle all environments, such a system is beyond current technology. For a long time to come, the technology necessary will generally be dependent on the application domain for which the VE is being built. Real-world simulation applications will be highly bound by the graphics and network protocols and by consistency issues; information visualization and scientific visualization applications will be bound by the computational performance and will involve issues of massive data management (Bryson and Levit, 1992; Ellis et al., 1991). Some applications, such as architectural visualization, will require photorealistic rendering; others, such as information display, will not. Thus the particular hardware and software required for VE implementation will depend on the application domain targeted. There are some commonalities of hardware and software requirements, and it is those commonalities on which we focus in our examination of the state of the art of computer hardware and software for the construction of real-time, three-dimensional virtual environments.

HARDWARE FOR COMPUTER GRAPHICS

The ubiquity of computer graphics workstations capable of real-time, three-dimensional display at high frame rates is probably the key development behind the current push for VEs today. We have had flight simulators with significant graphics capability for years, but they have been expensive and not widely available. Even worse, they have not been readily programmable. Flight simulators are generally constructed with a specific purpose in mind, such as providing training for a particular military plane. Such simulators are microcoded and programmed in assembly language to reduce the total number of graphics and central processing unit cycles required. Systems programmed in this manner are difficult to change and maintain. Hardware upgrades for such systems are usually major undertakings with a small customer base. An even larger problem is that the software and hardware developed for such systems are generally proprietary, thus limiting the availability of the technology. The graphics workstation in the last 5 years has begun to supplant the special-purpose hardware of the flight simulator, and it has provided an entry pathway to the large numbers of people interested in developing three-dimensional VEs. The following section is a survey of computer graphics workstations and graphics hardware that are part of the VE development effort.

Notable Graphics Workstations and Graphics Hardware

Graphics performance is difficult to measure because of the widely varying complexity of visual scenes and the different hardware and software approaches to computing and displaying visual imagery. The most straightforward measure is given in terms of polygons/second, but this only gives a crude indication of the scene complexity that can be displayed at useful interactive update rates. Polygons are the most common building blocks for creating a graphic image. It has been said that visual reality is 80 million polygons per picture (Catmull et al., 1984). If we wish photorealistic VEs at 10 frames/s, this translates into 800 million polygons/s. There is no current graphics hardware that provides this, so we must make approximations at the moment. This means living with less detailed virtual worlds, perhaps via judicious use of hierarchical data structures (see the software section below) or off-loading some of the graphics requirements by utilizing available CPU resources instead.

For the foreseeable future, multiple processor workstations will be playing a role in off-loading graphics processing. Moreover, the world modeling components, the communications components, and the other software components for creating virtual worlds also require significant CPU capacity. While we focus on graphics initially, it is important to note that it is the way world modeling effects picture change that is of ultimate importance.

Graphics Architectures for VE Rendering

This section describes the high-level computer architecture issues that determine the applicability of a graphics system to VE rendering. Two assumptions are made about the systems included in our discussion. First, they use a *z-buffer* (or depth buffer), for hidden surface elimination. A z-buffer stores the depth—or distance from the eye point—of the closest surface "seen" at that pixel. When a new surface is scan converted, the depth at each pixel is computed. If the new depth at a given pixel is closer to the eye point than the depth currently stored in the z-buffer at that pixel, then the new depth and intensity information are written into both the z-buffer and the frame buffer. Otherwise, the new information is discarded and the next pixel is examined. In this way, nearer objects always overwrite more distant objects, and when every object has been scan converted, all surfaces have been correctly ordered in depth. The second assumption for these graphic systems is that they use an application-programmable, general-purpose processor to cull the database. The result is to provide the rendering hardware with only the graphics primitives that are within the viewing volume (a perspective pyramid or parallelpiped for perspective and parallel projections respectively). Both of these assumptions are valid for commercial graphics workstations and for the systems that have been designed by researchers at the University of North Carolina at Chapel Hill.

The rendering operation is composed of three stages: per-primitive,

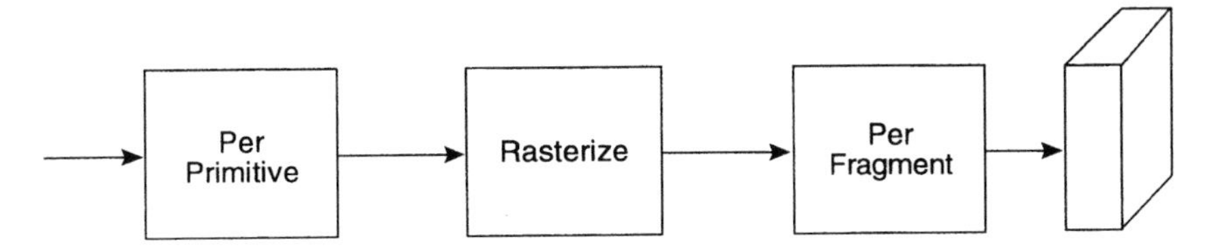

FIGURE 8-2 The graphics pipeline.

rasterization, and per-fragment (as shown in Figure 8-2). *Per-primitive* operations are those that are performed on the points, lines, and triangles that are presented to the rendering system. These include transformation of vertices from object coordinates to world, eye, view volume, and eventually to window coordinates, lighting calculations at each vertex, and clipping to the visible viewing volume. *Rasterization* is the process of converting the window-coordinate primitives to fragments corresponding to the pixels held in the frame buffer. The frame buffer is a dedicated block of memory that holds intensity and other information for every pixel on the display surface. The frame buffer is scanned repeatedly by the display hardware to generate visual imagery. Each of the fragments includes x and y window coordinates, a color, and a depth for use with the z-buffer for hidden surface elimination. Finally, *per-fragment* operations include comparing the fragment's depth value to the value stored in the z-buffer and, if the comparison is successful, replacing the color and depth values in the frame buffer with the fragment's values.

The performance demanded of such a system can be substantial: 1 million triangles per second or hundreds of millions of fragments per second. The calculations involved in performing this work easily require billions of operations per second. Since none of today's fastest general purpose processors can satisfy these demands, all modern high-performance graphics systems are run on parallel architectures. Figure 8-3 is a general representation of a parallel architecture, in which the rendering operation of Figure 8-2 is simply replicated. Whereas such an architecture is attractively simple to implement, it fails to solve the rendering problem, because primitives in object coordinates cannot be easily separated into groups corresponding to different subregions of the frame buffer. There is in general a many-to-many mapping between the primitives in object coordinates and the partitions of the frame buffer.

To allow for this many-to-many mapping, disjoint parallel rendering pipes must be combined at a minimum of one point along their paths, and this point must come after the per-primitive operations are completed. The point or crossbar can be located prior to the rasterization (the primitive crossbar), between rasterization and per-fragment (the fragment

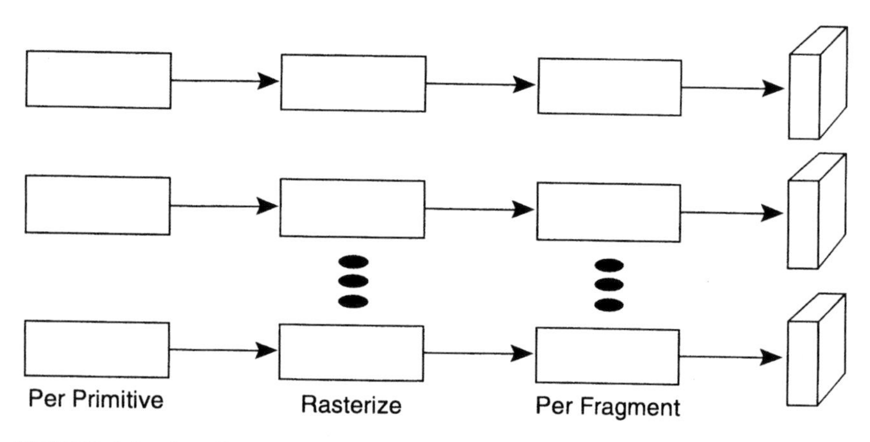

FIGURE 8-3 Parallel graphics pipelines.

crossbar), and following pixel merge (the pixel merge crossbar). A detailed discussion of these architectures is provided in the technical appendix to this chapter. There are four major graphics systems that represent different architectures based on crossbar location. Silicon Graphics RealityEngine is a flow-through architecture with a primitive crossbar; the Freedom series from Evans & Sutherland is a flow-through architecture with a fragment crossbar; Pixel Planes 5 uses a tiled primitive crossbar; and PixelFlow is a tiled, pixel merge machine.

Ordered rendering has been presented to help clarify a significant distinction in graphics architectures; however, it is not the only significant factor for VE rendering. Other primary issues for VE rendering are image quality, performance, and latency. Measured by these metrics, RealityEngine and PixelFlow are very effective VE machines architecturally. Freedom and Pixel Planes 5 are less suitable, though still useful.

Computation and Data Management Issues in Visual Scene Generation

Many important applications of VE require extensive computational and data management capabilities. The computations and data in the application primarily support the tasks taking place in the application. For example, in simulation, the computations may support the physical behavior of objects in the VE, while in a visualization application the computations may support the extraction of interesting features from a complex precomputed dataset. Such computations may require on the order of millions of floating point operations. Simulations currently demand only modest data management capabilities but, as the complexity of simulations increases, the data supporting them may increase. Visualization applications, in contrast, often demand a priori unpredictable access to gigabytes of data (Bryson and Gerald-Yamasaki, 1992). Other types of applications can have similar demands. As computer power increases, more ambitious computational demands will be made. For example, an application may someday compute a fluid flow solution in real time to high accuracy. Such computations can require trillions of floating point operations.

An Example: The Virtual Wind Tunnel

In this section, we consider the implications of the VE performance constraints on the computation and data management requirements of a VE system. An example of an application that is both computationally intensive and works with large numbers of data is the virtual wind tunnel (Bryson and Gerald-Yamasaki, 1992). A modest modern problem in the virtual wind tunnel is the visualization of a precomputed dataset that gives five values (one for energy, one for density, and three for the velocity vector) at 3 million points at a time, for 106 times. This dataset is a total of 5.3 Gbytes in size, with each time step being about 50 Mbytes. If the virtual wind tunnel is to allow the user to interactively control the time-varying visualization of this dataset, each time step must be loaded, and the visualizations must be computed. Assuming that 10 time steps must be loaded per second, a data bandwidth of 500 Mbytes per second is required. The computations involved depend on the visualization technique. For example, the velocity vector field can be visualized by releasing simulated particles into the flow, which implies a computation requiring about 200 floating point operations per particle per time step. A typical visualization requires thousands of such particles and hundreds of thousands of floating point operations. The computation problem expands further as such visualizations are combined with other computationally intensive visualization techniques, such as the display of isosurfaces. It is important to stress that this example is only of modest size, with the size and complexity of datasets doubling every year or so.

It is quite difficult to simultaneously meet the VE performance constraints and the data management requirements in the above example. There are two aspects to the data management problem: (1) the time required to find the data in a mass storage device (seek time), which results in delays, and (2) the time required to read the data (bandwidth). The seek time can range from minutes in the case of data stored on tape through a few hundred thousandths of a second in the case of data stored on disk, to essentially nothing for data stored in primary memory. Bandwidths range from a few megabytes per second in the case of tapes and disk to on the order of a hundred megabytes per second for RAID disks and physical memory. Disk bandwidth is not expected to improve significantly over the next few years.

Support is needed to meet the requirements of VE applications for real-time random access to as much as several gigabytes (Bryson and Gerald-Yamasaki, 1992). Whereas for some visualization techniques, only a small number of data will be addressed at a time, a very large number of such accesses may be required for data that are scattered over the file on disk. Thus the seek time of the disk head becomes an important issue. For other visualization techniques (such as isosurfaces or volume rendering), many tens of megabytes of data may be needed for a single computation. This implies disk bandwidths of 300 to 500 Mbytes/s in order to maintain a 10 Hz update rate, an order of magnitude beyond current commercial systems. For these types of applications, physical memory is the only viable storage medium for data used in the environment. Workstations are currently being released with as much as 16 Gbytes of memory, but the costs of such large amounts of memory are currently prohibitive. Furthermore, as computational science grows through the increase in supercomputer power, datasets will dramatically increase in size. Another source of large datasets will be the Earth Observing Satellite, which will produce datasets in the terabyte range. This large number of data mandates very fast massive storage devices as a necessary technology for the application of VEs to these problems.

Strategies for Meeting Requirements

One strategy of meeting the data management requirements is to observe that, typically, only a small fraction of the data is actually used in an application. In the above particle injection example, only 16 accesses are required (with each access loading a few tens of bytes) per particle per time step. These accesses are scattered across the dataset in unpredictable ways. The bandwidth requirements of this example are trivial if only the data actually used are loaded, but the seek time requirements are a problem: 20,000 particles would require 320,000 seeks per time step or 3.2 million seeks per second. This is two orders of magnitude beyond the seek time capabilities of current disk systems.

Another way to address the data size problem is to develop data compression algorithms. The data will be decompressed as they are used, trading off reduced data size for greater computational demands. Different application domains will make different demands of compression algorithms: image data allow "lossy" compression, in which the decompressed data will be of a slightly lower fidelity than the original; scientific data cannot allow lossy compression (as this would introduce incorrect artifacts into the data) but will perhaps allow multiresolution compression algorithms, such as wavelet techniques. The development of appropriate data compression techniques for many application domains is an open area of research.

Another strategy is to put as much of the dataset as possible in physical memory. This minimizes the seek time but restricts the number of data that may be investigated. This restriction will be relieved as workstation memories increase (see Figure 8-4). Datasets, however, are expected to grow radically as the available computational power increases.

Computational requirements can be similarly difficult to meet. The above example of injecting 20,000 particles into a flow requires 4 million floating point operations, implying a computational performance of 40 million floating point operations per second (or 40 Mflops) just to compute the particle visualization. Such an application will often use several such visualizations simultaneously. As more computational power becomes available, we may wish to include partial differential equation solvers, increasing the computational requirements by several orders of magnitude.

There are many ways in which supercomputer systems have attained very high computational speeds, but these methods typically work only for special computations. For example, Cray supercomputers rely on a vectorized architecture, which is very fast for array-type operations but is

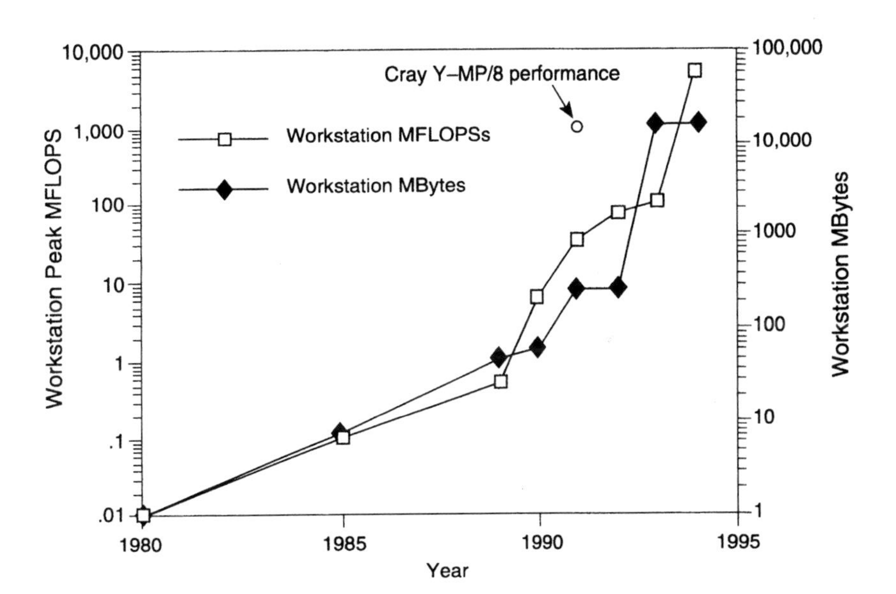

FIGURE 8-4 History of workstation computation and memory.

not so fast as for the particle example discussed above. Another example is the massively parallel system, which distributes memory and computation among many processors. Massively parallel systems are very fast for some applications, but are slow for computations that are not parallelizable or require large amounts of data movement. In a VE system, many kinds of computations may be required, implying that a unique computational architecture typically will be unsuitable. To maximize versatility, computations in VE systems should be based on a few parallel high-power scalar processors with large shared memory.

As Figure 8-4 shows, workstation computational power is increasing dramatically. It is expected that in 1994 workstations will be available that will match the computational power of the supercomputers of 1992.

The run-time software architecture of the VE is an area of serious concern. There are two run-time models that are currently common in computer graphics: the *simulation loop* model, in which all operations in the visualization environment (including interaction, computation, data management, and graphics) are performed in a repeated single loop; and the *event-driven* model, in which operations occur in response to various events (usually generated by the user). Neither model is attractive for large VEs.

The time required for a single loop in the simulation loop model may, due to the combination of data management, computation, and graphics, exceed the VE performance constraints. This is a particularly severe problem if these various operations are performed in sequence, drawing each frame only after the entire computation has been completed. This can lead to very low frame rates both with respect to display and interaction, which is unacceptable in a VE system. For multiprocessing systems, one answer is to put the computation and data management in one process while the graphics is in another, asynchronously running process. Then the graphics can be performed as fast as possible even though the computations may take much longer times. For multiprocessor systems, the computation can be parallelized as well, in which all computation takes place on as many processors as possible to reduce the overall time required for a computation. This parallel implementation of the computation is still a single loop. The time needed for execution will be determined by the slowest computation in that loop.

The event-driven model is unsuited for VE, as there are many events that may be generated at any one time (including repeated "compute the environment" events that amount to an effective simulation loop), and the time ordering and priority of these events are critical. For example, several user interaction events may occur simultaneously and the priority and meaning of these events will depend on their relationship to one another and their environment. Put more succinctly, the meaning of the events will be context-sensitive and will require the system to interpret the state of the user. This operation will be difficult to do on the basis of an event queue.

An alternative run-time model that is gaining popularity is the *concurrent* model, in which different operations in the environment are running simultaneously with one another, preferably on several processors. The example of the simulation loop broken into the two asynchronously running graphics and computation processes discussed above is a simple example of concurrency. In full concurrency, one may assign a process to each element of the VE. These processes should be implemented as threads or lightweight processes, which are regularly preempted to prevent a single process from taking too much time. Each process would be a small simulation loop, which repeatedly computes and draws its object. The concurrent model has the advantage that slow processes will not block down faster processes. It has the disadvantage that processes requiring very different time scales (fast streamlines versus slow isosurfaces in a visualization application, for example) will not always be in sync. This is a serious problem for time-dependent environments, in which a concurrent implementation may lead to the simultaneous display of, for example, the streamline from one time and the isosurface from another. One can constrain the various processes to stay in sync, but the result would be an environment in which all processes are executed in a time determined by the slowest process (in effect, a parallelized simulation loop).

The choice of run-time architecture will be closely tied to and constrained by the operating system of the computer platform running the VE. In order to allow the parallelization of graphics and computation described above, the operating system should support many lightweight, shared-memory processes, thus minimizing the time required for context switching and interprocess communication. The operating system should be capable of ensuring that high-priority processes (such as the processes handling user tracking) can be serviced at very short and regular intervals. In addition, a synchronous process capability could be provided for various types of simulation computations. A further capability of operating systems that would significantly facilitate the development of VE applications is facilities for time-critical computing and rendering. While it is probably unreasonable to ask the operating system to schedule the time-critical tasks by itself, these facilities should provide the ability for the developer to determine scheduling through tunable parameters. Looking farther into the future, we expect that distributed VE applications will become common. Developing operating systems that make such distribution transparent and easy to implement then becomes high priority.

Another strategy to meet the computation and data management requirements is to distribute the computation and data management to several machines. There are several possible models for such a system. One is to keep all data and perform all computations on a remote supercomputer (Bryson and Gerald-Yamasaki, 1992). This approach is motivated when the local workstation does not have a large amount of computing power or large memory. Another approach is to distribute the data and computations among several computers. In the virtual wind tunnel example, there would be a density machine, which would contain the density data and handle all visualizations of the density field, a velocity machine, which would contain the velocity vector data and handle all visualizations of the velocity vector field, and so on. The resulting visualizations would be collected by the workstation that is handling the graphics and driving the VE interface. These distributed architectures would require fast low-latency networks of the type discussed elsewhere in this document.

There are many occasions on which the computations required to support the VE cannot be done to full accuracy within the VE speed performance constraints. The trade-off between accuracy and speed is a common theme in the design of VE systems. There are occasions in which faster, less accurate computational algorithms are desirable over slower, more accurate algorithms. It is not known at this time how to design these trade-offs into a system in a way that can anticipate all possibilities. Research into how these trade-offs are made is therefore needed. A current strategy is to give users full control over these trade-offs. A related issue is that of time-critical computing, in which a computation returns within a guaranteed time. Designing time-critical computational architectures is an active area of research and is critical to the successful design of VE applications.

Extrapolating current trends, we expect that VE applications will saturate available computing power and data management capabilities for the indefinite future. Dataset size will be the dominant problem for an important class of applications in VE. In the near term, an effective VE platform would include the following: multiple fast processors in an integrated unit; several graphics pipelines integrated with the processors; very large shared physical memory; very fast access to mass storage; operating systems that support shared-memory, multiprocessor architectures; and very high-speed, low-latency networks.

Graphics Capabilities in PC-Based VE Systems

Small VE systems have been successfully built around high-end personal computers (PCs) with special-purpose graphics boards. Notable examples are the W Industries system from England, which uses an Amiga computer controlling auxiliary graphics processors. This system is capable of rendering several hundred polygons at about 15 Hz, and is used extensively in the Virtuality video arcade VE games. The Virtuality systems are networked and allow a few participants to play together in the same environment. Another common example is the use of an IBM-compatible personal computer with the Intel DVI graphics board, which is capable of rendering a few hundred textured polygons at 15-20 Hz.

PC-based VE systems are a natural consequence of the widespread availability of PCs. PC-based systems will provide the public with a taste of virtual reality that will eventually lead to demand for more capable computational and graphics platforms. It is anticipated that, by 1996, systems similar to the entry-level Indy machines from Silicon Graphics should replace the PC-based platforms as the total price of the PC system becomes comparable to that of the Indy. Already there are signs that computer graphics workstation companies are developing RISC CPUs with IBM PC compatibility nodes to simplify this transition.

SOFTWARE FOR THE GENERATION OF THREE-DIMENSIONAL VIRTUAL ENVIRONMENTS

There are many components to the software required for the real-time generation of VEs. These include interaction software, navigation software, polygon flow minimization to the graphics pipeline software, world modeling software (geometric, physical, and behavioral), and hypermedia integration software. Each of these components is large in its own right, and all of them must act in consort and in real time to create VEs. The goal of the interconnectedness of these components is a fully detailed, fully interactive, seamless VE. Seamless means that we can drive a vehicle across a terrain, stop in front of a building, get out of the vehicle, enter the building on foot, go up the stairs, enter a room and interact with items on a desktop, all without delay or hesitation in the system. To build seamless systems, substantial progress in software development is required. The following sections describe the software being constructed in support of virtual worlds.

Interaction Software

Interaction software provides the mechanism to construct a dialogue from various control devices (e.g., trackers, haptic interfaces) and to apply that dialogue to a system or application, so that the multimodal display changes appropriately. The first part of this software involves taking raw inputs from a control device and interpreting them. Several libraries are available both as commercial products and as "shareware" that read the most common interface devices, such as the DataGlove and various trackers. Examples of commercial libraries include World ToolKit by Sense8. Shareware libraries are available from the University of Alberta and other universities. These libraries range in sophistication from serial drivers for obtaining the raw output from the interface devices to routines that include predictive tracking and gesture recognition.

The second part of building interaction software involves turning the information about a system's state from a control device into a dialogue that is meaningful to the system or application, at the same time filtering out erroneous or unlikely portions of dialogue that might be generated by faulty data from the input device. The delivery of this dialogue to the virtual world system is then performed to execute some application-meaningful operation.

Interaction is a critical component of VE systems that involves both hardware and software. Interface hardware in VEs provides the positions or states of various parts of the body. This information is typically used to: (1) map user actions to changes in the environment (e.g., moving objects by hand, etc.), (2) pass commands to the environment (e.g., a hand gesture or button push), or (3) provide information input (e.g., speech recognition for spoken commands, text, or numerical input). The user's intent must be inferred from the output of the hardware as read by the computer system. This inference may be complicated by inaccuracies in the hardware providing the signal.

Existing Technologies

Although there are several paradigms for interaction in VEs, including direct manipulation, indirect manipulation, logical commands, and data input, the problem of realistic, real-time interaction is still comparatively unexplored. Generally, tasks in VEs are performed by a combination of these paradigms. Other paradigms will certainly need to be developed to realize the potential of a natural interface. Below we provide an overview of some existing technologies.

With direct manipulation, the position and orientation of a part of the user's body, usually the hand, is mapped continuously to some aspect of the environment. Typically, the position and orientation of an object in the VE is controlled via direct manipulation. Pointing in order to move is another example of direct manipulation in which orientation information is used to determine a direction in the VE. Analogs of manual tasks such as picking and placing require display of forces as well and therefore are well suited to direct manipulation, though more abstract aspects of the environment, such as background lighting, can also be controlled in this way.

When indirect manipulation is employed, the user performs direct manipulation on an object in the VE, which in turn controls some other aspect of the environment. This is an extension to VE of the concept of a widget, that is, a two-dimensional interface control used in graphics interface design. Thus one may directly manipulate a slider that controls the background color, while direct manipulation of another slider may control the volume of sound output. Several groups, including the University of North Carolina and the National Aeronautics and Space Administration (NASA), have developed this concept with generalizations of menus and sliders to VEs (Holloway et al., 1992; Jacoby, 1992; Conner et al., 1992). The term employed by these groups is *three-dimensional widget*. Creators of three-dimensional widgets go beyond the typical slider and checkboxes of traditional two-dimensional interfaces and attempt to provide task-specific widgets, such as the Computational Fluid Dynamics (CFD) widgets used in the virtual wind tunnel and surface modeling widgets (Bryson, 1992a). Indirect manipulation provides the opportunity to carry out many actions by using relatively few direct manipulation capabilities.

Logical commands detect the state of the user, which is then mapped to initiate some action by the environment. Logical commands are discrete events. The user's state that triggers the command may be detected via buttons, gestures as measured by haptic devices, voice commands, etc. The particular command triggered by a user state may depend on the state of the environment or on the location of parts of the user's body. For example, a point gesture may do different things depending on which virtual object happens to be coincident with the position of the user's hand. Logical commands can also be triggered via indirect manipulation using menus or speech recognizers.

Data or text input can be provided by conventional keyboard methods external to the VE. Within the environment, speech recognition may be used for both text and numerical input, and indirect manipulation of widgets provides limited numerical input.

There are high-level interfaces that should be explored. Research must be performed to explore how to use data measuring the positions of the user's body to interact with a VE in a way that truly provides the richness of real-world interaction. As an example, obvious methods of manipulating a virtual surface via a DataGlove have proven to be difficult to implement (Bryson, 1992b; Snibbe et al., 1992). This example demonstrates that research is needed to determine how user tracking data are to be applied as well as how the objects in the VE are to be defined to provide natural interaction.

In addition, research is needed on the problem of mapping continuous input (body movement) to discrete commands. There are significant segmentation and disambiguation problems, which may require semantic decoding. Since such decoding is application-dependent, the VE user interface cannot easily be separated from the application in the way that it can be with current two-dimensional WIMP (windows, icons, mouse, pointer) interfaces.

Design Approaches and Issues to be Addressed

A crucial decision in designing the interaction is the choice of conceptual approach. Specifically, should researchers focus on ways in which the existing two-dimensional technology might be enriched, or should the starting point be the special attributes and challenges of three-dimensional immersive environments? Some researchers are recreating the two-dimensional graphic user interface (GUI) desktop metaphor in three dimensions by placing buttons and scroll bars in the environment along with the user. While we believe that there is great promise in examining the very successful two-dimensional desktop metaphor as a source for ideas, we also believe that there are risks because of the different sets of problems in the two environments. Relying solely on extensions of our experience with two dimensions would not provide adequate solution approaches to three-dimensional interaction needs, such as flying and navigation or to issues related to body-centered coordinates systems and lines of sight.

Two of the more important issues associated with interacting in a three-dimensional environment are line of sight and acting at a distance. With regard to line of sight, VE applications have to contend with the fact that some useful information might be obscured or distorted due to an unfortunate choice of user viewpoint or object placement. In some cases, the result can lead to misinformation, confusion, and misunderstanding. Common pitfalls include obscuration and unfortunate coincidences.

- *Obscuration* At times, a user must interact with an object that is currently out of sight, hidden behind other objects. How does dealing with this special case change the general form of any user interface techniques we might devise?
- *Unfortunate Coincidences* The archetypical example of this phenomenon is the famous optical illusion in which a person stands on a distant hill while a friend stands near the camera, aligning his hand so that it appears as if the distant friend is a small person standing in the palm of his hand. Such devices, while amusing in some contexts, could under other circumstances, such as air traffic control, prove quite dangerous. Perhaps we should consider alternative methods for warning the user when such coincidences are occurring or for ensuring that the user has enough depth information via parallax to perceive this.

When the user is immersed in a three-dimensional environment, he or she is interacting with objects at a distance. Some are directly within arm's reach, others are not. In each case, there is a question of how to specify the arguments to a particular command—that is, how does a user select and manipulate objects out of the reach envelope and at different distances from the user (that is, in the same room, the same city, across the country)? Will the procedure for distant objects be different from those used in selecting and manipulating nearby objects? Some solutions to the selection problem involve ray casting or voice input, but this leaves open the question of specifying actions and parameters by means of direct manipulation.

Some solutions emphasize a body-centric approach, which relies solely on the user's proprioceptive abilities to specify actions in space. Under this scheme, there is no action at a distance, only operations on objects in close proximity to the user. This approach requires one of three solutions: translate the user's viewpoint to within arm's reach of the object(s) in question, scale the user so that everything of interest is within arm's reach, or scale the entire environment so that everything is within arm's reach.

The first solution has several drawbacks. First, by moving the user over significant distances, problems in orientation could result. Next, moving objects quickly over great distances can be difficult (moving an object from Los Angeles to New York would require that the user fly this distance or that the user have a point-and-click, put-me-there interface with a global map). Finally, moving close to an object can destroy the spatial context in which that move operation is taking place. The second and third solutions are completely equivalent except when other participants or spectators are also in the environment.

Perhaps the most basic interaction technique in any application is object selection. Object selection can be implicit, as happens with many direct manipulation techniques on the desktop (e.g., dragging a file to the Mac trash can), or it can be explicit, as in clicking on a rectangle in any common GUI drawing package to activate selection handles for resizing. It is interesting to note that most two-dimensional user interface designers use the phrase "highlight the selected object," to mean "draw a marker, such as selection handles" on the selected object. With VE systems, we have the ability to literally highlight the selected object. Most examples thus far have used three-dimensional extensions of two-dimensional highlighting techniques, rather than simply doing what the term implies; applying special lighting to the selected object.

The following list offers some potentially useful selection techniques for use in three-dimensional computer-generated environments:

• *Pointing and ray casting.* This allows selection of objects in clear view, but not those inside or behind other objects.

• *Dragging.* This is analogous to "swipe select" in traditional GUIs. Selections can be made on the picture plane with a rectangle or in an arbitrary space with a volume by "lassoing." Lassoing, which allows the user to select a space of any shape, is an extremely powerful technique in the two-dimensional paradigm. Carrying this idea over to three dimensions requires a three-dimensional input device and perhaps a volume selector instead of a two-dimensional lasso.

• *Naming.* Voice input for selection techniques is particularly important in three-dimensional environments. "Delete my chair" is a powerful command archetype that we should not ignore. The question of how to manage naming is extremely important and difficult. It forms a subset of the more general problem of naming objects by generalized attributes.

• *Naming attributes.* Specifying a selection set by a common attribute or set of attributes ("all red chairs with arms") is a technique that should be exploited. Since some attributes are spatial in nature, it is easy to see how these might be specified with a gesture as well as with voice, offering a fluid and powerful multimodal selection technique: all red chairs, shorter than this [user gestures with two hands] in that room [user looks over shoulder into adjoining room].

For more complex attribute specification, one can imagine attribute editors and sophisticated three-dimensional widgets for specifying attribute values and ranges for the selection set. Selection by example is another possibility: "select all of these [grabbing a chair]." All of the selection techniques described above suffer from being too inclusive. It is important to provide the user with an opportunity to express "but not that one" as a qualification in any selection task. Of course, excluding objects is itself a selection task.

An important aspect of the selection process is the provision of feedback to the user confirming the action that has been taken. This is a more difficult problem in three dimensions, where we are faced with the graphic arts question of how to depict a selected object so that it appears unambiguously selected from an arbitrary viewing angle, under any lighting circumstances, regardless of the rendering of the object.

Another issue is that of extending the software to deal with two-handed input. Although manipulations with two hands are most natural for many tasks, adding a second pointing device into the programming loop significantly complicates the programmer's model of interaction and object behavior and so has been rarely seen in two-dimensional systems other than research prototypes. In three-dimensional immersive environments, however, two-handed input becomes even more important, as individuals use both gestures and postures to indicate complex relationships between objects and operations in space.

If an interface is poorly designed, it can lull the user into thinking that options are available when in fact they are not. For example, current immersive three-dimensional systems often depict models of human hands in the scene when the user's hands are being tracked. Given the many kinds of actions that human hands are capable of, depicting human hands at all times might suggest to users that they are free to perform any action they wish—yet many of these actions may exceed the capabilities of the current system. One solution to this problem is to limit the operations that are possible with bare hands, specifying for more sophisticated operations the use of tools. A thoughtful design would depict tools that suggest their purpose, so that, like a carpenter with a toolbox, the user has an array of virtual tools with physical attributes that suggest certain uses. Cutting tools might look like saws or knives, while attachment tools might look like staplers. This paradigm melds together issues of modality with voice, context, and command.

Interaction techniques and dialogue design have been extremely important research foci in the development of effective two-dimensional interfaces. Until recently, the VE community has been occupied with getting any input to work, but it is now maturing to the point that finding common techniques across applications is appropriate. These common techniques are points of leverage: by encapsulating them in reusable software components, we can hope to build VE tools similar to the widget, icon, mouse, pointer (WIMP) application builders that are now widely in use for two-dimensional interfaces. It should also be noted that the progress made in three-dimensional systems should feed back into two-dimensional systems.

Visual Scene Navigation Software

Visual scene navigation software provides the means for moving the user through the three-dimensional virtual world. There are many component parts to this software, including control device gesture interpretation (gesture message from the input subsystem to movement processing), virtual camera viewpoint and view volume control, and hierarchical data structures for polygon flow minimization to the graphics pipeline. In navigation, all act together in real time to produce the next frame in a continuous series of frames of coherent motion through the virtual world. The sections below provide a survey of currently developed navigation software and a discussion of special hierarchical data structures for polygon flow.

Survey of Currently Developed Navigation Software

Navigation is the problem of controlling the point and direction of view in the VE (Robinett and Holoway, 1992). Using conventional computer graphics techniques, navigation can be reduced to the problem of determining a position and orientation transformation matrix (in homogeneous graphics coordinates) for the rendering of an object. This transformation matrix can be usefully decomposed into the transformation due to the user's head motion and the transformation due to motions over long distance (travel in a virtual vehicle). There may also be several virtual vehicles concatenated together.

The first layer of virtual world navigation is the most specific: the individual's viewpoint. One locally controls one's position and direction of view via a head tracking device, which provides the computer with the position and orientation of the user's head.

The next layer of navigation uses the metaphor of a virtual vehicle, which allows movement over distances in the VE greater than those distances allowed by the head-tracker alone. The position and orientation of the virtual vehicle can be controlled in a variety of ways. In simulation applications, the vehicle is controlled in the same way that an actual simulated vehicle would be controlled. Examples that have been implemented are treadmills and bicycles and joysticks for flight or vehicle simulators. For more abstract applications, there have been several experimental approaches to controlling the vehicle. The most common is the point and fly technique, wherein the vehicle is controlled via a direct manipulation interface. The user points a three-dimensional position and orientation tracker in the desired direction of flight and commands the environment to fly the user vehicle in that direction. Other methods of controlling the vehicle are based on the observation that in VE one need not get from here to there through the intervening space. Teleoperation is one obvious example, which often has the user specify a desired destination and then "teleports" the user there. Solutions have included portals that have fixed entry and exit locations, explicit specification of destination through numerical or label input, and the use of small three-dimensional maps of the environment to point at the desired destination. Another method of controlling the vehicle is dynamic scaling, wherein the entire environment is scaled down so that the user can reach the desired destination, and then scaled up again around the destination indicated by the user. All of these methods have disadvantages, including difficulty of control and orientation problems.

There is a hierarchy of objects in the VE that may behave differently during navigation. Some objects are fixed in the environment and are acted on by both the user and the vehicle. Other objects, usually virtual tools that the user will always wish to have within reach, will be acted on by the head transformation only. Still other objects, such as data displays, are always desired within the user's field of view and are not acted on by either the user or the vehicle. These objects have been called variously *world stable, vehicle stable,* and *head stable* (Fisher et al., 1986). Although most of the fundamental mathematics of navigation software are known, experimentation remains to be done.

Survey of Hierarchical Data Structure Techniques for Polygon Flow Minimization

Hierarchical data structures for the minimization of polygon flow to the graphics pipeline are the back end of visual scene navigation. When we have generated a matrix representing the chosen view, we then need to send the scene description transformed by that matrix to the visual display. One key method to get the visual scene updated in real time at interactive update rates is to minimize the total number of polygons sent to the graphics pipeline.

Hierarchical data structures for polygon flow minimization are probably the least well understood aspect of graphics development. Many people buy workstations, such as the Silicon Graphics, that promise 2 million polygons/s and expect to be able to create realistic visual scenes in virtual worlds. This is a very common misconception. Visual reality has been said to consist of 80 million polygons per picture (Catmull et al., 1984). Extending this to the VE need for 10 frames/s minimum, 800 million polygons/s are needed.

Today, as noted above, workstations are advertised to have the capability to process approximately 2 to 3 million polygons/s (flat shaded, nontextured). If textured scenes are desired, the system will run slower at approximately 900,000 textured polygons/s. We expect to see 10 to 25 percent of this advertised performance, or 225,000 textured polygons/s. At 10 frames/s, this is 22,500 polygons per frame or 7,500 textured polygons at 30 frames/s (7,500 polygons is not a very detailed world).

The alternatives are to live with worlds of reduced complexity or to off-load some of the graphics work done in the pipeline onto the multiple CPUs of workstations. All polygon reduction must be accomplished in less time than it takes just to send the polygons through the pipeline. The difficulty of polygon flow minimization depends on the composition of the virtual world. This problem has historically been approached on an application-specific basis, and there is as yet no general solution. Current solutions usually involve partitioning the polygon-defined world into volumes that can readily be checked for visibility by the virtual world viewer. There are many partitioning schemes—some of which work only if the world description does not change dynamically (Airey et al., 1990).

A second component of the polygon flow minimization effort is the pixel coverage of the object modeled. Once an object has been determined to be in view, the secondary question is how many pixels that object will cover. If the number of pixels covered by an object is small, then a reduced polygon count (low-resolution) version of that object can be rendered. This results in additional software complexity, again software that must run in real time. Because the level-of-detail models are precomputed, the issue is greater dataset size rather than level selection (which is nearly trivial).

The current speed of z-buffers alone means we must carefully limit the polygons sent through the graphics pipeline. Other techniques that use the CPUs to minimize polygon flow to the pipeline are known for specific applications, but those techniques do not solve the problem in general.

In a classic paper, Clark (1976) presents a general approach for solving the polygon flow minimization problem by stressing the construction of a hierarchical data structure for the virtual world (Figure 8-5). The approach is to envision a world database for which a bounding volume is known for each drawn object. The bounding volumes are organized hierarchically, in a tree that is used to rapidly discard large numbers of polygons. This is accomplished by testing the bounding volumes to determine whether they are contained or partially contained in the current orientation of the view volume. The process continues recursively until a node is reached for which nothing underneath it is in the view volume.

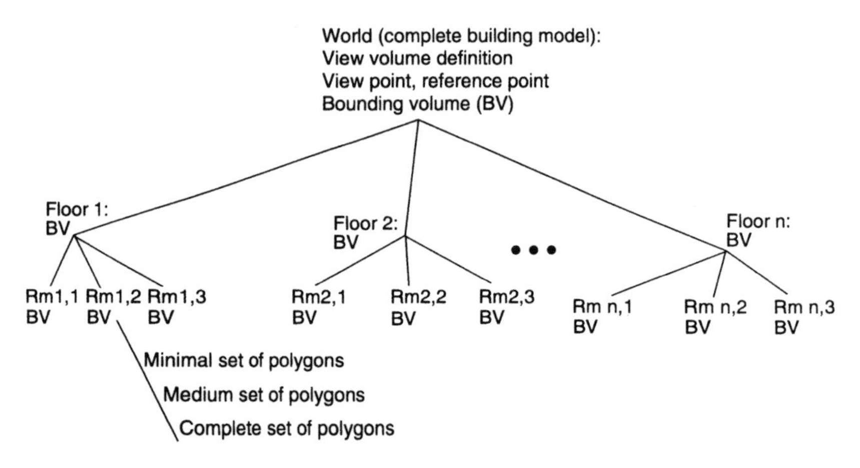

FIGURE 8-5 Hierarchical data structure for polygon flow minimization.

This part of the Clark paper provides a good start for anyone building a three-dimensional VE for which the total number of polygons is significantly larger than the hardware is capable of drawing.

The second part of Clark's paper deals with the actual display of the polygons in the leaf nodes of the tree. The idea is to send only minimal descriptions of objects through the graphics pipeline (minimal based on the expected final pixel coverage of the object). In this approach, there will be multiple-resolution versions of each three-dimensional object and software for rapidly determining which resolution to draw. The assumption of multiple-resolution versions of each three-dimensional object being available is a large one, with automatic methods for their generation remaining an open issue. Other discussions of this issue are found in DeHaemer and Zyda (1991), Schroeder et al. (1992), and Turk (1992).

Application-Specific Solutions

Polygon flow minimization to the graphics pipeline is best understood by looking at specific solutions. Some of the more interesting work has been done by Brooks at the University of North Carolina at Chapel Hill with respect to architectural walkthrough (Brooks, 1986; Airey et al., 1990). The goal in those systems was to provide an interactive walkthrough capability for a planned new computer science building at the university that would offer visualization of the internal spaces of that building for the consideration of changes before construction.

The walkthrough system had some basic tenets. The first was that the architectural model would be constructed by an architect and passed on to the walkthrough phase in a fixed form. A display compiler would then be run on that database and a set of hierarchical data structures would be output to a file. The idea behind the display compiler was that the building model was fixed and that it was acceptable to spend some 45 minutes in computing a set of hierarchical data structures. Once the data structures were computed, a display loop could then be entered, in which the viewpoint could be rapidly changed. The walkthrough system was rather successful, but it has the limitation that the world cannot be changed without rerunning the display compiler. Other walkthrough systems have similar limitations (Teller and Sequin, 1991; Funkhouser et al., 1992).

Real-time display of three-dimensional terrain is a well-researched area that originated in flight simulation. Terrain displays are an interesting special case of the polygon flow minimization problem in that they are relatively well worked out and documented in the open literature (Zyda et al., 1993a). The basic idea is to take the terrain grid and generate a quadtree structure containing the terrain at various display resolutions. The notion of the grid cell is used for reducing polygon flow by drawing only those objects whose cell is also to be drawn (assuming the grid cell on which the object lies is known). This strategy works well for ground-based visual displays; a more comprehensive algorithm is required for air views of such displays.

World Modeling

Models that define the form, behavior, and appearance of objects are the core of any VE. A host of modeling problems are therefore central to the development of VE technology. An important technological challenge of multimodal VEs is to design and develop object representation, simulation, and rendering (RSR) techniques that support visual, haptic, and auditory interactions with the VE in real time. There are two major approaches to the RSR process. First, a unified central representation may be employed that captures all the geometric, surface, and physical properties needed for physical simulation and rendering purposes. In principle, methods such as finite element modeling could be used as the basis for representing these properties and for physical simulation and rendering purposes. At the other extreme, separate, spatially and temporally coupled representations could be maintained that represent only those properties of an object relevant for simulating and rendering interactions in a single modality (e.g., auditory events). The former approach is architecturally the most elegant and avoids issues of maintaining proper spatial and temporal correlation between the RSR processes for each modality. Practically, however, the latter approach may allow better matching between modality-specific representation, simulation, and rendering streams. The abilities and limitations of the human user and the VE system for each of the modalities impose unique spatial (e.g., scale and resolution) and temporal (e.g., device update rate and duration) constraints. For example, it is likely that the level of detail and consequently the nature of approximations in the RSR process will be different for each of the modalities. It is unclear, therefore, whether these modality-specific constraints can be met by systems based on a single essential or core representation and still operate in real time.

The overwhelming majority of VE-relevant RSR research and development to date has been on systems that are visually rendered (e.g., Witkin and Welch, 1990). The theoretical and practical issues associated with either of the two RSR approaches or variants in a multimodal context, however, have received minimal attention but are likely to become a major concern for VE system researchers and developers. For example, geometric modeling is relevant to the generation of acoustic environments (i.e., room modeling) as well as visual environments, and the development of physical models is critical to the ability to generate and modulate auditory, visual, and haptic displays. Novel applications such as the use of auditory displays for the understanding of scientific data (e.g., Kramer, 1992; Blattner et al., 1989) require models that may not be physically based. In this section, we concentrate on the visual domain and examine the problems of constructing geometric models, the prospects for vision-based acquisition of real-world models, dynamic model matching for augmented reality, the simulation of basic physical behavior, and simulation of autonomous agents. Parallel issues are involved in the other modalities and are discussed in Chapters 3 and 4.

Geometric Modeling: Construction and Acquisition

The need to build detailed three-dimensional geometric models arises in computer-aided design (CAD), in mainstream computer graphics, and in various other fields. Geometric modeling is an active area of academic and industrial research in its own right, and a wide range of commercial modeling systems is available. Despite the wealth of available tools, modeling is generally regarded as an onerous task. Among the many factors contributing to this perception are sluggish performance, awkward user interfaces, inflexibility, and the low level at which models must typically be specified. It is symptomatic of these difficulties that most leading academic labs, and many commercial animation houses (such as Pixar and PDI), prefer to use in-house tools, or in some cases a patchwork of homegrown and commercial products.

From the standpoint of VE construction, geometric modeling is a vital enabling technology whose limitations may impede progress. As a practical matter, the VE research community will benefit from a shared *open* modeling environment, a modeling environment that includes physics. In order to understand this, we need to look at how three-dimensional geometric models are currently acquired. We do this by looking at how several VE efforts have reported their model acquisition process.

Geometric models for VEs are typically acquired through the use of a PC-based, Macintosh-based, or workstation-based CAD tool. If one reads the work done for the walkthrough project at the University of North Carolina (Airey et al., 1990), one finds that AutoCAD was used to generate the 12,000+ polygons that comprised the Orange United Methodist Church. In the presentation of that paper, one of the problems discussed was that of "getting the required data out of a CAD program written for other purposes." Getting the three-dimensional geometry out of the files generated by AutoCAD was not difficult, but there was a problem in that not all of the data required were present in the form needed for the VE walkthrough. In particular, data related to the actual physics of the building were not present, and partitioning information useful to the real-time walkthrough algorithms had to be added later "by hand" or "back fed in" by specially written programs.

The VPL Reality Built for Two (RB2) system (Blanchard et al., 1990) used a Macintosh II as its design station for solid modeling and an IRIS workstation as its rendering/display station. RB2 is a software development platform for designing and implementing real-time VEs. Development under RB2 is rapid and interactive, with behavior constraints and interactions that can be edited in real time. The geometric modeling function of RB2 was provided by a software module called RB2 Swivel and a data flow/real-time animation control package called Body Electric. RB2 has a considerable following in organizations that do not have sufficient resources to develop their own in-house VE expertise. RB2 is a turnkey system, whose geometric and physics file formats are proprietary.

In the NPSNET project (Zyda et al., 1992), the original set of three-dimensional icons used was acquired from the SIMNET databases. These models were little more than three-dimensional skins of the weapons systems known to SIMNET. As a result, project researchers have developed an open format for storing these three-dimensional models (Zyda et al., 1993a), added physics to the format (Zyda et al., 1992), and have rewritten the system to include object-oriented animation capabilities (Wilson et al., 1992). For example, at the Naval Postgraduate School, the NPSNET Research Group is currently using Software Systems' expensive and proprietary MultiGen CAD tool for the development of physics-free models for its SGI Performer-based NPSNET-4 system. Computer-aided design systems with retrofitted physics are beginning to be developed (e.g., Deneb Robotics and Parametric Technologies), but these systems are expensive and proprietary.

Many applications call for VEs that are replicas of real ones. Rather than building such models by hand, it is advantageous to use visual or other sensors to acquire them automatically. Automatic acquisition of complex environment models (such as factory environments) is currently not practical but is a timely research issue. Meanwhile, automatic or nearly automatic acquisition of geometric models is practical now in some cases, and partially automated interactive acquisition should be feasible in the near term (Ohya et al., 1993; Fuchs et al., 1994).

The most promising short-term approaches involve active sensing techniques. Scanning laser finders and light-stripe methods are both capable of producing *range images* that encode position and shape of surfaces that are visible from the point of measurement. These active techniques offer the strong advantage that three-dimensional measurements may be made directly, without the indirect inferences that passively acquired images require. Active techniques do, however, suffer from some limitations: because sensor-to-surface distances must be relatively small, they are not applicable to large-scale environments. Surfaces that are nonreflective or obliquely viewed shiny surfaces may not return enough light to allow range measurements to be made. Noise is enough of a problem that data must generally be cleaned up by hand. A more basic problem is that a single range image contains information only about surfaces that were visible from a particular viewpoint. To build a complete map of an environment, many such views may be required, and the problem of combining them into a coherent whole is still unsolved.

Among passive techniques, stereoscopic and motion-based methods, relying on images taken from varying viewpoints, are currently most practical. However, unlike active sensing methods, these rely on point-to-point matching of images in order to recover distance by triangulation. Many stereo algorithms have been developed, but none is yet robust enough to compete with active methods. Methods that rely on information gleaned from static monocular views—edges, shading, texture, etc.—are less effective.

For many purposes, far more is required of an environment model than just a map of objects' surface geometry. If the user is to interact with the environment by picking things up and manipulating them, information about objects' structure, composition, attachment to other objects, and behavior is also needed. Unfortunately, current vision techniques do not even begin to address these deeper issues.

Dynamic Model Matching and Augmented Reality

The term *augmented reality* has come to refer to the use of transparent head-mounted displays that superimpose synthetic elements on a view of the real surroundings. Unlike conventional heads-up displays in which the added elements bear no direct relation to the background, the synthetic objects in augmented reality are supposed to appear as part of the real environment. That is, as nearly as possible, they should interact with the observer and with real objects, as if they too were real.

At one extreme, creating a full augmented-reality illusion requires a complete model of the real environment as well as the synthetic elements. For instance, to place a synthetic object on a real table and make it appear to *stay* on the table as the observer moves through the environment, we would need to know just where the table sits in space and how the observer is moving. For full realism, enough information about scene illumination and surface properties to cast synthetic shadows onto real objects would be needed. Furthermore, we would need enough information about three-dimensional scene structure to allow real objects to hide or be hidden by synthetic ones, as appropriate. Naturally, all of this would happen in real time, in response to uncontrolled and unpredictable observer motions.

This sort of mix of the real and synthetic has already been achieved in motion picture special effects, most notably, Industrial Light and Magic's effects in films such as *The Abyss* and *Terminator 2*. Some of these effects were produced by rendering three-dimensional models and creating a composite of the resulting images with live-action frames, as would be required in augmented reality. However, the process was extremely slow and laborious, requiring manual intervention at every step. After scenes were shot, models of camera and object motions were extracted manually, using frame-by-frame manual measurement along with considerable trial and error. Even small geometric errors were prone to destroy the illusion, making the synthetic objects appear to float outside the live scene.

Automatic generation of augmented-reality effects is still a research problem in all but the least demanding cases. The two major issues are: (1) accurate measurement of observer motions and (2) acquisition and maintenance of scene models. The prospects for automatic solutions to the latter were discussed above. If the environment is to remain static, it would be feasible to build scene models off-line using interactive techniques. Although VE displays provide direct motion measurements of observer movement, these are unlikely to be accurate enough to support high-quality augmented reality, at least when real and synthetic objects are in close proximity, because even very small errors could induce perceptible relative motions, disrupting the illusion. Perhaps the most promising course would use direct motion measurements for gross positioning, using local image-based matching methods to lock real and synthetic elements together.

Physical Simulation for Visual Displays

In order to give solidity to VEs and situate the user firmly in them, virtual objects, including the user's image, need to behave like real ones. At a minimum, solid objects should not pass through each other, and things should move as expected when pushed, pulled, or grasped.

Analysis of objects' behavior at the scale of everyday observation lies in the domain of classic mechanics, which is a mature discipline. However, mechanics texts and courses are generally geared toward providing insight into objects' behavior, whereas to support VE the behavior itself is of paramount importance—insight strictly optional. Thus classic treatments may provide the required mathematical underpinnings but do not directly address the problem at hand.

Simulations of classic mechanics are extensively used as aids in engineering design and analysis. Although these traditional simulations do yield numerical descriptions of behavior, they still do not come close to meeting the needs of VEs. In engineering practice, simulation is a long, drawn-out, and highly intellectualized activity. The engineer typically spends much time with pencil and paper developing mathematical models for the system under study. These are then transferred to the simulation software, often with much tweaking, and parameter selection. Only then can the simulation actually be run. As a design evolves, the initial equations must be modified and reentered and the simulation rerun.

In strong contrast, a mechanical simulation for VEs must run reliably, seamlessly, automatically, and in real time. Within the scope of the world being modeled, any situation that could possibly arise must be handled correctly, without missing a beat. In the last few years, researchers in computer graphics have begun to address the unique challenges posed by this kind of simulation, under the heading of *physically based modeling*. Below we summarize the main existing technology and outstanding issues in this area.

Solid Object Modeling Solid objects' inability to pass through each other is an aspect of the physical world that we depend on constantly in everyday life: when we place a cup on a table, we expect it to rest stably on the table, not float above or pass through it.

In reaching and grasping, we rely on solid hand-object contact as an aid (as do roboticists, who make extensive use of force control and compliant motion). Of course, we also rely on contact with the ground to stand and locomote.

The problem of preventing interpenetration has three main parts. First, collisions must be detected. Second, objects' velocities must be adjusted in response to collisions. Finally, if the collision response does not cause the objects to separate immediately, contact forces must be calculated and applied until separation finally occurs.

Collision detection is most frequently handled by checking for object overlaps each time position is updated. If overlap is found, a collision is signaled, the state of the system is backed up to the moment of collision, and a collision response is computed and applied. The bulk of the work lies in the geometric problem of determining whether any pair of objects overlap. This problem has received attention in robotics, in mechanical CAD, and in computer graphics. Brute force overlap detection for convex polyhedra is a straightforward matter of testing each vertex of every object against each face of every other object. More efficient schemes use bounding volumes or spatial subdivision to avoid as many tests as possible. Good general methods for objects with curved surfaces do not yet exist.

In fact, checking for object overlaps at each update is not sufficient to guarantee noninterpenetration, because objects may have collided and passed through each other between the previous configuration and the new one. This is not merely an esoteric concern, because it means that rapidly moving objects, e.g., projectiles, may pass entirely through thin objects, such as walls, with no collisions ever being detected. Needless to say, large errors can result. Guaranteed methods have been described by Lin and Canny (1992) for the case of convex polyhedra with constant linear and angular velocity.

Collision response involves the application of an impulse and producing an instantaneous change in velocity that prevents interpenetration. The basics of collision response are well treated in classic mechanics and do not pose any great difficulties for implementation. Problems do arise in developing accurate collision models for particular materials, but many VE applications will not require this degree of realism.

To handle continuous multibody contact, it is necessary to calculate the constraint forces that are exchanged at the points of contact and to identify the instants at which contacts are broken. Determining which contacts are breaking is a particularly difficult problem, turning out, as shown by Baraff, to require combinatorial search (Baraff and Witkin, 1992; Baraff, 1989). Fortunately, Baraff also developed reasonably efficient methods that work well in practice.

Many virtual world systems exhibit rigid body motion with collision detection and response (Hahn, 1988; Moore and Wilhelms, 1988; Baraff, 1989; Baraff and Witkin, 1992; Zyda et al., 1993b). Baraff's system also handles multibody continuous contact and frictional forces for curved surfaces. These systems provide many of the essential elements required to support VEs.

Constraints and Articulated Objects In addition to simple objects such as rigid bodies, we should be able to handle objects with moving parts—doors that open and close, knobs and switches that turn, etc. In principle, the ability to simulate simple objects such as rigid bodies, together with the ability to prevent interpenetration, could suffice to model most such compound objects. For instance, a working desk drawer could be constructed by modeling the geometry of a tongue sliding in a groove, or a door by modeling in detail the rigid parts of the hinge. In practice, it is far more efficient to employ direct geometric constraints to summarize the effects of this kind of detailed interaction. For instance, a sliding tongue and groove would be idealized as a pair of coincident lines, one on each object, and a hinge would be represented as an ideal revolute joint.

The simulation and analysis of articulated bodies—jointed assemblies of rigid parts—have been treated extensively, particularly in robotics. In addition to classic techniques such as Lagrangian dynamics, streamlined recursive formulations have been developed, making it possible to simulate forward dynamics of a kinematic chain in linear time, rather than the N^3 time that Lagrangian dynamics requires. These methods only pay off for relatively long chains (N > 9, according to Featherstone) and in their original form do not readily handle closed loops in the graph of part connectivity. Building on the work of Lathrop, Schroeder demonstrated that it is nevertheless feasible to build a "virtual erector set" based on recursive formulations (Schroeder and Zeltzer, 1990).

Another approach to simulating constrained systems of objects builds on the classic method of Lagrangian multipliers, in which a linear system is solved at each time step to yield a set of constraint forces. This approach offers several advantages: first, it is general, allowing essentially arbitrary holonomic constraints to be applied to essentially arbitrary (not necessarily rigid) bodies. Second, it lends itself to on-the-fly construction and modification, an important consideration for VEs. Finally, the constraint matrices that form the linear system are typically sparse, reflecting the fact that everything is not usually connected directly to everything else. Using numerical methods that exploit this sparsity can yield performance that competes with recursive methods. Methods of this kind were used for animation by Platt and by Barzel and Barr (Platt and Barr, 1988; Barzel and Barr, 1988). Witkin et al. (1990) demonstrated a fully interactive snap-together construction system and showed how the constraint equations could be built on the fly and solved in a way that exploits sparsity.

Nonrigid Objects A vast body of work treats the use of finite element methods to simulate continuum dynamics. Most of this work is probably of limited relevance to the construction of conventional VEs, simply because such environments will not require fine-grained nonrigid modeling, with the possible exception of virtual surgery. However, interactive continuum analysis for science and engineering may become an important specialized application of VEs once the computational horsepower is available to support it.

Highly simplified models for flexible-body dynamics are presented by Witkin and Welch (1990), by Pentland and Williams (1989), and by Baraff and Witkin (1992). The general idea of these models is to use only a few global parameters to represent the shape of the whole object, formulating the dynamic equations in terms of these variables. These simplified models capture only the gross deformations of the object but in return provide very high performance. They are probably the most appropriate choice for VEs that require simple nonrigid behavior.

A special form of nonrigid modeling, constituting a potential VE application in itself, is interactive sculpting of free-form surfaces. The general idea is to use simulated flexible materials as a sculpting medium. Flexible thin sheets are employed by Celniker and Gossard (1992) and by Welch and Witkin (1992). Szeliski and Tonnesen (1992) uses clouds of oriented particles to form smooth surfaces.

Motivated by the obvious need in both computer graphics and engineering for realism and physically based environments that support various levels of object detail and interaction (depending on the application), Metaxas (1992, 1993; Metaxas and Terzopoulos, 1992a, 1992b, 1993; Terzopoulos and Metaxas, 1991) developed a general framework for shape and nonrigid motion synthesis, which can also handle rigid bodies as a special case. The framework features a new class of dynamic deformable part models. These models have both global deformation parameters that represent the gross shape of an object in terms of a few parameters and local deformation parameters that represent an object's details through the use of sophisticated finite element techniques. Global deformations are defined by fully nonlinear parametric equations. Hence the models are more general than the linearly deformable ones included in Witkin and Welch (1990) and quadratically deformable ones included in Pentland and Williams (1989). By augmenting the underlying Lagrangian equations' motion with very fast dynamic constraint techniques based on Baumgarte (1972), he adds the capability to compose articulated models (Metaxas, 1992, 1993; Metaxas and Terzopoulos, 1992b) from deformable parts, whose special case for rigid objects is the technique used by Barzel and Barr (1988). Moreover, Metaxas (1992, 1993) also develops fast algorithms for the computation of impact forces that occur during collisions of complex flexible multibody objects with the simulated physical environment.

Issues to be Addressed Most of the essential pieces that are required to imbue VEs with physical behavior have already been demonstrated. Some—notably snap-together constraints and interactive surface modeling—have been demonstrated in fully interactive systems, and others—notably the handling of collision and contact—are only now beginning to appear in interactive systems (recent work by David Baraff at Carnegie Mellon University involves an interactive 2.5-dimensional simulation of noninterpenetrating objects). The most immediate challenge at hand is one of integrating the existing technology into a working system, along with other elements of VE construction software.

Many performance-related issues are still to be addressed, for example, doing efficient collision detection in a large-scale environment (systems with from 500 to 300,000 players or parts) and further accelerating constrained dynamics solutions. In addition, many of the standard numerical techniques are not tuned to real-time systems. For example, the ratio of compute time to real time can vary by orders of magnitude in the simulation of noninterpenetrating bodies, slowing even further when complex contact situations arise. Maintaining a constant frame rate will require the development of new methods that degrade gracefully in such situations.

Autonomous Agents

The need for simulated autonomous agents arises in many VE application areas, such as training, education, and entertainment, in which such agents could play the role of adversaries, trainers, or partners or simply supernumeraries to add richness and believability. Although fully credible simulated humans are the stuff of science fiction, simple agents will often suffice. The construction of simulated autonomous agents draws on a number of technologies, including robotics, computer animation, artificial intelligence, and optimization.

Motion Control Placing an autonomous agent in a virtual physical environment is essentially like placing a robot in a real environment: the agent's body is a physical object that must be controlled to achieve coordinated motion. Fortunately, controlling a virtual agent is much easier than controlling a real one, since many simplifications and idealizations can be made. For example, the agent can be given access to full and perfect information about the state of the world, and many troubling mechanical effects need not arise.

Closed-loop controllers were used to animate virtual agents by McKenna and Zeltzer (1990) and by Miller (1988). More recently, Raibert and Hodgkins (1992) adapted their controller for a real legged robot to the creation of animation. Rather than hand-crafting controllers, Witkin and Kass (1988) solve numerically for optimal goal-directed motion, in an approach that has since been elaborated by Van de Panne et al. (1990) and by Cohen (1992).

Human Figure Simulation In many applications, a VE system must be able to display accurate models of human figures, possibly including a model of the user. Consider training systems, for example. Out-the-window views generated by high-end flight simulators hardly ever need to include images of human figures. But there are many situations in which personnel must cooperate and interact with other crew members. Carrier flight deck operations, small squad training or antiterrorist tactics, for example, require precise coordination of the actions of many individuals for safe and successful execution. VE systems to support training, planning, and rehearsal of such activities must therefore provide computational models of human figures.

We call a computer model of a human figure that can move and function in a VE a *virtual actor*. If the movement of a virtual actor is slaved to the motions of a human using cameras, instrumented clothing, or some other means of body tracking, we call that a *guided virtual actor*, or simply, a *guided actor*. Autonomous actors operate under program control and are capable of independent and adaptive behavior, such that they are capable of interacting with human participants in the VE, as well as with simulated objects and events. In addition to responding to the typed or spoken utterances of human participants, a virtual actor should be capable of interpreting simple task protocols that describe, for example, maintenance and repair operations. Given a set of one or more motor goals—e.g., pick up the wrench and loosen the retaining bolts, or put the book on the desk in my office—a virtual actor should be capable of generating the appropriate motor acts, including necessary and implicit tasks and motor subgoals.

Beyond the added realism that the presence of virtual actors can provide in those situations in which the participants would normally expect to see other human figures, autonomous actors can perform two important functions in VE applications. First, autonomous actors can augment or replace human participants. This will allow individuals to work or train in group settings without requiring additional personnel. Second, autonomous actors can serve as surrogate instructors. VE systems for training, education, and operations rehearsal will incorporate various instructional features, including knowledge-based systems for intelligent computer-aided instruction (ICAI) (Ford, 1985). As ICAI systems mature, virtual actors can provide personae to interact with participants in a VE system.

The required degree of autonomy and realism of simulated human figures will vary, of course, from application to application. However, at the present time, rigorous techniques do not exist for determining these requirements. It should also be noted that autonomous agents need not be literal representations of human beings but may represent various abstractions. For example, the SIMNET system provides for semiautonomous forces that may represent groups of dismounted infantry or single or multiple vehicles that are capable of reacting to simulated events in accordance with some chosen military doctrine. In the remainder of this section, we confine our discussion to simulated human figures, i.e., virtual actors.

In the course of everyday activity, we touch and manipulate objects, make contact with various surfaces, and make contact with other humans either directly, e.g., shaking hands, or indirectly, e.g., two people lifting a heavy object. There are other ways, of course, in which two or more humans may coordinate their motions that do not involve direct contact, for example, crew members on a carrier flight deck who communicate by voice and hand signals. In the computer graphics community, there is a long history of human figure modeling, but this work has considered, for the most part, kinematic modeling of uncoupled motion exclusively.

With today's graphics workstations, kinematic models of reasonably complex figures (say, 30 to 40 degrees of freedom) can be animated in real or near-real time; dynamic simulations cannot. We need to understand in which applications kinematic models are sufficient, and in which applications the realism of dynamic simulation is required.

Action Selection In order to implement autonomous actors that can function independently in a virtual world without the need for interactive control by a human operator, we require some mechanism for selecting and sequencing motor skills appropriate to the actor's behavioral goals and the states of objects— including other actors—in the VE. That is, it is not sufficient to construct a set of behaviors, such as walking, reaching, grasping, and so on. In order to move and function with other actors in a virtual world that is changing over time, an autonomous actor must link perception of objects and events with action. We call this process *motor planning*.

Brooks (1989) has developed and implemented a motor planning mechanism he calls the subsumption architecture. This work is in large part a reaction against conventional notions of planning in artificial intelligence. Brooks argues for a representationless paradigm in which the behavior of a robot is modulated entirely by interaction between perception of the physical environment and the robot's task-achieving behavior modules.

Esakov and Badler (1991) report on the architecture of a simulation-animation system that can handle temporal constraints for task sequencing, rule sets, and resource allocation. No on-line planning was implemented. Task descriptions were initially in the form of predefined animation task keywords. This keyword-based input constraint was subsequently relaxed to allow simple do-this/do-that commands, e.g., "The man should flip (switch) tglJ-1 with his left hand and the woman should move (switch) twF-1 to position 1." Most recently, reactive planning based on sensory perception has been used in locomotion control (Beckett and Badler, 1993), as well as real-time collision avoidance. A high-level task expansion planner (Geib, 1993) creates task-actions that are interpreted by an object-specific reasoner to execute animation behaviors. Recent work by Badler et al. (1991) also involves the exploration of natural language as a means of communicating task descriptions. Badler's new AnimNL project is deeply committed to high-level motion planning (Badler et al., 1993, 1991; Webber et al., 1993); various other motor planning issues have also been studied and published (Badler et al., 1993; Ching and Badler, 1992).

Magnenat-Thalmann and Thalmann (1990, 1991), and Rijpkema and Girard (1991) have reported some work with automated grasping, but their systems seem to be focused on key frame-like animation systems for making animated movies, rather than for real-time interaction with virtual actors. Maiocchi and Pernici (1990) describe the PINOCCHIO system, which is capable of animating realistic character movement derived from recorded human movements. Their system uses limited natural language for describing body configurations, e.g., dance motions. However, this has only limited use in describing interactions with objects in the environment.

Ridsdale (1990) describes the Director's Apprentice, which is intended to interpret film scripts by using a rule-base of facts and relations about cinematic directing. This work was primarily concerned with positioning characters in relation to each other and the synthetic camera, but it did not address the representation and control of autonomous agents. In later work, Ridsdale describes a method of teaching skills to an actor using connectionist learning models (Ridsdale, 1990).

Maes (1990) has developed and implemented an action selection algorithm for goal-oriented, situated robotic agents. Her work is an independent formalization of ideas discussed in earlier work by Zeltzer (1983), with an important extension that accounts for the continuous flow of activation energy among a network of motor skills. Routine, stereotypical behavior is a function of an agent's currently active drives, goals, and motor skills. As a virtual actor moves through and operates in an environment, motor skills are triggered by presented stimuli, and the agent's propensities for executing some behaviors and not others are continually adjusted. The collection of skills and the patterns of excitation and inhibition determine an agent's repertoire of behaviors and flexibility in adapting to changing circumstances.

Populating the World: NPSNET as an Example

One of the key aspects of a virtual world is the population of that world. We define population as the number of active entities within the world. An active entity is anything in the world that is capable of exhibiting a behavior. By this definition, a human-controlled player is an active entity, a tree that is blown up is midway between an active and static entity, and an inert object like a rock is a static entity. In the NPSNET system, all of the active entities have been divided into four general categories based on the control mechanism: expert system, scripting system, network entity, and driven entity. Recently, the term *computer generated forces* (CGF) has been developed to group all entities that are under computer control into a single category. In NPSNET, the entities controlled by both the expert system and scripting system are part of this category. The controlling mechanisms of the expert systems and autonomous players are briefly discussed below.

The expert system is capable of executing a basic behavior when a stimulus is applied to an entity. Within NPSNET it controls those entities that populate the world when there are an insufficient number of human or networked entities to make a scenario interesting. These added entities are called noise entities. The noise entity expert system has four basic behaviors: zig-zag paths, environment limitation, edge of the world response, and fight or flight. These behaviors have been grouped by the stimuli that causes the behavior to be triggered. The zig-zag behavior uses an internal timer to initiate the behavior. Environment limitation and edge of the world response are both dependent on the location of the entity in the database as the source of stimuli. The fight or flight behavior is triggered by external stimuli.

The purpose of an autonomous force is to present an unattended, capable, and intelligent opponent to the human player at the simulator. In NPSNET, the autonomous force is broken down into two components: an observer module that models the observation capabilities of combat forces and a decision module that models decision making, planning, and command and control in a combat force. The autonomous force system employs battlefield information, tactical principles, and knowledge about enemy forces to make tactical decisions directed toward the satisfaction of its overall mission objectives. It then uses these decisions in a reactive planning approach to develop an executable plan for its movements and actions on the battlefield. Its decisions include distribution of multiple goals among multiple assets, route planning, and target engagement. The autonomous force represented in this system consists of a company of tanks. The system allows for cooperation between like elements as well as collaboration between individuals working on different aspects of a task.

The observer module, described by Bhargava and Branley (1993), acts as the eyes and ears of the autonomous force. In the absence of real sensors, the observation module uses probabilistic models and inference rules to generate the belief system of the autonomous force. It accounts for battlefield conditions, as well as the capabilities and knowledge of individual autonomous forces, to determine whether and with how much accuracy various events on the simulated battlefield can be observed. The system converts factual knowledge about the simulated environment into a set of beliefs that might correspond to the beliefs that a real combat force might form under the given conditions. It does so by combining the agent's observations with evidence derived from its knowledge base and inference procedures.

Hypermedia Integration

If one considers three-dimensional VEs as the ideal interface to a spatially organized database, then hypermedia integration is a key technological component. Hypermedia consists of nonsequential media grouped into nodes that are linked to other nodes. If we embed such nodes into a structure in a virtual world, the node can be accessed, and audio or compressed video containing vital information on the layout, design, and purpose of the building can be displayed, along with historical information. Such nodes will also allow us to make a search of all other nodes and find related objects elsewhere in the virtual world.

We also envision hypernavigation, which involves the use of nodes as markers that can be traveled between, either over the virtual terrain at accelerated speeds or over the hypermedia links that connect the nodes. Think of rabbit holes or portals to information populating the virtual world. Hypermedia authoring is another growing area of interest. In authoring mode, the computer places nodes in the VE as a game is played. After the game, the player can travel along these nodes (which exist not only in space but also in time, appearing and disappearing as time passes) and watch a given player's performance in the game. Authoring is especially useful in training and analysis because of this ability to play back the engagement from a specific point of view. Some examples of the uses of hypermedia in virtual worlds are presented in the following paragraphs.

One application is the extension of hypermedia to NPSNET (Zyda et al., 1993a). Hyper-NPSNET combines virtual world technology with hypermedia technology by embedding hypermedia nodes in the terrain of the virtual world. Currently, hypertext is implemented as nonsequential text grouped into nodes that are linked to other text nodes. The NPSNET group also has embedded compressed video (QuickTime and Moviemaker) into its worlds. This video contains captured video of the world being represented geometrically. Thus it provides information not easily represented or communicated by geometry.

In another application, the University of Geneva has a project under way entitled "A Multimedia Testbed" (de Mey and Gibbs, 1993), in which an object-oriented test bed for multimedia is presented. This is a test bed for prototyping distributed multimedia applications. The test application of that software is a virtual museum. The museum is a three-dimensional geometric structure, the Barcelona Pavilion, in which is embedded various multimedia objects, compressed video, audio, and still imagery.

HARDWARE AND SOFTWARE ISSUES TO BE ADDRESSED

In all likelihood, the main short-term research and development effort and commercial payoff in the VE field will involve the refinement of hardware and software related to the representation, simulation, and rendering of visually oriented synthetic environments. This is a natural and logical extension of proven technology and benefits seen in such areas as general simulation, computer-aided design and manufacturing, and scientific visualization.

Nevertheless, the development of multimodal synthetic environments is an extremely important and challenging endeavor. Independent of the fundamental psychophysical issues and device design and development issues, multimodal interactions place severe and often unique burdens on the computational elements of synthetic environments. These burdens may, in time, be handled by extensions of current techniques used to handle graphical information. They may, however, require completely new approaches in the design of hardware and software to support the representation, simulation, and rendering of worlds in which visual, auditory, and haptic events are modeled. In either case, the generation of multimodal synthetic environments requires that we carefully examine our current assumptions concerning VE architectural requirements and design constraints.

In general, multimodal VEs require that object representation and simulation techniques now represent and support the generation of information required to support auditory signal generation and haptic feedback (i.e., rendering). Both of these modalities require materials and geometric (i.e., volume) information that is not typically incorporated into today's surface-oriented geometric models. Consequently, volumetric approaches may become more attractive at all three levels of information handling (i.e., representation, simulation, and rendering). Not only may volumetric approaches facilitate the representation of the information needed for objects in multimodal VEs but they may also lend themselves to local interaction models of physics that are elegant and straightforward to implement (Toffoli, 1983). In addition, hardware to support this form of physical simulation is starting to become available on such machines as the CAM-8 and the FX-1 from Exa Corporation. These approaches have been successfully employed in the modeling of fluid flow (Frisch, 1987) and may point the way for future VE representation, simulation, and rendering approaches.

In addition, the concept of frame rate, both in terms of update rate and acceptable lags previously discussed in this chapter, must be refined or altered. Display update rates, which may be perfectly adequate for visual-only synthetic environments, are wholly inadequate when we consider the auditory and haptic modalities. Auditory events not only require rendering at rates exceeding 40 kHz but also have a temporal extent that may be measured in seconds. Furthermore, in even moderately complex synthetic environments, several auditory events may need to be generated and spatialized at any given instant. Update rate requirements for haptic events are also problematic and may be viewed on two levels. The first level is that associated with rendering kinesthetic or gross position/force information. These events require an update rate approaching 20 Hz. At the second level, tactile information (e.g., texture) requires rendering rates measured in the hundreds of Hz.

From the perspective of temporal lag, the generation of VEs that have auditory and haptic displays also places unique burdens on computational elements. Visual, auditory, and haptic events must be displayed without unacceptable intermodal lags. Although currently an area of research for the three modalities, it is known that delays on the order of a few tens of milliseconds can cause undesirable perceptual effects (e.g., the decorrelation of visual and auditory stimuli such that they are perceived as belonging to separate events). It is likely that the temporal alignment of visual, auditory, and haptic stimuli will need to be on the order of 10 ms. This number can potentially place burdens on the computation system one order of magnitude greater than currently acceptable update rates of visual-only VE.

How these issues and the numerous others that are likely to be encountered with further exploration are to be handled by computational systems is still an open and important area for further research. Clearly, however, more attention must be given to these issues from a computer hardware and software perspective. Currently, auditory and haptic interactions are predominantly handled by devices outside the major computational (usually graphical) workstation. This approach makes the essential temporal correlation of trimodal stimuli even more problematic and costly. Current generations of computer workstations benefit tremendously from special-purpose hardware that supports the rendering of graphical information. Related hardware may also be needed to support the rendering of auditory and haptic events. In a like manner, special-purpose hardware to support the representation and simulation of objects within a multimodal VE may be beneficial. Multimodal VEs put an even higher premium on several issues and shortcomings associated with current computational systems, such as: (1) veridical, real-time, physically inspired, simulation technology; (2) high-bandwidth, low-latency, input-output capabilities; (3) multimodal representation and simulation information exchange formats and methodologies; and (4) short- and long-haul information networking technologies. Finally, an area that presents both a challenge and the promise of multimodal VEs is the proper use of the three modalities in the control of VEs and other computational environments. Certain metaphors (e.g., pull-down menus), albeit flawed, have served us well in two-dimensional, graphically oriented human-machine interfaces. Metaphors for three-dimensional, multimodal systems will require further research.

RESEARCH NEEDS

The state of the art in computer technology for the generation of VEs is constantly shifting. We have tried to define the edge of the currently accessible and available technology and some of the difficult problems yet to be solved. We now turn to a discussion of the hardware and software needed to address these problems and to move the field forward.

Hardware

As noted in the opening section of this chapter, advances in graphics and computer hardware are key to the full realization of VEs. The hardware capabilities available today have given researchers, entrepreneurs, and consumers just a taste of virtual worlds and a promise of possible applications. Because of the potentially wide appeal and the large variety of applications with differing performance requirements, it is important to continue hardware development at several levels from the high-end multimodal workstations to the low-end personal workstation with modest visual-only three-dimensional capabilities. The following paragraphs detail some of the key technical needs generated by VEs.

There are several computer hardware requirements needed to support high-end VE systems in the future. Computer architectures that provide for applications with high computational demands are devices for which we already have a requirement. These machines must have very large physical memories (> 15 Gbytes), multiple high-performance scalar processors, high-bandwidth (> 500 Mbytes/s), low-latency (< 0.03 s) mass storage devices, and high-speed interface ports for various input and output peripherals. Disk bandwidth is not expected to improve significantly over the next few years. (Disk bandwidth and size thus arise as limiting factors in video on demand and hypermedia integration in virtual worlds.) Current projections suggest that workstations capable of supporting 15 Gbytes of physical memory might be available sometime in 1994 but that the cost will be prohibitive for all but the most well-funded research groups.

Extensive computational and data management capabilities will also be required. Physical modeling and visualization computations will be the driving force behind this computational requirement. Machines capable of 40 Mflops or greater are needed now for some problems. If we wish to add more data points or develop finer resolution models, we could easily use all of this level of available computing power. It is important to encourage the actual production of required machines.

There is not just one computer architectural requirement. Different VE systems require different computer architectures. Some systems require a few parallel, high-power scalar processors with large shared memories; others require large numbers of CPUs operating in parallel (CPUs perhaps without scalar processing capabilities). This flexible requirement for CPU configuration may not be possible if the majority of the hardware fabricators move into the PC clone business.

Extrapolating current trends, we expect that VE applications will saturate available computing power and data management capabilities for some time to come. Dataset size will be the dominant problem for an important class of applications in VE. In the near term, an effective VE platform would include multiple fast processors in an integrated unit; several graphics pipelines integrated with the processors; very large shared physical memory; very fast access to mass storage; operating systems that support shared memory within multiprocessed architectures; and very high-speed, low-latency networks.

To ensure continued development, it is necessary to encourage both private- and public-sector participation. A key concern is that the number of serious research and development efforts associated with VE design and implementation are decreasing. For example, the number of commercial companies in the business of producing special architectures for high-performance graphics rendering systems has decreased since 1988 to only a few today. Furthermore, the University of North Carolina is perhaps the only significant university-based computer graphics hardware research group in the United States that is still working. These developments have three important implications. First, with a small number of participants it is possible that fewer ideas will be generated. Second, the pace of development may be slowed and the cost to consumers may remain high due to lack of competition. Third, those few companies currently producing hardware for VE research and application may turn their interests to other technology areas that for the moment might appear more lucrative, such as video games or the television of the future. Such could be the consequence if corporate America continues its trend toward high-yield, short-term investment rather than the lower yields over the long term.

Software

Most current SE systems are built using commercial workstations running some variant of the UNIX operating system (which was not originally designed to meet real-time performance requirements). Other approaches are based on using collections of more specialized embedded computational elements (possibly with a general-purpose workstation acting as the "front end") running operating systems that have been designed to support real-time, distributed computation. The latter approach has been extensively used in fields having hard real-time requirements, such as process control and telerobotics, and is discussed in some detail in Chapter 9 of this report.

From the perspective of using commercial workstations and their powerful graphical capabilities, however, the committee feels that what is needed is either a new operating system (OS) architecture specifically designed for synthetic environment (SE) applications or upgrades and enhancements of existing operating systems. The operating system capabilities required for virtual environments include: support of very large numbers of lightweight processes communicating via shared memory, support of automatic and transparent distribution of tasks to multiple computing resources, support of time-critical computation and rendering, and very high resolution time slicing and guaranteed execution for high-priority processes (to within 0.001 s resolution). Although not specifically addressing all of these concerns, the efforts of the Institute for Electrical and Electronics Engineers (IEEE) Posix standards committee are starting to bring real-time capabilities to the open system workstation environment. In particular IEEE standard 1003.4 (on real-time extensions to UNIX), standard 1003.4a (on threads extensions to UNIX), and standard 1003.13 (on application environment profiles for real-time applications support) are important to SE developers requiring some level of real-time control. Substantial subsets of the capabilities specified in these standards are now available on some graphics workstations (e.g., SGI workstations running the IRIX 5 version of the operating system). Supporting these capabilities in the operating system will significantly facilitate the development of many SE applications, especially larger, more ambitious efforts.

Government funding for the development of such a VE operating system or upgrade should be accompanied by a plan to shift this new system to commercial sponsorship. In the past, the federal government has funded a considerable amount of operating system research, much of which has never made the transition from university research project to commercial viability. To make sure that VE systems are written using an appropriate operating system, a real, financially sound transition plan for specially designed or upgraded operating systems must be formulated, funded, and executed.

A second critical need is for large, multiyear, basic science programs created for developing large-scale VEs. Current trends in computer science research funding are for small university research grants with a typical size of approximately $400,000 over three years. The great majority of VE researchers receive significantly fewer grant funds. In addition, the trend in research funding for most agencies is toward the funding of projects with firm deliverables and schedules. With these constraints we believe that the level of experimentation researchers are willing or able to engage in will decrease, and, as a result, we cannot hope to see major advances in the technology. An example of the problem is the tendency of some government agencies to divert funds from large software development projects to impressive technology demonstrations. Although such demonstrations are appropriate for the culmination of significant basic research projects, they have most recently been used for rushed presentations in which the software was pieced together over a few months rather than carefully planned and designed over three to four years. There is an important need for the funding of large software development projects in which the goal is the development of large, open-ended, networked VEs. It is critical to concentrate funding on the basics. An additional trend of concern in the research funding arena is that exemplified by the Technology Reinvestment Program (TRP). Much of the nation's research dollars are moving into the TRP, which requires that universities take on a corporate partner. Such moves lock up research results in proprietary agreements and diminish the likelihood of shared research results.

Interaction Software

The general problem of inferring user intentions so as to provide a natural interface for all tasks in a three-dimensional VE is an area requiring a great deal of further research. Because language, which provides much of the intentional information in the real world, is not currently available for use in virtual worlds, other options must be thoroughly explored and developed. Unfortunately, the size of current government research efforts for work on providing natural interfaces is small. One government agency recently indicated that their entire VE human computer interface budget was approximately $150,000; another program in VE interfaces was funded at $2.5 million, with the goal of supporting six universities over one year. The rationale for these limited efforts is the belief that much of the interface research will be funded and carried out by industry. However, it seems likely that, once the novelty wears off, industry interest will wane. Thus it is unlikely that the private sector will take on long-term development efforts in the absence of standards.

Nevertheless, high-level interface issues should be explored. Specifically, research should be performed to examine how to use data measuring the positions of the user's body for interaction with the VE in a way that truly provides the richness of real-world interaction. Critical concerns are how to apply user tracking data and how to define objects in VE to ensure natural interaction.

One of the major research challenges that has both hardware and software implications is the continued use of the RS-232C interface for control devices. Current workstation technology typically provides one or two such ports. Control devices are usually attached to these ports, with commands sent via the UNIX *write* system call. There is a speed limitation on the use of these ports, a limitation often seen as latency in input response. It is not uncommon to hear 70 ms touted as the fastest response from the time of input device movement to the reporting of the change back to the application running on the workstation. That 70 ms is too long a delay for real-time interaction, for which a maximum of 10 ms is more appropriate. And there is the additional problem with UNIX system software layers that must be traversed for events to be reported back to the concerned VE application.

Current workstation manufacturers do not focus on the design of such high-speed ports. Even within one manufacturer there is no guarantee that such ports will behave consistently across differing models of workstations. Real standards and highly engineered ports are needed for control devices. In fact, a revolutionary redesign and restandardization of the input port is required if control devices are to take off. In addition, we need to rethink the layers of VE system architecture.

Visual Scene Navigation Software

Given the current workstation graphics polygon filling capabilities and the extrapolation of those speeds into the future, software solutions will be needed to reduce the total number of graphics primitives sent through the graphics pipeline for some time to come. The difficulty of polygon flow minimization depends on the composition of the virtual world. This problem has historically been approached on an application-specific basis, and there is as yet no general solution. Current solutions usually involve partitioning the polygon-defined world into volumes that can readily be checked for visibility by the virtual world viewer. There are many partitioning schemes—some of which work only if the world description does not change dynamically. We need to encourage research into generalizing this software so that dynamically changing worlds can be constructed.

Furthermore, there is a need to encourage the funding of research to reach a common, open solution for polygon flow minimization. Current researchers who have tackled polygon flow minimization have closely guarded their developed code. In fact, most software source code developed under university research contract today in the United States is held as proprietary by the universities, even if that code was developed under government contract. This fact, coupled with the stated goal of federal agencies of recouping investments, is counterproductive and disturbing. The unavailability of such software increases the overall development time and cost of progress in technology, as researchers duplicate software development. These redevelopment efforts also slow the progress of new development.

There are additional technical issues in polygon flow minimization that are important. One of these issues, the generation of multiple resolution three-dimensional icons, is a closely related technological challenge. In much of the work of polygon flow minimization, it is assumed that multiple resolutions, lower polygon count, and three-dimensional icons are available. This assumption is a large one, with automatic methods for the generation of multiple-resolution three-dimensional icons an open issue. There is some work in this area, and it is recommended that a small research program be developed to encourage more (DeHaemer and Zyda, 1991; Schroeder et al., 1992; Turk, 1992). In fact, the development of such public software and a public domain set of three-dimensional clip models with geometry and associated behavior could go a long way toward encouraging the creation of three-dimensional VEs.

Modeling

Simulation Frameworks Research into the development of environments in which object behavior as well as object appearance can rapidly be specified is an area that needs further work. We call this area *simulation frameworks*. Such a framework makes no assumptions about the actual behavior (just as graphics systems currently make no assumptions about the appearance of graphical objects). A good term for what a simulation framework is trying to accomplish is *meta-modeling*. Such frameworks would facilitate the sharing of objects between environments and allow the establishment of object libraries. Issues to be researched include the representation of object behavior and how different behaviors are to be integrated into a single system.

Geometric Modeling Because geometric modeling is integral to the construction of VEs, its current limitations serve as limits to development. As a practical matter, the VE research community needs a shared *open* modeling environment that includes physical and behavioral modeling. The current state of the art in VE technology is to use available CAD tools, tools more suited to two-dimensional displays. The main problem with CAD tools is not in getting the three-dimensional geometry out of the CAD files but rather the fact that data related to the actual physics of the three-dimensional objects modeled by the CAD systems are not usually present in such files. In addition, the partitioning information useful for real-time walkthrough of these data usually has to be added later by hand or back fed in by specially written programs. CAD files also have the problem that file formats are proprietary. An open VE CAD tool should be developed for use by the VE research community. This tool should incorporate many of the three-dimensional geometric capabilities in current CAD systems as well as physics and other VE-relevant parameters (i.e., three-dimensional spatial partitioning embedded into the output databases). It should also capture parameters relevant to haptic and auditory channels.

Vision-Based Model Acquisition Although CAD systems are useful for generating three-dimensional models for new objects, using them can be tedious. Currently, modelers sit for hours detailing each door, window, and pipe of a three-dimensional building. VEs could be much more widely used if this painful step could be automated, perhaps via laser range finders and the right "surface generation to CAD primitive" software. Unfortunately, there is the very hard multiple view, laser range image correlation problem. Automatic model acquisition would be a good first step toward providing the three-dimensional objects for virtual worlds. However, the physics of the objects scanned would still need to be added. This technology has many uses beyond developing VEs. An additional application area of high interest is providing CAD plans for older buildings, structures designed and constructed before the advent of CAD systems.

Augmented Reality Real-time augmented reality is one of the tougher problems in VEs research. The two major issues are (1) accurate measurement of observer motions and (2) acquisition and maintenance of scene models. The prospects for automatic solutions to the scene model acquisition and maintenance were discussed above. The problems with measuring observer motion are more difficult and represent a major research area. Although VE displays provide direct motion measurements of observer movement, these are unlikely to be accurate enough to support high-quality augmented reality, in situations in which real and synthetic objects are in close proximity. Even very small errors could induce perceptible relative motions that could disrupt an illusion. Perhaps the most promising course would be to use direct motion measurements for gross positioning and to use local image-based matching methods to lock real and synthetic elements together.

TECHNICAL APPENDIX

Graphics Architectures for VE Rendering

The rendering operation has three stages: preprimitive, rasterization, and prefragment. Because of the performance demand, all modern high-performance graphics systems are run on parallel architectures. To allow the many-to-many mapping, the parallel rendering pipes must be combined at one point along their paths. The three possible locations for the crossbar are illustrated in Figures 8-6 through 8-8. The primitive crossbar (Figure 8-6) broadcasts window-coordinate primitives from the engines that transformed and lighted them to the one or more rasterization engines that correspond to frame buffer regions that each primitive intersects. Depending on the window-coordinate size of a primitive, it might be processed by just one rasterization engine, or by all of the rasterization engines. Thus this crossbar is really a one-to-many bus.

The fragment crossbar (Figure 8-7) is a true, one-to-one crossbar connection. Each fragment that is generated by a rasterization engine is directed to the one fragment processor that manages the corresponding pixel in the frame buffer. Thus the fragment crossbar is itself more easily parallelized than the primitive crossbar, allowing for the necessarily greater bandwidth of rasterized fragments over window-coordinate primitives. The primary disadvantage of the fragment crossbar com-

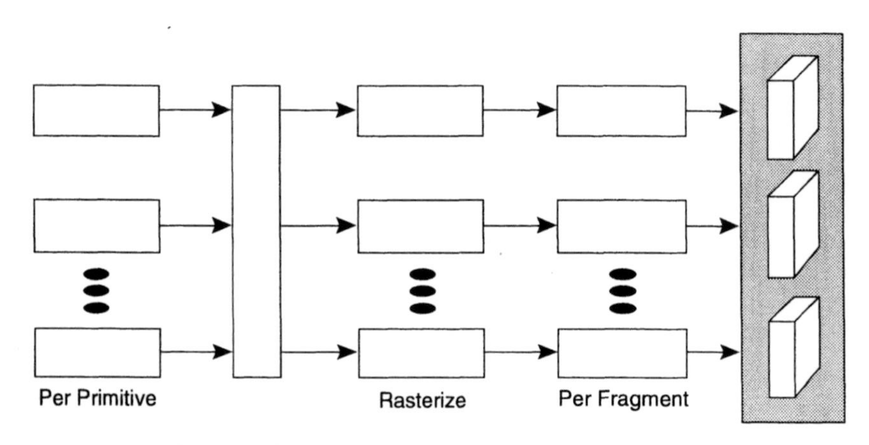

FIGURE 8-6 Primitive broadcast.

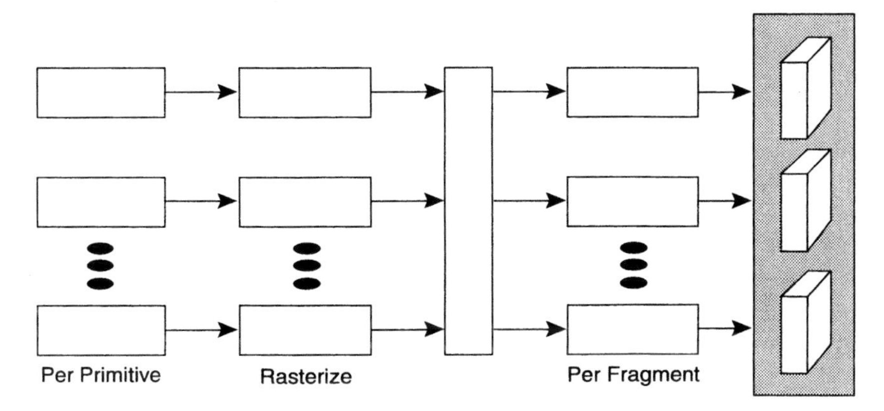

FIGURE 8-7 Fragment crossbar.

pared with the primitive crossbar is that fragment crossbar systems have difficulty rendering primitives in the order that they were presented to the graphics system, whereas primitive broadcast systems easily render primitives in the order presented.

Whereas the frame buffers in the primitive broadcast and fragment crossbar systems were disjoint, collectively forming a single, screen-size buffer, the frame buffers of a pixel crossbar system (Figure 8-8) are each complete, screen-size buffers. The contents of these buffers are merged only after all of the primitives have been rendered into one of the buffers. The primary advantage of such a system over primitive and fragment crossbars is that pixel merge, using the z-buffer algorithm to choose the final pixel value, is infinitely extensible with no performance loss. Again,

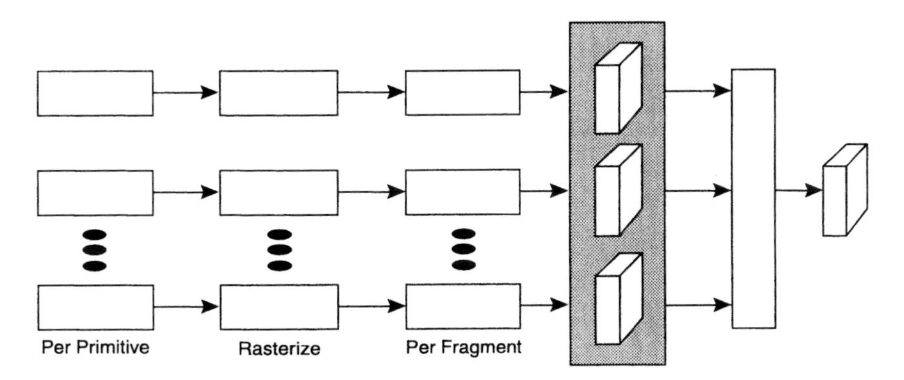

FIGURE 8-8 Pixel merge.

the term *crossbar* is misleading, since the pixel merge can be accomplished with one-to-one paths between adjacent buffer pairs.

The primary disadvantage of pixel merge systems is the requirement for large, duplicate frame buffers. A secondary disadvantage exists only with respect to primitive broadcast systems: the pixel crossbar, like the fragment crossbar, has difficulty rendering primitives in the order presented. (Each path renders the primitives presented to it in the order that they are presented, but the postrendering pixel merge cannot be done in order.)

The primary disadvantage of frame buffer size can be mitigated by reducing the size of each frame buffer to a subregion of the final, display buffer. If this is done, the complete scene must be rendered with multiple rasterization passes, with the subbuffers being merged into the final display buffer (which is full size) after each pass is completed. Application of such a multipass technique introduces the second differentiator of parallel graphics systems: whether the rendering is flow-through or tiled. Flow-through systems complete the processing of each primitive soon after that primitive is presented to the rendering system, in which "soon" is a function of the number of processing steps. Tiled systems accumulate all the primitives of a scene after the per-primitive processing is complete, then begin the rasterization and per-fragment processing. They must do this because frame buffer tiles are allocated temporally rather than spatially, and so are not available in the random sequence that the primitives arrive in. The primary disadvantage of tiled systems over flow-through systems is therefore one of increased latency, due to the serialization of the processing steps.

The third major differentiator is image quality: does the architecture support mapping images onto geometry (texture mapping), and is the sampling quality of both these images and the geometry itself of high quality (anti-aliasing)? This differentiator is less one of architecture than of implementation—primitive, fragment, and pixel crossbar systems, both flow-through and tiled, can be implemented with or without texture mapping and anti-aliasing.

The final differentiator is performance: the number of primitives and fragments that can be processed per second. Again this differentiator is less one of architecture than of implementation, although at the limit the pixel merge architecture will exceed the capabilities of primitive broadcast and fragment crossbar architectures.

Now we consider the architectures of four modern graphics systems, using the previously discussed differentiators. The Silicon Graphics RealityEngine is a flow-through architecture with a primitive crossbar. It therefore is able to efficiently render primitives in the order that they are presented and has low rendering latency. RealityEngine supports texture mapping and anti-aliasing of points, lines, and triangles and therefore is considered to have high rendering image quality. RealityEngine processes up to 1 million texture mapped, anti-aliased triangles/s, and up to 250 million texture mapped, anti-aliased fragments/s. It is able to generate 1,280 × 1,024 scenes of high quality at up to 30 frames/s.

Freedom series graphics from Evans & Sutherland use a flow-through architecture with a fragment crossbar. Thus Freedom machines also have low rendering latency, but are less able than the RealityEngine to efficiently render primitives in the order that they are presented. Freedom machines support texture mapping and can anti-alias points and lines, but they are unable to efficiently anti-alias surface primitives such as triangles. Hence the rendering quality of Freedom machines for full-frame solid images is relatively low. Although exact numbers for Freedom fragment generation/processing rates are not published, the literature suggests that this rate for texture-mapped fragments is in the tens of millions per second, rather than in the hundreds of millions. If that is the case, then the performance of Freedom graphics is not sufficient to generate 1,280 × 1,024 images at even 10 frames/s, the absolute minimum for interactive performance.

Pixel Planes 5, the currently operational product of the University of North Carolina's research efforts, uses a tiled, primitive crossbar architecture. Because the architecture is tiled, the advantage of ordered rendering typical of primitive crossbar systems is lost. Also, the tiling contributes to a latency of up to 3 frames, which is substantially greater than the single-frame latencies of the Freedom and the RealityEngine systems. The rendering performance, especially the effective fragment generation/processing rate, is substantially greater than either the Freedom or RealityEngine systems, resulting in easily maintained 1,280 × 1,024 30 frame/s image generation. However, Pixel Planes 5 cannot anti-alias geometry at these high rates, so the image quality is lower than that of RealityEngine.

Finally PixelFlow, the proposed successor to Pixel Planes 5, is a tiled, pixel merge machine. Thus it is unable to efficiently render primitives in the order in which they are received, and the rendering latency of PixelFlow is perhaps twice that of Freedom and RealityEngine, though less than that of Pixel Planes 5. PixelFlow is designed to support both texture mapping and anti-aliasing at interactive, though reduced rates, resulting in a machine that can produce high-quality, 1,280 × 1,024 frames at 30 or even 60 frames/s.

Silicon Graphics from the IRIS-1400 to the RealityEngine 2

Silicon Graphics, Inc., a computer manufacturer, creates visualization systems with some of the more flexible and powerful digital media capabilities in the computer industry, combining advanced three-dimensional graphics, digital multichannel audio, and video in a single package. Silicon Graphics systems serve as the core of many VE systems, performing simulation, visualization, and communication tasks. In such a role, it is critical that the systems support powerful computation, stereoscopic, multichannel video output, and fast input/output (I/O) for connectivity to sensors, control devices, and networks (for multiparticipant VEs). Textured polygon fill capability is also one of the company's strengths with respect to virtual worlds in that texturing enhances realism.

In support of this role, Silicon Graphics has engaged in the development of multiple processing, graphics workstations at the leading edge of technology since late 1983. A brief look at the graphics performance numbers of their high-end systems since that time is warranted (Table 8-1). Those systems comprise three generations, as described in the RealityEngine Graphics paper (Akeley, 1993). The 1000, 2000, and G are first generation, the GTX, VGX, and VGXT are second generation, and the RealityEngine and RealityEngine2 are third generation. Performance is listed for first-, second-, and third-generation operations for all these machines. Notice that the curve for first-generation performance falls off with second- and third-generation machines, because they are not optimized for first-generation rendering.

Onyx RealityEngine2

In January 1993, Silicon Graphics announced the Onyx line of graphics supercomputers, which incorporate a new multiprocessing architecture, PowerPath2, to combine up to 24 parallel processors based on the MIPS R4400 RISC CPU, which operates at 150 MHz. I/O bandwidth is rated at 1.2 Gbytes/s to and from memory, with support for the VME64 64-bit bus, operating at 50 Mbytes/s.

Onyx systems can utilize up to three separate graphics pipelines based on the new RealityEngine2 graphics subsystem. This new graphics system offers 50 percent higher polygon performance than the original RealityEngine introduced in July 1992. RealityEngine2 is rated at 2 million flat triangles/second and 900,000 textured, Gouraud shaded, antialiased, fogged, z-buffered triangles/s.

The optional MultiChannel board enables users to take the frame buffer and send different regions out to different display devices. Thus, a single 1.3 million pixel frame buffer could be used either as a 1,280 × 1,024 display or as four 640 × 512 displays. The MultiChannel option provides up to six separate outputs.

TABLE 8-1 Performance History for SGI Graphics

					Depth buffered, lighted, Gouraud shaded		Depth buffered, lighted, Gouraud shaded, anti-aliased, texture mapped	
System	Date	Points	Tris	Pixels	Tris	Pixels	Tris	Pixels
1000	1983	0.06	0.001	40	n/a	n/a	n/a	n/a
2000	1984	0.07	0.01	46	0.0008	0.1	n/a	n/a
G	1986	0.14	0.01	130	0.003	2.0	n/a	n/a
GTX	1988	0.45	0.135	80	0.135	40	n/a	n/a
VGX	1990	1.5	1.1	200	0.8	100	(0.08)	(10)
VGXT	1991	1.5	1.1	200	0.8	100	(0.08)	(50)
RE	1992	2.0	1.4	380	1.4	380	0.6	250
RE2	1993	3.0	2.0	380	2.0	380	1.0	250

NOTE: The first three columns are for generic rendering, no depth buffer, no lighting or shading, and no texture mapping or anti-aliasing. All values are in millions per second (points, triangles, or rendered pixels). Values in parenthesis are texture mapped but not anti-aliased. All these systems were introduced in the $50,000 to $100,000 range. Tris = triangle meshes; Pixels = pixel fill rate.

Evans & Sutherland Freedom 3000

Evans & Sutherland (E&S), an old line flight simulator company, has recently announced the Freedom Series of graphics accelerators targeted for the Sun Microsystems Sparc 10 line of workstations. The Freedom series offers a wide range of performance levels: from 500,000 polygons per second for the Freedom 1000 to 3 million polygons per second for the Freedom 3000. The Freedom series uses standard hardware and software interfaces to join seamlessly with the Sun Microsystems environment. The Freedom accelerators are programmable with Sun's standard interfaces and are software-compatible with workstations currently available from E&S and Sun.

The Freedom 3000 has 1,280 × 1,024, 1,536 × 1,280, and high-definition TV display formats. It also supports hardware texture mapping, including MIP-mapping, and resolutions up to 2,000 × 2,000. Additional features supported are: anti-aliased lines, dots, and polygons, alpha buffering, accumulation buffering, 128 bits per pixel, and dynamic pixel allocation.

The Freedom 3000 contains the following technology: five proprietary VLSI ASIC chips types using 0.8 μ CMOS, a parallel array of programmable high-speed microprocessors (DSPs), a very fast, proprietary graphics bus (G-bus) capable of speeds well beyond 3 million polygons/s, high-speed pixel routing interconnection, high-speed access to frame buffer for image processing (up to 100 million pixels/s), and a pixel fill rate of 95 million pixels/s.

Graphics Hardware from the University of North Carolina, Chapel Hill: PixelPlanes 4, 5, and PixelFlow

The University of North Carolina at Chapel Hill is one of the last schools still developing graphics hardware. Their efforts differ widely from what has been attempted in the commercial world, since their work is more basic research than machine production. Despite this research focus, the machines developed by Fuchs, Poulton, Eyles, and their colleagues have been close to the leading edge of graphics hardware at each of their prototypical stages (Fuchs et al., 1985, 1989). PixelPlanes4 had a 27,000 polygons/s capability in 1988, with a follow-on machine PixelPlanes5 shown first in 1991 with a 1 million polygons/s capability. The latest machine, PixelFlow, is still under development but shows great promise (Molnar et al., 1992). It is expected to be working in 1994.

Pixel Flow and its graphics performance scalability are an important part of the future of high-performance three-dimensional VEs. PixelFlow, an architecture for high-speed image generation, overcomes the transformation and frame-buffer access bottlenecks of conventional hardware rendering architectures (Molnar et al., 1992). It uses the technique of image composition, through which it distributes the rendering task over an array of identical renderers, each of which computes a full-screen image of a fraction of the primitives. A high-performance image-composition network combines these images in real time to produce an image of the entire scene.

Image composition architectures offer performance that scales linearly with the number of renderers. A single PixelFlow renderer rasterizes up to 1.4 million triangles/s, and an n-renderer system can rasterize at up to n times this basic rate. It is expected that a 128 renderer PixelFlow system will be capable of a polygon rate approaching 100 million triangles/s.

PixelFlow performs anti-aliasing by supersampling. It supports deferred shading with separate hardware shaders that operate on composite images containing intermediate pixel data. PixelFlow shaders compute complex shading algorithms and procedural and image-based textures in real time, with the shading rate independent of image complexity. A PixelFlow system can be coupled to a parallel supercomputer to serve as an intermediate-mode graphics server, or it can maintain a display list for retained-mode rendering.

9

Telerobotics

This chapter reviews issues and needs in telerobotics. A *telerobot* is defined for our purposes as a robot controlled at a distance by a human operator, regardless of the degree of robot autonomy. Sheridan (1992c) makes a finer distinction, which depends on whether all robot movements are continuously controlled by the operator (manually controlled teleoperator), or whether the robot has partial autonomy (telerobot and supervisory control). By this definition, the human interface to a telerobot is distinct and not part of the telerobot. Haptic interfaces that mechanically link a human to a telerobot nevertheless share similar issues in mechanical design and control, and the technology survey presented here includes haptic interface development.

INTRODUCTION

Telerobotic devices are typically developed for situations or environments that are too dangerous, uncomfortable, limiting, repetitive, or costly for humans to perform. Some applications are listed below:

Underwater: inspection, maintenance, construction, mining, exploration, search and recovery, science, surveying.

Space: assembly, maintenance, exploration, manufacturing, science.

Resource industry: forestry, farming, mining, power line maintenance.

Process control plants: nuclear, chemical, etc., involving operation, maintenance, decommissioning, emergency.

Military: operations in the air, undersea, and on land.

Medical: patient transport, disability aids, surgery, monitoring, remote treatment.

Construction: earth moving, building construction, building and structure inspection, cleaning and maintenance.

Civil security: protection and security, firefighting, police work, bomb disposal.

This chapter is divided into five sections, which represent one way of categorizing past developments in telerobotics:

(1) Remote manipulators
(2) Remote vehicles
(3) Low-level control
(4) Supervisory control
(5) Real-time computing

A recent survey including these and other topics is provided by Sheridan (1992a).

Relation to Robotics

Telerobots may be remotely controlled manipulators or vehicles. The distinction between robots and telerobots is fuzzy and a matter of degree. Although the hardware is the same or is similar, robots require less human involvement for instruction and guidance than do telerobots. There is a continuum of human involvement, from direct control of every aspect of motion, to shared or traded control, to nearly complete robot autonomy.

Any robot manipulator can be hooked up to a haptic interface and hence become a telerobot. Similarly, any vehicle can be turned into a teleoperated mobile robot. There are many examples in the literature of different industrial robots that have been used as telerobots, even though that was not the original intended use. For example, a common laboratory robot, the PUMA 560, has frequently been teleoperated (Funda et al., 1992; Hayati et al., 1990; Kan et al., 1990; Lee et al., 1990; Salcudean et al., 1992). There have also been a number of telerobots specifically designed as such, often with a preferred haptic interface. The design issues for robots, telerobots, and haptic interfaces are essentially the same (although Pennington, 1986, seeks to identify differences). Often telerobots have to be designed for hazardous environments, which require special characteristics in the design. Industrial robots have most often been designed for benign indoor environments.

Why don't we do everything with robots, rather than involve humans in telerobotic control? We can't, because robots are not that capable. Often there is no substitute for human cognitive capabilities for planning and human sensorimotor capabilities for control, especially for unstructured environments. In telerobotics, these human capabilities are imposed on the robot device. The field of robotics is not that old (35 years), and the task of duplicating (let alone improving upon) human abilities has proven to be an extremely difficult endeavor; it would be disturbing if it were not so. There is a tendency to overextrapolate from the few superior robot abilities, such as precise positioning and repetitive operation. Yet robots fare poorly when adaptation and intelligence are required. They do not match the human sensory abilities of vision, audition, and touch, human motor abilities in manipulation and locomotion, or even the human physical body in terms of compact and powerful musculature that adapts and self-repairs, and especially in terms of a compact and portable energy source. Hence in recent years many robotics researchers have turned to telerobotics, partly out of frustration.

Nevertheless, the long-term goal of robotics is to produce highly autonomous systems that overcome hard problems in design, control, and planning. As advances are made in robotics, they will feed through to better and more independent telerobots. For example, much of the recent work in low-level teleoperator control is influenced by developments in robot control. Often, the control ideas developed for autonomous robots have been used as the starting points for slave, and to a lesser extent, master controllers. Advances in high-level robot control will help in raising the level of supervisory control.

Yet the flow of advances can go both ways. By observing what is required for successful human control of a telerobot, we may infer some of what is needed for autonomous control. There are also unique problems in telerobotic control, having to do with the combination of master, slave, and human operator. Even if each individual component is stable in isolation, when hooked together they may be unstable. Furthermore, the human represents a complex mechanical and dynamic system that must be considered.

Relation to Virtual Environments

Telerobotics encompasses a highly diversified set of fundamental issues and supporting technologies (Vertut and Coiffet, 1985a, 1985b; Todd, 1986; Engelberger, 1989; Sheridan, 1992b). More generally, telerobots are representative of human-machine systems that must have sufficient sensory and reactive capability to successfully translate and interact within their environment. The fundamental design issues encountered in the field of telerobotics, therefore, have significant overlap with those that are and will be encountered in the development of veridical virtual environments (VEs). Within the virtual environment, the human-machine system must allow translation of viewpoint, interaction with the environment, and interaction with autonomous agents. All this must occur through mediating technologies that provide sensory feedback and control. The human-machine interface aspects of telerobotic systems are, therefore, highly relevant to VE research and development from a device, configuration, and human performance perspective.

Yet the real-environment aspect of telerobotics distinguishes it from virtual environments to some extent. Telerobots must:

- interact in complex, unstructured, physics-constrained environments,
- deal with incomplete, distorted, and noisy sensor data, including limited views, and
- expend energy which may limit action.

The corresponding features of virtual environments are more benign:

- Form, complexity, and physics of environments are completely controllable.
- Interactions based on physical models must be computed.
- Virtual sensors can have an omniscient view and need not deal with noise and distortions.
- The ability to move within an environment and perform tasks is not energy-limited.

Despite such simplifications, virtual environments play an important role in telerobotic supervisory control. A large part of the supervisor's task is planning, and the use of computer-based models has a potentially critical role. The virtual environment is deemed an obvious and effective way to simulate and render hypothetical environments to pose "what would happen if" questions, run the experiment, and observe the consequences. Simulations are also an important component of predictive displays, which represent an important method of handling large time delays. VE research and development promises to revolutionize the field of multimodal, spatially oriented, interactive human-machine interface technology and theory to an extent that has not been achievable in the robotics field. The two fields should therefore not be viewed as disparate but rather as complementary endeavors whose goals include the exploration of remote environments and the creation of adaptable human-created entities.

REMOTE MANIPULATORS

This section reviews remote manipulators from standpoints of kinematics, actuation, end effectors, and sensors. Specific examples of robots and telerobots in this review will tend to be drawn from more recent devices; some of the older telerobots are reviewed in Vertut and Coiffet (1985a). A review with similar categories is provided by Sheridan (1992a).

Kinematics

In this section we describe the number of joints and their geometrical layout. Some of the issues within kinematics are discussed below.

General Positioning Capabilities

A manipulator requires at least 6 degrees of freedom (DOFs) to achieve arbitrary positions and orientations. When a manipulator has exactly 6 DOFs, it is said to be general purpose. Examples include many industrial robots, such as the PUMA 560, as well as a number of commercial telerobots (Kraft, Shilling, Western Electric, ISE). The space shuttle's Remote Manipulator System (RMS), designed by Spar Aerospace, is another example.

If there are less than 6 DOFs, the device is said to be overconstrained. Often a task will require fewer DOFs, such as positioning ($x - y$) and orientating (a rotation θ about the normal z axis) restricted to a plane (a 3-DOF task). Another popular example is the SCARA robot geometry with 4 DOFs; the motions are planar with an extra translation in the direction normal to the plane. A modified SCARA robot, to which one joint was added, is being used for hip replacement surgery (Paul et al., 1992). Teleoperated heavy machinery usually is overconstrained; excavators have 4 DOFs (Khoshzaban et al., 1992).

An important subclass of mechanisms is a spherical joint, for which 3 rotations and no translations are required; this joint is useful in head-neck and head-eye systems. An example is the head-neck system described by Tachi et al. (1989). A 2-DOF pan-tilt system is presented in Hirose et al. (1992) and Hirose and Yokoyama (1992). Other pan-tilt systems are reviewed by Bederson et al. (1992), who also proposed a novel head-eye pan-tilt system employing a spherical motor. A 3-DOF parallel-drive head-neck system (Gosselin and Lavoie, 1993) has the potential for very fast motion, with some limitations in rotations.

Redundancies

When there are 7 or more DOFs, the mechanism is underconstrained. The extra DOFs may be used to fulfill secondary criteria (to general positioning), such as obstacle avoidance. There has been a lot of research in robotics addressing redundancy resolution. The human arm is a redundant 7-DOF mechanism (not counting shoulder shrug). Commercial examples include the Sarcos Dextrous Arm (Jacobsen et al., 1990a, 1990b, 1991), the Robotics Research Arm, and the Omnidirectional Arm (Rosheim, 1990). Laboratory examples include the Langley Laboratory Telerobotic Manipulator, the CESARm (Jansen and Kress, 1991), and the Anthropomorphic Tele-existence Slave Robot (Tachi et al., 1989, 1990a, 1990b). The Special Purpose Dextrous + Manipulator (SPDM) designed by Spar Aerospace will have 8 DOFs.

For direct control by the human arm, an exoskeleton master with 7 DOFs can be used to guide a slave 7-DOF manipulator. Hand-controllers (ground-based systems) are usually 6-DOF devices. To control a redundant arm, either the resolution is left to the discretion of the computer or an auxilliary control (such as a knob) must be manipulated.

Workspace

The term *workspace* describes the extent of the volume within which the manipulator may position the end-point, relative to the size of the manipulator. Certain manipulator geometries are known to offer superior workspaces (Hollerbach, 1985; Paden and Sastry, 1988; Vijaykumar et al., 1985); interestingly, these geometries are similar to the human arm geometry.

Serial Versus Parallel Mechanism

In a serial mechanism, joints are cascaded. The workspace is the union of motions of the joints, and hence is large. Because a proximal link must carry the weight of distal links, these arms may be slower, heavier, and less strong. Most manipulators (whether in robotics or telerobotics) are serial mechanisms, because a large workspace is often important.

In a parallel mechanism, several independent linkages meet at a common terminal link (the end effector). The workspace is the intersection of the independent linkages, and hence is small. One independent linkage does not carry the weight of the others, so these devices can be lightweight, strong, and fast. A prominent example of a parallel mechanism is the Stewart platform used in flight simulators. The human hand can also be viewed as a parallel mechanism; there are 5 independent 4-DOF linkages that can contact an object. A lot of recent research in robotics has focused on parallel mechanisms, to exploit their intrinsic advantages for specific tasks suited to their restricted workspaces. Examples include a 6-DOF parallel manipulator to be teleoperated for excavation (Arai et al., 1992), a 3-DOF microrobot based on beam bending (Hunter et al., 1991), and 6-DOF parallel hand controllers (Hayward et al., 1993; Long and

Collins, 1992). Landsberger and Sheridan (1985) have designed a cable-driven parallel mechanism. The parallel-drive hydraulic shoulder joint in Hayward et al. (1993) uses actuator redundancy to increase the workspace.

Kinematic Solvability

For serial mechanisms, the forward kinematics (find the end-point position given the joint angles) is easy, but the inverse kinematics (find the joint angles given the end-point position) is hard. The inverse kinematics is complicated unless the mechanism has a special structure: either a spherical joint or a planar pair (Tsai and Morgan, 1985). Almost all industrial robots have these special structures, but some for design convenience do not, such as the Robotics Research Arm, which has been used in teleoperation. Because of fast computers, iterative techniques to solve the nonlinear kinematics can work in real time.

For parallel mechanisms, the reverse is true: inverse kinematics is easy, but forward kinematics is hard (Waldron and Hunt, 1991).

Actuation

Actuation comprises the force or torque source (henceforth called the actuator) and any transmission element to connect to a joint or link. The actuation is the primary determinant of performance (speed, accuracy, strength). A survey of actuators for robotics is presented by Hollerbach et al. (1992).

Macrorobots

For macro motion control, standard actuators are electric, hydraulic, or pneumatic. For smaller robots (human size and less), electric actuators dominate. For larger robots (e.g., cranes), hydraulic actuators dominate.

Electric Motors and Drives Electric actuators are the most convenient, because the power source is an electric plug. For pneumatic or hydraulic systems, air supplies and hydraulic supplies often make them much less easy to install and maintain. Electric actuators, however, are weak relative to their mass; hence payloads are not great.

To amplify motor torque and couple high-rev motors to low-rev joints, nearly all electrical motor drives employ some form of transmission element, primarily gears. Thus nearly all commercial electric robots employ gears of some form. For space applications, the space shuttle RMS employs special high-ratio hollow planetary gears (Wu et al., 1993). These same gears will be employed in the two-armed SPDM (Mack et al., 1991).

The Flight Telerobotic Servicer (FTS) system produced by Martin Marietta employed harmonic drives (Andary and Spidaliere, 1993; Ollendorf, 1990), which are also commonly employed in industrial robots (e.g., the Robotics Research Arm, ASEA robots). Advanced spherical joint designs employing gears have been produced by Rosheim (1990).

Yet gears bring serious drawbacks: substantial friction, backlash, and flexibility. The performance consequences are poor joint torque control, poor end-point force control, reduced accuracy, and slower response. To overcome these drawbacks, some attempts are made to model the gear dynamics so that they may be compensated for in a controller (Armstrong-Helouvry, 1991). Other attempts include better gear designs that reduce friction and losses; examples include the Artisan arm (Vischer and Khatib, 1990) and the ROTEX manipulator (Hirzinger, 1987; Hirzinger et al., 1993), which is meant for space laboratory teleoperation.

Cable or tendon drives (including belts) are another common way to reduce arm weight, by remote location of the actuators. A number of master-slave systems have been designed using cable drives. More recent examples include the FRHC from JPL (Hayati et al., 1990; Kan et al., 1990) and the Whole Arm Manipulator (Salisbury et al., 1990) (both Salisbury's design). Nearly all multifingered robot hands employ tendon or cable routing; space constraints preclude direct mounting of actuators at joints (Jacobsen et al., 1986).

Another recent development is the integration of gears and actuators. For example, Rosheim (1990) has used miniature integrated lead screw mechanisms for finger joint-mounted actuators. Similar systems, developed originally for ROTEX (Hirzinger et al., 1993), are now commercially available (Wittenstein Motion Control GmbH). A related concept is the harmonic motor, in which the rotor rolls along the inside of the stator (Jacobsen et al., 1989). Analogous to harmonic drives, with harmonic motors, high effective gear ratios can be obtained.

To avoid problems with transmission elements, direct drive actuators and robots have become popular in the past decade to provide smooth and controllable motion (An et al., 1988; Asada and Youcef-Toumi, 1987). Examples of direct drive telerobots include the CMU DDArm II (Papanikolopoulos and Khosla, 1992) and MEISTER (Sato and Hirai, 1988); MEISTER is also an example of a direct drive master. Advances in electric motor technology continue, and particularization to robotics is a key to enhanced performance. One example of a new electric motor specifically designed for direct-drive robotics is the McGill/MIT Direct Drive Motor (Hollerbach et al., 1993).

An important new area of development in mechanism design is magnetic bearings and levitation, which seek to avoid problems of transmission elements, including bearings as structural members. In principle, devices with magnetic bearings should produce the smoothest motions. Hollis (Hollis et al., 1990) has designed a 6-axis magnetically levitated wrist, which can be used either as a hand controller or as a robot end effector. Salcudean (Salcudean et al., 1992) has developed a teleoperated robot, in which the master is a magnetically levitated wrist and the slave is an industrial 6-axis robot (coarse positioner) and the end effector is a magnetically levitated wrist. Because the wrist's range of motion is small, the hand controller is used in rate mode for large excursions and in proportional mode for fine motions. Although not magnetically suspended, a 2-axis force-reflecting mouse was developed by Salcudean using the same actuator elements (Kelley and Salcudean, 1993).

Another area under development, related to microelectromechanical systems (MEMS), is electrostatic actuators. All the electric motors mentioned above work by magnetic attraction. At small scales, the electrostatic effect is more favorable than the magnetic effect (Trimmer, 1989). By using small air gaps and many poles, powerful muscle-like actuators are conceivable. In terms of what has been realized on the macro scale, Higuchi has demonstrated lightweight but very strong linear electrostatic actuators (Niino et al., 1993).

Hydraulic Actuators Hydraulic actuators offer the most strength for the size. There are a number of commercial telerobot systems that are hydraulic, such as the Kraft arm, the Shilling arm, and the International Submarine Engineering (ISE) arm. Teleoperation of excavators (which are hydraulic) with hand controllers has also been pursued (e.g., Khoshzaban et al., 1992). To some extent, hydraulics have received a bad reputation due to concerns about leakage and controllability. Advances such as the Sarcos Dextrous Arm are a counterpoint to these concerns.

Pneumatic Actuators Pneumatic actuators are intermediate between electric and hydraulic drives, in terms of force produced for a given size and mass. There are very few high-performance robots powered pneumatically, because of control problems associated with the compressibility of air. Perhaps the most advanced example is the Utah/MIT Dextrous Hand Master (Jacobsen et al., 1986).

Micromotion Actuators

One of the more exciting new areas under development is micromotion robotics, in which (macro) robots are able to position precisely down to 1 nanometer (Dario, 1992). As a counterpoint to simulated molecular docking (Ouh-young et al., 1988), these robots would actually be able to manipulate molecules. The development of true microsize robots is still somewhere off in the future, but the new area of microelectromechanical systems (MEMS) holds promise for developing the proper components: structures, actuators, and sensors.

For micromotion control, piezoelectric actuators dominate. They are used in scanning tunneling microscopes (STMs) and atomic force microscopes (AFMs). Hollis has used a magnetically levitated hand controller to control an STM (Hollis et al., 1990). A stacked actuator consisting of a linear voice coil motor in series with a piezoelectric element was the basis for Hunter's telemicrorobot (Hunter et al., 1991). Hatamura (Hatamura and Morishita, 1990) has teleoperated a 6-axis force-reflecting nanorobot whose individual axes are flexure elements activated by piezoelectrics; the masters are two bilateral joystick mechanisms, and vision is provided by a stereo scanning electron miscroscope (SEM).

Shape memory alloy (SMA) actuators hold considerable promise as a compact but powerful actuator source. Various robotic mechanisms have been proposed that incorporate SMA actuators, including active endoscopes (Dario et al., 1991; Ikuta et al., 1988). A tactile stimulator employing cantilever arrays activated by SMA wires has been developed commercially (TiNi Alloy Company). At the moment, two major drawbacks of SMA are highly nonlinear dynamics and slow response speed. Recent developments by Hunter (Hunter et al., 1991) have sped up the response by 100 times and hold promise for the future.

End Effectors

Most end effectors on robots or telerobots are unremarkable, usually two-jaw grippers or special purpose tooling. Multifingered robot hands have been developed to provide robots with the same dexterity as the human hand. The major commercial examples are the Stanford/JPL hand (Salisbury, 1985) and the Utah/MIT Dextrous Hand (Jacobsen et al., 1986). Master gloves have been used to teleoperate particularly the Utah/MIT Dextrous Hand (Hong and Tan, 1989; Pao and Speeter, 1989; Rohling and Hollerbach, 1993; Speeter, 1992). Force-reflecting multifingered master-slave systems have been developed by Jacobsen et al. (1990a) and Jau (1992). A 3-DOF hand partly inspired by prosthetics is the end effector for the Sarcos Dextrous Arm.

Sensors

Sensor technologies for telemanipulators include sensors required to monitor the internal mechanical state of the arm (joint angle sensors and joint torque sensors), the external contact state (wrist force/torque sensors and tactile sensors), and proximity sensors. We do not cover visual sensors (cameras) and image processing. Position trackers and inertial sensors are reviewed in Chapter 5.

Joint Motion Sensors

A review of traditional joint motion sensors is provided by deSilva (1989). For rotary motion, common sensors are potentiometers, optical encoders, resolvers, tachometers, Hall-effect sensors, and rotary variable differential transformers (RVDTs). For linear motion, common sensors are linear variable differential transformers (LVDTs), linear velocity transducers (LVTs), and position sensitive detectors (PSDs).

Most of the rotary sensors listed above are analog sensors. Potentiometers are not that favored because of noise and sensitivity problems, and they are hard to make small. For use in robot fingers and hand masters, compact Hall-effect sensors are used in the Utah/MIT Dextrous Hand and in the EXOS Dextrous Hand Master. The resolution is not high (0.2 deg), but is adequate for these devices. The VPL DataGlove employs fiber optic sensing, but this effect is too coarse to be really useful and there have been many complaints about the accuracy of this system. Resolvers and tachometers are suitable for larger actuators and joints, such as robot shoulders and elbows. RVDTs are moderate in size but also have a moderate resolution.

The trend is increasingly toward digital sensors. Optical encoders offer the highest resolution; for example, Canon produces an incremental laser rotary encoder with 24 bits of resolution. The Sarcos Dextrous Arm has 18 1/2 bit incremental encoders. The trend is for rotary encoders to become less expensive while maintaining high resolution, to become more compact in size, and to provide high-resolution absolute joint angle readings (Steve Jacobsen, personal communication).

For linear transduction, LVDTs and LVTs are common. Resolutions in the range of 10-100 nm are possible for LVDTs. Linear PSDs have been reported to have resolutions in the range of 1-5 nm. Digital linear sensors are being developed with a resolution of 2.5 nm (Steve Jacobsen, personal communication). The ultimate in high-resolution linear measurement is of course interferometry, for which resolutions of 0.1 nm are possible. An additional consideration is the sampling rate while maintaining high resolution; Charette et al. (1993) reported a 1 MHz rate. Future developments should result in reduced size of such sensors and increased use in micromanipulation.

With increased resolution such as that provided by optical encoders, the calculation of joint velocity and acceleration from positional data will become more accurate. This calculation is required for precise control and calibration. Recent research has focused on how these derivatives are to be calculated (Belanger, 1992; Brown et al., 1992).

Joint Torque Sensors

Strain gauges are most commonly used for force and torque sensing; the review by deSilva (1989) is again relevant. Typically some flexible structure is attached in series with a joint axis; an example is the Sarcos Dextrous Arm, with torque sensors having a dynamic range of 1:2,000. Autonomous calibration of joint torque sensors was considered by Ma et al. (1994). Joint torque sensors have been retrofitted to PUMA robots by Pfeffer et al. (1989), to the Stanford Arm by Luh et al. (1983), and to a direct drive arm by Asada and Lim (1985). An instrumented harmonic drive for joint torque sensing was presented by Hashimoto et al. (1991).

Displacement sensors may also be employed to measure joint torque; for example, inductive sensors were employed by Vischer and Khatib (1990) and in ROTEX. A variable reluctance joint torque sensor is also discussed by deSilva (1989). Hall-effect sensors on a cantilever system are employed to sense tendon tension on the Utah/MIT Dextrous Hand. Optical joint torque sensors using light emitting diodes have been developed by Hirose and Yoneda (1989). Displacement sensors can have advantages over strain gauge sensors, such as lower cost and higher robustness, although the sensitivity is typically lower. The future will probably continue to see alternatives to, and a movement away from, strain gauge sensing as micro positional sensors improve.

Another trend is the use of improved electric motor models to predict torque accurately open loop. This may involve the design of new motors (Hollerbach et al., 1993; Wallace and Taylor, 1991) or the reverse engineering of existing motors (Newman and Patel, 1991; Starr and Wilson, 1992). When a transmission element is employed, one alternative is a careful characterization of joint friction to compensate for its effect (Armstrong-Helouvry, 1991).

Accurate knowledge of joint torque is very important for precise control and, in the context of teleoperation and haptic interfaces, for force reflection. Despite this importance, very few manipulators actually have this capability. The development of torque-controlled joints will be essential for higher performance devices in the future.

Wrist Force/Torque Sensors

An alternative, or a complement to, joint torque sensing is to employ multiaxis force/torque sensors, usually mounted at the wrist. Such sensors have also been used as haptic interfaces, such as the Trackball or

Spacemouse. Sensor technology is the same as that discussed under joint torque sensors, but the multiaxis characteristic offers substantial complications.

Most frequently, a 4-beam arrangement in a Maltese cross configuration has been employed with strain gauges; commercial examples include the JR3 sensor and the Assurance Technology sensor. A significant problem is cross-axis interference, due to nonlinear beam bending (Flatau, 1973; Hirose and Yoneda, 1990); this effect may produce errors up to 3 percent. Although more complex calibration may alleviate this effect, another approach is to use alternative structures. Nakamura et al. (1988) proposed a boxlike sensor. Yoshikawa and Miyazaki (1989) proposed a three-dimensional cross-shaped structure. Another possibility is a membrane suspension (Gerd Hirzinger, personal communication).

As an alternative to strain gauges, the use of optical sensing has been proposed (Hirose and Yoneda, 1990; Hirzinger and Dietrich, 1986; Kvasnica, 1992). The most precise multiaxis force/torque sensor built to date employs a magnetically levitated wrist and optical sensors (Tim Salcudean, personal communication). The wrist is servoed to a null position, and a motor model is employed to infer the forces and torques; hence there is no cross-coupling. This idea is similar to that of the Sundstrand accelerometers, the most accurate on the market.

There is considerable room for improvement in the market for commercial 6-axis force/torque sensors. Sometime in the future we can expect accuracies of around 0.1 percent, including cross-coupling effects; this would represent about an order of magnitude improvement over those currently on the market.

Tactile Sensors

There have been a number of reviews of tactile sensing technology (Dario, 1989; Dario and De Rossi, 1985; Hollerbach, 1987; Howe and Cutkosky, 1992; Nicholls and Lee, 1989; Pugh, 1986). Many different effects have been employed; piezocapacitance, piezoresistance, and piezoelectrics are some of the more common ones. Tactile sensors have also been produced using optical sensing (Maekawa et al., 1992). Very large scale integrated (VLSI) fabrication methods have also been employed to produce tactile sensors.

Commercially, piezoresistive tactile sensors were produced by the former Lord Corporation and by Barry Wright Controls Division. Piezoresistive inks have been employed in the Interlink Electronics tactile sensors. Very few other examples of commercially available tactile sensors may be found.

Hysteresis, sensitivity, and repeatability are often problems with piezoresistive sensors. Piezocapacitance sensors circumvent some of these problems. Piezoelectric sensors are temperature sensitive, often function only in an AC mode, and are hard to make very small because of cross-talk. Other sensor technologies are often too complicated or fragile to make useful devices.

The vast majority of tactile sensors sense only normal force. Multiaxis stress sensors have been proposed by De Rossi et al. (1993) and McCammon and Jacobsen (1990). In the context of teleoperation and tactile stimulation, tactile sensors for normal force are probably adequate because tactile stimulators are likely only to be able to produce normal force.

The bottom line is that, despite all the published work on tactile sensors, almost none is used on robots. The problems have to do with packaging, cost and complexity, response properties, robustness, and suitability for curved surfaces such as fingertips. This is a technology area that needs considerable development, although economic drivers may not be in place for it.

Proximity Sensors

Proximity sensors, intermediate between contact sensors and visual sensors, are used for docking maneuvers of a manipulator end effector with an object or target. They are particularly useful in teleoperation to account for model discrepancies and to compensate for obstructed vision. For example, proximity sensors in the end effector of the ROTEX manipulator play an important role.

Main technologies include electromagnetic, optical, and ultrasonic proximity sensors. Electromagnetic sensors (induction, capacitance) are limited in range and detectable materials (Novak and Feddema, 1992). Ultrasonic sensors are not useful for short-range measurements. Hence most proximity sensing has hinged on optical reflectance sensors. A review of such sensors is provided by Espiau (1987). A challenge for these sensors is to separate the effects of distance, surface orientation, and reflectance properties. Multiple detectors are one way of inferring surface orientation (Lee et al., 1990; Okada, 1982; Okada and Rembold, 1991; Partaatmdja et al., 1992). An optical proximity sensor based on the confocal principle has been reported by Brenan et al. (1993).

REMOTE VEHICLES

Remote vehicles, or mobile robots, encompass any basic transport vehicle that can be operated at a distance: indoor motorized carts, road vehicles, off-terrain vehicles, airborne or space vehicles, boats, and submersibles. Many mobile robots also will carry one or more remote manipulators. This section highlights mobile robots exemplifying the current state of the art, major issues involved in the development of mobile robotic systems, and remaining research and development challenges.

Systems

The arguably perfect mobile robotic system would: (1) be easy to control or program, (2) automatically transit an unstructured, highly complex, dynamic environment, (3) automatically perform general sensory and manipulative tasks, and (4) if required or desired, transmit detailed, easily interpreted, sensory information describing its environment and task state in real time. It would be capable of performing these tasks for long periods of time, over long distances, and would not require a physical tether for either power or data transmission. Unfortunately, such a system is not currently technically feasible or even physically achievable for some scenarios.

Environmental and physical factors attenuate data transmission bandwidth with distance; platform design places constraints on the environments that can be traversed; available energy systems limit endurance; sensors and sensor processing technology limit the type, form, and reliability of information about the environment available to the robot; the state of the art in high-level control limits the robot's autonomous capabilities; and available computational devices limit how much sensor processing and high-level control can be embedded in the remote system. In addition, reliability, volume, and cost issues exert a strong influence on the designs of current mobile robotic systems. In spite of these constraints, however, highly successful mobile robotic systems have been developed. These systems can be classified into four major physical configuration classes based on different weighting of endurance, maneuverability, automation, and cost attributes.

Class 1 Systems: Power and Data Tethered

The power and data tether that characterizes Class 1 vehicles allows these systems to be optimized for endurance and cost. Due to the use of a power and data tether, mission duration is essentially unlimited and telemetry can be very high-bandwidth and immune to noise, jamming, and occlusion. Range, however, is limited by tether length and the tether is subject to entanglement. In addition, combined power and data tethers tend to be bulky and can impart tremendous drag to the remote vehicle, thereby limiting maximum achievable speed. As a rule, Class 1 systems are teleoperated and have minimal on-board automation (usually limited to closed-loop servo-control of actuators). All major navigation and strategy decisions are made by a human operator using vehicle navigation, collision avoidance, and scene understanding sensors. Human-machine interfaces for these types of systems range from relatively simple collections of analog and symbolic interface devices to more sophisticated systems with stereoscopic video feedback and force-reflecting manipulator controllers. Simple short-range land vehicles and a majority of undersea vehicles are been exemplars of Class 1 vehicles.

Typical of Class 1 vehicles are the dozens of low-cost, commercially available, remotely operated vehicles developed for undersea inspection or light work tasks to depths of a few thousand feet. An example is the Hydrovision Ltd. Hyball undersea inspection system (Busby Associates, 1990). This small (.46 m × .51 m × .47 m), lightweight (39 kg) system has an on-board video camera on a pan and tilt and work lights. The Hyball can operate to 300 m and is operated using a simple video display and joystick. It can be outfitted with a scanning sonar, a low-light level camera and auto altitude system.

Representing the high end of Class 1 vehicles, the Advanced Tethered Vehicle (ATV) (Morinaga and Hoffman, 1991; Busby Associates, 1990) is a large (6 m × 3 m × 2.5 m, 5,000 kg) undersea work system developed by the Navy for general undersea repair and recovery tasks at full ocean depth. It currently holds the world depth record for a tethered vehicle (20,600 ft) and is capable of speeds to 2 kn. Its overall sensor and actuator complement includes: (1) four video cameras—a stereo pair, a single camera with zoom lens on pan/tilt devices, and a fixed camera for position reference; (2) three manipulators—two force-reflecting arms (6-DOF arm and 1-DOF gripper) and a simple grasping device; (3) a scanning forward looking sonar; and (4) depth and heading sensors. Navigation is augmented by a long-baseline acoustic positioning system. A sophisticated van-based control system includes a vehicle driver and a manipulator or work system operator console. The manipulator operator console contains a stereoscopic panel-mounted display and a pair of human-sized, replicate, force-reflective master controllers. The vehicle driver console has access to video, sonar, and navigation information. The ATVM-Us, which has a 1.2 inch diameter, 23,000 ft power and data tether, represents one of the project's major contributions to Class 1 underwater vehicles (Freund, 1986).

Class 2 Systems: Data Tethered

Class 2 systems are highly maneuverable yet cost effective. Remote vehicle power requirements and overall system costs are minimized by relying heavily on the human operator for sensory and control decisions rather than on-board automation systems. Access to remote vehicle actuation and sensory capabilities is provided by a tethered telemetry system. Data tethers, typically fiber optic cables, are much less bulky than power tethers. These cables can support very high-bandwidth, secure, nonjammable, non-line-of-sight telemetry to ranges of over 100 km without repeaters, and they do so without imparting significant drag to the remote vehicle, since they can be actively or passively payed out from it. Fiber optic cables cost approximately $1-2 a meter and, depending on the application, may not be reusable. Due to the possibility of cable entanglement or breakage, a nontethered, low-bandwidth, non-line-of-sight telemetry system is frequently employed as a minimal capability backup. Mission duration of Class 2 systems is limited due to the requirement to carry on-board energy sources. This class of mobile robots has historically received the most interest in the human-machine interface area, since they are aimed at being highly maneuverable and capable yet still posseses continuous high-bandwidth telemetry. Some air vehicles, more advanced undersea vehicles, and most land vehicles capable of executing realistic missions are exemplars of Class 2 systems.

An example of the latter is the Department of Defense Unmanned Ground Vehicle Program TeleOperated Vehicle (TOV) system, an exterior, off-road capable, surveillance robot (Aviles et al., 1990). The remote vehicle platform is based on the military's four-wheel drive utility vehicle, the High Mobility Multi-purpose Wheeled Vehicle (HMMWV). It is configured as a modular, remotely operated mobile platform with a fiber optic data link to provide high-bandwidth telemetry out to 30 km. In addition to making all basic vehicle functions remote, a stereoscopic camera pair and two artificial pinnae are mounted on a pan and tilt platform to provide feedback for remote driving. Navigation information is provided by a satellite-based navigation system that performs dead reckoning between satellite updates. Up to three add-on mission-specific subsystems (mission modules) can be added to the base vehicle. The usual mission module is a reconnaissance, surveillance, and target acquisition system that includes a low-light level video system, a forward-looking infrared sensor, and a laser range finder and designator. All these sensors are mounted on a pan and tilt platform on top of an extendable 15 ft scissors mast.

The TOV control station is mounted in a mobile shelter and is designed to provide the human operator with a control and sensory experience as similar as possible to normal, nonremote driving. The operator is provided with replicas of the HMMWV steering wheel, accelerator, brake, shifter, and ignition controls. Feedback for driving is provided primarily by a stereoscopic head-mounted display (HMD) with a binaural headset. The remote vehicle camera pan and tilt is slaved to the operator's head motions while wearing the HMD, and basic navigation information is overlaid onto the operator's visual scene. The TOV has been extensively tested in both on-road and off-road conditions to 55 km/hr and can be remotely driven to the limits of the basic platform.

Sometimes Class 2 systems, such as the XP-21, a modular undersea vehicle developed by Applied Remote Technology (Busby Associates, 1990), are used as test beds for autonomous control. The XP-21M-Us data tether allows use of powerful off-board computational systems and quick reconfigurability. The XP-21 is approximately 5 m in length and .5 m in diameter, has a maximum speed of 6 kn, and a 40 mi cruise range. It can also run on preprogrammed missions without the tether.

Class 3 Systems: Nontethered Telemetry

Class 3 systems fall into two major categories. The first contains mobile robots that have continuous, high-bandwidth, line-of-sight, nontethered telemetry systems. These mobile robots are equivalent to Class 2 systems but are limiting their range of operation in order to remove the disadvantages of a physical data tether. Sometimes a Class 2 system will be put in a configuration of this type for training purposes or short-range missions and use cable only for extended-range missions.

The second category of Class 3 systems represents a uniquely different approach to the development of mobile robots. On-board automation is emphasized in order to remove the physical data tether yet still be capable of performing long-range missions. Class 3 systems have a telemetry connection to their control station, but it is a low-bandwidth, non-line-of-sight, connection incapable of supporting direct manual control by the human operator. These systems typically exhibit at least supervisory-level control and can often perform reasonably complex behaviors autonomously. The human's role is one of a supervisor, giving high-level commands to the remote vehicle and monitoring its progress. This level of control is not only a goal for human/operator load reduction purposes but also is a requirement for stable control of the remote platform under telemetry-induced delays (Ferrell, 1965; Sheridan, 1970). High-resolution imagery is often selectively telemetered to the operator at very low frame rates (on the order of seconds per frame). Mission duration is still limited by on-board energy systems but through intelligent power management approaches this can be significantly extended. This class of systems has historically received reasonable interest in the human-computer interaction arena as relating to control partitioning and sharing (Chu and Rouse, 1979), but relatively little attention has been paid to remote presence approaches. Most air vehicles, more advanced undersea vehicles, some land vehicles, and planetary rovers have been exemplars of this type.

Rocky III (Wilcox, 1992; Desai et al., 1992) is a 15 kg, 6-wheel, planetary rover test bed developed at the California Institute of Technology's Jet Propulsion Laboratory. It has a 9,600 baud radio telemetry system, on-board computation, and a 3-DOF arm outfitted with a soft soil scoop. Rocky III's on-board batteries provide a 10-hour mission duration. Very simple navigation and collision avoidance sensors are used. Navigation is accomplished using a gimballed flux-gate compass and wheel encoders for dead reckoning. Collision avoidance information is provided by sensors connected to the front wheels and to a skid plate for objects that go between the wheels. Using the telemetry system, an operator designates a site to be sampled with the soft soil scoop and optional intermediate way-points. The rover then accomplishes its mission, including obstacle avoidance maneuvers, with no further communication. Rocky III's larger cousin, Robby (Desai et al., 1992), a 6-wheeled 3-body, 1,000 kg, articulated vehicle, has demonstrated semiautonomous navigation through a rough natural terrain at a rate of 80 meters per hour using both deliberative and reactive control paradigms on stereo-vision-provided data.

The Mobile Detection Assessment and Response System (MDARS) program, a joint effort of the U.S. Army's Armament Research Development and Engineering Center and the Navy's Naval Command, Control and Ocean Surveillance Center, has developed an interior, supervisory controlled, physical security robot as an adjunct to fixed security sensors (Everett et al., 1990, 1993; Laird et al., 1993). The 3-wheel drive, 3-wheel steered, remote platform is 6 feet tall and weighs 570 pounds. It is outfitted with a 9,600 bit per second bidirectional radio telemetry system, navigation sensors, and intruder detection sensors. Collision avoidance sensors include a 9-element ultrasonic array and bumper-mounted collision detectors. Navigation is accomplished using a hybrid navigation scheme that combines compass/encoder-based dead reckoning and a wall-following/reindexing system. The wall-following system updates and refines the robot's computed position using an a priori map of static features in the environment and readings from acoustic ranging sensors. Intruder detection sensors include a 360 deg, 24-element ultrasonic array, microwave motion detectors, passive infrared motion detectors, a video motion detection system with a near-infrared light source, and near-infrared proximity detectors. The mobile platform can automatically follow preprogrammed or random paths and performs automatic obstacle detection and avoidance maneuvers. It periodically stops to look for intruders and alert a human supervisor when an on-board security assessment system determines that an intruder is likely (Smurlo and Everett, 1992). The operator then has the option of (1) ignoring the alert and ordering the robot to continue its patrol, (2) asking the robot to get closer to the detected object for evaluation using the on-board video camera, or (3) taking over control in a reflexive teleoperated mode (automatic collision avoidance). Initial tests of the system in military warehouse environments have demonstrated probabilities of detection well in excess of 0.90 with a very low false alarm rate. The system is being extended to allow the supervision of multiple mobile platforms by one operator.

In the underwater environment, the Advanced Unmanned Search System (AUSS) (Walton et al., 1993; Uhrich and Walton, 1993), developed by the Navy, is a supervisory controlled, broad-area, undersea search system. The remote vehicle is 17 ft long and 31 inches in diameter, weighs approximately 2,800 lb, has an endurance of 10 hr and a maximum velocity of 5 kn, and can operate to depths of 20,000 ft. An acoustic link transmits compressed search data from the vehicle at 4,800 bits/s and sends high-level commands to the vehicle at 1,200 bits/s. The primary search sensor is a side-looking sonar. Electronic still and 35 mm film cameras provide imagery for identification. Depending on the amount of compression desired, sonar and video images take from 20 s to 2 min to transmit. On-board navigation sensors include a forward-looking sonar, a Doppler sonar, gyro-compass, depth sensor, attitude sensors, and rate sensors. In addition, bottom-deployed long-baseline acoustic transponders and ship-based short-baseline acoustic, Loran-C, and global position system (GPS) navigation systems can be used to update the remote vehicle navigation system and to allow the surface support craft to maneuver to maintain the acoustic telemetry link. In a typical scenario, the AUSS system operator commands the remote vehicle to execute a search path and supervises the system by monitoring vehicle position, status, and transmitted imagery. If an object of interest is detected by the operator, the vehicle can be ordered to automatically home in on the object and get higher resolution video imagery for evaluation.

Class 4 Systems: Nontethered, No Telemetry

The final class of systems represents the perceived high ground of mobile robotics research and development. The premium on on-board automation is extremely high, and the remote vehicle carries out its mission without requiring human monitoring or intervention. The human is involved only in programming or specifying the desired high-level behavior of the system and possibly in retrieving mission or sensory data after the mobile robot has returned from an excursion. This means that all sensor regard control, interpretation, and the reasoning required to transit within the environment without collisions must occur on board. Class 4 systems do not need a telemetry connection to their control station and therefore can be highly maneuverable and operate to long distances. Mission duration, like Class 2 and Class 3 vehicles, is still limited by on-board energy sources but again can be extended tremendously by intelligent power management. In addition to being an intellectual focus of the mobile robotics community, Class 4 systems can perform tasks for which maintaining telemetry would be problematic or impossible, such as excursions deep inside of structures. This class of systems has historically received minimal attention in the human-machine interface area, since the overall effort is to limit human involvement to goal specification at most. As yet, systems capable of rapid transit in general, unconstrained environments without a priori knowledge of that environment do not exist. Some very interesting systems that operate in more constrained environments have been constructed, however.

The Carnegie Mellon University (CMU) Navlab and Navlab II (Kanade, 1992; Mettala, 1992; Thorpe, 1990) mobile ground robot test beds have demonstrated impressive performance at road following and cross-country traversal. These vehicles navigate using sonar, gigahertz radar, and an ERIM laser rangefinder. The best runs to date over moderate off-road terrain have occurred at 6 mph. ALVINN, a neural network road-following system, has been used by the CMU researchers to drive the Navlab II up to 62 mph on highways and for a continuous distance of over 21 miles. The neural network is trained by observing a human driver. Image understanding for mobile robotic applications, as exemplified by the CMU work, is currently the focus of intense research sponsored by the Advanced Research Projects Agency (1992).

Technologies and Directions

Although all the major technology areas depicted in Figure 9-1 continue to be the focus of intense research and development, the most significant and relevant developments have been in the sensor, platform, actuator, high-level robotic control, and human-machine interface fields.

Sensor Systems

One of the classic problem areas constraining the development of mobile robotic systems has been the development of sensors supporting navigation and collision-free transit through the environment. The platform must be able to navigate from a starting position to a desired new location and orientation, avoiding any contact with fixed or moving objects en route. The difficulty can be directly related to the requirement for the platform to move at reasonable speeds and the unstructured nature of the operating environment.

Navigation Sensors and Systems Major techniques for determining vehicle position and orientation are dead reckoning, inertial navigation, beacons,

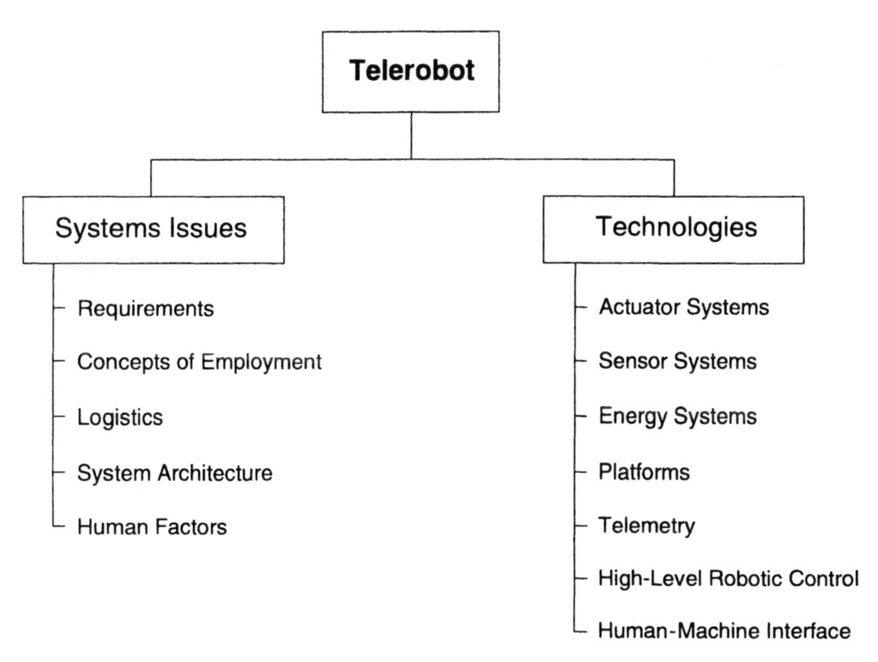

FIGURE 9-1 Mobile robot system and technology areas.

satellite navigation, and map matching. Recent developments in small, low-cost inertial linear accelerometers and angular rate sensors and the maturation of global positioning system (GPS) technology, however, are of particular import for mobile robot navigation and sensing. Inertial and GPS technologies are highly complementary. Inertial systems are very precise for short distances and short times but are subject to long-term drift. GPS systems, in contrast, provide somewhat less accurate but real-time updates over long periods of time and long distances without drift.

Standard GPS technology provides a location solution with an accuracy of about ± 30 m. The use of differential GPS techniques, through which correction factors from a fixed GPS receiver station are transmitted to the mobile platform and its GPS receiver, allow a positional accuracy of approximately ± 2 m. GPS receivers are now available in both hand-held and chip-set versions, such as the Rockwell NavCore V Positioning System Receiver Engine. In addition, multiantenna systems such as those developed by Trimble Navigation use signal phase time-of-arrival information to provide both orientation and more accurate position information.

Silicon-based miniature accelerometers are challenging the dominance of more traditional rate and acceleration sensors, such as quartz beam or fiber optic gyroscopes. In general, silicon-based microaccelerometers measure the displacement of a proof mass attached to a silicon chip in response to inertial force by piezoresistive, capacitive, or resonant means (Yun and Howe, 1992). For example, Triton Technologies Inc. has developed an accelerometer using capacitive sensing with 0.1 mg resolution, 120 dB dynamic range, and a cross-axis sensitivity of less than 0.001 percent (Henrion et al., 1990). Researchers at the Berkeley Sensor and Actuator Center have developed a capacitive accelerometer that integrates the sensor and readout electronics in a 2.5 mm × 5.0 mm area (Yun et al., 1992). Pointing to potential future performance enhancements, researchers at JPL's Center for Space are exploring the use of electron tunnel sensors for measuring acceleration. These systems hold the promise of surpassing the performance of capacitive sensors by four orders of magnitude (VanZandt et al., 1992). Magnetohydrodynamic angular rate sensors (Laughlin et al., 1992) also show promise with high precision (> 0.1 μrad) and dynamic range (> 100 dB) at low-power consumption (< 0.3 W). To date, however, sensors of this type capable of measuring constant input rates are still under development.

These advances in GPS and inertial technology are allowing the development of lightweight, low-cost, integrated GPS/inertial navigation systems specifically targeted for remote vehicle applications. A prototype system, built by Rockwell International, weighs 5.5 lb, draws 18 watts, and has a volume of 115 cubic inches (Griffin and Murphy, 1992). A NavCore V GPS chip set and tactical grade inertial sensors using piezoelectric bender crystals mounted on a motor shaft are utilized. Initial positional error specifications are 76 m SEP (spherical error probable) with future systems slated to have positional accuracies on the order of 15 m SEP by using the military version of GPS and solid-state accelerometers.

Collision Avoidance and Scene Understanding Sensors and Systems Acoustical, optical, and electromagnetic sensors using proximity, triangulation, time of flight, phase modulation, frequency modulation, interferometry, swept focus, and return signal intensity ranging techniques have been employed on mobile robots for collision detection and scene understanding purposes (Everett et al., 1992). Recent developments in millimeter wave radars and laser radars are especially relevant to the mobile robotics community.

Millimeter wave (MMW) radar utilizes that portion of the electromagnetic spectrum from wavelengths of approximately 500 μ to 1 cm. In theory, MMW-sensing systems can have much higher resolution and fit into smaller packages than conventional radar systems in the microwave portion of the electromagnetic spectrum (approximately 3 to 100 gHz) with some penalty in shorter operating distances and more attenuation by environmental factors. Although, MMW radar systems are currently not widely commercially available, several promising prototype systems have been developed. Examples of current MMW radar sensors are systems developed by Millitech, Battelle, and Kruth-Microwave. Millitech has developed two prototype sensors for robotic collision avoidance applications (Millitech, 1989). The first sensor has a 30 × 30 degree field of view, a maximum range of 10 m, a range resolution of 1 cm, and an update rate of 100 Hz. It is targeted at providing the range to the nearest obstacle. The second sensor is designed to scan 360 degrees in azimuth, track multiple targets, and determine range and bearing for each. It has a maximum range of 100 m, a 2 degree azimuth resolution, a range resolution of 10 cm, and an update rate of 40 Hz. Researchers at Battelle Memorial Institute have developed an MMW radar system for use as an automobile collision-avoidance sensor (Wittenburg, 1987). It allows range, velocity, amplitude of returned signal, and angle to be determined for multiple targets. In an envisioned configuration for automotive use, it would scan a 30 deg sector at a 5 Hz update rate with a 5 m range resolution out to 60 m and a 10 m range resolution out to 100 m. Finally, Kruth-Microwave Electronics Company has developed a low-cost, low-power prototype MMW radar system for unmanned system use (Kruth-Microwave Electronics, 1989). It has a theoretical range of 1 km and a range resolution of better than 1 m, can detect targets of approximately 0.1 m cross-section, and can be packaged in approximately one-third of a cubic ft volume.

Laser radar (LADAR) or laser scanner technology for terrestrial applications is now relatively mature and can be used for both obstacle detection and landmark identification (Besl, 1988; Hebert et al., 1988). Systems developed by the Environmental Research Institute of Michigan (ERIM) and Odetics exemplify the state of the art. The ERIM laser scanner, used by Carnegie Mellon University on its Navlab series of mobile robots, provides resolutions of 0.5 deg/pixel horizontal (80 deg field of view) by 0.3 deg/pixel vertical (30 deg field of view). It has a maximum unambiguous range of 20 m, a range resolution of 8 cm, and a 2 Hz update rate. The Odetics system provides 0.5 deg/pixel over a 60 deg vertical and horizontal field of view. It has a resolution of 1.8 cm out to a range of 9.4 m.

Along a similar front, laser-based imaging systems for the underwater environment are being developed (Fletcher and Fuqua, 1991). These systems typically use lasers capable of producing energy in the blue/green spectrum (wavelengths in the range of 460 to 560 nm), which corresponds to a window of minimum absorption in seawater. SPARTA Incorporated has developed a range-gated imaging system that has dimensions of approximately 12 × 27 in, a 12 deg field of view, a 10 Hz update rate; it weighs approximately 120 lb (neutral in water), and consumes 450 watts of power (Swartz, 1993). Westinghouse has continued development and run sea trials on a blue/green laser line scanner with fields of view from 15 to 70 deg (Gordon, 1993).

Platforms

In the area of platform design, research and development on legged robots is particularly important. In his overview of the field, Raibert (1990) suggests that legged locomotion research is well advised, since only half of the earth's land mass is navigable using current wheeled and tracked vehicles. In addition, not only is legged platform research attempting to develop platforms that can traverse difficult terrain but also it has served as a focus for understanding human and animal locomotion. This work has led to a reasonable understanding of gait under a variety of regimes (Song and Waldron, 1989) and to a variety of commercial and research platforms. Odetics Inc. has built a series of supervisory controlled hexapods. Its first legged platform, the ODEX I (Russell, 1983), weighed 370 lb, could lift 900 lb, and could walk onto a small truck bed. The Adaptive Suspension Vehicle, a 6000 lb, 6-legged, terrain-adaptive suspension vehicle developed at the Ohio State University (Waldron and McGee, 1986; Song and Waldron, 1989) has a single human operator and a 5 mph top speed and is capable of crossing 6 ft ditches, stepping over or onto obstructions over 4.5 ft high, and navigating 60 percent grades. Raibert at CMU and MIT has developed a variety of monoped, biped, and quadruped test beds including a planar biped capable of jumping through hoops and running at speeds of over 13 mph (Raibert, 1990). Legged platform research and development is particularly relevant to the design and control of figures and autonomous agents in virtual environments (see Badler et al., 1991).

High-Level Robotic Control

A relatively new and controversial approach to high-level control for mobile robotics is reactive control. This form of control and its subset, behavior-based control, rely heavily on the immediacy of sensory information, channeling it directly to motor behaviors without the use of an intervening symbolic representation of the world. More traditional strategies construct representations or models of the world and then reason based on these models prior to acting (Brooks, 1986, 1991; Jones and Flynn, 1993). Approaches relying on this traditional approach have typically been hampered by sensor-produced inaccuracies in the world model and the significant amount of computation required to reason about the abstract model. Reactive systems do not possess these two weaknesses because they bypass the creation of a world model altogether and directly couple sensory perceptual activities with action (Arkin, 1992). The ultimate limits on useful adaptive behavior achievable using reactive control approaches are unclear. A variety of systems have been developed that perform in dynamic, unstructured environments (Brooks, 1990; Connell, 1990; Anderson and Donath, 1991). For example, Brooks and his colleagues at MIT have built a series of mobile robots, including a small 6-legged robot (Brooks, 1990) that uses simple pitch and roll sensors, passive infrared sensors, and whiskers to exhibit robust autonomous walking, prowling, and directed prowling behaviors that avoid collisions. This level of competency is achieved without the generation of a symbolic representation of the world.

Human-Machine Interface

Human-machine interfaces for mobile robots range from simple analog or symbolic controls and displays maintaining minimal isomorphisms with the system to be controlled to highly immersive, spatially oriented, isomorphic, veridical human-machine interfaces engaging the visual, auditory, and haptic senses. Interfaces of the latter type have typically been designed in order to provide the human operator with some sense of remote presence or telepresence providing sensor feedback of sufficient richness and fidelity and controls of sufficient transparency that human operators feel as if they are present at the remote site. This approach is typically taken in order to engage the human's naturally evolved sensory, cognitive, and motor skills in the ways they are used in everyday tasks so as to minimize task completion times and the training required to operate the remote system (Pepper and Hightower, 1984). The definition and efficacy of remote presence, however, are currently a major research topic within both the robotics and virtual environment communities (Held and Durlach, 1992; Sheridan, 1992b). The type and form of the human-machine interface, although often not clearly separated, is orthogonal to the level of control (manual through autonomous) of the mobile robotic system. The type and form of the elements and level of information to be controlled, programmed, or monitored vary whether the mobile robot (or virtual entity in a VE) is manually or autonomously controlled, but similar human-machine interaction approaches, symbolic to immersive, can be applied.

Although a complete overview is beyond the scope of this paper, mobile robotics research and development on the human-machine system is highly relevant to the VE community. Not only have successful systems been developed that attempt to create a sense of remote presence

(Aviles et al., 1990; Morinaga and Hoffman, 1991) but a wide body of research outlining the time, speed, accuracy, and configuration trade-offs of different human-machine configurations exists (Sheridan, 1989, 1993; McGovern, 1990; Spain, 1992; Heath, 1992).

LOW-LEVEL CONTROL OF TELEOPERATORS

Teleoperators are complex systems composed of the human operator, master manipulator (joystick), communication channel, slave manipulator, and the environment (remote task). Since each of these systems is complex in its own right, in combination they create formidable analytical and design challenges. In particular, when slave contact force information is fed back to the operator through the master manipulator, the system becomes closed loop and thus stability is often a problem, even if each of the individual subsystems is stable in isolation. A related technology not considered here consists of "man-amplifiers" or "extenders" through which master and slave are effectively combined into one mechanism in direct contact with (or worn by) the human (Kazerooni and Guo, 1993).

The most common controller for robot manipulators in practice remains the classic proportional-integral-directive control (PID) compensator used on the individual joint positions. Experimental systems have become more sophisticated. For example, to accurately follow trajectories, robot controllers are greatly assisted by incorporating a dynamic model of the manipulator (Luh et al., 1980; Khosla, 1988).

Most teleoperators are used in tasks involving heavy contact with rigid or massive environments. Position and contact force cannot be simultaneously controlled because their relation is constrained by the environment. Thus, the task space can be segmented according to the contact constraints into subspaces in which position and force are individually controlled (Mason, 1981; Raibert and Craig, 1981). Alternatively, position error and force error can be related to each other through a stiffness constant (Salisbury, 1980) to generate actuator control signals in what is called *stiffness control*; for a review of these methods, see Whitney (1985). More generally, position and force can be related to actuator torque through a second-order dynamic model representing the desired mechanical impedance of the end effector (Hogan, 1985a, 1985b, 1985c).

Although force reflective teleoperators have been in existence for more than four decades (Spooner and Weaver, 1955; Sheridan, 1960; Goertz, 1964), there are still very few successful implementations of complex, high-DOF systems that satisfy what we envision as the ideal system. Although much of the performance limitation is due to physical hardware insufficiencies, great increases in performance can be achieved through proper design and implementation of the embedded control strategies used.

The ideal performance of a teleoperator has been described as a massless, infinitely stiff mechanical connection between the input device (master), and the effector (slave) (Biggers et al., 1989; Handlykken and Turner, 1980). Augmentation of the operator's sensory and motor skills, such as force magnification or displacement scaling, as well as compensation for environmental effects, such as gravity, are often additional design challenges (Dario, 1992; Flatau, 1973; Hatamura and Morishita, 1990; Hollis et al., 1990; Hunter et al., 1990; Vertut and Coiffet, 1985a). Desired characteristics of the coupled master-slave system include:

- low operator input impedance in free space
  - — inertia
  - — viscous drag
  - — friction
- high intersystem stiffness
- high-bandwidth force reflection
- stability for a wide range of contact impedances

These characteristics attempt to maximize the "transparency" of the overall system. An optimal system would be indistinguishable from direct operation on the environment itself, without the interposed machinery. However, given nonideal machinery, and given that there is still no clear consensus on an ideal remote manipulator, Sheridan et al. (1989) state that "the ideal manipulator is an adjustable one." These goals, however, have shown themselves to be formidable problems given the realities of limited-bandwidth actuation, limited sensory capability, and time delays in the communications and computation pathways.

Position-Based Teleoperation

The simplest form of teleoperator consists of a remote manipulator that is servo-controlled to follow the operator's position commands. The operator's intended motion is measured through a joystick or similar device. Trajectories for autonomous robot arms are mathematically defined for smoothness so that velocity and acceleration profiles are available as slowly changing inputs to the control system. In contrast, in teleoperation, only the position is measurable at a given time so that velocity and acceleration commands must be estimated and will therefore be noisier. Fortunately, most human movements are relatively smooth (Flash and Hogan, 1985).

The volume of space in which it is comfortable for the human operator to maintain hand position for extended periods is small compared with the total work volume of the human arm. Also, it is difficult to design hand controllers that can safely cover the entire work volume. As a result, many designs employ a much smaller work volume for the master hand controller than for the slave robot. To effectively use the slave robot then requires a method for selectively changing the offset between master and slave positions, usually referred to as *indexing*. This is accomplished in most systems with a finger button that momentarily breaks the connection between master and slave, during which time the operator repositions his or her hand. When the button is released, teleoperation resumes with motion increments referenced to the new master position.

The other significant form of control for remote manipulators is resolved rate control (Whitney, 1969). In this mode, human hand displacements are interpreted as a velocity command in an assigned Cartesian frame. This mode is used for example in the space shuttle Remote Manipulator System (RMS). One 3-DOF joystick is used for orientation commands and another for translation. A 6-axis hand controller has also been used as a rate controller for teleoperation (Bejczy et al., 1988). The command frame can be set arbitrarily so that commands can be referenced to the center of gravity of the held object, for example. One important requirement for rate control joysticks is a spring return. The spring return can be implemented with hardware springs or by computer-generated "software spring" force commands to an active joystick (Bejczy et al., 1988). This passive force feedback is essential for easily stopping the commanded motion.

A significant issue concerns which mode is better for teleoperators without force feedback. Kim et al. (1987) found that position control gave better completion times in simulated teleoperation, except for very slow simulated manipulators, for which rate control was slightly better. It is widely thought that, for tasks requiring large displacements, rate control can be superior because it eliminates the need for repeatedly indexing to perform motions larger than the master work volume.

Performance degradation occurs when there are significant rotations between the rate controller's frame and the frame defined by the user's body (Kim et al., 1993). In other words, if left-right hand motion causes end effector motion that is in a visibly different direction, confusion can result. This problem is particularly severe for the control of orientation when rate commands are referenced to rotation axes fixed to the robot hand. Feedback of contact forces in rate control has been tried by many laboratories without success. Parker et al. (1993) developed a novel control law that solved some of the problems by using a deadband. However, much work remains to show true performance benefits. Novel modes of rate control, transitions between rate control and position control, and relative performance between rate control and position control in teleoperation are unresolved issues that could have major impacts on application design.

Coupled Control of Manipulators for Force-Feedback Teleoperation

When mechanical master-slave manipulators were first made electrically remote, it was realized that force feedback to the master side was essential for good manipulation performance. The earliest systems used identical master and slave devices with decoupled controls of the individual joints in which joint torque for both master and slave was a function of position difference between them (Goertz and Thompson, 1954; Goertz, 1964). As improved computing power became available in the 1970s, it became possible to use kinematically different master and slave devices in which the master could be optimized for interfacing with the human operator, and the slave for its particular task (Bejczy and Salisbury, 1980, 1983). The computer performed the necessary coordinate transformations of the force and motion information and calculated the control laws in real time (Bejczy and Handlykken, 1981).

The details of these coordinate transforms depends on details of the master and slave manipulators. Human operator position is usually sensed indirectly through joint sensors on the master. These joint angle readings must be transformed through the forward kinematics of the master arm to derive the hand's position and orientation. Alternatively, the increments of joint motion can be transformed to Cartesian displacements through the manipulator Jacobian matrix. Similarly, in one popular force reflection architecture, slave motion, controlled and sensed in terms of joint torques, must be transformed into Cartesian coordinates, through the Jacobian transpose inverse of the slave and then into the joint coordinates of the master through the Jacobian transpose of the master. If a wrist-mounted force/torque sensor is used, the first of these transforms in not necessary. As with autonomous robot control, the performance of these operations is very sensitive to the numerical conditioning of the Jacobian matrix of the master and slave manipulators, since matrix inverses appear frequently in the relevant equations. Control methods have been developed for manipulators operating near singular configurations (Nakamura and Hanafusa, 1986; Wampler, 1986) but have not so far been applied to teleoperated robots.

Many teleoperator controllers have since been developed. Most of them have relied on existing position, stiffness, or impedance controllers on the slave and master manipulators (Jansen and Herndon, 1990; Yokokohji and Yoshikawa, 1992; Tachi, 1991; Goldenberg and Bastas, 1990; Strassberg et al., 1992). To link the master and slave and to provide kinesthetic feedback, these approaches choose one of the interaction variables (force or velocity) to send from master to slave (forward) and send its complement (velocity or force) from slave to master (feedback). A useful general representation of these models was developed by Fukuda et al. (1987), who formulated the controller in matrices that relate all measured variables to all actuated variables.

For practical reasons, many of these studies have been carried out using hardware designed for other purposes, with little capability for delicate force control. Often, for example, an industrial robot manipulator is used for the slave robot. Master manipulators (joysticks) are often highly geared mechanisms with joint limits or kinematic singularities significantly affecting the operator's motion. Mechanical limits are of key importance in determining teleoperator fidelity. Control technology can never fully overcome limitations imposed by friction, bandwidth, and actuator properties.

For example, stiction (static friction) imposes a lower limit on the magnitude of forces that can be displayed; actuator saturation provides an upper limit. Mechanisms capable of handling higher forces often have lower bandwidth and higher friction. Thus dynamic range of forces becomes important. Similarly, torque ripple in actuators will distort force commands to the operator or environment and costly force transducers in a closed-loop mode can only partially compensate. Many current studies are limited by a narrowness of approach in that they study only the control system. Major improvements in teleoperator performance will require attention to the underlying mechanisms as well as their control. This tightly integrated approach between control, actuation, and mechanization has been termed *mechatronics.*

The mechatronics of teleoperation needs further interdisciplinary study. However, research progress is currently limited by a lack of tools with which to quantify teleoperator performance. Although some work has been done (discussed below), few of the implementation studies have quantified the performance of their design. In the absence of standardized quantitative measures, an event at which researchers could bring their implementations together to allow comparative subjective testing could have a substantial impact on the field.

Modeling

The performance of a master-slave teleoperator has aspects that have been described qualitatively as "crispness," "viscosity," "stiffness," and "bandwidth." In the 1980s it was recognized that a useful analogy could be constructed between teleoperators with force feedback and 2-port electrical networks (Raju, 1988; Hannaford, 1989). This was an extension of the earlier treatment of robots in contact with their environments as 1-port impedances (Hogan, 1985a, 1985b, 1985c). The 2-port model describes and unifies the above qualitative descriptors into a multidimensional measure of transparency.

The 2-port model can be expressed in several forms by different matrices (impedance matrix, Z, admittance matrix, Y, hybrid matrix, H, and scattering matrix, S). Each of these matrices is a useful way to quantify the performance of force-reflecting manipulators. The elements of the various 2-port matrices are dynamic functions of frequency that quantify the stiffnesses, damping, and dynamic properties of the telemanipulators. Each 2-port matrix is a complete description of the teleoperator in a specific configuration, but each matrix is useful for different types of analysis.

Delay

A challenging issue arising in many applications is time delay between master and slave sites. This delay ranges from a few milliseconds in the case of computation delays to 10-100 ms delays induced by computer networks, to delays of seconds or more induced by multiple satellite communication links. These delays can induce loss of teleoperator fidelity and task performance as well as dangerous instability of operation (Hannaford and Kim, 1989; Kim et al., 1992). A breakthrough in this area occurred when Anderson and Spong (1988, 1989) developed an approach to making the two-way information channel between master and slave satisfy a passivity constraint regardless of time delay. Passivity quantifies the total energy output of a system and has been used before as a test for stability (Wen, 1988). With Anderson and Spong's approach used for the communication channel, if the master and slave controllers could be shown to be passive, then the system as a whole can be guaranteed passive and thus stable. This analysis assumes as well that the human operator and environment are passive.

The assumptions of passivity for the operator and environment are well accepted and appear to be consistent with the everyday experience of mechanical manipulators that are guaranteed passive and always stable under human control and actuation. Niemeyer and Slotine (1991) reformulated the passivity approach and explicitly normalized the relation between force and velocity by introducing the characteristic impedance. This work corrected a detail omitted by Anderson and Spong, who implicitly used a value of 1 for the characteristic impedance. Although passivity is an elegant method to prove stability, it allows no way to assess performance. Recent results (Lawn and Hannaford, 1993) have shown that the characteristic impedance of the simulated passive transmission line induces a trade-off between the series compliance from master to slave and the free motion damping. One or the other must be increased in proportion to time delay, which induces a signficant performance penalty. Therefore, the shortcoming of the passivity-transmission line theory is that performance of the system is not addressed.

Robustness

Recently, robust control concepts have been applied to teleoperator control, starting with earlier work in impedance control (Colgate and Hogan, 1988) and continuing by applying H-infinity optimization of control over a specified bandwidth (Andriot and Fournier, 1991; Kazerooni et al., 1993). This work allows the designer to optimize a controller to minimize distance measures between the 2-port model of the teleoperator and ideal transparency and passivity. This work seems especially promising since it is aimed at simultaneously achieving performance and stability for all loads over a specified bandwidth.

Scaling

An important extension of force-reflecting teleoperation is the idea of the scaling of mechanical energy between master and slave (Feynman, 1960; Flatau, 1973; Hannaford, 1990, 1991; Colgate, 1991). Recent implementations of this idea have encompassed an astonishing 10^{10} range of scale variations from 10^9 reduction of the operator's motion between powered joysticks and scanning tunneling microscopes (Hollis et al., 1990; Hatamura and Morishita, 1990), to approximately 10^1 increasing of the human motion in the control of large robots, such as construction or demolition manipulators. In between are applications such as muscle fiber manipulation (Hunter et al., 1990) and microsurgery (Colgate, 1991).

An important issue arises with scaling because of the scaling properties of dynamical systems. If an object is uniformly scaled in linear dimension by L, and some assumptions are made, then the inertia (which depends on volume) can be expected to scale with the cube of L, the friction (which depends on surface area) can be expected to scale with L^2, and the stiffness with L. This nonlinear scaling of the dynamic parameters results in qualitative changes in dynamic behaviors, such as natural frequency and damping ratio as size is changed. For example, if size is scaled down, natural frequencies will rise. Approaches to correcting this change in dynamics include time domain manipulations and frequency domain approaches (Colgate, 1991). Much research needs to be done in this area since the objects we will manipulate in the micro domain clearly will not be scaled versions of the objects in our own world.

Performance Evaluation

Finally, the performance of force-reflecting teleoperators is more difficult to assess than in unilateral systems with statically defined inputs and outputs. True system performance depends on visual display quality and manipulator and hand controller capabilities, in addition to the low-level controller's performance. The presence of the human operator introduces a source of variability that must be treated statistically at great cost in experiment time.

There have been a wide variety of measures used to characterize operator performance. Many of these measures, such as time of task completion, accuracy, and error, are similar to those used by production-line industrial engineers and motor-skill psychologists (Sheridan, 1992a). Other measures include Fitts information processing rate (Hill, 1979), peak force, variance in force, and sum of squared forces (Hannaford et al., 1989, 1991). Classes of tasks to which these measures have been applied include:

- Calibration tasks, such as tracking a specified path;
- Elementary tasks, such as stacking blocks, putting pegs in holes, and threading nuts on bolts; and
- Actual tasks, which depend on an application such as assembly of components of a machine.

In many cases, the measures are experiment-dependent and, due to the large variety of experiments humans have been subjected to, there is a large variety of measures. Even in the performance measurement of similar tasks with similar variables sensed, there is a lack of a standard metric (Sheridan, 1992a). For example, a variety of metrics that are a function of the force/torque of interaction have been used to characterize operator performance (Das et al., 1992; Hannaford et al., 1991; McLean et al., 1994). In addition, the tasks themselves need to be standardized so that comparisons can be made between alternative studies.

Related to the measure of human performance from experimental data is the analysis of the data. Statistical evaluation of data in teleoperation experiments has lacked standardization. Often, visual inspection of the means and standard deviations of data sets is used to make conclusions. There are a number of statistical methods available for the processing of data, and other fields of experimental research have adopted standards. It is encouraging to note that this trend is occurring in human performance studies in teleoperation.

In terms of published results, Kim et al. (1987) compared position control versus rate control, taking into account the joystick type (isotonic or isometric), display mode (pursuit or compensatory), three-dimensional task (pick-and-place or tracking), and manipulator workspace. They found that, regardless of joystick type, display mode, or task, when the workspace is small, position control is superior to rate control, by measures of completion time and accuracy. For larger workspaces or slow manipulators, rate control becomes superior to position control. Das et al. (1992) also found that position control is superior to rate control. In contrast, Zhai and Milgram (1993a) examined a six-dimensional task and concluded that isometric rate control can be as good as isotonic position control.

Isometric (pure force) versus isotonic (pure position) control are the extreme ends of a continuum of variable-stiffness hand controllers: infinitely stiff versus infinitely pliant. For intermediate stiffnesses, Zhai and Milgram (1993b) found that elastic rate controllers are better than isometric rate controllers, especially for less well rehearsed tasks. Mention should be made again of a comparable study by Jones and Hunter (1990), first discussed in Chapter 1, who found that increasing the stiffness of a manipulandum decreases the response time. A contrary effect is that high stiffnesses decrease accuracy, and there is an optimal value of low stiffness for best accuracy. More recently, they found that increasing the viscosity of the manipulandum decreases the delay and the natural frequency of the human operator (Jones and Hunter, 1993).

The effect of other mechanical properties of manipulanda, i.e., bandwidth, Coulomb friction, and backlash, were investigated by Book and Hannema (1980) using a Fitts' Law paradigm. In this paradigm, a motion with accuracy constraints is segregated into a gross motion (approach to a target) and a fine motion (accurately attaining the target). Both Coulomb friction and backlash increased the fine motion time but kept the gross motion time unchanged; backlash was the hardest to handle. Decreasing bandwidth increased the gross motion and fine motion times about the same.

Another issue is force reflection versus position control for teleoperators. In general, it is found that force reflection is significantly better than position control (Das et al., 1992; Hannaford et al., 1991).

SUPERVISORY CONTROL

The term *supervisory control* derives from the analogy between a supervisor's interaction with subordinate human staff members in an all-human organization and a person's interaction with intelligent automated subsystems. A supervisor of humans gives directives that are understood and translated into detailed actions by staff subordinates. In turn, subordinates collect detailed information about results and present it in summary form to the supervisor, who must then infer the state of the system and make decisions for further action. The intelligence of the subordinates determines how involved their supervisor becomes in the process. Automation and semi-intelligent subsystems permit the same sort of interaction to occur between a human supervisor and the computer-mediated process (Sheridan and Hennessy, 1984; Sheridan, 1992a).

In the strictest sense, supervisory control means that one or more human operators are intermittently programming and communicating to the computer information about goals, constraints, plans, contingencies, assumptions, suggestions, and orders relative to a remote task, getting back integrated information about accomplishments, difficulties, and concerns and (as requested) raw sensory data. In addition, the operator continually receives feedback from a computer that itself closes an autonomous control loop through artificial effectors and sensors to the controlled process or task environment.

In a less strict sense, supervisory control means that one or more human operators are continually programming and receiving information from a computer that interconnects through artificial effectors and sensors to the controlled process or task environment, even though that computer does not itself close an automatic control loop. The strict and not-strict forms of supervisory control may appear the same to the supervisor, since he or she always sees and acts through the computer (analogous to a staff) and therefore may not know whether the computer is acting in an open-loop or a closed-loop manner in its fine behavior.

Once the supervisor turns control over to the computer, the computer executes its stored program and acts on new information from its sensors independently of the human, at least for short periods of time. The human may remain as a supervisor or may from time to time assume direct control (this is called *traded control*), or may act as supervisor with respect to control of some variables and direct controller with respect to other variables (called *shared control*).

Supervisory control has been applied wherever some automation is useful, but the task is too unpredictable to trust it to 100 percent automation. This includes essentially all of what are called "robot" applications currently. Among the reasons to employ supervisory control, are:

(1) improved task performance (both speed and accuracy) and, in the special case of loop time delay, and avoidance of instability.

(2) human safety, when the work environment is hazardous.

(3) convenience, when the human must attend to other tasks while the automation is working and assuming the task does not require continuous monitoring.

(4) a means to construct, control, and continuously modify intelligent systems, so as to better appreciate the relation of people to machines.

Generic Paradigm for Supervisory Control

Figure 9-2 illustrates the generic supervisory control paradigm. At the bottom the tasks represent various material processing machines and transfer devices, each controlled by its own computer. All the computers are controlled from a central control station in which the human supervisor cooperates with a computer to coordinate the control of the multiple automatic subsystems. The supervisor's functions are five, which in turn can be subdivided as shown in the upper part of the diagram by the upper case labels. For each of the main supervisory functions, the computer provides decision-aiding and implementation capabilities, as represented by the separate blocks. These five functions are:

(1) *Plan*, consisting of (a) modeling the physical system or process to be controlled, (b) deciding on the objective function or trading relation among the various goal states that might be sought, and (c) formulating a strategy, which consists of scheduling and devising a nominal mission profile.

(2) *Teach*, which means (a) selecting the control action to best achieve the desired goal and (b) selecting and executing the commands to the lower-level computers to make this happen.

(3) *Monitor*, which means (a) allocating attention appropriately among the various subsystems to measure salient state variables, (b) estimating the current state (arriving at the current belief state), considering in addition to measurements the predicted response based on the previous belief state and the previous control actions, and (c) detecting and diagnosing an abnormality.

(4) *Intervene*, which means (a) to make minor parameter adjustments to the automatic control as it continues in effect, (b) to reprogram if the system has come to a normal stopping point and is awaiting further instruction, (c) to take over manual control if there has been a failure of the automation, and (d) to abort the process in the case of a major failure.

(5) *Learn* from experience so as to do better next time.

Status of Research in Supervisory Control

Computer-Based Planning of Telerobot Actions

Many new computer-based aids for planning (for collision avoidance, energy minimization, and other criteria of satisfaction) operate "what-would-happen-if—" trials on system models with hypothetical inputs. Typically, these provide the capability for graphical entry of test commands or trajectories and indicate the projected quality of the results. The recent system of Park (1991) is an example. Park's system allows the

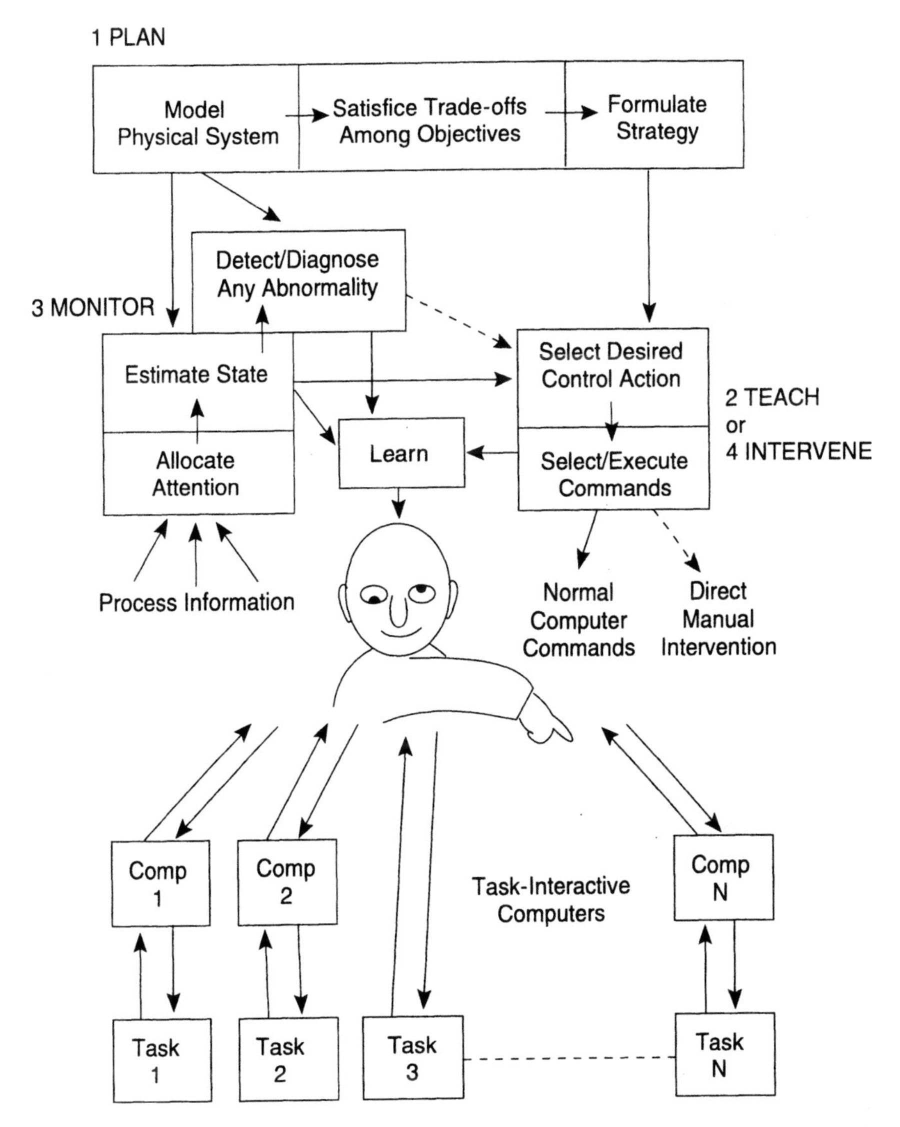

FIGURE 9-2 Generic Supervisory Control Framework. At the bottom are subtasks, all of which may be automated for short periods through a task-interactive microcomputer. At the top are the supervisory functions (*plan, teach, monitor, intervene*) subdivided into elements, all of which must be attended to by the human supervisor, but any of which may be aided by a computer-based expert system or on-line decision aid. In addition, off-line and not shown is the supervisory function *learn*. The supervisor must allocate his or her attention among all the boxes (Sheridan, 1992).

human operator to indicate, on a computer graphic model of the environment (as best it is known), a series of subgoal positions to serve as intermediate points along a multiple-straight-line-segment path for motion of the end-point of the teleoperator. The computer graphic model provides depth cues in the form of lines from the underside of the object to the floor as described above. Another algorithm checks for collisions with the environment by any part of the teleoperator or objects carried by it (again, as best known). As necessary, the algorithm makes small modifications to avoid such collisions. Areas that cannot be seen, or are not known from other modeling information, are considered virtual objects. If an initial trajectory turns out to be unsatisfactory, the operator can try another. The operator can thus evaluate a trajectory before ever committing to an actual motion. If the video camera is moved for a better view, the computer can reduce the virtual objects to more closely conform to actual objects. If, upon actual motion, collisions occur, the model can be updated with the operator's assistance.

Teaching the Telerobot What to Do

Machida et al. (1988) demonstrated a technique by which commands from a master-slave manipulator can be edited much as one edits material on a video tape recorder or a word processor. Once a continuous sequence of movements had been recorded, it can be played back either forward or in reverse at any time rate. It could be interrupted for overwrite or insert operations. Their experimental system also incorporated computer-based checks for mechanical interference between the robot arm and the environment.

Funda et al. (1992) extended the Machida et al. work and the earlier supervisory programming ideas in what they call *teleprogramming*. Again, the operator programs by kinesthetic demonstration as well as visual interactions with a (virtual) computer simulation. That is, commands to the telerobot are generated by moving the teleoperator master while getting both force and visual feedback from a computer-based model slave. However, a key feature of their work is that instructions to be communicated to the telerobot are automatically determined and coded in a more compact form than record and playback of analog signals. Several free-space motions and several contact, sliding, and pivoting motions, which constitute the terms of the language, are generated by automatic parsing and interpreting of kinesthetic command strings relative to the model. These are then sent on as instruction packets to the remote slave. The Funda et al. technique also provides for error handling. When errors in execution are detected at the slave site (e.g., because of operator error, discrepancies between the model and real situation and/or the coarseness of command reticulation), information is sent back to help update the simulation. This is to represent the error condition to the operator and allow him or her to more easily see and feel what to do to correct the situation.

Computer Assistance in Monitoring

There are various ways the computer can assist the human supervisor in monitoring the effectiveness of the automation, getting a better viewpoint and detecting and diagnosing any abnormality that may occur.

Das (1989) devised an algorithm for providing a "best" view for teleoperator control within a virtual model. His technique computes the intersection of the perpendicular bisector planes to lines from the end-effector to the immediate goal and to the two nearest obstacles. Such a point provides an orthogonal projection of the distances to the obstacles and to the goal. In evaluation experiments it proved useful to have the viewpoint thus set automatically, unless the operator needed to change it often, in which case the operator preferred to set it.

Hayati et al. (1992) and Kim et al. (1993) report a computer-aided scheme for performing telerobotic inspection of remote surfaces. The scheme involves both automation in scanning and capture of video images that appear to reveal flaws as well as graphic overlays and image maneuvering under human control.

Tsach et al. (1983) report a technique for automatically scanning and detecting a discrepancy in dynamic input-output relations relative to a multisubsystem normative model. Actual process measurements of upstream variables are fed to the appropriate model subsystems and corresponding downstream measurements are compared with outputs of the model. The advantages of their technique is that failures can be determined even during process transients, and failure locations can be isolated.

Intervening and Learning

If the automation fails, if the programmed actions end, or for other reasons, occasionally the human supervisor must intervene to reprogram or take over manual control. Criteria for doing this and which takeover mode is best tend to be context dependent.

Supervisory learning can be accomplished by keeping track of conditional frequencies (probabilities) and weighting by outcomes, by steepest ascent methods, by building up fuzzy rules and "growing an expert system" in terms of given fuzzy linguistic variables, calling for more evidence in state space regions in which membership is poorest, or by more rigorous neural net techniques. There is a dearth of research in both the intervening and learning roles of the supervisor; the reasons are elusive.

Predictor Displays to Cope with Time Delay

Control engineers are familiar with the destabilizing effect of loop time delays that occur in space teleoperation due to the limited speed of electromagnetic radiation, or similarly in underwater teleoperation when sonic communication is used (Sheridan, 1993). Predictor displays have been demonstrated (Noyes, 1984; Hashimoto et al., 1986; Cheng, 1991) to reduce task execution time by as much as 50 percent under ideal conditions, namely when: (a) the kinematics and dynamics of the object being controlled can be reasonably modeled; (b) there are minimal interactions with external uncontrollable objects or disturbances that cannot be modeled; (c) movements of the real object can be superimposed on the model movements and both can be seen in the same image plane; and (d) actions occur at a pace slower than human movement and perception. Condition (b) does not obtain in assembly tasks in which external objects are small enough to be moved. However, much as an athlete apparently does in running over rough ground, catching a ball, etc., it is possible to build and calibrate a model of the salient external objects by vision or optical proximity sensing and to use it to predict interaction with the end-effector or object under direct control.

Although the dynamics of the computer model used for prediction and planning must be synchronized to the dynamics of the actual task, the use of that model by the human operator need not be synchronized. In other words, an easy maneuver in free space may require no attention by the operator, and as that easy maneuver is occurring he or she may wish to focus attention on a more complex maneuver coming up later, even running the prediction of the complex maneuver in slow time or repeating it. Conway et al. (1987) called this "disengaging time control synchrony using a time-clutch" and "disengaging space control synchrony using a position-clutch." They built and demonstrated such a system.

Visual and Haptic Aids

Predictor displays may be considered to provide visual and haptic aids to the operator. Whether or not there are delays, other forms of visual and haptic aids may enhance operator performance. Brooks and Ince (1992) suggest a variety of display enhancements:

- In zone displays, areas of the visual field are highlighted that place restrictions on the operator, such as slave reachability (due to its own kinematics and to limits in the hand controller range) and imminent collisions. The reachability limits of the master can be changed by reindexing.

- Depth cues can be provided by superimposing a perspective grid with markers dropped onto the grid from the end effector and object workpoint, by virtual placement of the camera in a more useful location or by superimposing graphs representing end effector position and orientation relative to the task.
- View enhancements include delineating edges and surfaces that may be poorly visible due to lighting or similar textures, image distortions to highlight positioning errors, and status indication, such as coloring a gripper depending on the grasp stability and hot spots in the environment sensed with a temperature probe on the end effector.

Milgram et al. (1993) also propose virtual pointers to mark points in displays, virtual tape measures to show the distance between points, and virtual tethers between slave endpoint and target point to help guide a motion.

An issue is registration of the graphical image with the real image (Oyama et al., 1993). In addition, the graphical simulation has to be periodically corrected to correspond to the real display because of modeling errors; this process has been called stimulus-response reconciliation (Brooks and Ince, 1992). Rasmussen (1992) emphasized the importance of aligning the visual and kinesthetic reference frames for best performance; concern was expressed for laparascopic procedures in which vision via an overhead monitor is badly aligned with manipulation (Tendick et al., 1993).

Haptic aids are not as well developed as visual aids. Sayers and Paul (1993) propose the concept of synthetic fixtures, whereby parts mating operations are facilitated by force attractors. Similarly, force-reflecting mice for graphical interfaces have been proposed to mechanically enforce window boundaries, simulate push buttons, and act as attractors (Hannaford and Szakaly, 1992; Kelley and Salcudean, 1994; Ramstein and Hayward, 1994).

A hypothetical scenario for the use of predictor displays and sensory aids is presented in Chapter 12 for hazardous operations.

Performance Measurement, Learning, and Modeling

Any task performed under supervisory control requires time for the human to program (teach) the operation and then time to monitor while the operation is being executed by the computer. Each of these components takes more time as the complexity of the task increases. For very simple tasks, one might expect direct control to be quicker because instruction of a machine, as with that of another person, requires some minimum time. For very complex tasks, instructing the computer is likely to be extremely difficult and therefore supervisory control is essentially impossible. In between, as has been shown, the sum of the two times can be significantly less than that for direct human control.

There have been models of supervisory control, but none has captured all of the elements, including those of planning and setting objectives. The Baron et al. (1980) PROCRU model includes continuous sensorimotor skill, attention sharing among displays, decision making, and procedure following and has been applied successfully, for example, to the final approach of aircraft.

REAL-TIME COMPUTING

In a virtual environment or teleoperation system, the user senses a synthetic environment through visual, auditory, and haptic displays and then controls the synthetic environment or telerobot through a haptic or auditory interface. These input/output (I/O) operations must be sufficiently fast to be effective, that is to say, for control of a device or for presentation of human stimuli. How to organize a computer system to handle the computational demands of the many components of a VE or teleoperation system, especially those of the I/O operations, is a challenge for real-time computing.

At the most basic level, computers simply have to be faster and cheaper to be used to compute the necessary algorithms in real time. Such requirements have already been discussed for graphic displays, auditory displays, and information visualization. A more particular requirement is for real-time I/O: often powerful computer systems can compute fast, but they are unable to perform real-time I/O for control and sensing of external devices. For example, the elaborate operating systems of workstations typically do not permit fast and reliable interaction with the external world. Given the diverse components of a VE or teleoperation system, a decentralized computing architecture with an individual computer attending to just one component, such as a haptic interface, seems more appropriate than a monolithic supercomputer. Software and operating systems have to facilitate the programming and interaction of such networks of computers, walking a delicate line between efficiency and features.

Since World War II, real-time computing has been particularly associated with the development of control systems, simulators, and input-display systems, which are the three primary ingredients of VE and teleoperation systems. Many approaches and issues have been addressed in robotics and telerobotics. In fact, telerobotics puts greater demands on real-time computing than does robotics, because, in addition to the control of a robot, sampling and control of the human interface and the real-time computation of a predictive display are required. That VE and teleoperation systems must comply with the requirements of the human beings with whom they interact puts lower bounds on any measure of performance.

Requirements

VE and teleoperation systems might include a head-mounted visual display, eye trackers, various limb trackers, haptic interfaces, sound generators, and remotely controlled robots. Each of these components may itself be a complicated system; for example, a telerobot can have many sensors and actuators and complicated end effectors. Not counting the visual data, it can be estimated that such system should be able to sustain, at the bottom layer of a real-time control hierarchy, several hundreds of thousands of transactions per second with latencies measured in a few microseconds. These numbers may not seem so high by today's standards in terms of communication systems, but, unfortunately, because of the short latency requirements (ideally a few tens of microseconds for a rate of a thousand of samples per second) currently commercially available equipment either barely can keep up or simply is inadequate, because often this equipment is designed for quite different purposes.

Levels of Control

A telerobot requires a hierarchical control system, with servos and higher-level control operating at several levels. A force-reflecting haptic display or a locomotion display is also essentially a robot with similar control system requirements. The real-time requirements vary considerably, depending on the level of the hierarchy in question. At the lower levels, speed, precision, responsiveness, and predictability are the features that are sought from the real-time control; at the higher levels, flexibility will be the key, along with facilities normally provided by high-level operating systems.

Low-Level Control

At the lowest level is joint control. In the simplest case, the control design may be approached by considering each joint of a complex robot (or haptic interface) as a single-input/single-output system, with joint torque as input and joint position as output. Manipulators and haptic interfaces are made of a collection of actuated and passive joints organized into a kinematic structure. Position, force, and relationships between these quantities and their derivatives for the entire manipulator or haptic interface will then become the variables of interest. From a systems perspective, the manipulator or the haptic interface will then be viewed as a multiple-input/multiple-output system, as seen from a higher level in a control hierarchy.

For a medium-size manipulator, servo rates will be on the order of a fraction of to several thousand samples per seconds, which translates into 50 kflops to 5 Mflops, depending on the control algorithm (Ahmad, 1988). Feed-forward dynamic compensation, for medium-size manipulators, can be used effectively with rates as low as 20 updates per second; nevertheless, the computational burden is significant (Hollerbach, 1982; Izaguirre et al., 1992). For micro manipulators, these numbers are much higher: on the order of Gflops in extreme cases (Yoshichida and Michitaka, 1993).

Intermediate Level Control

Higher still in the hierarchy, not only will the state of the system at one instant in time be considered, but also entire state trajectories. The calculation of these trajectories will in turn require an additional computational demand, which depends on the number and the nature of the constraints that need to be enforced. As a general rule, each control system at a level of a control hierarchy must handle subsystems, which grow more complex as we move up the control hierarchy, but if the design is correct, their complexities will be hidden by the lower levels.

High-Level Control

At the higher levels, the control of the system is described in terms of general directives that cover a complete task or subtask. Here the time scale is measured in seconds, minutes, or hours. At such a level, the real-time control requirements are described only in general terms. The higher up in the hierarchy, the more the features offered by conventional operating systems found on engineering workstations will supply the required services for VE and teleoperation systems. The requirements are expressed in terms of powerful programming environments, languages, and ordinary system services, such as data storage and retrieval (Hayward and Paul, 1984).

Latency Versus Update Period

The latency is affected not only by calculation time, but also by input/output transactions between computing units and peripheral devices. These operations may include analog to digital conversion, data transfer from peripheral devices, data formatting and conversion, safety checks, and so forth.

The capital rule of digital control systems is the minimization of latency, defined as the time elapsed between the time a measurement is taken, the relevant calculations are performed, and the signal is fed back to the system to be controlled (in general, to the actuators). In fact, it can be readily realized that a digital control system with a latency larger than the basic response time of the system being controlled will fail at its task: at each update, the new control signals will be based on measurements made in the past, thus reflecting a state of the system that no longer is relevant.

The update period is simply the time between two successive outputs of control signals. Notice that parallel processing can always be used to achieve a very small update period, but the effect on latency is limited by the minimum number of sequential operations that have to be performed on the inputs to calculate the outputs (Lathrop, 1985). An experimental study (Rizzi et al., 1992) demonstrates the effect of a 40-fold increase in either latency or update period on the performance of a tracking controller. The increase in update period from 1 ms to 40 ms has a very minor degrading effect on tracking performance and stability. The 40-fold increase in latency completely destabilizes the system.

One way to counteract a large latency is for the control system to predict future states of the system based on previous measurements. Of course, precise prediction of the state is computationally demanding, and this will result in even larger control latency, possibly leading to worse performance than a simpler controller well matched to the response of the controlled system. The conclusion is: if latency can be avoided, it must be—even if it means simpler control algorithms.

For low-level control, latency requirements for manipulators and haptic interfaces are identical to the sampling period, which is of the order of a few tens to hundreds of microseconds. For intermediate-level control, latency requirements range from a few milliseconds to a few seconds. For high-level control, latency requirement can vary from a fraction of a second to a few seconds.

System Architecture

By system architecture for real-time computing is meant the type and number of processors, the connectivity and communications between processors, the input/output interfaces, the operating systems, and the development tools.

Processors and Computing Platforms

Major processor architectures can be classified as CISC (complex instruction set computers), RISC (reduced instruction set computers), VLIW (very long instruction word), Superscalar, and DSP (digital signal processors). Table 9-1 lists some of the main commercial examples. At the moment, certain of these processors are more prevalent and dominant than others; Juliussen (1994) discusses their relative strengths and future commercial viability. Some of the CISC and RISC processors have been designed specifically for personal computers (PCs), such as the 486s and the Pentium; others are incorporated into both PCs and workstations, such as the Alpha, the PowerPC, and MIPS. Processors such as Transputers and the Texas Instruments TMS320C40 (C40) DSP chips are not meant for general-purpose computing platforms, but for fast numerical computation and I/O. The scenario of some general-purpose computing platform as a front end to a network of Transputers or C40s that do the I/O and real-time computing has become popular in robotics and may well serve as an example for real-time VE computing as well.

There is a tendency to discuss only workstations for VE and teleoperation systems, primarily because of graphics and networking, but PCs are approaching workstations in power and are already heavily used in real-time computing. The outcome of the much-discussed battle, or convergence, between PCs and workstations will have a major impact on the nature of real-time systems in general and on VE and teleoperation systems in particular. Because of the generally lower costs of PCs and associated software, their development will facilitate the spread of VE and teleoperation systems.

In turn, the development of PCs is threatened from below by the advances of more powerful video game players and set-top interactive TV players. Planned game players from Sega and Nintendo will soon have comparable computational power to the fastest desktop computers. The addition of typical PC applications software such as spreadsheets, word processing, and electronic mail are likely in the near term and will convert such TV players into more general-purpose computers. With the combination of video, sound, input devices, and fast computation, these TV players could represent a formidable computational force in VE. Given

TABLE 9-1 Categorization of Some Advanced Processors

CISC	RISC	VLWI	DSP
Alpha	Sparc	AMD Bit Slice	TMS320C40
68040	RS/6000		M960002
486	PowerPC		AT&TDSP32C
Pentium	MIPS		
	PA-RISC		
	Transputer		

the huge market forces driving these developments, costs of such systems would run at several hundred dollars and severely threaten PC- or workstation-based VE systems.

Parallel Processing and Communications

To run processes faster and meet real-time constraints, parallel processing can be advantageous up to a point. That point is defined by losses in throughput and increases in latency due to communication, and by limitations in the ability to parallelize a computation. When there are 10 or more processors to coordinate, it can become a formidable task to avoid conflict and to ensure timely performance. In such cases, it may be best to upgrade to a faster processor to reduce the number required.

The main approaches toward connecting processors include a common bus, point-to-point links, and crossbars. A common bus is the most traditional approach; industry standards with a roughly equivalent performance of 40 to 80 Mbytes/s include VME Bus, MultiBus II, FutureBus, EISA, NuBus, SBUS, and PCI Bus. Each computing unit is equipped with fast local memory, and a global memory bank is made available to all computing units (Bejczy, 1987; Chen et al., 1986; Clark, 1989; Narasimham et al., 1988). The advantages of such an architecture include modularity, easy expansion for more performance, and the possibility to mix hardware from different sources (Backes et al., 1989). The disadvantages obviously lie in the bottleneck created by the two shared resources: the global memory and the system bus.

Point-to-point links circumvent the bottleneck problem of bus-based architectures to some extent: communication between connected points will be fast, but that between distant nodes in a graph will be slow. Such built-in links are used in most engineering workstations for graphic processors, sound processors, memory access, disk controllers, etc. The advantages include design for minimum cost, but then the system, once configured, offers few possibilities for expansion.

Some developers have proposed high-performance central processing units (CPUs) equipped with built-in high-speed point-to-point links (1-20 Mbytes/s). Following this philosophy, Inmos from Europe (now a division of SGS-Thomson) introduced the Transputers series (T800, T9000), which can loosely be classified as RISC computers. These computers are each fitted with four communication channels, which permit the user to design a variety of coarsely parallel architectures (Buehler et al., 1989; Zai et al., 1992). Recently Texas Instruments combined the Transputer concept with its experience in DSP design leading to the C40 with six high-speed links, thus competing head-on with the latest Inmos Transputer T9000. The principal disadvantage of point-to-point communication systems, such as those made available with Transputers or C40s, is the lack of industry standards in terms of signal conversion and data acquisition hardware. Whereas such hardware is abundantly available for most bus standards, it is only beginning to become so for point-to-point communication systems, requiring for system implementers much custom electronic design.

The recent availability of these powerful processors, which lend themselves to parallel processing, together with commercially available development and run-time software, has virtually eliminated the need to build custom computing platforms and develop custom multiprocessing environments. In the past, this has been necessary and has consumed substantial financial investment and personnel. Examples are the CHIMERA II (Stewart et al., 1989) and Condor (Narasimham et al., 1988) systems. Even though most of these custom systems relied increasingly on standard commercial boards, the efforts in upgrading, to stay compatible with the host operating system as well as the latest run-time boards, constituted a major ongoing investment that fewer and fewer labs are willing to undertake. The same situation of the availability of powerful processors applies to VE and teleoperation systems. This is very fortunate, since it means that all the custom development efforts can be focused on the critical I/O interface, a current bottleneck that may be better served by commercial suppliers in the future.

The most generic architecture consists of a full N × M crossbar switching network connecting a set of CPUs and register files totalling N inputs and M outputs. This architecture has the advantage of being capable of reaching the theoretical optimum performance, which minimizes latency and maximizes throughput. An example is a system for control of a complete microrobot system (Hunter et al., 1990; Nielson et al., 1989). The disadvantage is the massive complexity and high cost of the system.

Although considerably slower, it is also possible for processors to communicate across networks. Researchers have become interested in running robotic experiments or virtual environment setups across a computer network with nodes located in different cities or even in different continents. In one example, robotic sessions are run among a network of sites across the United States (Graves et al., 1994); in another, a telerobotic experiment has been performed between Japan and the United States (Mitsuishi et al., 1992). This type of work is bound to be increasingly investigated given the high potential for applications. In effect, with such systems, the services of specialists in any area could be requested and put to use without asking the specialists to travel where they are needed. Clearly, the demand on communication networks will follow the same route that is being taken by low-level real-time control systems: high volumes of data transactions, low latency and time precision. It must be expected that the performance levels required at the scale of a laboratory site or an application site will need to become available across entire networks.

Operating Systems and Development Environments

A large number of real-time operating systems are currently in competition to provide the services needed to implement high-performance real-time applications (Gopinath and Thakkar, 1990). These operating systems (OS) may be classified as embedded, full-featured real-time, or hybrid.

An embedded operating system is targeted at original equipment manufacturer (OEM) applications, and as such provides only essential facilities, like a real-time executive (real-time multitasking, memory management, and device I/O). It is usually a simplified system offering no support for a file system and must be used as a cross-development tool. Yet most offer hard real-time features essential to VE and teleoperation systems. In this category, SPOX (Spectron Microsystems) has emerged as one of the industry standards.

A full-featured real-time OS resembles the more standard UNIX but with additional real-time features. An example of this category is HELIOS by Distributed Software Ltd., which offers an eight-level priority real-time scheduler, virtual message routing, X and Microsoft Windows graphic support, as well as SUN and PC host interfaces. Clearly, processing nodes with microsecond interrupt latency requirements can—and must—be stripped of the inherently slow high-level functionality.

In between embedded and full-featured real-time operating systems are the hybrids, the most popular of which is probably VxWorks by Wind River Systems. These typically are real-time operating systems running on dedicated shared-memory VME processor boards and provide a transparent interface to the host UNIX operating system. Despite its high cost ($10,000-20,000) VxWorks has been attractive for its convenient interface to existing UNIX hosts, debugging and development facilities, and the availability of a large number of supported processor and I/O VME cards.

Nevertheless, many researchers are quite dissatisfied with the services provided by commercial operating systems, usually for reasons of cost and inefficiency. Because these operating systems are designed with general purpose requirements in mind, the few requirements essential to VE and teleoperation systems are not well addressed, and the resulting systems are cumbersome to use and clumsy in their design. For these reasons, countless custom operating and development systems have been designed and implemented, with their scope limited to one or to a handful of laboratories.

Experience shows that no single operating system (and no single CPU technology) will satisfy the varied needs of VE and teleoperation systems. This is why real-time extensions to the UNIX systems will most likely remain limited to the higher levels of the hierarchy (soft real time). For low-level control and data acquisition, no single run-time environment has been shown to be satisfactory to the community. HELIOS is probably one of the most suitable operating systems for VE and teleoperation systems to date. It provides all the UNIX-like services on a distributed real-time network, allowing one at the same time to "peel away" the higher levels of the operating system on part of the network dedicated to time-critical code. In practice, the response is mixed: some users report sufficient control at the device level, some others don't (Poussart, personal communication). Systems like HELIOS are clearly going in the right direction and might be the approach of choice for VE and teleoperation systems in the future. To obtain full processor control on a network of only a few processors, enhanced C compilers, like 3L C by 3L Ltd. are minimal systems, adequate for providing standard host I/O, memory management functions, and multiprocessing facilities (network loading, communication management, load balancing, message routing, micro-kernel for multitasking).

RESEARCH NEEDS

A major trend in telerobotics is toward higher-level supervisory control, that is to say, toward autonomous robotics (robotics for short). Major drivers in this direction include: (1) communication problems (e.g., delays, low bandwidth, low resolution), especially in space and undersea, (2) burden lifting from the operator (e.g., partially automating repetitive tasks), and (3) performance enhancements (e.g., using local sensing when vision is obscured, assisting in obstacle avoidance). Hence the needs in telerobotics are often the same as the needs in robotics. The line between supervisory control and robotics is quite blurred.

At the same time, manually controlled telerobots remain important. One reason is safety, particularly in space and medicine applications, due to uneasiness about the loss of control. For example, developers of the planned space station are concerned that the telerobot not damage the space station structure. Another reason is the inability of robots to handle unstructured environments, which obviates the possibility of even partially automating a task—the reason for having telerobots in the first place.

In the control of telerobots, the issues of haptic interfaces, computer-generated environments, and real-time systems are important. Although predominantly covered in other sections of the report, these issues are addressed here to some extent in discussing various needs in telerobotics.

Handling Communication Delays

In space applications, the sentiment for ground-based teleoperation is increasing, for a number of reasons. First, for the planned space station, the actual amount of time spent by astronauts in orbit will be relatively short, and of that time the fraction that could be devoted to teleoperation is shorter yet. Because internal vehicular robots (IVRs) and external vehicular robots (EVRs) will be present the whole time, they would be used more efficiently if operated from the ground when astronauts are not attending. Second, humans are much less effective in space than on the ground; weightlessness seriously affects concentration and attention. Third, sending humans into space is expensive and dangerous.

To communicate from the ground, several satellite and computer systems must be traversed. The result is variable delays on the order of 5 s. There is also a need to funnel commands through a central computer at the Johnson Space Flight Center, not only to share the communication channel but also to verify the commands. Again, safety is an overriding concern. The trend is to devolve ground control from Houston to other sites; for example, the recent ROTEX experiment (Brunner et al., 1993; Hirzinger et al., 1993) involved control of this German-built telerobot on the space shuttle from Karlsruhe. The Canadian Space Agency is interested in controlling Canadian-built telerobots from its main facility in St. Hubert, Quebec. Other groups will wish to perform scientific experiments in space from their home locations on the ground; the word *telescience* has been coined to describe this activity. This devolution will accentuate the delays.

Time delays are also important in underseas teleoperation. To avoid power and data tethers, communication with submersibles must rely on relatively slow sonar signals. For example, at a distance of 1,700 m, sound transmission imposes a round-trip delay of 2 s (Sheridan, 1992a).

There are two main approaches to handling teleoperation with time delays: (1) control theory approaches that incorporate time delays and (2) predictive displays. In general, the greater the delays, the lower the system gains may be (e.g., stiffness and viscosity) in order to avoid instability. Thus the system response slows down and performance suffers, to an extent that depends on the controller. Recently developed controllers based on passivity approaches guarantee stability but seem to suffer in performance because their gains are too low (Lawn and Hannaford, 1993). Exactly how to formulate a controller that addresses both stability and performance, under various time delays, needs further work.

Force reflection under delays is more problematic than position control. For delays approaching a second or more, position control becomes superior to force control (Lawn and Hannaford, 1993). In fact, humans are sensitive to extremely small delays in terms of fidelity of force reflection (Biggers et al., 1989). One solution is to implement a force controller, such as impedance control, on the slave side and a position controller on the master side. Thus there is no force feedback to the operator, and the slave force controller is presumed capable enough to handle interaction forces. This requires that the slave force controller be appropriately tuned from a knowledge of the environment and the task. This knowledge can either exist from a priori information about the task or from measurements made on the environment. It is therefore important to get several kinds of information: (1) determine whether to use position control or force reflection as a function of delay and tasks; (2) determine how to configure the slave manipulator's force controller as a function of a priori knowledge of the task; and (3) identify the environmental characteristics through slave robot sensing, if a priori knowledge is not available. The last two needs are essentially robotics problems.

For delays greater than 1-2 s, direct manual control becomes ineffective without the aid of predictive displays (Sheridan, 1992a). The main issue then becomes how well the remote manipulator and environment are simulated. For known structured environments, good simulations of the geometrical aspects are quite feasible. Force control can also be simulated, but difficulties are present due to closed-chain dynamics and arbitrary contact conditions. In ROTEX, the predictive displays for force were found to be quite close to actual experimental forces.

Even in the case of ROTEX, there are small misalignments and inaccuracies, which are accommodated by local sensing. Force, proximity, and tactile sensing are fused to correct such deviations. For less structured problems, a model of the environment must be built using vision and other sensors. This generic problem in robot vision and mobile robots is far from solved. One envisions that a robot would spend some time mapping its immediate environment, from which a simulation model is extracted; the operator then controls the robot through this simulation model.

Accurate, Real-Time Simulations

Besides predictive control, other uses for telerobotic simulations are training and mission development. Training simulators prevent damage from being done in an operator's learning phase (e.g., learning a surgical procedure), free expensive telerobotic equipment for actual use (e.g., teleoperation of forestry harvesting equipment), and prepare trainees for equipment that is available only at the remote site (e.g., underwater robots) or that cannot function under normal conditions (e.g., space robots that cannot lift themselves against gravity). Mission development involves extensive simulation to work out operational scenarios; this is particularly important in space.

Three needs, which are similar for virtual environments, are reemphasized here for telerobotics.

(1) *Construction of a simulated environment.* This involves the software tools for representing real environments, creating a particular virtual environment, and sharing VE modules. As mentioned above, it would also be desirable to "reverse engineer" a real environment into a simulation, by using visual and other sensory recognition and representation methods (see, e.g., Oyama et al., 1993). This is a hard problem, especially if object attributes other than geometry (space occupancy) are to be addressed, such as mass and surface properties.

(2) *Accurate representation of task dynamics.* Although the laws of physics have been around for a long time, often we don't understand detailed mechanical interactions between objects in sufficient detail for simulation. A very basic example is motion of three-dimensional objects under friction, such as sliding, and in collisions (Mason, 1989). The field of robotics is just beginning to enumerate how such objects are expected to behave and to develop appropriate concepts.

Another difficulty is representing closed-chain systems under arbitrary contact conditions. Examples are locomotion on different surfaces, the behavior of the operator's hand tissues in the interaction with the haptic interface, and the behavior of the remote manipulator with such compliant environmental substances as human tissues (in surgical applications) and rock or soils (in mining applications). Furthermore, there is an issue of experimentally verifying such task dynamics. The particular robot sensing also needs to be simulated (Brunner et al., 1993).

In the continuous domain, accurate finite element models are required, which may be nonlinear. They will require that the constitutive equations for real objects be known or measured. Abrupt changes in boundary conditions, such as simulating the cutting of tissue, are difficult to represent.

(3) *Real-time computation.* In addition to generating graphical images, there are substantial difficulties in real-time computation of a simulated environment. Often classical multibody simulation is more concerned with accurate long-term integration of initial value problems than with computational efficiency. Other intensive computations are simulating and detecting collisions and real-time calculation of finite element models. Needs in real-time computing are discussed in greater detail below.

Real-Time Computing

The multitude of challenges in real-time computing for VE and teleoperation systems can be successfully addressed only with computing solutions offering high performance, ease of use, reconfigurability, expandability, and support for massive and fast I/O. Coarse-grained parallel systems, based on the most powerful computational nodes available, which communicate via high-bandwidth, integrated communication links, are the most promising contenders to satisfy all requirements. This has been recognized by academia and industry alike and explains the early strong support for the Transputer processor family and recently the sweeping popularity of C40-based systems.

In this setting, a high-speed intercommunication standard would be highly desirable, in a similar fashion to existing bus standards. This would offer users an immense degree of flexibility, since computational nodes could be mixed and matched across vendors and different processor types. Naturally, this communication standard must go hand in hand with a physical module standard, akin to Texas Instrument's TIM-40 modules.

Currently, code development is done on traditional host systems connected to dedicated run-time systems. This results typically in a bottleneck at the interface between the two systems. In addition, communicating between run-time code on the dedicated architecture and host programs is typically problematic, due to the difference in development tools, architecture, and response time. Based on the emerging paradigm of coarse grained parallelism based on standardized communication, the boundary between development and run-time hardware could become transparent. Since the host architecture itself could be based on the same paradigm, communication between host (development system) and controller (run-time system) could be fast and flexible, as there could exist numerous communication links between the two. Furthermore, the boundary could be laid dynamically, depending on the requirements of each system.

If the input/output devices for control, teleoperation, and VE hardware adhere to the same communication standard, much effort in system development could be eliminated and progress toward powerful real-time teleoperation and VE systems would substantially accelerate. It is likely that, with the increased interest in telerobotics and VE, there will be sufficient thrust to finally address the I/O standardization issue in addition to reducing cost through volume. It should be reemphasized that, in contrast to I/O system interfacing, custom software developments for real-time multiprocessing systems are no longer necessary, and the effort should rather go toward identifying the most suitable commercial products and focus on industry-academia collaboration to develop one or several compatible standards.

Better Robot Hardware

A goal in teleoperation is that performing a task with a remotely controlled manipulator should feel nearly indistinguishable from directly performing the task with one's own limbs. Shortfalls in this goal are primarily due to the mechanical hardware, both on the master and slave sides. Often, limited-performance masters are hooked up to limited-performance industrial robots. Although comparisons of different master-slave control strategies have been published for these systems, it is unlikely that the results generalize beyond the evaluation of these specific low-performance systems.

On the slave side, robot arms and hands are required with sufficient dexterity and responsiveness to match that of the human arm. To achieve such devices requires improvements in actuators, sensors, and structures. Some particular needs are discussed below.

Multiaxis, High-Resolution Tactile Sensors

The robot must have a comparable sense of touch to a human, even before considering how to transmit the sensation to an operator. Tactile sensors continue to be a problem in robotics: hardly any robots have them, and those that do sense touch only coarsely. Although interesting designs have been proposed, few have yet been realized in a usable form. Some robotics researchers are working with tactile physiologists to look for correspondences between physiological and robotic designs, for example, the use of accelerometers to act like Pacinians and piezoelectric strips to sense shear (Cutkosky and Hyde, 1993). Attention to robot skin mechanics is important, for example, the existence of fingerprint-like ridges as stress amplifiers.

Robust Proximity Sensors

As a step toward supervisory control, small adjustments in grasping or an approach to a surface should be performed with local sensing. The intelligence required is much less than for the general case of autonomous control. A highly sophisticated gripper on ROTEX (Hirzinger et al., 1993), incorporating proximity, tactile, and force sensing, was the key to the success of the predictive display control. Proximity sensors have typically been ultrasonic, electromagnetic, or optical. All have limitations in range, accuracy, and sensitivity to different surface conditions. Visual time-of-flight systems are not yet practical because of complicated electronics, but they could be a future solution.

Multiaxis Force Sensors

Multiaxis force sensors would typically be mounted on a robot wrist, to measure the net force and torque exerted on the end effector. Miniature force sensors would also be useful on finger segments, to control fingertip force accurately. Key issues at the moment are cost, robustness, and accuracy. For wrist sensors, inaccuracies of a few percentage points due to cross-coupling effects are typical and are problematic. A possible solution is based on magnetic levitation.

High-Performance Joints

Robot limbs should be strong and fast yet should be able to interact gracefully with the environments. Better actuator and transmission designs are the key. Improvements in electric, hydraulic, and pneumatic drive systems are required. Novel actuators such as shape memory alloy and polymeric actuators look promising in terms of force to weight; the result might be lightweight, responsive limbs that accurately track human motion commands and faithfully reflect back contact conditions.

Hardware considerations on the master side are very similar. The requirements of a human wearing or interacting with a haptic interface are even more demanding and involve concerns of safety, convenience, and bulk. If reasonably performing devices cannot be obtained at reasonable costs, the spread of such systems will be limited.

Improved Telerobotic Controllers

Even without time delays, there are unresolved issues about how to best implement controllers for master-slave systems. One reason is limitations in hardware, as mentioned above. Another reason is a lack of tools with which to quantify and evaluate teleoperator performance. There are deficiencies in taxonomies of tasks: we don't understand well enough at a detailed level what we want these systems to do. At a basic level, we don't completely understand human sensorimotor abilities, for example, the discrimination of arbitrary mechanical impedances (stiffness, viscosity, and inertia) (Jones and Hunter, 1992). Clearly this knowledge is necessary to set design goals for telerobotic systems. We also need to understand how more complicated tasks decompose into more basic tasks, which can then be measured and used as discriminators between telerobotic controllers. As mentioned earlier in this chapter, robotics also has this goal.

As an example, it is not fully understood when to apply rate control versus position control, or how to include force feedback into rate control.

Other examples were mentioned in the section on low-level control. When is force reflection more advantageous than the position control of a locally force-controlled robot? If force reflection is used, what is the exact form of controller that best meets goals of robustness and stability? There is still considerable work to be done in this area.

A new set of issues arises from scaling: teleoperation of very small and very large robots. The mechanical behavior of objects in the micro domain is very different than in the macro domain. One problem is that the dynamics will be very fast; somehow movements will have to be slowed down for the operator.

As robot autonomy improves, so will the level of supervisory control. A number of functions could be increasingly automated:

(1) *Path planning and collision avoidance.* The main issues here are efficient routines and obtaining geometric descriptions of the environment (Latombe, 1991).

(2) *Trajectory specification.* Any trajectory has to stay within system constraints of joint limits, actuators, and safety concerns.

(3) *Grasping.* The robot should be relied on to obtain a stable grasp and to regrasp as necessary.

(4) *Intermittent dynamic environments.* Trajectories should be modified in real time subject to changes in the environment. For example, a robot may swerve to avoid hitting someone entering its workspace. Some forms of hand-eye coordination, such as catching or hitting, may require a speed of response not possible with teleoperation.

(5) *Force control.* With more sophisticated abilities to interact with the environment and to complete such tasks as the generic peg-in-hole problem, the need for force reflection will diminish.

A step toward such autonomous control capabilities would be a higher-level transfer of skills between the operator and the telerobot. The idea is to program by kinesthetic demonstrations: the human makes a movement, this movement is measured, and the telerobot extracts symbolic information about how to accomplish the task (Funda et al., 1992; Ikeuchi, 1993). This differs from direct manual teleoperation, in that an exact trajectory is not being commanded, but rather a strategy for completing a task. Difficulties particularly present themselves in transferring force control skills.

10

Networking and Communications

Computer networks offer real-time interaction among people and processes without regard to their location. This capability coupled with virtual environments (VE) makes telepresence applications, distance learning, distributed simulation, and group entertainment possible. It has been suggested that the most promising use of VE and networks will be in applications in which people at different locations need to jointly discuss a three-dimensional object, such as radiologists using a VE representation of a CT (computer-assisted tomography) scan (Pausch, 1991) or aeronautical engineers using a distributed virtual wind tunnel (Bryson and Levit, 1992). Another exciting concept is that of virtual libraries, like the one being developed at the Microelectronics Center of North Carolina. This project, Cyberlib, will allow patrons to enter "the information space independently" or go to a "virtual reference desk" from anywhere across the United States via the Internet (Johnson, 1992). We already have virtual newspapers. The *San Jose Mercury News* publishes its entire text (including classified advertisements) via the American Online Service using graphically based software for the Macintosh and IBM personal computers.

VE and high-speed networks are the tools that will allow us to explore Mars and the earth's oceans. The National Aeronautics and Space Administration's Ames Research Center is examining the use of VE to control robot explorers. Scientists at the Monterey Bay Research Institute are integrating such diverse technologies as computer simulations, robotics, and VE with a sophisticated undersea local-area network to explore the nation's newest marine sanctuary.

The Department of Defense's Advanced Research Projects Agency (ARPA) has also recognized the importance of networks with regards to VEs in one of its seven science and technology thrusts. Since 1984 ARPA has funded the simulation network (SIMNET) system, which enables simultaneous participation of approximately 300 players in simulated aircraft and ground combat vehicles located in Europe and the United States on the same virtual battlefield. ARPA is currently working on a much larger technology demonstration to "create a synthetic theater of war" using the Defense Simulation Internet (DSI). This network will link thousands of real and simulated forces all across the United States (Reddy, 1992).

It is anticipated that, in the future, high-speed networks will allow VE systems to take advantage of distributed resources, including shared databases, multimedia sources, and processors, by providing the required computational power for building the most demanding applications. High-speed networks will provide VE applications with access to huge datasets generated by space probes, dynamic climatic information from weather models, and real-time imaging systems such as ultrasound.

One less serious but rather lucrative combination of VE and networking that is currently in use is multiuser interactive VE games offered by Genie and Sierra Online services. These games are provided over slow telephone lines with limited graphics. Other upcoming cooperative arrangements are planned between VE and cable television. Some examples of these are the promise of multiuser games by Sega and Nintendo and the development of virtual Walmart department stores by the Home Shopping Network.

STATUS OF THE TECHNOLOGY

Distributed VE will require enormous bandwidth to support multiple users, video, audio, and possibly the exchange of three-dimensional graphic primitives and models in real time. Moreover, new protocols and techniques are required to appropriately handle the mix of data over a network link. The technologies providing these gains in performance blur the traditional distinction between local-area and wide-area networks (LANs and WANs). There is also a convergence between networks that have traditionally carried only voice and video over point-to-point links (circuit-switching) and those that have handled only data such as electronic mail and file transfers (packet-switching).

Wide-Area Networks

The fabric of our national telecommunications infrastructure is being radically altered by the rapid installation of fiber optic cabling capable of operating at gigabit speeds for long-haul traffic. The change has come so fast that AT&T wrote off $3 billion of equipment in a single year while replacing its analog plant with digital systems. Long-distance carriers have been quietly installing synchronous optical network (Sonet) switches to support those speeds. Sonet is a U.S. and international standard for optical signals; it is the synchronous frame structure for multiplexed digital traffic and the operating procedure for fiber optic switching and transmission systems. Sonet allows lower-speed channels such as OC-3 (155 Mbit/s) to be inserted and extracted from the main rate. Also, Sonet defines data transmission speeds to 2.4 Gbit/s and the major carriers believe that this can be extended to 10 Gbit/s for a single fiber link (Ballart and Ching, 1989). It is likely that data rates will go much higher in the next century. Japan's telephone company, NT&T, has announced transmission of a 20 Gbit/s data stream over 600 miles of fiber, and it is working to increase throughput to 100 Gbit/s using soliton technology. Both MCI and Sprint have announced that they will have Sonet completely deployed by 1995.

The new switches will also incorporate asynchronous mode technology (ATM). ATM (2.4 Gbit/s) provides fast variable-rate packet-switching using fixed-length 53 byte cells. This permits ATM networks to carry both asynchronous and isochronous (video and voice) transmissions at Sonet speeds. ATM also supports multicasting, and the Consultative Committee on International Telegraph and Telephony has written a standard for interfacing ATM with Sonet. AT&T has announced that it will provide WAN ATM services in 1994.

Two other high-speed services are being offered today: switched multimegabit data service (SMDS) and frame relay. SMDS, based on the Institute for Electrical and Electronics Engineers 802.6 MAN standard, is connectionless, uses frames and fixed-length cells, and offers speeds up to 34 Mbit/s with plans to upgrade to 155 Mbit/s. It is currently being offered only in local metropolitan areas. Frame relay is connection-oriented (dial-up) and offers speeds up to 1.544 Mbit/s. Although neither of the services are considered well suited for voice or video applications, they are likely to reduce the data transmission cost of WAN services.

The major carriers are not alone in this effort to push wide area networking to faster speeds. The backbone of the Internet, NSFnet, has been completely upgraded to T-3 (45 Mbit/s) and will transition to OC-3 (155 Mbit/s) by 1994. The backbone rate is expected to go to OC-12 (622 Mbit/s) by 1996. The National Science Foundation is responsible for this effort as part of the overall National Research and Education Network (NREN) project, which is one of the four components in the U.S. High Performance Computing and Communications (HPCC) Program, established by Vice President Gore. Part of the project involves the installation of OC-12 networks at several regional test beds: Aurora, Casa, Blanca, Nectar, and Vistanet. These test beds will be pursuing "Grand Challenge" applications ranging from medical imaging to interactive visualization using ATM, SMDS, and Sonet technologies (Johnson, 1992).

The Internet Engineering Task Force, foreseeing future increases in WAN speeds, has been conducting experiments over the Internet multicast backbone network to develop a standard for sending multimedia applications over packet-switching networks. The Real-Time Transport protocol that is currently being tested supports packet video and audio and could be used with ATM as it becomes more pervasive.

At the local loop (the telephone line between the central office and customers), intense competitive pressures in the cable and telephone industries are spurring the development of new technologies to allow the currently installed copper lines to operate at megabit speeds without expensive repeaters. The high-bit-rate digital subscriber loop is an encoding scheme being used now to deliver duplex T1 service. Another scheme that is in the trial stage, asymmetric digital subscriber loop, provides 1.5 Mbit/s in one direction and 16 kbit/s in the other. With the use of new compression standards such as those of the Consultative Committee on International Telegraph and Telephony (H.261—P × 64) and the Motion Pictures Experts Group (MPEG), ADSL-II, a follow-on technology with 3 to 4 Mbit/s transport capability, could carry real-time video, audio, and VE data (Hsing, 1993).

The rewiring of the local loop has also begun. Tele-Communications Inc. (TCI), the nation's largest cable company, has announced that it intends to upgrade by 1996 the broadband lines to more than 90 percent of its customers with fiber. TCI is doing this in order to support increased channel capacity, high-definition television (HDTV—which, when uncompressed, requires 1.2 Gbit/s bandwidth), and VE services such as games from Sega/Genesis. TCI also will try to counter the threat of the telephone companies entering the lucrative market. AT&T has had a test bed for developing the fiber optic local loop for several years near Pittsburgh. Bell Atlantic, a regional phone carrier, is conducting tests in its employees' homes of a system that delivers movies over the telephone line. The Regional Bell Operating Companies recently proposed to the Clinton administration a plan to rewire the local loop with fiber optic cable within 10 years in exchange for permission to enter the information services market and manufacture telecommunications equipment. A bill that failed to pass Congress in 1993 would have permitted the regional Bell operating companies to enter the cable business.

Local-Area Networks

The connection from a three-dimensional graphics workstation to high-speed WANs is most likely to come from a LAN. (Table 10-1 presents LAN capabilities.) Most LANs use Ethernet (10 Mbit/s), which is inadequate for the high-performance demands of VE and multimedia. Several companies have endorsed the proposal of standards for 100 Mbit/s Ethernet. In addition, through the use of switching hubs (which are collapsed backbones with gigabit-speed backplanes), a workstation can use all the bandwidth available on an Ethernet segment.

FDDI (Fiber Distributed Data Interface—100 Mbit/s) is used extensively in supercomputer centers. However, most host interfaces operate in the 20-50 Mbit/s range. A new standard for FDDI over unshielded twisted pair wiring should make FDDI more affordable for general computing. Unfortunately, both FDDI and Ethernet technologies are not ideal for isochronous data because there is no guaranteed data rate or prioritizing in the protocols. The American National Standards Institute (ANSI) has developed FDDI-II to address this problem by dynamically allocating bandwidth to isochronous applications. ANSI is working on FDDI Follow-On with plans for completion to be finished in the middle of the decade. FDDI Follow-On is likely to be designed for speeds up to 1.25 Gbit/s. In addition IEEE has established a working group that issued a final draft of a standard, the Integrated Services LAN Interface, which defines a LAN that carries voice, data, and video traffic over unshielded twisted pair wires.

The high-performance parallel interface (HPPI) is an ANSI standard supporting 32 and 64 bit interfaces that run at rates of 800 and 1,600 Mbit/s, respectively. It is a switched architecture and operates over a distance of 25 m on copper cables connecting supercomputers and their peripheral devices. A serial version of HPPI for fiber optic cables has been proposed to extend the range to 10 km. The NREN Casa test bed researchers from

TABLE 10-1 Local Area Network (LAN) Capabilities

LAN Technology	Capacity Mbit/s	Year of Final Standard	Status
Ethernet	10	1985	In use
FDDI	100	1989	In use
HPPI	800/1600	1992	In use
Fiber Channel	132.8 - 1064.2	1993	Some products available
ATM	45-622	1993	Some products available
FDDI-FO	1250	Approx. 1995	Not available

Los Alamos, California Institute of Technology, the Jet Propulsion Laboratory, the San Diego Supercomputing Center, and the University of California, Los Angeles, are developing HPPI-Sonet interfaces to connect supercomputers over multiple OC-3 circuits, providing 1.2 to 2.5 Gbit/s bandwidth (Cattlett, 1992).

Fiber Channel is a proposed ANSI standard for very-high-speed LANs. It is designed to connect more than 4,000 computers and peripherals over several kilometers at data rates up to 1,062.4 Mbit/s. Fiber Channel will provide a number of upper-layer network services that HPPI does not, and it has the backing of IBM and Sun Microsystems. Another proposed standard, Scalable Coherent Interface, has a potential speed of 8 Gbit/s (Cattlett, 1992).

ATM has also been deployed for local area networks. Several vendors, including Fore Systems Inc. and Adaptive Corp., are selling ATM switches for LANs. Fore Systems sells interfaces card for SGI, DEC, and Sun workstations. Each workstation is linked via fiber optic cable to a switch at 140 Mbit/s. The Aurora and Nectar test beds are investigating the use of ATM host interfaces for supercomputers (Cattlett, 1992). The allure of ATM is that it might eliminate the distinction between wide-area and local-area networks, providing high-speed connectivity from desktops across the United States.

Issues to be Addressed

Despite the apparent momentum in the development of network technology, there are still many problems that impede its use for VE. First, the hardware technology is emerging so rapidly that standards are in flux. For example, ATM is not completely specified because the standards bodies representing the system users and developers have not agreed on all the protocol requirements. How ATM maps to the upper-layer software protocols for transport services and routing is not yet clear (Cavanaugh and Salo, 1992). Other protocols, like FDDI-II, probably will not be implemented widely unless ATM does not succeed.

Operating at gigabit speeds presents a new set of problems for networking. New methods of handling congestion are required because of the high ratio of propagation time to cell transmission time (Habib and Saadawi, 1991). By the time a computer in New York sends a message telling a host in San Francisco to stop sending data, a gigabit of information will have been transmitted. Latency also becomes a major concern. Much as a jet aircraft can be severely damaged by a small bird, short delays can cause major disruptions for high-speed networks and VE applications that demand real-time performance.

The most likely bottlenecks identified in the Aurora project at the

University of Pennsylvania will be in the network interfaces, memory architectures, and operating systems of the computers on either end. For example, the Fore Systems ATM interface for the SGI Indigo can handle only 20 Mbit/s of data, even though the media can deliver 140 Mbit/s. The slow progress in increasing the interface performance of FDDI is an example of the lag in technologies we may see as high-speed networks are fully deployed. Nor have memory speeds kept up with the strides made in central processing unit and network performance.

At the operating system level, most VE applications are built on commercial versions of UNIX—a system that is not designed for real-time performance. Additional problems are introduced by questions surrounding the adequacy of current transport protocols, like Transmission Control Protocol (TCP), that provide an interface between the operating system and the network. Furthermore, there is strong debate concerning the efficiency of the new generation of interface protocols, like Versatile Message Transaction Protocol (VMTP) and Xpress Transfer Protocol (XTP—designed to be implemented in silicon) (Rudin and Williamson, 1989). Other protocols, such as ST-II developed by Bolt Beranek and Newman (BBN) for the Defense Simulations Internet and the OSI suite, are challenging the Internet Protocol (IP) as networked applications like VE demand a host of integrated network services, such as multicast support and resource information for dynamic bandwidth allocation.

BBN has also developed dead-reckoning techniques to abstract data from simulators, thus reducing the communications loads on the network (Miller et al., 1989). Developers of distributed VE should follow this example and examine how to balance network bandwidth requirements with better efforts at preprocessing the data.

Finally, perhaps the greatest impediment to distributed VE is the lack of overall standards in the field, extending from file formats for three-dimensional models, to graphic and video images, to audio, to application interfaces. The IEEE standard for Distributed Interactive Simulation applications is in its infancy and is incomplete. Although it does not define a standard applicable to the diverse requirements of VE, it marks a milestone because it shows a widening of understanding of the relationship between VE and networking technology.

A related concern is networking architecture that will accommodate the connection of large numbers of communicating devices of different kinds that are delivering and using different kinds of services. Standards are needed for implementing architectural concepts as well as for achieving interoperability. A recent report of the National Research Council (1994) points out the importance of an open data network architecture (based on shared technical standards that enable users and providers to interact freely and that permit technology upgrades) in fulfilling the promise of the National Information Infrastructure. This lesson also is applicable to distributed VE.

COMMUNICATIONS SOFTWARE

Communications software passes changes in the virtual world model to other players on the network and allows the entry of previously undescribed players into the system. It is quite easy to exceed the capabilities of a single workstation when constructing a virtual world, especially if one expects multiple players in that world. As we move to a networked environment, we go beyond graphics and interface software issues to a much more complicated system involving database consistency issues. A standard message protocol between workstations is needed to communicate changes to the world. For small systems, it is important to ensure that all players on the network have the same world models and descriptions as time moves forward in the VE action.

In systems with fewer than 500 players, each node in the virtual world has a complete model of the world. The current SIMNET system looks like this and uses Ethernet and T1 links. For systems with more players (1,000 to 300,000, as envisioned for the Defense Department's Louisiana Maneuvers project), it is not reasonable to propagate complete models of the world but rather to consider rolling in the world model just as aircraft simulators roll in terrain. Although only a few researchers are examining this problem, there is at least an abstraction that might be relevant in the work of Gelernter (1991), entitled "Mirror Worlds." Mirror Worlds presents the notion of information tuples (like a distributed blackboard) and tuple operations: publish, read, and consume. Such an abstraction allows flexibility in communicating any type of information throughout a large, distributed system; flexibility that is necessary for constructing large virtual worlds. Although the abstraction appears appropriate, efficient and real-time implementations are an open research problem.

SIMNET, which is a simulation of entities on a battlefield, is currently the largest communications network in any VE (Pope, 1989; Thorpe, 1987). It is a standard for distributed interactive simulations developed under ARPA auspices that began running in 1988. *Distributed* means that the processing of the simulation can take place on different hosts on a network. *Interactive* means that the simulation can be dynamic and guided by human operators. SIMNET is also a network protocol with a well-defined set of communication packets called Protocol Data Units (PDUs). In addition to packet definitions, SIMNET also defines algorithms for the dead reckoning of vehicles whose velocities and directions can be predicted. The purpose of the dead reckoning is to minimize traffic on the computer network. SIMNET currently uses T1 links for long distance and

Ethernet for local communications. The protocol does not use TCP/IP, multicasting or any other network services. SIMNET sits on top of the device driver/link layer and, consequently, it requires that processes reading/writing SIMNET packets run as root privileged. SIMNET is limited in that it is only capable of engagements of a maximum of 300 players. Also the SIMNET protocol does not allow for generalized information transfer.

Distributed interactive simulation (DIS) is the latest standard for communicating Defense Department simulations (IST, 1991), with a latency of less than 100 ms in situations in which players are expected to interact. This is to minimize human perception of lag, which can induce simulator sickness. It has common goals with SIMNET but, as its replacement, is far more ambitious. DIS is intended to overcome the limitations imposed by SIMNET and to include packet definitions not present in SIMNET. DIS uses the Defense Simulation Internet (DSI) as its support network. DSI is currently being installed in more than 150 sites throughout the United States. The NPSNET project is connected to this network and currently displays records propagated by the DIS 2.0.3 draft standard. DIS uses IP multicasting services and does not require that its processes run under root privilege. DIS is planned to grow, with the eventual communicating player load expanding to 10,000 to 300,000 players. A DIS-compliant version of NPSNET, NPSNET-IV, was recently demonstrated at SIGGRAPH '93; approximately 50 players located in Washington, D.C.; Anaheim, California; Dayton, Ohio; and San Diego and Monterey, California, were connected. The hardware communications medium was Ethernet and a gateway machine to a T1 WAN.

There are additional players in the development of networked virtual environments; a good source is the Networked Virtual Environments and Teleoperation special issue of *Presence* (1994).

RESEARCH NEEDS

Advances in network hardware and communication software are key to the full realization of virtual environments. The following paragraphs detail some of the key network and communications needs generated by virtual environments.

Hardware

Although high-speed networks will provide the required computational power to build large-scale VE systems, there are several problems to be addressed. First, the actual cost of WANs may present a problem for the development of large scale VEs. The current price of a one-year T1 (1.5 Mbit/s) link is beyond the budget of most VE research groups, and we already know T1 is too slow. This is an issue of accessibility. Establishing a subsidized, nationwide, open VE network may be one way to eliminate this bottleneck to networked VE development.

Second, network host interface slowdown problems are another significant problem for the large-scale VEs of the future. We have high-speed interfaces today, but the layers of UNIX system software make these interfaces hard to utilize at their full rates.

Third, because latency across long distances is a permanent problem for VE systems (i.e., the speed of light), there is a need to build software mechanisms to cope with latency—such as dead reckoning or predictive modeling. Single-packet DIS transfers require approximately 300 ms between Monterey, California, and Washington, D.C., today using the NPSNET and UNIX software layers and Ethernet/T1 links. While NPSNET dead-reckoning algorithms for vehicles provide some predictive capabilities, these algorithms are not generalizable to other player paradigms. In addition, there is a long way to go to reach the 10 ms necessary for two participants on opposite ends of a long-distance network to work together cooperatively in real time.

Fourth, network hardware technology is evolving so rapidly that standards are in flux, and we are currently at risk of having them set by the large entertainment conglomerates. At the present time, there is only one governmental effort in network standard setting, the DIS packet format, IEEE standard 1278—and this standard can be considered a stopgap. DIS was developed as a standard of communication for military vehicle simulators and operates with Ethernet/T1 links and a particular software architecture for virtual environments. It was not designed for generalized information exchange between large-scale virtual environments; it is not even general enough to handle the articulations necessary for an animated, walking human figure. Packet-switching standards should be set that take into consideration the requirements of distributed VE systems with huge databases and potentially large numbers of participants.

Software

The problem of generalized information distribution in a large-scale, virtual environment of more than 300 participants is not yet solved. There are interesting abstractions but no real solutions (Gelernter, 1991). For example, the solutions offered by the Department of Defense's DIS standard are limited to 300 participants or less and are too specific and complex to be useful to the general VE development community. In the near future, both the military and civilian communities will have requirements for environments capable of handling as many as 10,000 to 300,000 simultaneous participants. There are several fundamental infrastructure and software problems associated with the research to develop the communications software needed to solve the technical issue of interaction among thousands of participants.

One of the primary infrastructure problems is that only a few university research groups are working in networking large-scale distributed VEs, and they are constrained in at least two ways. First, the networks are operating within extremely limited bounds—the Department of Defense (DIS) bounds; and second, they are very expensive to run. DIS is an applications protocol developed under ARPA and U.S. Army contract as the networking protocol for Department of Defense simulators. Although DIS is known to have significant problems (i.e., it is limited in capability and too large for what it needs to do), it has been made an IEEE standard. The Department of Defense is now putting all of its development resources into using this protocol for moving from the SIMNET-sized limit of approximately 300 participants (using Ethernet/T1 links) toward the 10,000 to 300,000 participant level. What is needed is a major research initiative to investigate DIS alternatives—alternatives that will allow a generalized exchange of information between the distributed participants of large-scale virtual environments. The new protocol needs to be extensible, a feature that does not appear to be a part of DIS.

A second infrastructure issue is the high cost of research into large-scale networked virtual environments. There are very few universities that can afford to dedicate the T1 lines (with installation expenses of $40,000 and operating costs of $140,000 per year) needed to support these activities. At the present time, only two universities in the United States have such dedicated resources: the University of Central Florida and the Naval Postgraduate School. Various approaches, such as an open VE network and the necessary applications protocol, should be considered for providing research universities with access to the needed facilities. Unless costs are significantly reduced, a concerted development effort on software solutions for networked VE cannot begin.

A critical ingredient in the development of large-scale networks for VE is the interest of the entertainment industry in introducing telecomputer and interactive video games into the home. To date, cooperative financial arrangements have been made between manufacturers of video games and large corporations already in the telecommunications business. The focus of these arrangements is to provide low-end, relatively inexpensive systems for large numbers of participants with an eye to making a profit. Like the Department of Defense, the video game industry is not interested in generalizability of information transfer, nor is it interested in openness and accessibility. The danger is that the video game industry will set the networking protocol standards at the low end and the Defense/DIS community will set the standards at the high end. Neither of these standards is general enough for the widespread VE application development we would like to see.

11

Evaluation of Synthetic Environment Systems

Many questions arise about design, performance, usability, and cost-effectiveness as a new system progresses from inspiration to realization. A carefully chosen program of evaluation studies, conducted throughout the development and operational life of a system, can help provide the answers to such questions, as well as reduce development time and minimize the need for expensive design changes. To the extent that synthetic environments (SE) systems are based on a family of new technologies configured in new ways to perform new functions, the need for evaluation studies becomes especially important. In this chapter we first outline a variety of approaches to evaluation and identify some key issues in the evaluation of systems in general, whether or not they are SE-based. We then comment on special issues in the evaluation of SEs, including the current tendency to ignore, or at least minimize, the evaluation problem.

GENERAL ISSUES IN SYSTEM EVALUATION

There are many practical reasons why evaluation studies should be conducted. At the outset of development, they can be used to refine the requirements for a system and to compare design concepts. (For simplicity, and unless stated otherwise, we use the term *system* here to refer to components and subsystems of SE systems as well as to complete SE systems.) Once a design has been chosen, evaluation studies can be used to diagnose problems and suggest alternative approaches. If appropriate, the results of these formative evaluations can also be used to facilitate communication with a sponsor or a customer. To be useful, the findings from such formative studies must be timely; this may require that scientific rigor and precision be traded for speed. Accordingly, rapid prototyping and simulations are often used to provide representations of the elements to be examined in formative evaluations.

Once a system has been developed, a summative evaluation can be used to measure the capability of the system to fulfill its intended function, to compare its performance with that of alternative systems, and to assess its acceptance by intended users. For example, the training effectiveness of a virtual environment (VE) training system might be compared with that of a conventional training simulator. Quantitative measures of training performance and training transfer, together with pooled ratings of experts and judgmental information about the friendliness of the system, all would have a role in such a summative evaluation. Taken together, formative and summative evaluations provide a critique of a system over its entire life-cycle. Finally, evaluation studies can be used to estimate the cost-effectiveness of an SE system in performing a particular application function. The results of these studies can then be used to inform policy decisions about investment and production.

The specific type of evaluation to be conducted will depend, of course, on the characteristics of the system to be evaluated as well as the purpose of the evaluation. One dimension along which evaluations vary, mentioned in the preceding paragraph, concerns the extent to which the purpose of the evaluation is to guide system development or to determine various characteristics of a system that has already been developed (e.g., related to overall performance, cost-effectiveness). A second dimension concerns the amount and type of empirical work involved. The empirical component of the evaluation can be restricted to pure observation (of how the system performs, how the user behaves, how the market reacts); it can involve surveys in which system users and other individuals affected by the system are asked questions; or it can involve highly structured and controlled scientific experiments. Under certain conditions, evaluations can be conducted without any empirical work at all and be based solely on theoretical analyses, for example, when appropriate models are available for describing all relevant components and performance can be reliably estimated solely by human or machine computation. A third dimension concerns the extent to which the item being evaluated constitutes a whole system or just one of the components in the system of interest. As one would expect, in most cases, the evaluation of a system component is much simpler than the evaluation of the whole system (particularly when the whole system involves a human operator). A fourth dimension concerns the extent to which the evaluation is analytic, in the sense of providing information on how the performance of the whole system is related to the performance of various components and subsystems. Obviously, analytic evaluations play an exceedingly important role in guiding the development of improved systems. A fifth dimension concerns the distinction between static and dynamic tests. Whereas static evaluation methods focus on nonperformance attributes of a system, dynamic methods focus on performance attributes. General background on types and methods of evaluation can be found in Meister (1985).

SPECIAL ISSUES IN SE EVALUATION

As discussed throughout this book, the creation of SE systems draws on previous work in, and provides research and development challenges to, a wide variety of established disciplines, including computer science, electrical and mechanical engineering, sensorimotor psychophysics, cognitive psychology, and human factors. In each discipline, the requirements associated with creating cost-effective SE technology raise new questions that call for evaluations within the context of research and development. In general, evaluation studies of SEs and SE technology are needed to help ensure that: (1) the perceptual and cognitive capabilities and limitations of human beings, as well as the needs of the specific tasks under consideration, are being used as driving criteria for system design; (2) hardware and software deliver SEs in a cost-effective manner; and (3) SE applications represent a significantly better way of doing old things or of doing new things that have never before been possible.

Despite the clear need for evaluation in the SE area, the types and amounts of evaluation currently taking place in this area are rather limited (a brief review of the limited work on performance measurement in teleoperator systems is provided in Chapter 9; three highly experimental studies on VE are described in Chapter 12). This is undoubtedly due, at least in part, to the high level of enthusiasm that exists about what the technology is likely to be able to accomplish, as well as the belief among many individuals in the SE field that no special evaluation efforts are required. According to this belief, the informal evaluations that take place more or less automatically as one is developing a system and the evaluation evidenced by the degree of acceptance in the marketplace (i.e., a system is good or bad according to whether it is used or not used) are sufficient. Although these forms of evaluation are necessary, the committee does not agree that they are sufficient; the cost-effectiveness of the research and development is likely to be significantly increased if the task of evaluation is taken more seriously. Although many existing evaluation tools can be adapted for use with SEs, a variety of new tools will need to be developed in order to evaluate the unique properties of this technology. In the following paragraphs, we comment briefly on some of the considerations that are relevant to the design of an analytic evaluation of an SE system. We discuss evaluation of SE system characteristics and issues that arise in observing and measuring human behavior in SE systems.

Evaluation of System Characteristics

Perhaps the first general evaluation task consists of measuring the physical characteristics of the SE system and considering how these characteristics relate to those of the prospective human user. Thus, for example, the characteristics of the displays and controls (dynamic ranges, resolutions, time lags, distortions, internally generated noises) should be measured and compared with the sensorimotor capabilities of humans as determined from psychophysical studies. Ideally, some kind of metric of physical fidelity should be developed that takes account not only of the fidelity that exists in all the relevant sensorimotor channels, but also of the extent to which this fidelity falls short of the maximally useful fidelity and of the implications of this shortfall for overall system performance. Another portion of the evaluation effort should focus on an analysis of the task to be performed by the SE system and an examination of how well the system is designed to perform this task. Such an evaluation would take account of physical fidelity and how such fidelity is expected to influence performance on various task components, as well as more central issues, such as the degree to which the SE system has been designed to anticipate the user's intention by examination of, and extrapolation from, the control signals. A further set of related issues can perhaps be best grouped under the heading of "cognitive fidelity." Such issues arise in connection with interface design and the interaction metaphors employed in this design, as well as the structure and function of the machine (computer or telerobot) to which the human is interfaced. The classic notion of stimulus-response compatibility is a pale and restricted version of the cognitive fidelity factors that require consideration in an analytic evaluation of an SE system.

Observation and Measurement of Human Behavior in SEs

In addition to evaluating basic system characteristics, it is important to evaluate overall system performance and how well the system performs the tasks for which it was designed. Less obvious, but also of considerable importance, is the need to examine the behavior of the human operator in the SE system. The most obvious method for accomplishing this involves storing all the signals that occur as part of the SE operation (i.e., all the display signals and all the control signals flowing in and out of the interface) and then studying this set of signals. In order to make such a procedure meaningful and efficient, however, procedures must be developed for filtering and transforming this mass of data in a manner that addresses well-defined evaluation questions. One such question, for example, might relate to how the user's behavior (as measured by the relation of the control signals to the previous display signals) compares with the behavior that would have been generated under the same circumstances by some model operator (e.g., an operator that is ideal according to some well-defined criterion).

Beyond examining stored SE records, evaluation of human behavior in the SE can make use of supplementing information derived from external observations and measurements. Such information could be obtained, for example, from direct observation of the subject by the evaluator, from video or audio recordings or physiological measurements of the subject, or from administering questions to the subject after termination of the SE experience.

Further information can be obtained by performing experiments during the SE experience. For example, the evaluator can intentionally degrade the system in some fashion to probe the effects of degradations that are likely to occur during field use. Similarly, the evaluator can introduce special signals into the displays and observe the subject's actions in response to these special signals (e.g., to test alertness).

Additional supplementary information can be obtained by having the evaluator enter the SE in which the subject is operating and making observations and measurements of subject behavior from within the SE. This can be done passively and unobtrusively (in the sense that the subject's environment remains identical to that which would have existed if the evaluator had remained outside) or the evaluator can intentionally interact with the subject or the subject's environment. In general, observations and experiments can be performed either from outside the SE or from inside the SE.

Among the special features to consider when designing an SE evaluation program are those related to measurements of the sensation of presence and the sopite syndrome. Although it has not yet been demonstrated that the sense of presence is an important variable for predicting objective performance (it has not yet even been adequately defined), it seems likely that interest in this variable will continue. Also, it is clear that the extent to which an SE system elicits the sopite syndrome is of major importance. Thus, it is important to consider measuring both these variables in any comprehensive SE system evaluation. (Information on both the sense of presence and the sopite syndrome are available in recent issues of *Presence.*)

Special considerations arise in connection with assessing system usability and acceptance. For example, in addition to choosing individual test subjects who truly constitute a representative sample of the anticipated population of users as test subjects, because many envisioned SE applications involve simultaneous multiple users communicating and working together on a common task, attention must be given to appropriate sampling of groups of users as well as to observations and measurements of relevant group processes.

Finally, because of the immersive character of SEs, special attention in SE evaluation must be given to possible negative long-term psychological and social effects. Illustrative questions in this category include the following: To what extent, if any, will individuals begin to confuse occurrences in SEs with occurrences in the real world? How will an individual's self-image be influenced by spending large amounts of time in SEs that seriously transform the individual's interactions with the environment? What impact will widespread use of networked SEs have on various types of social institutions? Fundamental psychosocial questions of this type are not likely to be addressed adequately (if at all) by the developers of SE technology. However, it is important that they be seriously addressed by some group.

In general and as a consequence of the many special features associated with SE evaluation, as well as the current tendency to ignore evaluation in the SE field, the committee believes that it would be extremely useful to develop a special evaluation tool kit for this field. Such a tool kit could serve to educate people in the field, to provide a more or less standardized set of evaluation tools for the field, and eventually to help provide a cumulative and shareable database that would constitute both a current snapshot of accomplishments in the field and a guide for future research and development.

III
APPLICATIONS

As the chapters of this book have noted, synthetic environment (SE) systems present a number of highly challenging scientific and technological research problems that, in an ideal and unconstrained funding environment, could be supported solely on the basis of intrinsic scientific interest. However, the funding environment is hardly unconstrained. Policy makers are more concerned with the promise of applications to areas of national need or economic significance than they are with the purely scientific or technical promise or potential in a given area of investigation. Cognizant of this priority, researchers are often tempted to promise a cornucopia of applications even when the state of the art does not permit realization of these applications. Avoiding the hyperbole that leads to a later fall is thus of paramount importance.

Both virtual environments (VE) and augmented reality pose challenging intellectual problems and also evidence considerable potential for a wide variety of applications; however, with few exceptions (one of which is entertainment), serious commercial applications (as opposed to research demonstrations of concept feasibility and promise) are likely to be realizable only in a long-term time frame—perhaps 5 to 10 years. This is not to say that meaningful progress cannot be demonstrated in a shorter time frame—only that according to the metric of commercial viability, major economic and social benefits are not expected to be demonstrated in the near future.

Teleoperation, in contrast, has already been used extensively in a variety of activities, including handling nuclear materials, operating heavy machinery, exploring space, performing underwater inspections, and removing hazardous waste. Furthermore, there are a number of experimental programs in which the use of teleoperation is being explored for surgery, patient monitoring, and delivery of remote treatment. Although teleoperation technology can provide many potential benefits, it has not, to date, been demonstrated to have a high commercial value.

That said, the scientific and technical study of SE is entirely compatible with a tight integration between research and applications development. The history of SE suggests that many interesting research problems in the field have arisen from difficulties faced by applications developers concerning, for example, perception, motion sickness, and software development for real-time interactive systems.

Given that applications-oriented work in areas of national need is appropriate, it is important to choose judiciously which applications should be singled out for near-term attention. An important consideration in this choice is the fact that the entertainment industry has been the primary driver of nonfederal work in VE. For all practical purposes, this industry has supported the development of low-cost, low-performance proprietary hardware and software. This suggests that federal efforts should focus more appropriately on the development of medium- and high-end technology that can be readily shared. However it does not specify which among the various possible application areas should be most strongly supported. The committee has developed five criteria for determining appropriate application areas. Specifically, the applications should be:

(1) of demonstrable national importance;
(2) intellectually challenging;
(3) realizable in a relatively short time;
(4) an area to which SE technology can make unique contributions; and
(5) an area in which success can be achieved through modest application of additional federal efforts.

Using these criteria, the committee focused on the areas of design, manufacturing, and marketing; medicine and health care; hazardous operations; training; education; information visualization; and telecommunications and teletravel. Our selection of these seven areas should not be construed to mean that these are the only areas worthy of special interest. A different committee with different expertise and interests could well have chosen another set.

Important omissions to this list are entertainment, art, and national defense. Entertainment is not discussed in detail because it is receiving significant commerical support and because the development of technology for this purpose was not deemed a pressing national need. Art is not disucssed separately because VE in its current stage of interface technology development provides only marginal opportunities for artistic expression over those offered by more traditional computer-generated art. However, as the interface technology improves, we expect a significant increase in the use of VE for art. National defense is not treated separately because it includes functions and tasks that are represented in many of the other application domains discussed. We do, however, briefly mention certain facets of these areas because of their important roles in technology development and implementation.

Progress in the entertainment industry regarding virtual environments is of interest because its contribution to the field will probably generalize to other application areas. Moreover, entertainment can be expected to be a primary driver and test bed for certain aspects of the technology. According to the popular press, since 1992, many joint ventures have been created among video game companies, computer graphics companies, motion picture studios, and telecommunication conglomerates to use VE technology as a new medium for entertainment, education, and artistic expression. Although several of these ventures are occurring in the United States, large programs are also under development in Japan and Europe (Fisher, 1993). To date, the entertainment industry's efforts in VE have been proceeding on several fronts, ranging from low-end systems for home use to arcade games, location-based entertainment, and theme parks.

At the low end of the technology, several companies, including Sony, Reflection Technology, Olympus, and Sega, are developing inexpensive VE displays for use in the home with interactive, three-dimensional games. In arcades, VE action games involving one or more players are now appearing. Those currently in existence offer motion platforms and realistic interaction, but the visual quality remains poor. Location-based entertainment differs from arcade games in that it provides several interactive systems on a common theme. Such systems usually involve several players sharing a virtual space over a local area network. In the near future, Paramount Communication will be introducing a location-based Star Trek game in which players enter the bridge of a starship, become one of the characters, and interact with other characters. As the industry moves from passive viewing and interactive two-dimensional environments to the realism of three-dimensional environments in which multiple participants actively play in a fantasy world, serious questions are raised about the effects of the content of these worlds on human behavior. Of particular concern is the use of such worlds to depict violence and sex; both of these areas have been heavily hyped in newspapers and magazines.

VE also offers a new medium for artistic expression that has only begun to be explored (see Loeffler and Anderson, 1994; *Leonardo*, 1994). A number of individuals are designing highly imaginative video games. A few artists have produced special VE experiences, and some programs are beginning to emerge that encourage artists to create VE art pieces; however, these efforts are in their infancy. Although VE has the potential to support the creation of a wide variety of aesthetic experiences, it is too early in the development process to know how the technology will influence the techniques, effects, genres, and content that will emerge to define the medium. In the case of film, another artistic medium, the process of discovering new techniques (e.g., cuts, pans, fades, slow motion, close-ups and telephoto shots) and using them to create new experiences for the viewing public has taken many decades to evolve.

A few organizations have developed VE art programs or exhibitions over the last few years. The Banff Centre for the Arts in Canada has had an ongoing program allowing artists to create VE art pieces since 1991, and about a dozen pieces have been created so far (Moser, 1991). The annual exhibition Ars Electronica in Linz, Austria, has included VE art pieces in recent years (Hattinger et al., 1990). Since 1991 the annual

SIGGRAPH computer graphics conference has sponsored an exhibition including VE demos and art pieces (Lineham, 1993).

Currently, most efforts are directed toward building the hardware and software that can effectively generate realistic VEs. As adequate equipment becomes more widely available and affordable, we expect to see an increasing number of artists exploring its potential. Although it is difficult to make reliable predictions, it seems fairly obvious that future VE art will make extensive use of observer participation and interactivity of various kinds. Precisely how these features, as well as the other special features of VE, are used to create truly artistic experiences, however, remains to be seen. In order to encourage the use of VE for purely artistic purposes, it will undoubtedly be necessary for the art world to create appropriate supporting attitudes, programs, and funding opportunities.

Finally, as noted above, national defense is not treated as a separate application area because its scope intersects with functions in all the applications discussed in Chapter 12. For example, information visualization and distributed collaboration are critical to strategic and tactical engagement planning; hazardous operations relates to the use of technology in handling unsafe materials or remotely operating vehicles in hostile environments; and telemedicine and the eventual promise of remote surgery are important to the rapid provision of medical support to soldiers on the battlefield or to those located in isolated or inaccessible locations. The two most relevant applications to military concerns may be training and design and manufacturing.

According to a recent draft strategic plan (Thorpe, 1993), the most ambitious application of SE technology in the military is the integration, management, development, and acquisition of new systems such as tanks or aircraft. In this application, the technology will be used to model and test alternative systems in a variety of synthetic exercises. The most promising configuration will then be designed by a computer-aided design (CAD) system, manufactured in a virtual factory, and tested in an SE. Once the appropriate technology is available, the entire system can be selected, designed (with manufacturing specifications, time, and cost included) and tested without the need for physical prototype development. Moreover, individuals and units can receive training on the new system using virtual war games before the system has been produced. The discussion of manufacturing in Chapter 12 uses the design of an aircraft to describe many of the processes involved in the acquisition of a new system.

In the area of training, the Department of Defense (DoD) is currently using VE technology to cover a range of instructional experiences, from those of the individual soldier and small team to theater-level synthetic battlefields in which more than 10,000 participants interact in real time over distributed networks. For all levels of training, VEs can be used to safely provide exercises in hostile and dangerous environments. Moreover, in the future, computers will have the capability to generate VEs as replacements for and extensions of traditional equipment-bound simulators. A critical issue discussed in the training application section of this chapter is the problem of conducting evaluation studies that demonstrate the impact of one method or another on training success as defined by how well the knowledge and skills acquired in training transfer to performance on the job.

Of particular interest to DoD is the development of networking hardware and software for large-scale distributed training. The first large network simulation of realistic battlefield engagements was SIMNET, developed by DoD's Advanced Research Projects Agency. In SIMNET, as many as 300 soldiers in tanks and aircraft simulators located at different military bases can engage in a realistic battle against an intelligent enemy on a common battlefield. Each participant in the battle views the portion of the terrain and the action that would be visible to him if he were present. As a battle unfolds, the scene changes in real time. Once a battle is completed, it can be replayed as an after-action briefing in which trainees can zoom in on various portions of the engagement or take the perspective of any tank or aircraft.

More recently, DoD has used its newest software—Distributed Interactive Simulation (DIS)—to develop a detailed, true reconstruction of the 73 Eastings battle that occurred during the Persian Gulf war. This fully interactive simulation is based on the events in an actual battle; as a result, it can be used as a benchmark to examine what-if training scenarios when enemy or friendly weapon system capabilities are changed. The projections are that DIS will provide many of the networking capabilities needed by the military for both training and system acquisition.

The final chapter of this report provides a more detailed description of the use of SE technology for each of the application areas selected for review.

12

Specific Applications of SE Systems

DESIGN, MANUFACTURING, AND MARKETING

Computer technology generally, and synthetic environment (SE) technology more specifically, are potentially important drivers for future developments in design, manufacturing, and product marketing. Trends already under way suggest a movement toward the development of manufacturing systems in which production processes are integrated with all elements of the product life-cycle from concept through sales, including quality, cost, schedule, and the determination of user requirements. This process, known as concurrent engineering, provides for parallel development across product life-cycle activities through the use of technologies such as computer-aided design/computer-aided manufacturing (CAD/CAM) and computer-integrated manufacturing (CIM). Using shared databases, customers, designers, and production managers can simultaneously evaluate a proposed product design. As a result, the design of the product, as it evolves, can incorporate the requirements of the user, special needs for marketing, and any limitations of the production process (Krishnaswamy and Elshennawy, 1992). Once developed, advanced visualization technologies such as virtual environments (VE) may provide valuable extensions to current practices.

In April 1993, manufacturing was named as one of six national initiatives to be administered by the Federal Coordinating Council for Science, Engineering, and Technology (FCCSET). The primary focus of the FCCSET mission on manufacturing is to assess special opportunities for technology to change the way manufacturing will be carried out in the next century.

Rationale

Although the ultimate goal of all manufacturing is to produce a tangible object or component, information—in the form of plans, specifications, and processes—plays a most important role. Thus, it should be expected that information technology, including VE, could have a meaningful role in manufacturing by enabling people to generate and manage such information more effectively. Consider the following points in the life-cycle of a manufactured product:

- *Developing design requirements.* VE could be used as a medium in which a customer's mental image of a product can be fashioned into a virtual image of the product. That image could be subsequently manipulated or even used as the basis for production specifications. Examples: architectural walkthroughs of spatial designs, such as proposed buildings, rooms, and aircraft interiors.
- *Undertaking detailed design.* VE could provide designers with the ability to reach inside the design and move elements around, to test for accessibility, and to try out planned maintenance procedures. The designer could thus have a comprehensive view of how changes made in the design or placement of one component could affect the design of other system components.
- *Producing the artifact.* Virtual pilot lines could simulate both human and machine processes on the production line. Such a virtual pilot line could be used to predict performance and to diagnose the source of faults or failures. Plant management could be improved as engineers are given the capability of reviewing and modifying various plant layouts in virtual space.
- *Marketing the artifact.* By providing potential customers with the ability to visualize various uses of an artifact, VE could be used for marketing an array of completed product designs to customers prior to their production.

Specific Manufacturing Applications

Building Prototypes Electronically

Building prototypes electronically provides a number of advantages, including the opportunity for sharing data across manufacturing functions and the ability to modify designs with greater ease than in a physical mock-up. A further advantage is the ability to incorporate stress and durability test data into the design process without physically performing each test. VE promises to enhance the value of prototyping electronically by offering customers, sales staff, and engineers the ability to walk around the product and manipulate it in virtual space, much the same way as they would explore a physical mock-up in real space. A long-range goal is to create VE systems that can be extended to provide groups of individuals in different locations with the capability to work together in a shared virtual space.

Researchers at the University of North Carolina (Airey et. al., 1990) have worked on the development of software for creating interactive virtual building environments. This software can be used to present architectural walkthroughs of buildings that have not yet been constructed. In touring a virtual building, an individual will be provided with changing views and lighting that are consistent with his or her position relative to the building space. Such software can be useful for design of any interior spaces, including industrial buildings, hospital operating rooms, churches, homes, and aircraft passenger compartments, to name a few.

Electronic Configuration and Management of Production Lines

Another potentially important area for the application of VE technology is in the design and testing of the processing, fabrication, and assembly lines. Virtual pilot lines might be developed instead of real pilot lines to simulate human and machine tasks and make predictions about potential problems for human performance and safety as well as estimating the probability of failure and the line's expected operating efficiency. The promise is that virtual pilot lines will be far easier to modify in response to diagnosed problems than a physical pilot line, and they will provide the opportunity to introduce information on manufacturing efficiency early in the product design process. In addition, a virtual line could be run in parallel with an operating line for purposes of diagnosing failures, retooling for new products, or changing human-machine interface designs or procedures at points in the process at which errors or problems are occurring.

Although VE technology provides more of a promise than an existing capability for industry, several forces within various government and manufacturing enterprises will push for its development and use. From the industry perspective, VE technology has the potential to make the manufacturing process (from planning through sales) more flexible and economical. The aerospace, automobile, and textile industries are pursuing VE technology as a means for speeding development and making product modification easier. Chrysler, Ford, and General Motors have formed a VE consortium with the U.S. Army vehicle center, the automotive division of United Technologies, the University of Michigan, and several small companies. In a recent proposal to the Advanced Research Projects Agency (ARPA), the consortium predicted that VE technology would lead to improved product design, a better market response, and reductions in time or cost (Adam, 1993).

Exemplar Industries

In the following sections we provide a discussion of the potential for VE in the textile and aerospace industries. The selection of these industries was not based on an exhaustive or systematic search of industries and applications; however, both industries offer some interesting illustrations of VE technology that transfer broadly to other industries.

Textiles

VE may have very important applications in the marketing and manufacture of clothing. The concept is that customers could shop for apparel in a VE in which they would see virtual clothes on virtual images of their own bodies and feel how the clothes would fit. On the basis of this experience, customers would select and order outfits that would be fabricated on demand and sent out to them within a short time period. The result would be to significantly reduce financial losses associated with fabric waste during apparel production and with product markdown and liquidation. Moreover, the customer would be provided with a greater range of choice and an improved made-to-measure fit. This approach appears to be a natural extension of the current market trends of increased shopping through catalogs and home shopping networks and the accompanying decrease in retail outlet shopping.

Industry Efforts VE technology has captured the interest of the textile industry (Steward, 1993). In 1993, a collaborative research and development program, the American Textile Partnership (AMTEX), was initiated between the Department of Energy (DOE), the DOE national laboratories, and the fibers, textiles, and apparel industry to improve the competitiveness of the U.S. textile industry through the application of technology. The national laboratories plan to work together and coordinate with industry through major industry-supported research and technology transfer facilities. Matching funds for the partnership are to be provided by government and industry. The first joint project between the national laboratories and the industry will involve the creation of an industry model for integrating hardware and software in a system to provide Demand Activated Manufacturing Architecture (DAMA). One aspect of this effort will involve research on the uses of VE technology (Hall and Walsh, 1993).

The U.S. textile industry, which includes fiber producers, textile weavers, apparel makers, and retailers, employs over 1 million workers (10 percent of the manufacturing work force in the United States) and includes 26,000 companies. It is the largest producer of nondurable goods, experiences annual consumer sales of approximately $200 billion, and contributes $53 billion to the U.S. gross national product (Hall and Walsh, 1993; Steward, 1993). Each year the industry fails to realize revenues of approximately $25 billion due to inventory markdowns and liquidation.

Most companies are small, with profit margins of 2 percent or less, and so are not in a position to conduct or support research. Almost all the research in the industry is conducted by five large research centers based in universities and jointly funded by industry and government. One of these centers, the Apparel CIM Center, was established in 1988 with the goals of removing barriers to adopting proven CIM technology, establishing CIM standards, providing assistance to state industry, and conducting broad-based research and development to keep the industry competitive.

The primary charge of the Apparel CIM Center is to investigate applications of VE to clothing as seen, examined, and purchased by the retail customer. A second charge is to apply VE technology to represent the internal view of a textile manufacturing plant, including the position of machines, the air conditioning, the noise level, and the lighting. The goal of the project is to facilitate the reorganization of a manufacturing plant by providing engineers and factory workers with the ability to walk through a virtual plant; to move machines around on the basis of requirements to produce new lines of apparel (seasonal changes); to examine spacing, lighting, and noise to ensure good human factors practices; and to assess the effects of various equipment configurations on work flow.

Technology Requirements The technology required for implementing the internal plant layout includes: (1) building an object database of all equipment needed in the plant, (2) creating the capability to determine light and ventilation, (3) providing noise levels based on the combination and spacing of machines, (4) matching lighting requirements and noise levels against federal requirements, (5) integrating new software capabilities with existing simulations of work flow through the plant for manufacturing different products, and (6) developing an interface for engineers that is easy to use and acceptable. According to Steward (1993), all of these activities are under way. There are many technologies, including VE, contributing to this application—some existing and some in development. As these activities evolve, there will be a need for VE technology to rely on and interface with other developing information technologies.

Developing the technology required to fully implement a VE system for marketing clothing is a long-term effort. One area for development is body measurement technology. Currently the concept is to have the customer don a body stocking and be electronically scanned. The linear and volumetric dimensions from the scan would be stored on a card that the customer would use when entering the virtual shopping space. When a customer's dimensions change, he or she could be scanned again. Cyberware has built and demonstrated effective full body scanners. However, this technology is produced on an individual basis and is expensive to acquire.

A second area of development is the technology for accurately representing material draping. A critical factor in deciding to purchase a garment is appearance: how the jacket hangs, how the folds appear, how the fabric moves when the individual wearing it moves, etc. Thus, the draping of virtual clothes on a virtual customer must appear real.

Other areas requiring technology development include providing accurate colors in the virtual world, giving customers the opportunity to "feel" fabrics, and providing customers with a sense of how the garment "fits." All of these factors are important to customers in selecting clothing. Colors must be accurate so that different parts of an outfit can be matched; feel and fit are critical to comfort and style. Of all the research and technology development issues identified above, the most complex and long range will be developing the tactile feedback needed to create a sense of fit.

Aerospace

The aerospace industry is expected to be a major user of VE technology in the future. Companies such as Boeing and Rockwell International have long-range plans to develop VE systems that will provide all interested parties with the ability to view and interact with three-dimensional images of prototype parts or assemblies of prototype parts. Currently, both companies are using CAD tools to create electronic prototypes of parts in lieu of physical mock-ups.

Industry Efforts Staff at Rockwell International, through its Virtual Reality Laboratory (Tinker, 1993), are working on virtual prototypes and mock-ups; virtual world human factors assessment for proposed task environments; and training for manual factory workers, maintenance personnel, and equipment operators. These efforts are in the early stages of implementation. Proprietary software has been developed to read CAD data into a virtual reality database. The long-range goal is to provide the ability for multiple participants to work together in a shared virtual space interacting with high-resolution CAD data in real time.

At Boeing, the design of the 777 aircraft is being accomplished without a physical mock-up; all of the 6.5 million parts are being prototyped electronically using CAD tools. As a result, designers, engineers, and possibly customers see only models of parts or assemblies of parts on a computer screen. Although this approach provides for shared databases among design, manufacturing, and sales components and adds significant flexibility to the design process, it takes away the ability to walk around, explore, and manipulate parts. As a result, Boeing is working toward the development of a VE system that would give these capabilities to designers, engineers, customers, and marketing personnel.

According to Mizell (1993), the plans for Boeing's VE project include: (1) giving designers the ability to reach in and move parts assemblies around, (2) conducting human factors tests in virtual space, using human models, to determine whether maintenance can be accomplished and control operations can be easily performed, and (3) providing customers with a variety of customized aircraft interiors to walk through and make modifications in real time. All of these applications feed directly back into the design process. Cabin layout modifications made by customers influence the placement of wiring, ventilation, windows, seats, etc. Mizell believes that Boeing would consider the work in virtual reality a success if its only use was to provide customers with the ability to walk through and experience various configurations of aircraft interiors.

Implementation of the project is in its initial stages. But Boeing already has software that reads CAD data into a VE preprocessed database. A limiting factor at this point is the computing and graphics power needed to represent the CAD database in a three-dimensional virtual space so that real-time interaction and a feeling of presence can be facilitated. It is anticipated that VE technology will begin to contribute to Boeing's productivity in the next two years, but development will probably need to be continued over a 15-year time period (see the more detailed discussion of implementation issues in the section on technology requirements).

A second major project area at Boeing is the application of augmented reality to various parts of the manufacturing process. This project seeks to eliminate the need for complex assembly instructions or manually manipulated templates by creating a system in which computer-produced diagrams are superimposed onto work pieces. The technology to accomplish this goal involves a head-mounted see-through display and a head-position-sensing and real-world registration system. The augmented-reality project is being developed to assist factory workers in performing many complex, manual, skill-based tasks that rely heavily on human perception and decision making and therefore are not easily automated. Currently, guidelines are presented to workers in the form of overlays, templates, or written instructions for each step in the process. When parts or processes are modified by designers in the CAD system, a substantial amount of time may be required to reflect the appropriate changes in the manufacturing documentation. The augmented-reality system would link manufacturing instructions with the CAD system and superimpose these instructions in the form of diagrams on work pieces. The diagrams would appear to be painted on. For each step, a new diagram is projected. The link with the CAD system would make it possible to show changes in design or procedures to the worker immediately. A more detailed description of this project is provided by Caudell and Mizell (1992).

Technology Requirements In Boeing's augmented-reality project, a prototype system has been developed and tested. The primary technology needs are for a comfortable head-mounted color display with a field of view wider than 30 degrees. Another goal is a position-tracking system that will leave the worker untethered.

In order to implement Boeing's vision for using VE, several areas require technology development. One critical problem is the lack of graphics and computing power. The CAD database for the 777 aircraft contains between 5 and 10 billion polygons. Even though only a fraction of the database may be needed at any one time, the existing graphics hardware limits the ability to create a scene that is interactive in real time, particularly because of the complexity of the geometry in the CAD database. The problems created by the size of the database, the inadequate hardware, and the requirement for a VE that looks real and behaves in predictable ways underscore the need for research on real-time scheduling, assigning reduced workload areas, and developing heuristics to accomplish graceful degradation.

Another goal requiring technology development is providing engineers with the ability to interact with objects in virtual space. Currently, Boeing is working with a mannequin developed by Norman Badler at the University of Pennsylvania (Badler et al., 1993) that can be put inside the CAD geometry and changed in size or shape. Similar technology has been used by the automobile industry for several years. The next major step is to develop the capability for an engineer to inhabit the mannequin in a virtual space, to move around inside the CAD geometry, perform maintenance checks, and in general, feel present inside the scene while others monitor from a third person. Particularly important development areas include the need for collision detection and the requirement to give the individual in the virtual space some sense of force feedback, especially when testing the difficulty of performing various maintenance operations. Developments in this area, particularly those involving haptic feedback, are at least 10 to 15 years in the future.

Creating architectural walkthroughs of customized aircraft interiors is another important area for development. These models would provide customers with the opportunity to see and experience the aircraft they are purchasing before it is actually built.

MEDICINE AND HEALTH CARE

Rationale

The knowledge base of medicine has exploded in the past 30 years, and it continues to expand at a staggering rate. As a result, medical practitioners have difficulty in keeping pace with changes in practice, and medical students and residents have difficulty in assimilating the information presented in their medical educations.

As in other information-intensive disciplines, computer and communications technologies have important roles to play in reducing the cognitive demands on medical practitioners and students by helping to manage, filter, and process multiple sources of information. The following subset of medical knowledge and skill is well-suited to management and handling by VE, augmented reality, and teleoperator systems:

- *Anatomical relations of various organs and systems.* Knowing that a particular organ is located underneath another organ is an essential part of anatomical knowledge. The ability to "walk through" the body and to see anatomy in its natural state with all of the interrelations of various organs and systems would greatly facilitate the acquisition of certain important pieces of anatomical knowledge.
- *Development of manipulative skills involving precise motor control and hand-eye coordination.* Surgical trainers can be particularly useful in acquiring needed skills.
- *Image interpretation.* Although various imaging devices are common in medicine, their effective use depends on the skill of the viewer to identify often small differences between normal and abnormal images.
- *Telemedicine through teleoperation.* Medical expertise is often unavailable in remote areas. Telemedicine—whether through consultation or through remote manipulators that enable teleoperation—offers some potential to place medical expertise in locations that might not otherwise have access to such expertise. Teleoperation also enables one to effectively transform the sensorimotor system of the physician (diagnostician or surgeon) to better match the task.

Specific Medical Applications

The following discussion focuses on six applications of VE technology to medicine: medical education, accreditation, surgical planning, telepresence, telesurgery, and rehabilitation. Each subsection addresses a long-range vision of how VE might assist in these applications and a description of possible near-term demonstrations.

Preservice and Continuing Medical Education

Medical education has changed little in the last 30 years, despite enormous advances in knowledge. Most medical schools emphasize learning facts by rote. Information is provided in a lecture format, and students study outlines for endless hours in the library. Little effort is expended to place the information into a context or framework that might help to structure and organize seemingly disparate facts. As a result, students must use their own, perhaps incomplete experience to begin assimilating the data and creating a logical, integrated framework of anatomy, physiology, biochemistry, genetics, and the myriad springs of subspecialized knowledge from contemporary medical research.

Teaching Anatomy

The teaching of anatomy is illustrative, and the application of VE and augmented reality to such teaching has great potential. The static, transparent, two-dimensional overlays typical of anatomy textbooks could someday be replaced by a virtual human. Indeed, today the National Institutes of Health is funding the Visible Human, a project to develop a complete static digital representation of an adult human. Once the data are collected, a student would be able to operate a VE system for anatomy that would illustrate the spatial interrelationships of all body organs relative to each other, selectively enabling or suppressing the display of selected body subsystems (e.g., displaying only the digestive system, viewing the complete image without the circulatory system).

A much more sophisticated version of the Visible Human would be a dynamic model that could illustrate how various organs and systems move during normal or diseased states, or how they respond to various externally applied forces (e.g., the touch of a scalpel). Thus, a student could view the heart in normal and diseased states pumping blood, or observe how the stomach wall moved while cutting it.

Today, several virtual worlds have been developed to demonstrate basic anatomy and as rudimentary models of training simulators. One is a model of the optic nerve created by VPL (VPL, Inc., 1991). This model illustrates, in three dimensions, the path of the optic nerve from the retina to the optic cortex. By pointing a finger one can fly along this path, looking to either side at adjacent structures. In this way, less effort is expended in constructing a three-dimensional image in the individual's mind and more effort is channeled into learning the anatomical relationships.

A second model is a rudimentary simulator for the abdomen created by Satava (1993b). With this simulator, one can travel from the esophagus throughout the intestine, taking side trips through the biliary system and the pancreas. It is a unique instructional tool that describes anatomy from the inside of the intestines rather than from the outside. It is of considerable benefit in training individuals to perform colonoscopy and esophagogastroduodenoscopy, as well as teaching students the true anatomic relationships of intraabdominal structures. Basically, one is able to fly around various organs and experience their actual relationships—the model provides the learner with the ability to interdigitate between organs and behind them without destroying their relationships to one another in the process.

Another educational tool is an augmented-reality system that allows the user to see virtual information superimposed over real structures. See-through displays provide the user with a view of the surrounding environment, along with an image displayed on goggles. Investigators in Boston and at the University of North Carolina (UNC) have created see-through displays using computer-assisted tomography (CT) scan, magnetic resonance imaging (MRI), or ultrasound technology as imaging techniques (Bajura et al., 1992).

The work on augmented reality at UNC (Bajura et al., 1992) is based on the images from an ultrasound that delineates abdominal structures in three dimensions. Specifically, the investigators created a graphic of a three-dimensional model and projected it through the head-mounted display (HMD), as an overlay onto the user's view of the abdomen. This program, used on a pregnant woman, allows the operator to "open a window" into the abdomen and view the fetus in a three-dimensional manner without incising the skin. Although the application of such programs to view a developing fetus is limited, the technology raises the possibility of visualizing other intraabdominal structures.

See-through models can be used to teach surgeons where an organ is located and show its relation to surrounding tissues. Novice surgeons often have difficulty visualizing the location of the gallbladder and the cystic duct in relation to the common bile duct. Despite extensive anatomy instruction, the first few operations are difficult because structures in a living body appear different from the illustrations in an anatomical atlas. A see-through display gives surgeons in training an opportunity to develop their own internal three-dimensional map of living organs, rather than having to operate without one.

Surgical Training

A variety of surgical training applications are plausible as well. Surgeons know there is no substitute for hands-on practice and training. Consider laparoscopic procedures, which involve surgery performed through very small incisions in the body. The advantage of such procedures is that patient recovery time is greatly reduced over conventional surgery, because of the smaller trauma to the body. However, manipulation of tools through the incision, hand-eye coordination, and understanding the spatial relationships of the tools relative to the organs within the body place high cognitive demands on the surgeon.

Today, a surgeon wishing to learn laparoscopic procedures may attend a one- or two-day course designed to teach necessary skills. In the course, the surgeons-in-training begin with laparoscopic cholecystectomy trainers to familiarize themselves with the technique. These trainers are quite rudimentary, consisting of a black box in which endoscopic instruments are passed through rubber gaskets. Trainees use these instruments to practice such tasks as tying knots, grasping structures, and encircling plastic arteries. The current technology of velour-and-plastic organ models are stiffer and harder to manipulate than normal tissue; arteries are not easily transected or avulsed and do not bleed; and damaged organs do not ooze. As a result, the experience is far from realistic. Following work with the trainer, the surgeons-in-training begin practicing these techniques on pigs—the approximation to humans is better, but the anatomy is dissimilar, and the amount of experience is limited by both the cost and the availability of the pigs.

An appropriately constructed VE simulator could obviate some of these problems. For example, the abdominal simulator developed by Satava (1993b) includes several laparoscopic tools, making it possible for the surgeon to practice some (primitive) endoscopic surgical techniques. The current drawbacks of this system are the low-resolution graphics, the lack of realistic deformation of organs with manipulation, and the lack of tactile input and force feedback. However, the model sets the framework for further investigation into surgical simulators. Such a simulator could be used for initial training, as well as for the additional follow-up training, which has been shown to significantly reduce the incidence of post-surgical complications in patients operated on by surgeons with such training (William et al., 1993).

A second application to surgical training is the use of the see-through augmented-reality model to support novice surgeons in performing their first few appendectomies, cholecystectomies, or arthroscopies. After gaining confidence in their knowledge of anatomy, these surgeons could proceed without the aid of the display (Haber, 1986).[1]

Close on the horizon are hybrid programs that will combine technology from current surgical simulators with VE technology (Stix, 1992). The pressure and tactile feedback provided to the surgeon could be improved by using the actual tools of endoscopic procedures, instrumented to act as interface devices, to train with a simulator. If the VE program did not have to generate the instruments, computer power could be reserved for producing more intricate displays.

VE simulators might also serve as mentors to teach residents, interns, and medical students the basics of surgical practice, such as suturing, ligating blood vessels, and the basic tenets of dissecting. Operating procedures might be stored in a VE library, ready to demonstrate a Billroth I, a choledochojejunostomy, the reconstruction of an uncommon craniofacial anomaly, pathologic rarities, and uncommonly encountered trauma scenarios.

Building on the groundwork of what has been created to date, Robert Mann's vision of "the ultimate simulator" will be achieved: "a computer environment in which a surgeon will not only see but 'touch' and 'feel' as if actually performing surgery" (Mann, 1985).

Accreditation

The purpose of medical accreditation is to ensure that physicians have a minimum level of skill and knowledge adequate to serve the public. The knowledge and skills acquired by physicians-in-training depend very much on the idiosyncrasies of the patient population and the epidemiology of diseases to which they are exposed. Thus, whereas residents of different programs have extremely disparate bases of knowledge and different levels of skill, even residents in the same program may acquire very different funds of knowledge. For recently developed surgical techniques, training programs may differ even more strongly, lacking standardized training methods and standardized accreditation procedures (see Bailey et al., 1991).[2] How are physicians at various points on a spectrum of experience, without any unified teaching, and without exposure to a uniform patient population, to be judged against a national standard?

---

[1] This training protocol is analogous to the head-up displays often used by fighter pilots during their initial flights. The pilot's head-up display gives cues about the height of objects or ground contours while the pilot develops his or her own internal approximations. Once the pilot has developed a scale of altitude approximation, the cues can be deleted from future flights.

[2] In some cases, surgeons have published proposed guidelines for such accreditation.

Today the answer is board certification. Knowledge can be tested by means of paper-and-pencil examinations, but proficiency in operative skills cannot be demonstrated in this way. As a result, individual residency programs are left to judge a physician competent in the operating room.

VE simulators offer some hope for standardizing the surgical accreditation process. For example, a more sophisticated version of the abdominal simulator developed by Satava (1993b) could offer a set of standardized tests for laparoscopic procedures. The ability to track the instruments would enable the National Board of Medical Examiners (or a similar governing body) to monitor performance during a given procedure and to document the types and frequency of errors made, thus providing some uniform means of assessing surgical skills across programs.

Surgical Planning

An actual medical operation is never performed in the abstract—it is performed on a specific individual whose precise physical dimensions are unique and whose anatomy almost certainly deviates from those found in anatomy textbooks. Thus, to a certain extent, a surgeon confronts surprises every time he or she undertakes an operation.

VE-based surgical planning aids offer a way to reduce the uncertainty. In principle, imaging data for the patient could be used to update a generic digital human model, allowing the surgeon to understand more fully the specifics of the individual. The surgeon could then explore freely various approaches to solving a surgical problem on the VE simulator and could practice the operation if required.

Such an application is a promise for the long term. However, Jolesz and his colleagues (Gleason, 1993) have developed an augmented-reality system for video registration of brain tumors to aid in the planning and performance of surgical resection. In this system, the mass is imaged using either CT or MRI, a three-dimensional construct is created, and the image is projected over the patient's head to plan the optimal site of skin incision and bone flap to expose the tumor. This program is then taken into the operating room to provide a reference map for the surgeons during the resection. Thus, the surgeon is able to consult the image at any time to assess the remaining tissue and the extent of further excision.

A model of the lower extremity envisioned by Mann (1985) and designed by Delp (Delp et al., 1990) is an example of using virtual reality to test various procedures. This model allows the surgeon to "perform" the planned surgery, and then simulate a number of years of walking or other normal activity. The altered model can be reanalyzed at the end of the simulated activity period and the outcome of the procedure evaluated.

This allows surgeons to refine decision-making skills before ever operating on a real patient.

The augmented-reality system designed by Bajura et. al (1992), described above, could also be used for preplanning complex abdominal procedures, such as complex tumor resections. Liver masses, pancreatic pseudocysts, and edematous gallbladders could be viewed in three dimensions, making it easier to estimate size and interrelations with other abdominal organs and to plan and perform invasive procedures.

Telemedicine

Telemedicine is the technology that would allow physicians to interact directly from locations thousands of miles apart. They would be in the same virtual "room" when discussing a case or when performing a procedure and could refer, simultaneously, to paramedical data, reducing the likelihood that a miscommunication would occur, that information would be missed, or that results would be misdirected or lost.

The Medical College of Georgia in Augusta has developed and introduced the first stages of a statewide telemedicine system. The system uses interactive voice, video telecommunication, and biomedical telemetry to link rural health care facilities and primary care physicians with large medical centers. As a result, primary care physicians and their patients can consult with specialists without leaving their communities. In these consultations, the participants would see each other and share common diagnostic data and images. The idea also extends to providing remote assistance to surgeons in rural hospitals during surgery. According to Sanders and Tedesco (1993), the system will eventually have five hubs, each serving several clinics and rural facilities. Currently, a test is being conducted with one hub—the Medical College of Georgia—serving five sites. The ultimate goal is to deliver the quality health care available at major medical centers to all underserved areas in the state. Early results suggest that the proposed system, when fully developed, will not only make high-quality care more accessible but will also reduce costs. Telemedicine networks are also under development in Iowa, West Virginia, and Colorado.

Telesurgery

Developments are also under way in telesurgery. One example is the Green Telepresence system, created at SRI International (Palo Alto, Calif.) by Phillip Green. The system based on technology used by the National Aeronautics and Space Administration (NASA) for remote manipulation consists of a separate operative worksite and surgical workstation (Rosen et al., 1994). The operative site has a remote manipulator with surgical instruments, a stereoscopic camera, and a stereophonic microphone to transmit the environment to the surgical workstation. The workstation has a three-dimensional display on polaroid glasses and an interface for the surgical tools. It also has the capability of accepting digital information from CT, MRI, and vital signs monitors.

Hunter and his colleagues at McGill (1993) have developed a prototype teleoperated microsurgical robot and an associated virtual environment that are intended to allow a surgeon to perform remote microsurgery on the eye. The helmet worn by the surgeon controls a remote camera in the operating room. Images from the camera are relayed back to the helmet for the surgeon to view. Tools shaped like a microsurgical scalpel and attached to a force-reflecting interface are provided to the surgeon. As the surgeon moves the tools, he or she causes the microsurgical tool held by the microrobot to move proportionally. The forces exerted by the microrobot are reflected back to the surgeon. These forces are amplified, thus providing the surgeon with an experience of cutting that would, under normal conditions, be imperceptible. The fact that the master and slave computers communicate by optical fiber connection will make it possible for the surgeon and the microsurgical robot to be located at different sites once the system is implemented.

VPL has developed a system, RB2, in which two operators can interact in virtual space (VPL Inc., 1989). Using this program, two surgeons could operate on a virtual patient, one as the initiate, the other as a mentor. Ultimately, a surgeon in New York could help a surgeon in London to do an operation without setting foot on a plane. A data highway rather than a commercial airline would make possible the transmission of complex knowledge (Satava, 1993a).

There is an enormous amount of interest in the promise of telesurgery; however, all current work is at the research and development stage.

Rehabilitation

There are several ongoing programs in which the use of VE technology is being tested as a means to assist in the rehabilitation of physically and mentally challenged persons. Some of this work is focused on building better human-computer interfaces that take into account the specific physical limitations of the disabled individual. These studies are examining alternative methods, such as eye movements or flexion of facial muscles, for sending bioelectric signals to a computer, which will in turn enable the individual to perform a desired task or see a desired display. At Loma Linda University Medical Center, researchers are developing such interfaces using what they call a "biocybernetic controller" (Warner et al., 1994). Their next series of studies will involve the use of bioelectric signals based on muscle activity to provide disabled children with the capability to play Nintendo-type computer games now being played by the general population. Other centers working on the same problem, such as the Children's Hospital in Boston, have proposed studies to examine the motor output capacity of disabled individuals and to match that capacity with control interfaces making use of multimodal outputs.

Loma Linda researchers are also engaged in having disabled persons work with virtual objects generated by a computer as a means to begin rehabilitating their motor skills. For example, an individual could practice manipulating a virtual object at a weight they could handle. It is believed that this practice is useful even if the virtual object weighs less than the real one. In related work, Greenleaf Medical Systems (Greenleaf, 1994) is working toward using VEs to enable individuals to perform tasks that they could not perform in the real world. An example provided by researchers at Greenleaf is creating a VE in which a cerebral palsy sufferer could operate a switch board.

Weghorst et al. (1994) are experimenting with using virtual objects to treat walking disorders associated with Parkinson's disease. According to the authors, objects placed at the feet of these patients may serve to stimulate a walking response. Since using real objects is not a particularly practical approach, virtual objects are being tried with some success. Specifically (p. 243):

> Near-normal walking can be elicited, even in severely akinetic patients, by presenting collimated virtual images of objects and abstract visual cues moving vertically through the visual field at speeds that emulate normal walking. The combination of image collimation and animation speed reinforces the illusion of space-stabilized visual cues at the patient's feet.

In another example, Greenleaf Medical Systems is in the process of adapting the VPL DataGlove and DataSuit for use in measuring the functional motion of a disabled person and recording progress over time. Moreover, researchers are working on developing a gesture control system designed to enable individuals wearing the DataGlove to perform complex control activities with simple gestures and on creating a system that will recognize personalized gestures as speech signals. In the latter example, the DataGlove would receive the hand gestures and translate them into signals sent to a speech synthesizer, which then "speaks" for the individual wearing the DataGlove. All of these products are in the early development stage.

In yet another example, VE technology is being used by architects to design living and working spaces for the disabled. The technology now enables an individual in a wheelchair wearing a head-mounted display and a DataGlove to travel through a space testing for access, maneuverability, counter heights, and reach distances for doors and cabinets.

Finally, the use of VE is being explored for behavioral therapy. Hodges et al. (1993) report on a project at the Graphics, Visualization, and Usability Center at the Georgia Institute of Technology that makes use of VE technology to provide acrophobic patients with fear-producing experiences of heights in a safe situation. Advantages of this approach are that VE provides the therapist with control over the height stimulus parameters and with the ability to isolate those parameters that generate a phobic response in the individual. In another project, Kijima et al. (1994) have been exploring the use of VE technology to create a virtual sand box to be used for virtual sand play. Sand play, a technique used in diagnosing an individual's mental state and providing psychotherapy, involves patients creating landscapes and populating them with bridges, buildings, people, animals, and vegetation. An important advantage of using VE for sand play is that the patient's actions are recorded and can be viewed several times by several trained observers.

Issues to be Addressed

Although the long-term potential of VR for medical applications is suggested by an extrapolation from current demonstrations, a number of research problems must be addressed to fulfill this potential.

- *Simulations of higher fidelity.* The simulated organs and body structures seen by the simulator user are far from realistic, with graphics that are primitive and cartoon-like. Tactile and force feedback, an important consideration in simulating the actual "feel" of a surgical procedure, is mostly absent. Changes in visual perspective are not seen by the user in real time.
- *Realistic models of organs and other body structures.* In current simulations, organs and body structures do not morph as real tissue does; for example, they do not deform with gravity or change shape with manipulation. Blood vessels should bleed; bile ducts should ooze; hearts should pump.
- *Better image registration techniques for augmented reality.* In many cases, a VR surgical or diagnostic aid will require the superposition of images acquired from a number of different modalities (e.g., a CAT scan coupled with an ultrasound image). See-through displays must superpose artificially generated images on the real image through the user's eyes. Techniques for aligning these images so that the right parts correspond to each other remains a considerable intellectual problem.

- *Appropriate data representation schemes.* How can patient-specific data best be combined with generic models of humans in a computationally efficient manner?
- *Reduction in wide-area network delays.* Delays must be significantly reduced to provide for coordinated long distance work.

There are several social issues that will also need to be addressed if VE technology is to achieve wide acceptance in medicine:

- *Acceptability to care providers.* Physicians generally practice from a perspective of conservatism, refraining from the use of techniques that may be unproven, such as the opportunities for surgical training and performance offered by VE technology. Accepted practice, to be encoded in the future as practice guidelines, is widely regarded as a way to ensure that the well-being of patients is not placed at undue risk.
- *Public opinion.* For understandable reasons, the public may feel uncomfortable with the perception that a robot is undertaking surgical operations. Negative public reactions to the recently tested robot used in hip replacement surgery are a case in point. A cultural shift that acknowledges that automation can help care providers do their jobs more effectively will be necessary. Moreover, the necessary shift is bidirectional: patients will have to change the way they regard their care providers, and care providers will have to change the way they present themselves to patients.

To gain physician and public acceptance, convincing demonstrations will be necessary. These systems will have to demonstrate that their use will result in better outcomes, fewer complications, and ultimately even less invasive, less debilitating procedures.

In the current health care environment, a third social consideration must be addressed: cost. Explicit cost-benefit analyses for various technologies are likely to become increasingly common, and VE will be no exception.

TELEOPERATION FOR HAZARDOUS OPERATIONS

Rationale

Any activity that involves performing tasks in environments that are unsafe for humans, or in which safety is too costly, is a potential application of a teleoperated system. Chapter 9 provides a review of the technology underlying teleoperation; in this section we discuss ways in which this technology has been applied in hazardous environments.

Survey of Major Applications

There are many examples of teleoperation applied to hazardous environments. We provide a short survey of some major applications. In the first two, toxic environments and space operations, detailed scenarios are presented to illustrate critical tasks. These scenarios contain examples that can be considered prototypical of all hazardous environments.

Toxic Environments

To avoid human exposure to radioactivity or toxic chemicals, teleoperators have been used in handling radioactive and chemically toxic materials, maintaining nuclear power plants, and disposing of hazardous wastes. Handling radioactive materials was the first application for which teleoperators were developed in the 1940s by Raymond Goertz at Argonne National Laboratory near Chicago. Glove boxes are used when more dexterity is required than is afforded by current teleoperators, but advances in telerobotic dexterity should eventually replace them.

Maintenance and cleanup of nuclear power plants typically requires mobile telerobots as well as large manipulators. A variety of wheeled or tracked mobile robots have been designed that carry sensors and manipulators through power plants (Fogle, 1992; Trivedi and Chen, 1993). Large manipulators are required to reach inside reactor vessels for maintenance (Munakata et al., 1993; Rolfe, 1992). More generally, telerobots are required to identify and handle toxic materials at accident sites (Stone and Edmonds, 1992).

The disposal of hazardous wastes is a major public concern. For example, the U.S. Department of Energy has an extensive program to retrieve low- and medium-level nuclear and chemically toxic wastes of nuclear weapon fabrication from desert disposal sites and to place these wastes into improved containers and sites (Harrigan, 1993). At these desert disposal sites, most of the waste is stored in large concrete tanks or buried in 100 gallon drums whose location is known only approximately. Removing these wastes (and any soil that may have been previously contaminated through leaks) without causing further environmental damage and without endangering human life is a challenging task.

Another challenging task is disposing of the high-level nuclear waste from commercial nuclear reactors and government facilities; high-level waste is currently stored under water in storage pools. As these facilities fill up, current plans call for placing the wastes in deep underground storage facilities. Moving the material from storage pools to a processing facility where they can be more easily handled (e.g., a glass vitrification plant) to the final storage facility will require the assistance of remote handling operations.

In order to discuss the application of hazardous waste disposal in more detail, we take as a scenario the removal of hazardous waste from an underground storage tank, such as at Hanford, Washington (Harrigan, 1993), which has been demonstrated in part in laboratory settings. Sensors and manipulators are needed that are immune to the chemical, radiation, and thermal stresses that would destroy ordinary video camera lenses and sensing elements, manipulator lubricants and mechanisms. Moreover, the hazardous nature of the materials being handled places a premium on high reliability. Manipulators for gaining access to storage vats must be large yet have the ability to work in confined spaces. Such systems will be extremely expensive. Since there are many waste sites to handle, advances in supervisory control are required to speed up operations.

The first problem is that the contents of each tank is not known. Hence vision and proximity sensors have to be inserted by a manipulator to map the tank's contents. In this mapping phase, the manipulator is manually controlled; proximity sensors are used to approach surfaces without collisions using local control. In this operation, it is critical that the tank sides not be contacted to avoid rupture. From these data, a representation of the tank's contents is constructed with operator assistance. For recognizable objects such as pipes, an object model can be specified. For waste surfaces and nonrecognizable objects, a three-dimensional surface representation is employed.

The model of the tank's contents is then used to generate a graphical image and to perform simulations of operations. Tasks and robot motions are planned through real-time graphical simulations: sequences are developed, operations are dynamically simulated, and collisions and joint limits are checked. Just prior to a real operation, a simulation can be run to verify the expected result.

For the actual task, the operator relies heavily on the graphical display and simulation to control the telerobot because of poor visibility in the tank. Sensing during the operation is employed to update and modify the graphical model. As a motion develops, the simulation is run simultaneously to check for emerging problems such as collisions or other unsafe operations. Approaches to surfaces and objects will be conducted in local mode, using proximity and contact sensors in the end effector. Wherever possible, autonomous control will be employed to speed up such operations as grasping objects, cutting pieces of the structure loose, and conveying waste from the tank.

Space Operations

Achievement of remote manipulation from the earth to the moon dates from the 1967 United States Surveyor and the Russian Lunakhod, which followed shortly thereafter (both were unmanned lunar roving vehicles). Canadian-built remote manipulators have served the space shuttle project on many separate missions for loading and unloading the cargo bay in space and to provide a stable base for astronauts performing extravehicular tasks. The deep-space probes have been the most impressive teleoperators, e.g., the Viking was controlled from earth even when at the very edge of the solar system, several light-hours away.

The planned space station construction is motivating a great deal of space telerobotics work, particularly in the United States, Germany, Japan, and Canada. These considerable efforts have resulted in some of the most advanced telerobotic systems, which are developing the kinds of generic capabilities that will be required in applications other than space. Uses for space telerobots include construction and maintenance of the space station, satellite servicing and repair, and space laboratory work.

For the present scenario we concentrate on space laboratory robotics, motivated in part by the success of the German ROTEX effort (Brunner et al., 1993; Hirzinger et al., 1993). There are two main reasons for conducting laboratory and manufacturing operations in space with robots. First, humans are not aseptic enough for clean-room applications such as crystal growing. Second, control of the robot from the ground allows full utility of the space laboratory, because operations do not require the active participation of astronauts.

A main challenge therefore is teleoperation under variable time delays on the order of several seconds. Predictive displays and local autonomous control are the approach that has been taken (Sheridan, 1993b). Space station environments, unlike those of hazardous waste disposal, are likely to be well known in advance, so a detailed simulation of the workspace can be fashioned. The main difficulty is to model the dynamics of the manipulator and objects in the contact tasks; the goal is to accurately simulate the contact forces for the predictive display.

The predictive display is used in the first instance for operator training. An operator interfaces to the real-time simulation via a stereo graphical display and force-reflecting hand controllers. Display enhancements are added to aid the operator (see discussion in Chapter 9).

In addition to training, the predictive display is employed for task planning and preview (Kim and Bejczy, 1993; Kim et al., 1993). Prior to sending a motion command to the remote robot, the operator can rehearse a task to ensure its outcome. The resultant motion can be replayed or relayed to the slave robot; otherwise the operator may directly control the robot after feeling confident about the outcome.

During an actual space laboratory task, the operator may view not only the stereo graphical display but also a time-delayed real image superimposed with a phantom display. The phantom display is an outline of the graphical image of the robot superimposed on the real image. This phantom display responds immediately to an operator's commands, indicating where the robot will go from its current position.

Enhancements on the real image can be performed to improve visibility, such as removing bright specular reflections and increasing the contrast in dimly lit areas. Remote cameras automatically track the end effector to keep the manipulator in view or to focus on the task. Voice commands are employed to switch a camera onto a monitor or to zoom in on a known location. A separate graphical display provides a more global view for operator orientation.

Commands to the slave robot include not only position specifications, but also nominal sensory patterns such as for force. Differences between the real and simulated worlds are accommodated by local sensing and intelligence of the robot; the degree of autonomy is controllable and restricted, because any adjustments are small. The local sensing in the robot's end effector includes six-axis wrist force sensing, optical proximity sensing (both long and short range), and tactile sensing (Hirzinger et al., 1993). This local sensing can also be employed in dynamic tasks, such as catching a free-floating object, which might not be possible with long time delays.

Many elements of the scenario described above were successfully completed by the German Space Agency's ROTEX telerobot (Brunner et al., 1993; Hirzinger et al., 1993), during a flight of the space shuttle Columbia in April and May 1993.

Teleoperated Heavy Machinery

Tractors, diggers, cranes, and dump trucks have been controlled remotely for construction, demolition, earth moving, forestry, farming, and mining. Excavators and cranes are now being fitted with hand controls and sensors so that the end-point may be directly controlled instead of the arm joints (Wallensteiner et al., 1988). Force feedback is being supplied to the operator to accommodate the hardness of the ground. Teleoperated bulldozers and diggers have been employed to remove contaminated soil (Fogle, 1992; Potemkin et al., 1992). Grapple yarders and log loaders have been teleoperated in the forestry industry (Sauder et al., 1992). In mining, teleoperation is being applied for haulage and drilling (Kwitowski et al., 1992); low-bandwidth communications through mine shafts offer major difficulties.

Off-the-Ground Environments

Manipulators mounted on teleoperated cherry pickers have been applied in remote power line maintenance, tree trimming (Goldenberg et al.,

1992), and firefighting. In the last five years, several companies in Japan and the United States have developed teleoperator systems capable of changing insulators and performing other repairs on active high-tension power lines. Current methods require the use of rubber gloves (on low voltage lines) or several-foot insulated "hot sticks" for high voltage lines. The human line workers are boosted to within close proximity to the active wires in buckets on the ends of long hydraulic arms. However, due to instability of these devices or human errors, workers have inadvertently been burned and electrocuted. Newer manipulator devices allow workers to stay much farther away in the buckets or to operate from the ground using remote video.

Windows and outside surfaces of buildings must be cleaned, painted, and repaired, but the cost of doing so is high and dangers significant. Many modern buildings have platforms that can be raised and lowered by motor drives and moved laterally around the building. Yet since the geometry of the buildings is known in detail, it should be possible to perform all the necessary tasks by teleoperation. For inspection of structures, experimental vehicles have been demonstrated that scale walls (Hirose, 1987; Nishi et al., 1986).

A more unusual application of teleoperation is the inspection and cleaning of the insides of pipes. Pipes are essential elements of most buildings and plants (especially in nuclear and chemical plants with high pressure pipes), yet they are often difficult to inspect or clean. Experimental prototypes have been developed, but no large-scale effort has yet occurred (Fujiwara et al., 1993; Fukuda et al., 1987; Okada and Kanade, 1987). Eventually such devices could also be used in water, gas, and even sewer pipes, many of which are aging and leaking. Some teleoperators for this application drag cables behind them, although developers have envisioned autonomous self-powered devices that make surveillance journeys for significant distances before returning with their findings.

Firefighting and Security

Although teleoperated fire rescue and firefighting operations are not currently undertaken, it is clear that firefighters could use a remote vehicle capable of entering a burning building. Development of such a fire rescue teleoperator would make an excellent national demonstration project. There is no requirement that is not technologically feasible. A remote firefighting vehicle would be capable of crawling up one or more flights of stairs under remote human control; entering various rooms; and allowing the human operator to look around (with better vision than a human eye and most likely with camera pan and tilt controlled by a head-mounted display) to find and to give instructions to persons who are found. It would include fresh air breathing masks, an insulated compartment into which a person could crawl and be brought out to safety, and foam dispensers and other fire-extinguishing gear. Furthermore, the vehicle would probably be battery-operated so as not to have to drag a power cord and risk getting snagged, although conceivably it could unroll a cord behind it to avoid entanglement.

Several firms are already marketing simple wheeled sentry robots capable of guiding themselves along paths laid out with magnetic or optical markers, listening for unexpected sounds, or looking for unexpected people or objects. Future police teleoperator capability should include radio-controlled stair-climbing teleoperators equipped with low-light-level cameras capable of seeing in the dark, which are able to dispense tear gas or even to fire weapons.

Explosive ordnance disposal is an area that is very well matched to teleoperations, and prototype teleoperators have been developed for this application. A teleoperator could approach a suspected bomb with the intent of first inspecting it, then disarming it if possible, covering it with something to limit damage from detonation, or carrying it to another location for detonation.

Finally, a number of teleoperated systems have been developed for military surveillance, sabotage, and warfare. Some take the form of aircraft, such as cruise missiles, and remotely piloted vehicles and helicopters. Some are land vehicles, capable of laying down kilometers of optical fiber for high-bandwidth surveillance and control. Still others are undersea vehicles, capable of performing surveillance, sabotage, and weapons delivery. Some are manually controlled; others are preprogrammed telerobots.

Bulk Transportation Environments

Loading and unloading operations—whether within plants or to or from shipping docks, ships, trucks, trains or aircraft—tend to be unique with respect to the relative positioning movement to be made (location of receiving vehicle relative to initial location of container or product). Some bulk materials such as liquids, coal, and grain can be handled by pouring or pumping, so that remote manipulations are primarily for the purpose of locating the transfer pipe or duct relative to the receiving container or vehicle. Sometimes this manipulation is dynamically tricky, as in air-to-air refueling and cargo transfer between two ships on rolling seas. By sensing the (continuously varying) relative position of the receiving vehicle with reference to the sending vehicle, the teleoperator can be commanded to null out the difference in platform positions, while the human operator controls the cargo transfer as though the two platforms were fixed relative to one another. When massive objects are being transferred, manual control must be executed slowly because of large inertias. Computer-aided control can make use of predictor displays to anticipate the effects of control actions before they are imposed, thus adding lead stabilization.

Ocean Environments

Two widely publicized deep ocean-teleoperations were used in the recovery of the accidentally air-dropped hydrogen bomb off the Polomares Islands in 1967 and the discovery and exploration of the ship Titanic. In the former, the Navy vehicle CURV (cable-operated underwater remote vehicle), equipped with a pressure-sealed camera, lights, and manipulator, was dragged until the parachute attached to the bomb could be found and grasped. In the second case, the Woods Hole Oceanographic Institution submersible vehicle Argo was passively towed at the end of a several-mile-long cable, and its cameras, lights, and powerful sonar were used to discover the Titanic (Whitcomb and Yoerger, 1993). The small ROV (remotely operated vehicle) Jason then explored inside the Titanic at the end of its power and signal umbilical. Deep-ocean teleoperators have also found a number of other sunken ships, historic artifacts, and buried treasure, as well as discovering the existence of hydrothermal vents deep on the ocean floor.

Deep-ocean vehicles, sensors, and manipulators have been used to survey the ocean bottom topographically, geologically, and biologically. It has been estimated by some biologists that more than 90 percent (volumetrically) of the earth's ecosphere has yet to be explored—namely, the oceans below the surface—and it is already clear that the oceans are full of creatures at all depths. Teleoperators appear to be an ideal way to perform this exploration.

A more commercial application of teleoperation in ocean environments has been in the area of off-shore oil exploration. Many petroleum wellhead preparations have been performed by teleoperators. Still other operations that might have been done by teleoperation have been neglected, such as inspection and repair of the legs of oil platforms (some of which have broken up in heavy seas because their welds cracked) and inspection of outflow pipe lines (some of which have burst for similar reasons). Many associated underwater robotic technologies still need much development, such as cleaning, inspection of welds, and rewelding.

Infrastructure and Research Needs

The usefulness of teleoperation in a particular hazardous domain may not be at issue, but its costs may well be. In many cases, it is simply more economical to use tried and true practices (and accept but attempt to minimize their attendant human losses) than to invest in risky new technologies. A good example is that of firefighting, an area in which there is little national infrastructure to encourage application of advanced technology research and development, and in which local communities buy apparatus from a myriad of suppliers of conventional trucks and equipment with funds derived from local taxes and bond issues.

The technology needs for teleoperations for hazardous operations can be divided into a number of categories:

- Manipulators must today be developed on an application-by-application basis that is well-matched to the task in question. Such manipulators must be made sufficiently precise and sensitive, producing enough feedback that the human operator can control the manipulations appropriately. In addition, the range of tasks that can be performed by manipulators should be expanded so that unanticipated contingencies can be handled with greater ease.
- The survivability of manipulators and sensors is obviously important in hostile environments. As an example, radioactive environments can lead to the breakdown of lubricants, the deterioration of electronic components, and the darkening of glass lenses. The dirt and grime of mining environments are hard on optics and other sensors and on actuator joints and the bearings of robotic arms.
- Cost continues to be a major issue. As noted above, the cost-effectiveness of teleoperated systems remains to be demonstrated, and until it is, the cost of building and testing teleoperated systems will remain high and demand will remain low.
- Using sensors to construct a model of a disposal tank's content is challenging, particularly when dynamics as well as geometry are to be simulated.
- During actual operations, the incoming sensory information has to be fused with the current model of the tank.
- Higher-level supervisory control and partial autonomy are required to speed up the operations and ensure safety.

TRAINING

Requirements for training are ubiquitous in today's world. They range from general childhood education to highly specialized training of military special forces in preparation for specific critical missions. Our purpose here is to consider the application of VEs for training, that is, for the acquisition of special skills for specific purposes.

The natural precursor to VE, technology-based task simulation, came into being in the late 1930s. The archetype for such training devices was the Link Trainer, essentially a plywood box on a gimballed stand. The inside of the box was furnished like the cockpit of an airplane, complete with a functioning instrument panel. When occupied by a trainee, a large hood was dropped over the canopy area so the trainee's view of the external world was cut off. Activation of the instruments was controlled electronically by an instructor, and each trainee action was recorded in the same manner. By operating the simulator, the trainee could practice instrument navigation skills in ways that could not or would not be duplicated in the real world because of their hazardous qualities. While having relatively modest fidelity compared with present-day flight simulators, the Link Trainer incorporated nearly all of the training concepts that are featured in the most advanced training devices (Williams and Flexman, 1949). After World War II, the ideas of dynamic task simulation were extended into other task environments: air defense (Chapman et al., 1959), air traffic control (Fitts et al., 1958), submarine combat (McLane and Wolf, 1965), and the operation of surface vehicles such as tanks (Denenberg, 1954).

Meanwhile, computer programs were devised that could represent contexts as varied as multifirm commercial markets (Kibbee et al., 1961) and the processes of municipal governments (Guetzkow, 1962). Developers formulated interactive video representations of such complex situations as the emergency room in a large hospital in a way that the trainee would see the patient, see and hear the actions and questions of other members of the health care team, and, most dramatically, experience the consequences of his or her own decisions in the form of medical outcomes.

The advantages of relatively low cost, low hazard, speed, repeatability, and good transferability of acquired skills were the driving force for progress in simulation-based training. These qualities are likely to be even more prominent in mature VE-based training facilities. Indeed, the qualities that VE might bring to the whole enterprise of dynamic task simulation are those explicitly called for in the most comprehensive historical review of the use of simulation for pilot training (Orlansky and String, 1977).

The idea of using VE technologies for training is a very natural extension of the use of simulation for training. Given the committee's view of VE as simply an extension of the concepts of simulation to include closer coupling between the participant and the technology supporting the creation of the artificial world, it is only natural that training, a subject that has benefited greatly from advances in simulation, should be a prime candidate for exploration of VE technology.

Rationale

Virtual environments have the potential to extend the scope of circumstances that can be satisfactorily simulated and to extend the advantages of using simulation versus real-world training. When the visual world is presented to each individual on a personally mounted visual display, when the auditory environment is re-created with perceptually compatible virtual sound sources, when sensorimotor interaction is accomplished through an effector interface that simulates the touch and feel of live interaction without requiring a physical hardware mock-up, then the definition and scope of what can be represented is limited not by what can be presented in a physical mock-up, but only by the quality of the sensors and effectors, the speed and processing power of the supporting computer, and the bandwidth of the transmission channels.

The issue of what can be trained with simulation has therefore moved from the physical world to the potential worlds that can be created with software and general-purpose hardware. At this point in time we are not only uncertain of exactly what those boundaries are, but also confident that they are expanding with each new technological development in the field. When VE technology is mature, a suite of hardware will be able to be converted from a nuclear physics training simulation to a mission rehearsal application or a surgical simulator, simply with the change of software application.

VE-based training also has the potential for being better than conventional training. For example, realistic training for hazardous, risky, and dangerous emergency situations, in which the trainee feels that he or she is present in the simulated environment, can be undertaken in ways not possible with conventional training. Artificial cues, not realizable dynamically in the physical world, may be utilized to augment the training effectiveness of VE worlds.

Lastly, VE-based training may be more cost-effective. Current experience with simulators is instructive. Simulators are very expensive because they require the fabrication or acquisition of detailed mock-ups of the participants' actual operating environments. Nevertheless they are extremely cost-effective when compared with equivalent training in the corresponding real environment. (One example—the cost of a 747 flight approaches $10,000 per hour, whereas today's 747 simulator can be run for less than a tenth of that cost.) If research demonstrates that VE-based training is as good as other methods, then the economics of running simulators versus mock-ups and test equipment and vehicles may be such that the former is much cheaper than the latter.

The Future

It is presumptive to assume that serious applications of VE to training will await the clear experimental demonstration of cost-effectiveness. Most training techniques in place today did not have the benefit of such evaluation before they were adopted. There is little doubt that VE training will be advocated and perhaps even adopted on the basis of its face validity alone. However, the scientific community should not be satisfied simply with proof from the marketplace.

The committee anticipates that training will be a powerful and useful early application of VE. Its success in limited applications, such as passive simulation scenarios, involving three-dimensional demonstrations that are difficult to create in a two-dimensional world, will be the first areas of successful application. Then, as haptic interfaces emerge, at first with physical manipulators and later with general-purpose earth-referenced devices or with exoskeletons, the full potential for cost-effective applications to training will emerge. At this point we should be able to forecast applications for which VE technology will be successful and those for which it will not. In the long range, we envision a full range of interactive portable training simulators both for operational activities and for maintenance activities.

Specific Applications

Commercial aircraft simulators substitute for real aircraft in training flight crews with respect to standard and emergency procedures in their current aircraft and in qualifying them to transfer to a new aircraft type. Every nuclear power plant in the United States is required to have a simulator on site for regular training of the control-room crews. Aspiring doctors practice surgery on cadavers, a medical simulation of a real patient.

Although there is a massive literature on the use of simulation for training, the extension to VE technology has only begun. For now let us address the experimental literature on VE training. We are aware of one as yet unpublished and two published experiments; however, we can expect to see many more in the near future.

Wes Regian had conducted a series of experiments on the use of VE training for two applications: (1) learning to navigate an unfamiliar building and (2) learning to operate a simple control panel (Regian and Shebilske, 1992; W. E. Regian, personal communication, 1993). He has found conclusively that VE training is as good as training in the actual environment when the performance measures are success in finding one's way in the building unaided by encumbrances or artificial cues or operating a physical realization of the control panel. In further experiments in this series, as yet unpublished, Regian is investigating what can be accomplished with a similar training paradigm when a two-dimensional display is used as a control condition.

Peter Hancock and his students at the University of Minnesota (Kozak et al., 1993) conducted an experiment that purported to evaluate the usefulness of virtual environments for training simple pick and place movements in the laboratory. The task he chose for training was the movement of five soda cans arranged in a row, one by one, to a position about 6 inches to the rear of their prior position and then to return them to their original position. The subjects in one control condition practiced moving real cans while a second control condition involved no training. The experimental condition practiced using a VPL Eyephone system and data glove to move virtual cans. The transfer condition required moving the real cans as rapidly and accurately as possible.

Not surprisingly, early in the transfer trials they found that the virtual training condition was worse than training with the real cans when real movement of the cans was the criterion task. The resolution and viewing angle of the virtual display were inferior to those involving the real cans. The data glove representation of the hand projected in the virtual world provided degraded positioning accuracy compared with a real hand and real cans. And the visual-motor coordination lagged behind the actual hand movements under the conditions that used as the criterion the time for speeded movement. The task required during virtual training was simply very different from the criterion task.

The conditions of this experiment were a poor choice for demonstrating successful VE training. The task of picking up and moving soda cans is so highly overlearned even before beginning the experiment, that all that is left to learn is the very context-specific features of the situation. Nevertheless, the experiment illustrates one of the biggest barriers to the application of VE to training. The fidelity of the visual-motor representation was so poor that the training condition did not correspond closely enough to the conditions required for transfer to even hope for positive transfer.

Lochlan Magee and his colleagues of the Canadian Defense and Civil Institute of Environmental Medicine have designed a VE system for training naval officers in the piloting tasks of maneuvering ships in formation on the open sea. The officers currently practice on real ships at considerable cost of operation. Magee conducted an experiment, as yet unpublished, that compared land-based VE training (using a head-mounted display of the visual scene and Polhemus head sensing to stabilize the imagery) with training at sea. Both groups of approximately 13 junior officers transferred to performance at sea. He allowed the instructors to utilize the VE system in the same way that they used the real ships. They made no attempt to augment the training with artificial cues or special conditions that could be accomplished only in the VE system. His data analysis is still incomplete, but preliminary results show that the instructor ratings during the transfer condition for the simplest maneuvers studied were significantly better with training using the VE system than with training at sea, and the two training conditions were not significantly different for the two more difficult classes of maneuvers. Because of the very great cost differences between the two training methods, showing equivalence is more than good enough to support use of the VE system.

Training Transfer

The major concern in designing any training experience is how well the knowledge and skills acquired in the training environment transfer to the working or operational environment. The two theories that have dominated the thinking about transfer since the turn of the century grew out of the behaviorist tradition. The earliest of these was proposed by Thorndike and Woodworth in 1903 (Thorndike, 1903). This theory, known as the theory of identical elements, is based on the notion that the degree of transfer is a function of the identity of stimulus-response pairs between the original (training) task and the transfer task. That is, the task similarity is determined by the number of shared elements. This model can be used to give qualitative predictions of positive transfer; however, it does not address negative transfer.

The second theoretical formulation that was offered by Osgood in 1949 proposed that the amount of transfer is a function of the degree of similarity between stimuli and responses in the original and the training tasks. In this theory the predictions are qualitative and continuous for both positive and negative transfer. For VE training, the qualitative predictions based on Osgood's model are: (1) maximum positive transfer will be obtained with stimulus and response identity between original and transfer learning, (2) if there is complete response identity, negative transfer cannot be generated, and (3) even with maximum identity of stimuli between original and transfer learning, negative transfer can still be obtained if the responses are antagonistic or in opposition to one another.

A drawback of these approaches is the difficulty of identifying and defining similar elements and in determining the amount of contribution each element makes to transfer. Clearly, both theories are most appropriate for the training of discrete, concrete, and simple tasks that involve the acquisition of motor skills.

More recently, with the shift to cognitive approaches, interest has begun to focus on the process by which learning occurs and on the development of models to explain that process. Many of these models include not only a representation of the knowledge to be acquired but also a set of rules for using the knowledge. For the past 10 years, Anderson (1993) has been using such a model to build intelligent tutors to teach content and procedural knowledge in algebra, geometry, and computer programming.

There is some indication from transfer of training research that single theories of transfer will not hold for both cognitive and motor tasks (Schmidt and Young, 1987; Ritchie and Muckler, 1954). For complex simulations of the type represented by VE technology, in which complex tasks may be learned and transferred, Hays and Singer (1989:319) suggest starting with fidelity analysis: "the first step in the fidelity analysis should be to determine the major emphasis (either cognitive or psycho-motor) of the task. If the task is cognitively oriented, it is likely that the training systems should emphasize functional fidelity. On the other hand, if the task has strong psycho-motor requirements, physical fidelity should be emphasized." Functional fidelity refers to the accuracy of representation of the system's procedures and procedural sequences; physical fidelity refers to the accuracy of representation of the physical design and layout of the system on which the individual will receive training.

VE may be particularly suited to increasing the probability of transfer because of its flexibility, its feedback capabilities, and its potential for motivating the learner. The promise of flexibility may make it possible to design individually tailored training experiences that take into account individual differences in the skill and knowledge level of trainees as they enter and proceed through training. For example, a particular training program could be made more compatible with the specific motor skills of each trainee; as proficiency was gained, the training scenarios could be modified accordingly. Further, VE technology is better suited than previous technology to augment feedback to the trainee by adding special cues or by providing multimodal stimulation (e.g., haptic, visual, and auditory). This may be most useful as a reinforcment strategy for training lower-skilled students. Finally, VE technology has the potential to furnish an intrinsically interesting and motivating training environment through the presentation of special sound effects and interesting visual patterns.

Currently there appears to be no way to predict qualitatively or quantitatively the kinds of transfer that will result from VE. As a result, evaluation will be needed during all stages of system design, development, and implementation. The following section discusses the problems associated with quantitative demonstrations of transfer of training effectiveness in the laboratory and the field.

Issues to be Addressed

There are two limitations to demonstrating success in VE training. First are limitations inherent in experimental demonstration of success in the training of almost anything. That is not to say that simulation has not been a successful training medium. It has, as measured in the number of simulators sold, the number of disciplines that have committed to simulation as an alternative to training in the real environment, and the number of organizations that have enthusiastically adopted it as the training medium of choice. Commercial pilots routinely use simulators to maintain their currency in selected aircraft. NASA used virtual reality to train astronauts and other personnel in preparation for the Hubbell telescope repair mission that involved extensive extravehicular activity. What has been difficult to provide are *quantitative demonstrations* of transfer of training effectiveness in either the laboratory or the field.

Let us examine some of the issues that must be considered when conducting meaningful training evaluations. Since in this section we are concerned with the use of VE to support training rather than performance, we presume that, after training is complete, the trainees will begin work on the same or similar tasks in a different, non-VE environment. The issue therefore is transfer of training from the VE conditions to the non-VE conditions.

For purposes of illustration, let us assume that we are training firefighters. We place a VE helmet-mounted display of the visual scene of a fire on the trainee, together with a treadmill that allows us to simulate walking into a fire. The heat of the fire is represented by adjustable infrared lamps. The trainee then practices the procedures associated with fighting fires. After a period of training, we observe the trainee's performance fighting real fires. We hope that, for most of the required skills, there is positive transfer, that is, that trainee performance after practicing in the VE is better at fighting real fires than it was before training. We would expect that for some skills there would be no transfer, perhaps because they were skills for which the VE system could not provide practice, such as climbing ladders or manipulating hoses. However, there is always the risk of negative transfer, that is, that selected firefighting skills will be performed more poorly after practice in the VE. We would expect negative transfer under conditions in which students learn the wrong response associations to the stimuli to which they are exposed. For example, suppose the trainee approaches closer to the flames in the VE than he or she should because the VE fire leaves out the risk of getting burned. To the extent that the trainee gets closer than she should to the flames in a real fire for the same heat stimulus, we would say that the training exhibited negative transfer. A bad response was learned.

It is not enough simply to show that the training improved performance without asking, "Compared with what?" Firefighters are trained today in the classroom, using controlled fires in mock-ups or real buildings, or both. Unless we can show that the VE training is cost-effective compared with the current methods, we have not accomplished a useful result. This is usually accomplished by training a control group using the standard method and comparing the transfer of training to the real environment to that obtained with the experimental method.

It is likely in this example that there are some firefighting skills for which it is cost-effective to train in an VE, such as communication and coordination among personnel in firefighting teams, whereas others, such as the physical skills of using hoses and climbing ladders, are best trained in mock-ups.

It is very expensive and time-consuming to conduct the kind of transfer of training experiment that is described here. Efficient experiments require that each trainee-subject perform the task twice, once using the experimental method and once using the control method. But there is a dilemma: it is not sensible to compare performance with the experimental method with performance when the subject is trained again on the same task. Human variability being what it is, large numbers of different subjects must be tested in each condition in order to obtain statistically reliable results. Furthermore, because we would expect some skills to show transfer and others not, it is important to develop detailed performance measurement at the level of individual skills so that those that produce positive transfer are detectable.

Because of the cost and difficulty of conducting such studies, government-sponsored experiments of this kind are needed to evaluate scientifically the benefits of VE training.

A second limitation of the current state of virtual environment technology is the problem of successfully representing the real world with satisfactory fidelity to achieve a training goal. There is a trade-off between the field of view and the resolution of helmet-mounted displays. There are limitations to the technology supporting the computation and display of virtual images. There are limitations in the acceptability of computational delays that result from dynamically generated, rapidly changing, complex scenes. The hardware and software supporting the haptic interfaces, which allow the user to touch and feel the objects with which he or she is interacting, have until now been limited to data gloves that can sense only approximately the position of the hands and fingers and provide no sensory feedback about the objects with which the user in interacting. Technology problems also include the technology supporting the computation and display of virtual images.

In thinking about other barriers, it is possible to ask why existing simulation techniques are not more widely used in training. Despite the number of successful applications cited above, it is surprising that the use of simulation is not more general. Although our review does not support a definitive answer, we can speculate that, until the last few years, the cost of computers and of the special hardware required to support realistic displays, together with the inflexibility embodied in special-purpose hardware, made them not cost-effective for many training applications. For example, until recently, driving simulators may have been too expensive to be cost-effective when compared with the cost of training in a real vehicle. This is now changing, and there is growing interest in this and related applications.

At this stage in the development of VE, the lack of demonstration of cost-effectiveness is a clear barrier, but this stems as much from the lack of demonstrated effectiveness as from high cost.

Research Needs

There is a need to mature the state of VE technology in order to broaden the range of training applications for which it is cost-effective. We run the serious risk of trying to test potential training applications and failing, not because an application is bad, or even one that would not ultimately show usefulness, but because the state of the technology was not yet mature enough to support it effectively.

We also have a need for improved methodologies for evaluating cost-effectiveness. We can just as easily fail because we did not conduct an experiment that was sensitive enough to reliably detect the potential gains that were presumed to be there. This can happen for reasons of inadequate control of the experiment or because of an injudicious choice of performance measures.

Finally, since research of this kind has just begun, there is a need to develop a taxonomy of application areas that promise high leverage in VE training effectiveness. It seems likely that tasks requiring spatial learning or awareness are good candidates. For example, in isolating faults in the avionics of highly complex aircraft, the technician has to conduct the logical fault-isolation task and also map the location of the fault on a schematic diagram to the spatial location of the box containing the appropriate circuit on the actual aircraft. For today's aircraft, that is a very difficult task. It also seems likely that tasks for which enhanced artificial and spatially distributed cues could enhance learning could be important applications. The oft-referred to demonstration of illustrating the molecular forces associated with the position of atoms in a carbon-chain molecule could provide a compelling lesson in molecular structure. Such a taxonomy of application areas should not be based on speculation, however; it should emerge from a review of a large number of empirical tests of the success of VE training, tests that have yet to be accomplished.

EDUCATION

The problems of American education are too well known to need elaboration here. Our children are not sufficiently engaged in learning the skills and concepts presented to them by our schools, and the learning opportunities afforded are not always well matched with our society's future needs (Secretary's Commission on Achieving Necessary Skills, 1991). Synthetic environments may offer opportunities to address both concerns, within the public education system as well as in other arenas, such as home-based entertainment and communications. For students these opportunities include:

- visiting simulations of ancient India or Greece, the Paleozoic era or the inner ear, to gather data for presentations, plays, and virtual world-building of their own;
- observing ("job-shadowing") adults working in information space, such as CAD artists and researchers using databases (such as medical librarians, historians, etc.);
- remotely experiencing such phenomena as the eruption of an underwater volcano or the live birth of an elephant at the zoo in Beijing;
- experimenting with simulations constructed by experts (e.g., physicists, ecologists, social and econometric model builders);
- working with other students of different ages and cultures, at different sites internationally, on a daily basis, to improve one another's language skills, or on tasks like those described above;
- building improved tools for their own use and use by others, such as libraries of images, elemental simulations, stories, local history, and demographic and geographic data.

In essence, students using virtual reality would be able to do what we would like students to be doing today—but with vastly expanded ability to access information in the larger world, to experiment, visualize, and understand, and to interpret the information to their own ends.

Rationale

Education is an application area that cuts across subject-specific domains; to the extent that a person can learn something using a VE system developed for any specific area, VE is being used for educational purposes. For example, a scientific visualization of a computer simulation that teaches a researcher something new about nature is arguably an educational application. However, for purposes of this discussion, educational applications of VE will be focused on the potential use of VE in grades K-12, in which improvements are of demonstrable and pressing national concern.

Most commentators on the goals of K-12 education would agree that it should develop a student's capacity to think independently; increase a student's desire and motivation to learn; and increase the extent to which a student learns and retains specific skills and knowledge. In contrast, there is no unanimity about how to create an environment in which these things happen. For specificity, the discussion that follows is guided by the philosophy that people learn best when they can integrate what they are learning into the broader context of other things they know and care about; that they are more highly motivated when they can and have reason to influence the course of their own learning; and that they learn to think independently when they are given substantial opportunities for doing so. Much of this educational philosophy has been characterized as constructionism.

Constructionism is a theory of instructional design, based on constructivist theory. Constructivism is a school of thought among developmental psychologists (Carey, 1987; Piaget and Inhelder, 1967) that concerns the way in which children develop models of the world. The idea is that the essential steps toward a mature understanding of a particular subject include a series of differentiations and reintegrations of experiences involving the dissection and reconstruction of internal models. Within this philosophical framework, computer and other information technologies, specifically VE, may have important roles to play in improving education. In particular:

- VE is a potential vehicle through which the range of experience to which students are exposed could be vastly increased.
- VE can provide immersive and interactive environments that provide macro contexts in which interesting intellectual problems naturally arise.[3]
- VE potentially provides micro worlds in which students can exercise the skills and use the knowledge they learn.

[3] Theories of *situated cognition* (Brown et al., 1989) suggest that learning takes place most effectively in contexts that are meaningful to the learner. Such pedagogical approaches contrast sharply with those that assume that the learner learns a general skill that can then be applied to all relevant situations. One instructional design technique based on situated cognition is referred to as *anchored instruction* (Cognition and Technology Group at Vanderbilt, 1993). Anchored instruction consists of providing a rich story line or macro context, within which VE environments may be able to provide macro contexts that are both rich and controllable.

- VE potentially expands the peer group among which collaborative learning experiences are possible.[4]

Specific Applications

The technology of VE, augmented reality, and telepresence is too new to have any real educational applications, if *real* is taken to mean an application that is fully integrated into the intellectual substance of a non-experimental curriculum. The following discussion focuses on a number of applications of VE technology to education that are suggested by preliminary experiments and field trials to date: field trips and telepresence, spatial relations (real space or phase space), playrooms to build things, micro worlds, simulations of things that are too complex or expensive to experience or experiment with, and new conceptualization tools for traditional subjects.

Each section below addresses a long-range vision of how VE might assist in these applications and a description of possible near-term demonstrations.

Simulated Field Trips and Telepresence

Schools often use field trips to expose students to unfamiliar physical environments (e.g., an inner-city class may visit a farm). However, cost, convenience, and safety can limit such opportunities for students. VE technology could provide immersive display systems that would enable students to experience exotic environments—in museum dioramas, in microscopic worlds (see Taylor et al., 1993), and in remote and hazardous surroundings.

One example, the Jason Project (see Tyre, 1989; Ullman, 1993), a Massachusetts nonprofit enterprise developed by ocean explorer and geologist Robert Ballard, is built around a remotely controlled submarine (Jason). Live video images of grey whales in the ocean off Mexico and hydrothermal vents in the ocean floor were broadcast from Jason to sites in schools and universities, where nearly 1 million students could see underwater exploration as it was occurring and experience a sense of immediacy and involvement. The Jason project also permitted 23 students to participate directly and interactively. Specifically, these partici-

[4] The value of collaborative work is suggested by a number of projects in which students collaborate via electronic mail, for example, the National Geographic Society's KIDSnet, Bank Street College's Earth Lab, and AT&T Learning Circles systems. VE could enable collaborating students to share environments, thus reducing or eliminating the difficulty of using words to describe objects or phenomena that they collectively need to reference.

pants were directly involved in the control of the robot's motion in real time, thus creating an experience of telepresence.

Spatial Relations

In the summers of 1991 and 1992, the Pacific Science Center in Seattle, Washington, sponsored a Technology Academy Program in which groups of students ages 9 to 15 were introduced to some aspects of SE. Six one-week day camps were held each summer. Participation consisted of seeing videotapes and demonstrations and working with CAD software (Swivel 3D) on Macintosh computers to construct graphical models for use in a VE. These models were taken to the Human Interface Technology Laboratory at the University of Washington for installation by graduate assistants into the laboratory's SE system, which incorporated gloves, helmet, and sound.

Project managers Meredith Bricken and Chris Byrne (1992) reported that students showed a high degree of comprehension and rapid learning of computer graphics concepts such as Cartesian coordinates and three-dimensional modeling. Students indicated that they would much rather work and play with SE than with conventional video games or TV. However, the workshops were only one week long, and the students' experience of immersive SE consisted of parts of one day in which they saw their three-dimensional models in the SE system.

In spring 1992, Intel Corporation sponsored the creation of an exhibit at the Boston Computer Museum, which featured a two-participant virtual world. This world was based on Sense8's WorldToolKit software and Intel's DVI graphics cards, running in 486 PCs; the actual demonstration was constructed by Sense8's Ken Pimintel and Brian Blau at the University of Central Florida (Pimentel and Blau, 1994). Hundreds of people experienced the rather simple virtual world, in which it was possible to grasp building blocks using a three-dimensional wandlike pointing device and slide the blocks around to assemble one's own toy house.

The Boston Computer Museum invented a kind of video swivel chair (for which a patent application has been submitted). The chair carries a 13-inch TV monitor and allows the participant to look around in the virtual world. The imagery rotates to correspond to the chair's rotation. This approach avoids the sanitary problems and expense of providing stereo head-mounted displays for visitors. This kind of technology will probably also be useful in school settings.

The AutoDesk Cyberspace project has supported several experiments concerning multimedia tools in the public schools of Novato, Calif. In 1990, Mark Merickel of Oregon State University conducted experiments in the Olive Elementary School. One team used AutoSketch and

AutoCAD to develop three-dimensional models; another group explored Cyberspace—AutoDesk's immersive virtual world (Merickel, 1990).

Playrooms to Build Things

Some education researchers feel that learning is enhanced when students can build their own simulations (Harel and Papert, 1991), but the construction of virtual worlds with today's tools is technically challenging. Bowen Loftin at NASA/Johnson and the University of Houston has constructed a virtual physics laboratory, built on NASA's SE tools (Yam, 1993). Loftin is in the process of extending his previous work on intelligent tutoring (Loftin et al., 1991) into a richer virtual environment for science education.

Micro Worlds

Interactive, text-based, role-playing environments and games have been developing for several years on both the Internet and on bulletin board systems (Bruckman, 1992-1993). Known as MUDs, MOOs and MUSEs, these environments have gained substantial popularity with a certain segment of the population. MUD stands for multi-user dungeon, reflecting the genre's origins in role-playing games; MOO stands for MUD, object-oriented, which contains objects that can be manipulated. MUSE stands for multi-user simulation environment.

In a typical MUD or MOO, a participant types LOOK and receives a textual description of the room or place currently occupied. Any other players at that place can type messages, which are immediately echoed on the screens of all the others at that place. Objects can be created, picked up, dropped, and used. New places can be added to the universe.

MUDs provide a crude sense of space and a lively interaction with other participants. They thus prefigure some of the kinds of interactions that can be expected in fully immersive VEs. Some proponents of MUDs regard them as instances of virtual reality. The popularity of MUDs as educational tools is rapidly growing. They require only a PC, a modem, and access to the Internet. Pantelidis (1993) has provided a substantial bibliography on educational uses of MUDs and other VE systems in education.

Simulations

Interactive modeling and simulation is being pursued as an educational tool on many fronts (Fuerzeig, 1992). Simulations of physical, biological, and social phenomena can have substantial pedagogic value, especially when the systems being simulated are otherwise inaccessible to students. Many simulations have been successfully implemented on PCs and used in both K-12 and university environments (see White, 1984; Maxis, 1991; Glenn and Ehman, 1987; Schug and Kepner, 1984). As a research example, Tom DeFanti and collaborators at the University of Illinois at Chicago and the Center for Supercomputing Applications are designing educational applications of the CAVE display system (Cruz-Neira et al., 1992). This system projects images on three walls and the floor and tracks a principal viewer's head position to determine the view direction and content. Other viewers in the CAVE are "along for the ride," which can sometimes be disconcerting. The curricular topics under consideration at present include scientific visualization and non-Euclidean geometry. These topics are at present of more interest to university than to K-12 educators.

Issues to be Addressed

The discussion above is admittedly speculative. Educators and their constituents will have to address issues of serious concern in three areas.

Desirability

Perhaps the most fundamental question is that of desirability. Under what circumstances does it make sense—even given a technically perfect VE educational application—to use that application in the classroom? The question arises because it is all too easy to imagine a classroom in which the amount of interaction that students have with real blocks, real people, and real situations is reduced in favor of simulated experiences.

Put differently, concrete, manipulable objects enhance children's ability to handle abstractions, as repeatedly demonstrated in Montessori schools. Field trips enhance the realism and relevance of lessons. Why should children be deprived of these tools, in favor of virtual building blocks and virtual field trips? Moreover, unlike real blocks and environments, the rules that govern simulated systems are limited only by the system developer's imagination, and as such are essentially arbitrary. Why is it necessarily desirable for students to have more experience and interaction with such systems?

Ultimately, the intellectual challenge is likely to be learning to see VE as one tool of many in the tool chest of responsible educators. VE can obviously provide some experiences that would not otherwise be possible for students to have, but when a "real" alternative is available, it may make more sense for the latter to have priority. Some judicious mix of hands-on learning, VE experience, and book learning—varying from classroom to classroom and subject to subject—will demonstrate that these tools can complement each other. Such an approach is suggested by the Vanderbilt group (Cognition and Technology Group at Vanderbilt, 1993); they argue that hands-on and anchored instruction techniques are complementary. They use videotape and videodisk segments to establish the story line, and then use hands-on projects for the students to construct their own understanding. Such a paradigm can also be explored in VE, with or without virtual replacement of the hands-on activities.

Effectiveness and Feasibility

The VE hypothesis for education is that, for certain purposes, well-integrated VE systems will achieve results superior to the use of conventional capabilities. The hallmarks of such success would be (1) significant improvement in students' learning and retention of specific skills and concepts, compared with their response to similar content presented without VEs and (2) significant increases in students' voluntary use of such systems, compared with their response to similar content presented in other ways.

Some reports suggest that the successful introduction of education technology results in a sustainable increase in the enthusiasm of students, which increases the overall chances for educational success (Office of Technology Assessment, 1988). Indeed, the popularity of video games and other such technologies among K-12 students outside the educational context strongly indicates that engagement with technology is not likely to be a significant problem.

However, the eagerness with which students embrace technology should not be taken as an unqualified endorsement of immersive graphics for education. Rather, some part of the enthusiasm may come from the novelty of the medium. It is hard to test the kinds of learning achieved from field trips to exotic VE labs for one-shot viewings of one's geometric models or virtual submarine voyages. What is genuinely problematic is the following issue: To what extent can educators design engaging VE systems (a relatively easy task) that also result in what reasonable people would regard as learning (a much harder one)? To answer this question, much more empirical study is necessary.

Practicality

The concern about practicality is dominated by cost. A number of studies have established in specific contexts that computers can be cost-effective compared with other means of delivering instruction (Levin et al., 1984; Office of Technology Assessment, 1988). However, a number of other factors intervene to complicate their wider adoption by schools.

First, the introduction of technology seldom *decreases* costs—at least in public education. Thus, even if the provable incremental performance provided by the technology is cheaper per unit than other possible improvements, political will and funding may not be forthcoming.

Second, cost-benefit analysis for technology must necessarily analyze the teaching of skills and concepts that can be taught by other means. Certain skills (e.g., accessing remote on-line databases) are meaningless without the technology. Most parents want their children to have up-to-date skills.

Schools are not yet able to afford well-supported state-of-the-art personal computers in meaningful numbers, and only recently is there broad acceptance of the idea that computers are useful in education (Office of Technology Assessment, 1988). In 1993 a typical computer purchased for school use cost, with software, about $3,000—essentially the cost of a well-equipped Apple II in 1979. Increased performance expectations have offset the decrease in component prices, and so the cost of low-end personal computers has tended to remain about constant, even as their capabilities have increased.

An entry-level PC-based VE system can be purchased in 1993 for approximately $24,000. If current trends continue, such systems should be available for around $3,000 in three to five years. Current developments in entertainment electronics may advance that time frame somewhat, but display technology is not likely to advance as fast as image generation and simulation technology. We believe that VE will not begin to penetrate schools until its utility for specific purposes generates sufficient interest and desire, and when an individual station of acceptable performance costs $3,000 or less.

Assuming that costs have been adequately addressed, it appears that public opinion would be generally supportive of educational VE applications. But long-time educators have seen education fads come and go; they will need to be convinced that VE gives them *usable* capabilities that can enhance education. Thus, costs necessarily include those related to training teachers to use new technologies effectively and how to judge when the use of new technologies is appropriate. These costs are high, perhaps comparable to those related to deployment of the hardware itself.

Infrastructure and Research Needs

The development of high-quality software followed the arrival of the PC by more than a decade. Only recently has commercial software become available that meets real education needs beyond drill-and-practice. This is due both to the maturation of a generation of lesson-builders and their firms and to the arrival of mature, well-reasoned national agendas such as those provided by the National Council of Teachers of Mathematics (1989). The generational shift from emphasizing knowledge to emphasizing competence, represented by the National Council for Teachers of Mathematics curriculum and its siblings, has finally provided educational computing with appropriate subject matter and philosophical focus.

A similar evolution will be necessary for VE in education. A critical mass of competent VE programmers must develop and educators will need to work with the hardware and software over a period of years before clear directions will emerge.

On the technology side, no special issues stand out. This is partly because the specific requirements of VE systems in education cannot be articulated until we have a better understanding of the goals of those systems.

Although the technology research agenda for education is not separate from that of VE in general, a great many questions and issues for education research remain:

- The identification and characterization of skill and subject matter domains for which VE-based immersion can be demonstrated to provide clear didactic advantages over equivalent nonimmersive presentations.
- The relationship between immersion and nonimmersive representations *within* a given educational environment, as tools to help students understand their own learning process. Immersive presentations may be more engrossing and lead to intuitive understandings. Two-dimensional or schematic presentations may lead to better abstract understandings.
- The development of a variety of means (user interfaces, languages, CAD tools) whereby VE environments can be easily expressed and constructed by lesson designers.
- The development of concepts and tools (e.g., telepresence) that can be used by students to facilitate their own model-building within those environments and the educational significance of their use (e.g., as indicated by their ability to embody objective knowledge about the processes of science, economics, history, etc. in their models).
- The educational value of role-playing adventure games, anchored learning, and other shared simulation experiences in fostering the development of analytical skills such as problem formulation.
- The extent to which various features of the user interface can substitute for or enhance the interpersonal interactions among co-located team members.[5]

---

[5] For example, convolved sound, touch, and other sensory modalities and transmodal transpositions (e.g., a keyboard-driven speech synthesizer) may make it possible to get more involvement from relatively nonverbal people than would be possible in traditional in-person teamwork. Disabled learners could be involved on a more equal basis, when everyone is learning new metaphors for motion and control.

- The role of third-party mentors, either explicitly present and visible in the simulation or behind the scenes, in enhancing the learning experience.

With some relatively small modifications, all of these questions and issues can and should be asked of any information technology in an education context. But the three-dimensional immersive environment affords a much richer space than a two-dimensional screen in which to ask and address such questions. Indeed, the richness may well be a significant differentiating feature if answers to these questions depend on the specific technology being considered.

INFORMATION VISUALIZATION

With the rapidly increasing amount, types, and sources of information being produced for scientists, engineers, business executives, and the public, there is a pressing need to develop better presentation techniques and formats to support everyday tasks of exploration, understanding, and decision making. The information explosion, coupled with advances in computer technology, offers some promising opportunities to develop new approaches to visualizing information. Currently we are presented with information in a variety of forms, including text and images in two and three dimensions that can be static or dynamic. Interaction techniques are limited to point-and-click window interfaces that can require as many as 10 to 50 steps to accomplish a single task.

Possibilities for the future include augmented reality, in which a virtual image is superimposed on the real world (e.g, view the pipes inside a wall), and VEs in which the user is immersed in the information and interacts with it in real time. Effectively designed environments should provide individuals with easier access to the critical elements in the information; the ability to view and explore interactions among multiple problem dimensions simultaneously; and the opportunity to examine and test relationships that cannot be presented on a two-dimensional display. As an example, well-presented information may assist decision makers in a manufacturing plant to understand more readily the effects of several variables, such as current economic conditions, environmental impact, materials flow, staff training, and marketing forecasts, on the feasibility and cost of producing a particular product. In another example, VE technology could be used to visualize a proposed factory operation before it is developed. Furthermore, creative use of the technology to present financial data could assist investors in making more informed decisions.

Designing useful visualizations for different types of information in support of different user tasks is an enormous undertaking. One major area of required work is the development of computational software to manage large and diverse databases in ways that allow users to explore alternatives and make discoveries. This issue is discussed in more detail in Chapter 8, on computer generation of virtual environments and in the section below on scientific visualization.

A second area of required work is the determination of what information should be provided, how it should be formatted, and how we expect the user to interact with it. Researchers have worked for many years to create dynamic two-dimensional information displays for such tasks as monitoring the status of a nuclear power plant, flying a jet aircraft, controlling aircraft traffic, and analyzing complex data. This work has provided some insight into how much information can be absorbed at one time, how it should be organized on the screen, and how frequently it can be updated before the limits of human information processing are exceeded (Ellis, 1993). In addition, cognitive scientists have been exploring the relationship between types of tasks and the most appropriate types of information to support those tasks (Palmiter and Elkerton, 1993).

Although some of the knowledge about human information processing, learning, and problem solving gained when using two-dimensional displays will be of value in designing information displays in three-dimensional environments, we will need to mount a substantial research effort to determine how to use the capabilities of three-dimensional environments effectively. An integral part of these research efforts will be a determination of the most user-friendly and efficient interaction techniques. Other chapters of this book provide a discussion of these issues. Of particular importance will be research into the use of sensory modalities other than vision in increasing or modifying the comprehension of information.

Currently, little progress has been made in the use of virtual or augmented reality for the purposes of information visualization. However, some investigators have begun to explore various aspects of visualization for scientific purposes. A brief description of results in this area are reported below. Many of the problems raised will be pertinent to the design of information presentation for other types of activities.

Scientific Visualization

Scientific visualization (McCormick et al., 1987) is the use of computer graphics to create visual images that aid in the understanding of complex, often massive, numerical representations of scientific concepts or results. Such numerical representations, or datasets, may be the output of numerical simulations as in computational fluid dynamics and or molecular modeling, recorded data as in geological and astronomical applications, or constructed shapes as in visualizing topological arguments. These simulations may contain high-dimensional data in a three-dimensional volume, and they often vary in time. Different locations in such datasets can exhibit strikingly and interestingly different features, and difficulty in specifying locations will impede exploration. Scientific insight into complex phenomena depends in part on our ability to develop meaningful three-dimensional displays.

Rationale

Traditionally, scientific visualization has been based on static or animated two-dimensional images that have generally required a significant investment in time and expertise to produce.[6] As a result, severe limits have been placed on the number of ways in which a dataset can be explored. That is, an explorer does not know a priori what images are unimportant, but when the effort to produce a visualization is large, there will understandably be a hesitation to produce a picture that is likely to be discarded.

Other problems arise with traditional scientific visualization techniques because they are not well suited to the computational datasets associated with modern engineering simulations. These datasets may be inherently complex, consisting of a time series of three-dimensional volumes, with many parameters at each point. Also, scientists are often interested in behavior induced by these data (i.e., streamlines in a vector field) rather than the data values themselves. Under these circumstances, real-time interactive visualization is likely to pay off, due to the complexity of phenomena that can occur in a three-dimensional volume.

VE technology is a natural match for the analysis of complex, time-varying datasets. Scientific visualization requires the informative display of abstract quantities and concepts, rather than the realistic representation of objects in the real world. Thus, the graphics demands of scientific visualization can be oriented toward accurate, as opposed to realistic, representations. Furthermore, as the phenomena being represented are abstract, a researcher can perform investigations in VE that are impossible or meaningless in the real world. The real-time interactive capabili-

---

[6] In the early days of visualization, it was rather difficult for the researcher to produce visualizations beyond conventional drawings and plots. Familiarity with computer graphics programming was required to do more sophisticated visualization, a need that was addressed through the creation of "visualization shops," in which a visualization was produced to order. A researcher provided data to a visualization programmer, who then produced a high-quality image or animation. Thus, there was a significant investment involved in the production of visualization. This served the purpose of visualization as a presentation medium, but it hindered the use of visualization as an exploratory medium.

ties promised by VE can be expected to make a significant difference in these investigations, with the potential to provide: the ability to quickly sample a dataset's volume without cluttering the visualization; no penalty for investigating regions that are not expected to be of interest; and the ability to see the relationship between data nearby in space or time without cluttering up the visualization. In short, real-time interaction should encourage exploration.

Just as important, a natural, anthropomorphic three-dimensional VE-based interface can aid the unambiguous display of these structures by providing a rich set of spatial and depth cues. VE input interfaces allow the rapid and intuitive exploration of the volume containing the data, enabling the various phenomena at various places in that volume to be explored, as well as providing simple control of the visualization environment through controls integrated into the environment.

A properly constructed VE-based interface will require very little of the user's attention; it would be used naturally, using pointing and speech commands and directions rather than command-line text input. Someone using such an interface would see an unambiguous three-dimensional display. This would contrast with the current interaction paradigm in scientific visualization, which is based on text or two-dimensional input via graphical user interfaces and two-dimensional projections of three-dimensional scenes.

Specific Applications

VE systems for scientific visualization are in many ways like software packages for graphing: tools for displaying and facilitating the interpretation of large datasets. But it is too early to describe a single general-purpose VE system for scientific visualization. At the same time, a number of projects have demonstrated that VR does have significant application potential.

- *Aeronautical Engineering*: The virtual wind tunnel (Bryson and Levit, 1992; Bryson and Gerald-Yamasaki, 1992) uses virtual reality to facilitate the understanding of precomputed simulated flow fields resulting from computational fluid dynamics calculations. The visualization of these computations may be useful to the designers of modern high-performance aircraft. The virtual wind tunnel is expected to be used by aircraft researchers in 1994 and provides a variety of visualization techniques in both single-user and remotely located multiple-user environments.
- *General Relativity*: Virtual Spacetime (Bryson, 1992) is an extension of the virtual wind tunnel in which curved space-times, which are solu-

tions to Einstein's field equations of gravitation, are visualized using particle paths in virtual reality.

- *Molecular modeling*: Molecular docking studies using a VE that included a force-reflecting manipulation device have been performed with the GROPE system at the University of North Carolina at Chapel Hill (Brooks et al., 1990). Investigators employed a head-tracked stereo display in conjunction with the force-feedback arm to investigate how various molecules dock together. These studies have implications for the design of pharmaceuticals.
- *Scanning Tunneling Microscopy*: A VE coupled with a telerobot for the control and display of results from a scanning tunneling microscope called the Nanomanipulator has been developed at the University of North Carolina at Chapel Hill (Taylor et al., 1993). This system uses a head-tracked stereo display in conjunction with a force-feedback arm to display a surface with molecular resolution via graphics and force reflection based on data obtained in near-real time from a scanning tunneling microscope. In addition, there is the ability to deposit very small amounts of material on the surface via direct manipulation by the user.
- *Medical visualization*: Medical visualization systems using augmented reality (e.g., Bajura et al., 1992) have been developed at several sites. The primary difficulty with medical visualization at this time involves the very large amounts of graphic data being displayed. Bajura et al. are designing a system that will map ultrasound imagery in real time onto the physician's view of the real patient, allowing the location of the features shown in the ultrasound imagery to be quickly and intuitively located in the patient.
- *Astrophysics*: A system to investigate cosmic structure formation has been implemented at the National Center for Supercomputing Applications (Song and Norman, 1993). This system visualizes structure arising from simulations of the formation of galaxies in the early universe.
- *Circuit Design*: The Electronic Visualization Laboratory at the University of Illinois, Chicago, has implemented several scientific visualization applications in a virtual environment setting. For descriptions of the individual projects, see Cruz-Neira et al. (1993a, 1993b).

Issues to be Addressed

Experience with VE-based scientific visualizations has shown that in order to sustain usable interaction and to make the user feel that a series of pictures integrates into an insightful animation, a number of criteria must be met. First, the system must provide interactive response times to the user of approximately 0.1 s or less. Interactive response time is a measure of the speed with which the user sees the results of actions; if the interactive response time is too slow, the user will experience difficulty in precisely placing visualization tools (Sheridan and Ferrill, 1974). Second, effective systems for scientific visualization must have animation rates of at least 10 frames/s. Animation rate is a measure of how fast images are presented to the user; this rate is particularly relevant with respect to viewing control and for time-varying datasets. If the rate is too slow, the images will be perceived as a series of still pictures rather than a continuous evolution or movement. These two parameters are psychologically and perceptually related, albeit computationally distinct. Some VE systems may separate the computation and visualization processes, so that they run asynchronously. We are at the beginning of understanding the potential of this technology for scientists. Research is needed to answer such questions as when are continuous images more useful than discrete images for scientific insight.

Scientific visualization also makes particular demands on virtual reality displays. The phenomenon to be displayed in a scientific visualization application often involves a delicate and detailed structure, requiring high-quality, high-resolution, and full-color displays. Experience has shown that displays with 1,000 × 1,000 pixel resolution are sufficient for many applications. In addition, a wide field of view is often desirable, as it allows the researcher to view how detailed structures are related to larger, more global phenomena.

Lastly, user acceptance criteria suggest that few researchers would be willing to invest the time required to don and doff head-mounted displays available at the time of this writing. Furthermore, many researchers have expressed distaste for donning helmets or strapping displays onto their heads.

TELECOMMUNICATIONS AND TELETRAVEL

Rationale

As facilitators of distributed collaboration, the applications of telecommunications and teletravel cut across all of the other applications discussed in this chapter. For manufacturing activities of the future, it is anticipated that virtual images of products will be simultaneously shared by geographically dispersed design engineers, sales personnel, and potential customers, thus providing the means for joint discussion and product modification. In health care, there are several examples of distributed collaboration, including remote surgical practice and remote diagnostic consultation among patients, primary care physicians, and specialists who may all be viewing common data or three-dimensional images. An example of the latter is the development of a telemedicine system in Georgia, in which medical center expertise is shared over a network with rural doctors and their patients.

For education and training, there are many instances in which distributed collaboration may be useful. One example is the use of a shared virtual battlefield for mission planning, rehearsal, and training. Another potential use is offering students from several schools around the country the opportunity to come together through network technology to share a common virtual world—such as a reconstruction of a historic site that no longer exists. Finally, in hazardous operations, distributed collaboration is a central feature of humans and telerobots working together is the same remote environment.

The discussion in this section focuses on the increasingly collaborative nature of modern business and the potential contribution of SE technology to facilitating this collaboration in a cost-effective manner. Specifically, we discuss telecommunication and teletravel. Both of these processes use technology to reduce unproductive travel time to and from meeting sites or regular work sites.

Today, many of those who are knowledge workers already use technology to avoid travel. The nature of a knowledge worker's job is such that by working at home or in satellite locations using personal computers, modems, and the telephone network, these telecommuting workers can perform their jobs reasonably well. Greater understanding and acceptance of this phenomenon in the workplace is illustrated by the response of the Los Angeles work force to the earthquake of January 1994. In the aftermath of the disaster and the consequent disruption of customary commuter routes, telecommuting increased dramatically. However, most workers in the United States do not have jobs that can be performed using only a screen, a keyboard, and a mouse. Most sales people must interact face-to-face, and others, such as craftspeople, work on solid objects with their eyes and hands. Even telecommuting knowledge workers need face-to-face meetings for discussions involving more than two people, job interviews, and many other work situations in which gestures, facial expression, and eye contact are critical components of the interchange. These added task requirements open the door to the next step in the use of VE technology.

Background

The historical evolution of distributed collaboration provides a useful context for this discussion. Study of the paths followed in the earlier efforts in both research and system development can reveal the current robust status of telecommunication facilities, as well as the potential consequences of expanding such facilities to include VE capabilities.

Distributed collaboration first emerged as a product of advanced technology in the form of multiperson telephone conversations. Such services were provided by AT&T during the 1950s with economic benefits for both users and providers. Routine use of this technological capability came in the 1960s, along with expansion of the concept to include secure conferences between remote participants. A separate and special technology, telecommunication, came to be institutionalized as a consequence. Several major systems were developed at this time to serve the needs of the federal government. Among such systems were the first models of AUTOSEVOCOM, a secure network that could support remote conferences between top-level military commanders from their stations around the world (Sinaiko and Belden, 1965). This system also cemented the integration of computers into the multistation communication network. In these early instances, the computer was used only as a switching device and a tool for signal encryption. However, its presence in the system was a definite harbinger of things to come.

In addition, the design effort for military systems set off programs of research intended to explore the potentials, both positive and negative, of computer-mediated communication. For example, studies were initiated to determine the feasibility of using a computer-mediated network to link North Atlantic Treaty Organization member heads of state for purposes of joint crisis resolution. Problems ranging from how to implement rules of diplomatic protocol to overcoming language barriers were explored. The outcomes of these research programs revealed some of the limitations on communication effectiveness imposed by an absence of visual information to augment the direct voice transmissions.

Digitization, packet switching, and optical fibers, among other technological advances, began to open more vistas in the 1970s. It was then that the first instances of teletravel began to appear (Fordyce, 1974; Craig, 1980) after having been forecast several years previously. Entire new areas of economic advantage began to be apparent, such as the possible savings in gasoline use and pollution reduction. Negatives such as lowered productivity due to a lack of supervisory presence were played down by early enthusiasts but have come to be treated as significant matters in present-day application instances. In any case, telecommuting, as a form of distributed collaboration, has become an accepted option for some workers in some organizations (Shirazi, 1991).

The other critical ingredient in the evolution of distributed collaboration was the rapid adoption of small but powerful computers by workers in many different occupations. Computerized networks began to become widespread in the 1980s. In a sense, the computer became an actual participant in multiperson collaborations that were performed on the network. The computer provided an information storage and retrieval capability that far exceeded what the humans could contribute. Also, the computer could provide a dynamic color graphics capability that is not available by any other means. Object-oriented collaborations, such as designing electronic circuitry, have been quite successful when these technologies have been employed (Sheridan, 1993a; Fanning and Raphael, 1986).

In summary, distributed collaboration is not a particularly new idea. Working in this manner has gradually expanded over the past four decades, as people have accustomed themselves to the concept and its ramifications and as the technology has progressed to a point at which it supports new modes of activity at affordable costs. Now the question becomes: What can or will VE add to the process? Will VE provide the means to take a few more incremental steps in the further expansion of distributed collaboration—or will VE provide the basis for major change in how the concept is actualized?

Teletravel and Virtual Environments

VE offers the possibility for one to participate in a meeting in which all the other attendees were present in the form of virtual images. Each participant in the virtual meeting would see and hear the other participants through lightweight, see-through VE goggles that resemble eyeglasses, while his or her own appearance was captured by a video camera for broadcast to all the others. Different communications channels would support both group communications and communications to selected individuals—the equivalent of whispering to someone.

The social feasibility of virtual meetings goes beyond the technology. The enormous difficulty of scheduling conference phone calls for more than four busy people suggests that a new set of social norms would be needed before virtual meetings could be called routinely. For example, people would have to feel that is was unacceptable to remain in a virtual meeting and attend to other business simultaneously.

Shared workspaces refer to the real or virtual gathering of people at a specific location for the purpose of interacting with an artifact or an object. With the appropriate technology for automatic model generation, any physical space—and the relevant artifacts and objects—could be turned into a virtual meeting place. Thus a group could travel to any location where sensors existed and meet while observing events taking place in that real-world location.

VE also offers the possibility that one could be part of collaborative communities at a distance. For example, Xerox PARC is currently experimenting with technologies that virtually bring together people working in different offices. Although the current effort is limited to small-screen video and audio, VE-based collaborative communities could offer illusions of physical presence so real that offices of collaborators and colleagues could be geographically dispersed much more than they are today.

To facilitate the illusion of shared office spaces, office workers might wear glasses that could give them the feeling that their offices were part of a much larger common area in which many other participants were present. Each participant would see the other participants sitting at desks in their own offices.

In situations in which a traveler is unable to physically go to the target site, virtual travel may be useful. An example would be to enable inmates of correctional facilities to hold jobs or be trained in the outside world while still being controlled and monitored physically. Police might travel virtually to the middle of a jail riot to gather further intelligence. Researchers, regulators, and site planners might virtually meet in a hazardous environment, such as a nuclear dumping site, to examine conditions and plan operations for the future. It is conceivable that close observation and review of site conditions could provide superior input to planning based on video or still pictures.

Teleoperation and Remote Access

Teleoperated systems controlled through a VE interface would enable an individual or a group to go beyond mere passive observation. For example, a group on a virtual visit to a nuclear power plant could be authorized to open and shut valves and make other changes to the physical status of the plant. Since the valves and other actuators in such a plant are electronically controlled from a control panel, it is not too far-fetched to imagine the sensor data and control of the operation of the plant being part of the virtual world inhabited by the plant's human operators. The operators might be more aware of the status of the plant in emergencies if they could virtually travel through its radioactive corridors.

Any device that was connected to the communications grid could be controlled by a virtually present human. A home security system, for example, could summon home its owner when an alarm went off. (For an example based in telecommunications prior to VE technology, see Taylor, 1980.) With appropriate cameras and sensors, the owner could travel through the house (virtually) to see if intruders were present. With certain actuators present in the home and linked to the network, the owner might either drive off to escape confrontation or try to capture the intruders in the home.

A very general means of making changes to the world would be for a distant person to occupy a telerobot to perform some task. Discussions of telerobotics tend to treat the link between human operator and telerobot as semipermanent. But if telerobots were common, they might be treated more like telephones—that is, known locations into which one could project one's eyes and hands.

Research Needs

A key aspect of a virtual meeting is the ability to see body movements and facial expressions, a feature difficult to achieve with current video conferencing systems. In real meetings in real places, the participants perceive themselves to be in a place, surrounded by its walls. They are able to observe the positions of other participants within that space and hear their voices coming from specific directions. These perceptual possibilities are not available from video telecommunication systems. Therefore, the use of directional sound in virtual meetings will be especially important. With as few as two pairs of people having simultaneous independent conversations, the conversations will be disrupted without the ability of the hearers to filter out the unwanted sounds. This is done in real life (the "cocktail party effect") by the human ability to selectively filter sounds based on their directionality, and the fact that sounds from more distant sources are less loud. Both of these properties of real sounds can be supported in shared virtual worlds.

To capture an image of a user's facial expression while allowing the user to view the shared virtual world, several display methods are available. One way to do it is for the user to wear a see-through HMD, resembling eyeglasses. A video camera could be trained on the user's face. The see-through capability of the HMD in this case is primarily useful for allowing the camera to see though to the user's face, rather than for allowing the user to see the real surroundings.

Another viewing method is to put the user in an immersive viewing station, similar to the CAVE, a room whose walls, ceiling, and floor surround a viewer with projected images (Cruz-Neira et al., 1992). Since the CAVE uses polarized glasses for stereo, a camera is needed that can see through to collect images of both eyes of the user through the polarization. Both of these viewing setups are adequate to allow a group of people, each at a viewing station, to see and be seen by the other members of a group of people occupying a virtual meeting space.

The two-dimensional image of face and body that would be collected by a normal video camera is adequate but not optimal. The two-dimensional face texture could be mapped onto a virtual mannequin representing the person in the virtual world, and the same could be done for the two-dimensional body image. This would provide a very flat person to the virtual world, but it would still have advantages over video telecommunication, which does not show the locations of the different participants very effectively.

A more elaborate body image capture method would be to use range-imaging techniques to capture a three-dimensional model of the body and face. Such automatic three-dimensional model acquisition is needed by other branches of the SE field, and various prototype systems exist for range imaging. A three-dimensional image of the body of each participant in a virtual meeting begins to make such meetings sound like they could actually come close to duplicating the perceptual feel of being physically present at a real meeting.

These technologies, though still immature, offer the possibility of electronically projecting oneself, as easily as one currently makes telephone calls, into virtual worlds inhabited by other distant human users, with whom one can have face-to-face interactions both one-on-one and in groups. These shared multiperson virtual worlds create a shared space, in which each human participant has a position, a body image resembling his or her own real appearance, and a viewpoint from which to observe the behaviors and facial expressions of the other people engaged in the transaction.

References

OVERVIEW

Catmull, E., L. Carpenter, and R. Cook

1984 Private and public communication. Communication in reference to the number of polygons required to render reality, making certain assumptions about depth complexity and display resolution.

Larijani, L.C., ed.

1994 *The Virtual Reality Primer.* New York: McGraw-Hill, Inc.

Thompson, J., ed.

1993 *Virtual Reality: An International Directory of Research Projects.* Aldershot, U.K.: JBT Publishing.

Thorpe, J.

1993 Synthetic environments strategic plan. Draft 3B. Alexandria, Va.: Defense Research Projects Agency.

CHAPTER 1: SOME PSYCHOLOGICAL CONSIDERATIONS

Akin, D.L., M.I. Minksky, E.D. Thiel, and C.R. Kurtzman

1983 Space Applications of Automation, Robotics, and Machine Intelligence Systems. ARAMIS phase II, Vol. 3. National Aeronautics and Space Administration.

Anderson, J.R., A.T. Corbett, K.R. Koedinger, and R. Pelletier

1993 Cognitive Tutors: Lessons Learned. Unpublished paper. A copy may be obtained by contacting John R. Anderson at Carnegie Mellon University via electronic mail: ja0s@andrew.cmu.edu

Bach-y-Rita, P.

1972 *Brain Mechanism in Sensory Substitution.* New York: Academic Press.

Bach-y-Rita, P.
1992 Invited address: Tactile displays. *International Symposium: Digest of Technical Papers.* Vol: 23, May. Boston, Mass.: Society for Information Display.

Baron, S., C. Feehrer, R. Muralidharam, R. Pew, and P. Horwitz
1982 An Approach to Modelling Supervisory Control of a Nuclear Power Plant. Technical Report NUREG/CR-2988, ORNL/Sub/81-70523/1, Oak Ridge National Laboratory.

Bartlett, F.C.
1932 *Remembering: A Study in Experimental and Social Psychology.* Cambridge, England: Cambridge University Press.

Bejczy, A.K.
1982 Manual control of manipulator forces and torques using graphic display. Pp. 691-698 in *IEEE International Conference on Cybernetics and Society.*

Benthin, A., P. Slovic, and H. Severson
1993 A psychometric study of adolescent risk perception. *Journal of Adolescence* 16(2):153-168.

Bertsche, W., K. Logan, A. Pesch, and C. Winget
1977 Operator Performance in Undersea Manipulator Systems: Studies in Control Performance with Visual Force Feedback. Technical Report WHOI-77-6, Woods Hole Oceanographic Institution.

Bregman, A.S.
1990 *Auditory Scene Analysis.* Cambridge, Mass.: MIT Press.

Bryson, A.E., and Y. Ho
1975 Applied Optimal Control. *Hemisphere,* Washington, D.C.

Cesarone, B.
1994 Video Games and Children. ERIC Digest. ERIC Clearinghouse on Elementary and Early Childhood Education, Urbana, Ill., January.

Chi, M., P. Feltovich, and R. Glaser
1981 Categorization and representation of physics problems by experts and novices. *Cognitive Science* 5:121-152.

Coldevin, G., et al.
1993 Influence of instructional control and learner characteristics on factual recall and procedural learning from interactive video. *Canadian Journal of Educational Communication* 22(2):113-130.

Coombs, N.
1993 Global empowerment of impaired learners. *Educational Media International* 30(1):23-25.

Dubriel, R.
1993 All things being equal. *Vocational Education Journal* 68(8):28-69.

Durlach, N.I., L.A. Delhorne, A. Wong, W.Y. Ko, W.M. Rabinowitz, and J. Hollerbach
1989 Manual discrimination and identification of length by the finger-span method. *Perception and Psychophysics* 46(1):29-38.

Elkind, J.I.
1956 Characteristics of Simple Manual Control Systems. Technical Report 111, Lincoln Laboratory, Massachusetts Institute of Technology.
Fasse, E.D., B.A. Kay, and N. Hogan
1990 Human haptic illusions in virtual object manipulation. *Proceedings of Annual Conference of IEEE Engineering in Medicine and Biology Society* 12(5):1917-1918.
Fling, S., L. Smith, T. Rodriguez, D. Thornton, et al.
1992 Videogames, aggression, and self-esteem: A survey. *Social Behavior and Personality* 20(1):39-45.
Fontaine, G.
1992 The experience of a sense of presence in intercultural and international encounters. *Presence: Teleoperators and Virtual Environments* 1(4):482-490.
Funk, J.
1992 Video games: Benign or malignant? *Journal of Developmental and Behavioral Pediatrics* 13(1):53-54.
Garner, W.R.
1962 *Uncertainty and Structure as Psychological Concepts*. New York: Wiley.
1974 *The Processing of Information and Structure*. New York: Wiley.
Glaser, R., A. Lesgold, and S. Gott
1991 Implications of cognitive psychology for measuring job performance. Pp. 1-26 in A.K. Wigdor and B.F. Green, Jr., eds., *Performance Assessment for the Workplace, Vol. 2: Technical Issues*. Committee on the Performance of Military Personnel, National Research Council, Washington, D.C.: National Academy Press.
Heeter, C.
1992 Being there: The subjective experience of presence. *Presence* 1(2):262-271.
Held, R., and N. Durlach
1991 Telepresence, time delay, and adaptation. In *Pictorial Communication in Virtual and Real Environments*. London: Taylor and Francis.
1992 Telepresence, spotlight on: The concept of telepresence. *Presence* 1(1):109-112.
Hogan, N., B.A. Kay, E.D. Fasse, and F.A. Mussa-Ivaldi
1990 Haptic illusions: Experiments on human manipulation and perception of 'virtual objects.' *Cold Spring Harbor Symposia on Quantitative Biology* 55:925-931.
Howard, I.P., and W.B. Templeton
1966 *Human Spatial Orientation*. London: John Wiley.
Jex, H.R.
1971 Problems in modeling man-machine control behavior in biodynamic environments. In *7th Annual Conference on Manual Control, NASA SP-281*.
Jones, L.A., and I.W. Hunter
1990 Influence of the mechanical properties of a manipulandum on human operator dynamics: 1. Elastic stiffness. *Biological Cybernetics* 62:299-307.

1992 Human operator perception of mechanical variables and their effects on tracking performance. In *ASME Winter Annual Meeting: Issues in the Development of Kinesthetic Displays for Teleoperation and Virtual Environments*, Anaheim, Nov. 8-13.
1993 Influence of the mechanical properties of a manipulandum on human operator dynamics: 2. Viscosity. *Biological Cybernetics* (in press).

Kearney, R.E., and I.W. Hunter
1987 System identification of human joint dynamics. *Critical Reviews in Biomedical Engineering* 18(1):55-87.

Kidd, J.S.
1958 Social influence phenomena in a task-oriented group situation. *Journal of Abnormal and Social Psychology* 56(1):13-17.

Kleinman, S., D.L. Baron, and W.H. Levinson.
1974 A control theoretic approach to manned-vehicle systems analysis. *IEEE Transactions on Automatic Control* AC-16:824-832.

Kok, J.J., and R.A. Van Wijk
1977 A model of the human supervisor. Pp. 210-216 in *Thirteenth Annual Conference on Manual Control*, Moffett Field, Calif.

Kuipers, B.J.
1975 A frame for frames. In D. Bobrow and A. Collins, eds., *Representation and Understanding: Studies in Cognitive Science*. New York: Academic Press.

Lajoie, S.P., and A.M. Lesgold
1992 Dynamic assessment of proficiency for solving procedural knowledge tasks. *Educational Psychologist* 27(3):365-384.

Lesgold, A., S.P. Lajoie, D. Logan, and G.M. Eggan
1990 Cognitive task analysis approaches to testing. Pp. 325-350 in N. Frederiksen, R. Glaser, A. Lesgold, and M. Shafto, eds., *Diagnostic Monitoring of Skill and Knowledge Acquisition*. Hillsdale, N.J.: Lawrence Erlbaum.

Loomis, J.M., and S.J. Lederman
1986 Tactual perception. Chap. 31 in K.R. Boff, L. Kauffman, and J.P. Thomas, eds., *Handbook of Perception and Human Performance: Vol. I - Sensory Processes and Perception*. New York: John Wiley.

Loomis, J.M.
1992 Distal attibution and presence. *Presence* 1(1):113-19.

Massimino, M.J., and T.B. Sheridan
1993 Sensory substitution for force feedback in teleoperation. *Presence* 2(4):344-352.

McRuer, D.T., D. Graham, W.S. Krendel, and W. Reisener, Jr.
1965 Human Pilot Dynamics in Compensatory Systems—Theory, Models and Experiments with Controlled Element and Forcing Function variations. Technical Report AFFDL-TR-65-15, Air Force Flight Dynamics Laboratory, Wright-Patterson Air Force Base, Ohio.

Mead, G.H.
1934 *Mind, Self, and Society*. Chicago: University of Chicago Press.

Minsky, M.
1975 A framework for representing knowledge. In P.H. Winston, ed., *The Psychology of Computer Vision.* New York: McGraw-Hill.

Morrison, H., J. McClure, and C. Alewise
1993 The impact of portable computers on pupil's attitudes to study. *Journal of Computer Assisted Learning* 9(3):130-140.

Neuman, D.
1989 Computer based education for learning disabled students: Teacher's perceptions and behaviors. *Journal of Special Education Technology* 9(3):156-166.

Newell, A., and H.A. Simon
1972 *Human Problem Solving.* Englewood Cliffs, N.J.: Prentice-Hall.

Novelli, J.
1993 There's never been a better time to use technology. *Instructor* 103(3):34 ff.

Pang, X.D., H.Z. Tan, and N.I. Durlach
1991 Manual discrimination of force using active finger motion. *Perception and Psychophysics* 49(6):531-540.

Pepper, R.I., and J.D. Hightower
1984 Research Issues in Teleoperator Systems. Presented at 28th Annual Human Factors Society, San Antonio, Tex.

Quastler, H.
1955 *Information Theory in Psychology: Problems and Methods.* Glencoe, Ill.: Free Press.

Rasmussen, J.
1983 Skills, rules and knowledge: Signals, signs and symbols, and other distinctions in human performance models. *IEEE Transactions on Systems, Man and Cybernetics* SMC-133:257-267.

Rasmussen, J.
1986 *Information Processing and Human-Machine Interaction.* New York: Elsevier Press.

Reed, C.M., and N.I Durlach
1994 Note on Information Transfer Rates in Human Communication. Submitted to *Presence.*

Reed, C.M., N.I. Durlach, and L.D. Braida
1982 Research on tactile communication of speech: A review. ASHA Monograph 20, American Speech-Language-Hearing Association, Rockville, Md.

Reed, C.M., N.I. Durlach, L.A. Delhorne, W.M. Rabinowitz, and K.W. Grant
1989 Research on tactual communication of speech: Ideas, issues, and findings. *The Volta Review* (91):65-78.

Schloerb, D.W.
1994 A quantitative measure of telepresence. *Presence,* in press.

Schwartz, J.
1994 A terminal obsession. *The Washington Post.* March 27:F1.

Sheridan, T.B., and W.R. Ferrell
1974 *Man-Machine Systems: Information, Control and Decision Models of Human Performance.* Cambridge, Mass.: MIT Press.

Sheridan, T.B.
1976 Toward a general model of supervisory control. In T.B. Sheridan and G. Johansen, eds., *Monitoring Behavior and Supervisory Control.* New York: Plenum Press.
1992a Defining our terms. *Presence* 1(2):272-74.
1992b Musings of telepresence and virtual presence. *Presence* 1(1): 120-25.
1992c *Telerobotics, Automation and Human Supervisory Control.* Cambridge, Mass.: MIT Press.

Sjoberg, L., and G. Torell
1993 The level of risk acceptance and moral valuation. *Scandanavian Journal of Psychology* 34(3):223-236.

Stelmach, G.E.
1978 *Information Processing in Motor Control and Learning.* New York, Academic Press.

Steuer, J.
1992 Defining virtual reality: Dimensions determining telepresence. *Journal of Communication* (42): 73-93.

Tan, H.Z., X.D. Pang, and N.I. Durlach
1992 Manual resolution of length, force and compliance. Pp. 13-18 in H. Kazerooni, ed., *Advances in Robotics, ASME DSC Vol. 42.*

Tan, H.Z., N.I. Durlach, Y. Shao, and M. Wei
1993 Manual resolution of compliance when work and force cues are minimized. Pp. 99-104 in H. Kazerooni, J.E. Colgate, and B.D. Adelstein, eds., *Advances in in Robotics, Mechatronics and Haptic Interfaces, ASME DSC Vol. 49.*

Vasu, M.L., and E.S. Vasu
1993 Computer usage in a research methods course in the social sciences and education: A conceptual framework. *Collegiate Microcomputer* 11(3):177-182.

Warren, D.H., and E.R. Strelow
1985 *Electronic Spatial Sensing for the Blind.* Dordrecht, Netherlands: Martinus-Nijhoff.

Welch, R.B.
1978 *Perceptual Modification: Adapting to Altered Sensory Environments.* New York: Academic Press.

Zeltzer, D.
1992 Autonomy, interaction, and presence. *Presence* 1(1):127-32.

CHAPTER 2: THE VISUAL CHANNEL

Aukstakalnis, S., and D. Blatner
1993 *Silicon Mirage: The Art and Science of Virtual Reality.* Berkeley, Calif.: Peachpit Press.

Benton, S.A.
1982 Survey of holographic stereograms. Pp. 15-19 in J.J. Pearson, ed., *Proceedings of SPIE 367: Processing and Display of Three-Dimensional Data.*

1991 Experiments in holographic video imaging. In *Proceedings of the SPIE Institute on Holography*, Bellingham, Wash.

Boff, K.R., L. Kaufman, and J.P. Thomas, eds.
1986 *Handbook of Perception and Human Performance.* New York: John Wiley.

Bridgeman, B.
1991 Separate visual representations for perception and for visually guided behavior. In S.R. Ellis, M.K. Kaiser, and A.C. Grunwald, eds., *Pictorial Communication in Virtual and Real Environments.* London : Taylor & Francis.

Dallas, W.J.
1980 Computer-generated holograms. In R. Frieden, ed. *The Computer in Optical Research.* Berlin, West Germany: Springer-Verlag.

DiZio, P., and J.R. Lackner
1992 Altered Loading of the Head Exacerbates Motion Sickness and Spatial Disorientation. Paper presented at the Aerospace Medical Association Meeting, Miami, Fla.

Earnshaw, R.A., et al., eds.
1993 *Virtual Reality Systems.* London: Academic Press.

Graham, N.S.
1989 *Visual Pattern Analyzers.* Vol. 16 in the Oxford Psychology Series. D.E. Broadbent, S. Kosslyn, J.L. McGaugh, N.J. Mackintosh, E. Tulving, and L. Weiskrantz, eds. New York: Oxford University Press.

Gregory, R.L.
1991 Seeing by exploring. In S.R. Ellis, M.K. Kaiser, and A.C. Grunwald, eds., *Pictorial Communication in Virtual and Real Environments.* London: Taylor & Francis.

Haber, R.N., and M. Hershenson
1980 *The Psychology of Visual Perception*, 2nd ed. New York: Holt, Rinehart & Winston.

Hallett, P.E.
1986 Eye movements. In K.R. Boff, L. Kaufman, and J.P. Thomas, eds., *Handbook of Perception and Human Performance.* New York: John Wiley.

Held, R., and N. Durlach
1991 Telepresence, time delay and adaptation. In S.R. Ellis, M.K. Kaiser, and A.C. Grunwald, eds., *Pictorial Communication in Virtual and Real Environments.* London: Taylor & Francis.

Held, R., and S.J. Freedman
1963 Plasticity in human sensorimotor control. *Science* 54:33-37.

Holmgren, D.E., and W. Robinett
1994 Scanned laser displays for virtual reality: A feasibility study. *Presence* 2(3):171.

Hood, D.C., and M.A. Finkelstein
1986 Sensitivity to light. In K.R. Boff, L. Kaufman, and J.P. Thomas, eds., *Handbook of Perception and Human Performance.* New York: John Wiley.

Kalawsky, R.S.
1993 *The Science of Virtual Reality and Virtual Environments.* Workingham, England: Addison-Wesley.

Kobayashi, T., et al.
1984 Polyaniline film-coated electrodes as electrochromic display devices. *Journal of Electroanalytical Chemistry* 161:419-423.

Kohler, I.
1964 The Formation and Transformation of the Perceptual World. (Translated by H. Fiss.) International Universities Press. Originally Über Aufbau und Wandlungen der Wahrenehmungswelt. Vienna, Austria: R.M. Rohere (1951).

LaLonde, J.
1990 Warp speed into cyberspace. *The Seattle Times.* June 18.

Latour, P.L.
1962 Visual threshold during eye movements. *Vision Research* 2:261-262.

MacDowall, I.E., M. Bolas, S. Pieper, S.S. Fisher, and J. Humphries
1990 Implementation and integration of a counterbalanced CRT-based stereoscopic display for interactive viewpoint control in virtual environment applications. In J. Merritt, ed., *Proceedings of SPIE Stereoscopic Displays and Applications,* San Jose, Calif.

Matin, L.
1975 Eye movements and perceived visual direction. Pp. 1-76 in D. Jameson and L.M. Hurvich, eds., *The Handbook of Sensory Physiology,* Vol. 1. New York: Academic Press.

Matin, E., A. Clymer, and L. Matin
1972 Metacontrast and saccadic suppression. *Science* 178:179-182.

McKenna, M., and D. Zeltzer
1992 Three dimensional visual display systems for virtual environments. *Presence* 1(4):421-458.

Merritt, J.
1991 Innovations in Optics. Paper presented at the Southeast C41 Conference, Tampa, Fla.

Meyer, K., H.L. Applewhite, and F.A. Biocca
1992 A survey of position trackers. *Presence* 1(2):173-200.

Mittelstaedt, H.
1991 Interactions of form and orientation. In S.R. Ellis, M.K. Kaiser, and A.C. Grunwald, eds., *Pictorial Communication in Virtual and Real Environments.* London: Taylor & Francis.

Miyashita, T., and T. Uchida
1990 Causes of fatigue and its improvement in stereoscopic displays. *Proceedings of the Society for Information Display* 31(3):249-254.

Oyama, E., N. Tsunemoto, T. Susumu, and Y. Inoue
1993 Experimental study on remote manipulation using virtual reality. *Presence* 2(2):112-124.

Piantanida, T., D.K. Boman, and J. Gille
1993 Human perceptual issues and virtual reality. *Virtual Reality Systems* 1(1):43-52.

Picart, B., et al.
1989 Visualization of VLSI integrated circuits by means of ferroelectric liq-

uid crystals. *Liquid Crystal Chemistry, Physics, and Applications SPIE* 180:131-139.

Pirenne, M., and F. Mariott
1962 Visual functions in man. Pp. 3-217 in H. Davson, ed., *The Eye. Vol II. The Visual Process.* New York: Academic Press.

Roufs, A.J., and I.M.J. Goossens
1988 The effect of gamma on perceived image quality. *IEEE Proceedings: IDRC*:27-31.

Sekuler, R., and R. Blake
1990 *Perception.* New York: McGraw-Hill.

Soltan, P., J. Trias, W. Robinson, and W. Dahlke
1992 Laser based 3D volumetric display system. In *Proceedings of SPIE*, San Jose, Calif., February 9-14.

Spain, E.H.
1992 Effects of Camera Aiming Technique and Display Type on Unmanned Ground Vehicle Performance. Final Report to the Department of Defense Unmanned Ground Vehicle Joint Program Office, Huntsville, Alabama.

Spottiswoode, R.
1967 *A Grammar of the Film: An Analysis of Film Technique.* Berkeley and Los Angeles: University of California Press.

St. Hilaire, P., S.A. Benton, M. Lucente, M.L. Jepsen, J. Kollin, H. Yoshikawa, and J. Underkoffler
1990 Electronic display system for computational holography. In *SPIE Proceedings—Practical Holography IV*, Vol. 1212, Bellingham, Wash.

Tricoles, G.
1987 Computer generated holograms: An historical review. *Applied Optics* 26(20):4351-4360.

Westerink, J.H., and A.J. Roufs
1989 Subjective image quality as a function of viewing distance, resolution, and picture size. *SMPTE Journal* (Feb):113-119.

Westheimer, G.
1986 The eye as an optical instrument. In K.R. Boff, L. Kaufman, and J.P. Thomas, eds., *Handbook of Perception and Human Performance.* New York: John Wiley.

Wiegand, T.E.V., D.C. Hood, N.vS. Graham
1994 Testing a computational model of light-adaptation dynamics. Submitted to *Vision Research.*

CHAPTER 3: THE AUDITORY CHANNEL

Allen, J.B., and D.A. Berkley
1979 Image method for efficiently simulating small-room acoustics. *Journal of the Acoustical Society of America* 65(4):943-950.

Applied Acoustics
1992 Special Issue on Auditory Virtual Environments and Telepresence. 36(3-4).

1993 Special Issue on Computer Modelling and Auralization of Sound Fields in Rooms. 38(2-4).

Blattner, M.M., D.A. Sumikawa, and R.M. Greenberg

1989 Earcons and icons: Their structure and common design principles. *Human-Computer Interaction* 4:11-44.

Blauert, J.

1983 *Spatial Hearing*. Cambridge, Mass.: MIT Press.

Boff, K.R., and J. Lincoln, eds.

1988 Engineering Data Compendium: Human Perception and Performance. Henry G. Armstrong Aerospace Medical Research Laboratory, Wright Patterson Air Force Base, Ohio.

Boff, K.R., L. Kaufman, and J.P. Thomas, eds.

1986 *Handbook of Perception and Human Performance*, Vols. 1 and 2. New York: John Wiley.

Borin, G., G. Depoli, and A. Sarti

1992 Algorithms and structures for synthesis using physical models. *Computer Music Journal* 16(4):30-42.

Borish, J.

1984 Extension of the image model to arbitrary polyhedra. *Journal of the Acoustical Society of America* 75:1827-1836.

Bregman, A.S.

1990 *Auditory Scene Analysis*. Cambridge, Mass.: MIT Press.

Buxton, W., W. Gaver, and S. Bly

1989 The use of non-speech audio at the interface. (Tutorial No. 10). In *CHI'89, ACM Conference on Human Factors in Computing Systems*. New York: ACM Press.

Cadoz, C., and C. Ramstein

1991 Capture, representation, and "composition" of the instrumental game. *ICMC Proceedings 53-56*, Glasgow, Scotland.

Cadoz, C., A. Luciani, and J. Florens

1984 Responsive input devices and sound synthesis by simulation of instrumental mechanisms: The Cordis systems. *Computer Music Journal* 8(3):60-73.

1993 CORDIS-ANIMA—A modeling and simulation system for sound and image synthesis—the general formalism. *Computer Music Journal* 117(1):19-29.

Cadoz, C., L. Lisowski, and J. Florens

1990 A modular feedback keyboard design. *Computer Music Journal* 14(2):47-51.

Carterette, E.C., and M.P. Friedman

1978 *Handbook of Perception*, Vol. 4, Hearing. New York: Academic Press.

Chandler, D.W., and D.W. Grantham

1992 Minimum audible movement angle in the horizontal plane as a function of stimulus frequency and bandwidth, source azimuth, and velocity. *Journal of the Acoustical Society of America* 91:1624-1636.

Chowning, J.
1973 The synthesis of complex audio spectra by means of frequency modulation. *Journal of the Audiological Engineering Society* 21(7):526-534.
Chowning, J., and D. Bristow
1986 FM Theory and Applications by Musicians for Musicians. Yamaha Music Foundaton, Tokyo, Japan.
Cohen, M., and L.F. Ludwig
1991 Multidimensional audio window management. *International Journal of Man-Machine Studies* 34:319-336.
Colburn, H.S., and N.I. Durlach
1978 Models of binaural interaction. Pp. 467-518 in E.C. Carterette and M.P. Friedman, eds., *Handbook of Perception: Vol. 4, Hearing*. New York: Academic Press.
Colburn, H.S., P.M. Zurek, and N.I. Durlach
1987 Binaural directional hearing—impairments and aids. Pp. 261-278 in W.A. Yost and G. Gourevitch, eds., *Directional Hearing*. New York: Springer-Verlag.
Cook, P.R.
1993 Spasm, a real-time vocal tract physical model controller—and singer, the companion software synthesis system. *Computer Music Journal* 17(1):30-44.
Cook, P.R., S. Hirschman, and J.O. Smith
1991 Physical Modeling. Presentation at the International Computer Music Conference, CCRMA.
Cooper, D.H., and J.L. Bauck
1989 Prospects for transaural recording. *Journal of the Audio Engineering Society* 37:3-19.
Cruz-Neira, C., D.J. Sandin, and T.A. De Fanti
1993 Surround screen projection-based virtual reality: The design and implementation of the CAVE. Pp. 135-142 in *Computer Graphics*. Proceedings of SIGGRAPH.
Dannenbring, G.L.
1976 Perceived auditory continuity with alternately rising and falling frequency transitions. *Canadian Journal of Psychology* 30:99-114.
Djoharian, P.
1993 Generating models for modal synthesis. *Computer Music Journal* 17(1):57-65.
Dolson, M.
1986 The phase vocoder: A tutorial. *Computer Music Journal* 10:14-27.
Durlach, N.I.
1991 Auditory localization in teleoperator and virtual environment systems: Ideas, issues, and problems. *Perception* 20:543-554.
Durlach, N.I., and H.C. Colburn
1978 Binaural phenomena. Pp. 365-466 in E.C. Carterette and M.P. Friedman, eds., *Handbook of Perception, Vol. 4: Hearing*. New York: Academic Press.

Durlach, N.I., and X.D. Pang
1986 Interaural magnification. *Journal of the Acoustical Society of America* 80:1849-1850.
Durlach, N.I., R. Held, and B. Shinn-Cunningham
1992a Super auditory localization displays. In *Proceedings of the Society for Information Displays*, Boston, Mass.
Durlach, N.I., R.W. Pew, W.A. Aviles, P.A. DiZio, and D.L. Zeltzer
1992b *Virtual Environment Technology for Training (VETT)*. Report No. 7661. Cambridge, Mass.: Bolt, Beranek, and Newman.
Durlach, N.I., A. Rigopulos, X.D. Pang, W.S. Woods, A. Kulkarni, H.S. Colburn, and E.M. Wenzel
1992c On the externalization of auditory images. *Presence* 1(2):251-257.
Durlach, N.I., B.G. Shinn-Cunningham, and R.M. Held
1993 Supernormal auditory localization: I. General background. *Presence* 2(2):1-103.
Eargle, J.
1986 Stereophonic Techniques: An Anthology of Reprinted Articles on Stereophonic Techniques. Audio Engineering Society.
Fay, R.R., and A.N. Popper, eds.
1994 *Handbook of Auditory Research*. New York: Springer-Verlag.
Foster, S.H., E.M. Wenzel, and R.M. Taylor
1991 Real-time synthesis of complex acoustic environments. In *Proceedings of the IEEE Workshop on Applications of Signal Processing to Audio and Acoustics*, New Paltz, N.Y.
Gaver, W.W.
1993 Synthesizing auditory icons. Pp. 228-235 in *Proceedings of NTERCHI'93, Conference on Human Factors in Computing Systems*, Amsterdam, Netherlands.
Gaver, W.W., R.B. Smith, and T. O'Shea
1991 Effective sounds in complex systems: The ARKola simulation. Pp. 85-90 in *Proceedings of CHI'91, ACM Conference on Computer-Human Interaction*.
Gierlich, H.W., and K. Genuit
1989 Processing artificial-head recordings. *Journal of the Audio Engineering Society* 37:34-39.
Gilkey, R., and T. Anderson, eds.
1994 *Binaural and Spatial Hearing*. Hillsdale, N.J.: Lawrence Erlbaum Associates.
Grantham, D.W.
1986 Detection and discrimination of simulated motion of auditory targets in the horizontal plane. *Journal of the Acoustical Society of America* 79:1939-1949.
1989 Motion aftereffects with horizontally moving sound sources in the free field. *Perception and Psychophysics* 45:129-136.
1992 Adaptation to auditory motion in the horizontal plane: Effect of prior exposure to motion on motion detectability. *Perception and Psychophysics* 52:144-150.

Grey, J.M., and J.A. Moorer
1977 Perceptual evaluations of synthesized musical instrument tones. *Journal of Acoustical Society of America* 63:454-462.

Griesinger, D.
1989 Theory and design of a digital audio signal processor for home use. *Journal of the Audio Engineering Society* 37:40-50.

Handel, S.
1989 *Listening: An Introduction to the Perception of Auditory Events*. Cambridge, Mass.: MIT Press.

Hartman, W.M.
1988 Pitch perception and the segregation and integration of auditory entities. In G. Edelman, W. Gall, and W. Cowan, eds., *Auditory Function*. New York: John Wiley.

Howard, J.H.
1983 Perception of simulated propeller cavitation. *Human Factors* 25:643-656.

Jaffe, D., and J. Smith
1983 Extensions of the Karplus-Strong plucked-string algorithm. *Computer Music Journal* 7(2):43-55.

Kay, L.
1974 A sonar aid to enhance spatial perception of the blind: Engineering design and evaluation. *Radio and Electronics Engineer* 44:40-62.

Keefe, D.H.
1992 Physical modeling using digital waveguides. *Computer Music Journal* 16(4):57-73.

Kendall, G.S., and W.L. Martens
1984 Simulating the cues of spatial hearing in natural environments. In *Proceedings of the 1984 International Computer Music Conference*, Paris, France.

Killion, M.C.
1981 Earmold options for wideband hearing aids. *Journal of Speech and Hearing Disorders* 46:10-20.

Killion, M.C., and T.W. Tillman
1982 Evaluation of high-fidelity hearing aids. *Journal of Speech and Hearing Research* 25:15-25.

Kramer, G., ed.
1994 *Auditory Display: Sonification, Audification, and Auditory Interfaces*. Santa Fe Institute Studies in the Sciences of Complexity, Proceedings Volume 18. Reading, Mass.: Addison-Wesley.

Krokstadt, A., S. Strom, and S. Sorsdal
1968 Calculating the acoustical room response by the use of a ray tracing technique. *Journal of Sound and Vibrations* 8:118.

Lehnert, H.
1993a Systematic errors of the ray-tracing algorithm. *Applied Acoustics* 38:207-221.

1993b Auditory spatial impression. Pp. 40-46 in *Proceedings of the 12th International AES Conference*, Copenhagen, Denmark.

Lehnert, H., and J. Blauert
1989 A concept for binaural room simulation [Summary]. In *Proceedings of the IEEE ASSP Workshop on Applications of Signal Processing to Audio and Acoustics*, New Paltz, N.Y.
1991 Virtual auditory environment. *Proceedings of the 5th International Conference on Advanced Robotics* IEEE-ICAR 1:211-216.
1992 Principles of binaural room simulation. *Applied Acoustics* 36:325-333.
Li, X., R.J. Logan, and R.E. Pastore
1991 Perception of acoustic source characteristics. *Journal of the Acoustical Society of America* 90:3036-3049.
Liberman, A.M., F.S. Cooper, D.P. Shankweiler, and M. Studdert-Kennedy
1968 Why are speech spectrograms hard to read? *American Annals of the Deaf* 113:127-133.
Loomis, J.M., C. Hebert, and J.G. Cicinelli
1990 Active localization of virtual sounds. *Journal of the Acoustical Society of America* 88:1757-1764.
Ludwig, L.F., N. Pincever, and M. Cohen
1990 Extending the notion of a window system to audio. *Computer* 23: 66-72.
MacCabe, C., and D.J. Furlong
1994 Virtual imaging capabilities of Surround Sound Systems. *Journal of the Audio Engineering Society* 42:38-49.
Massimino, M.J., and T.B. Sheridan
1993 Sensory substitution for force feedback in teleoperation. *Presence* 2(4):344-352.
Mathews, M.V., and J.R. Pierce
1989 *Current Directions in Computer Music Research*. Cambridge, Mass.: MIT Press.
McKinley, R.L., and M.A. Ericson
1988 Digital synthesis of binaural auditory localization azimuth cues using headphones. *Journal of the Acoustical Society of America* 83: S18.
Mills, A.W.
1958 On the minimum audible angle. *Journal of the Acoustical Society of America* 30:237-246.
1972 Auditory localization. Pp. 301-345 in J.V. Tobias, ed., *Foundations of Modern Auditory Theory*, Vol. II. New York: Academic Press.
Moore, B.C.J.
1989 *An Introduction to the Psychology of Hearing*, 3rd edition. New York: Academic Press.
Moore, F.R.
1990 *Elements of Computer Music*. Englewood Cliffs, N.J.: Prentice- Hall.
Moorer, J.A.
1978 The use of the phase vocoder in computer music applications. *Journal of Audiological Engineering Society* 24:717-727.
Morrison, J.D., and J.M. Adrien
1993 MOSAIC - A framework for modal synthesis. *Computer Music Journal* 17(1):45-56.

Morse, P.M., and K.U. Ingard
1968 *Theoretical Acoustics.* New York: McGraw-Hill.
Patterson, R.R.
1982 *Guidelines for Auditory Warning Systems on Civil Aircraft.* Paper No. 82017. London: Civil Aviation Authority.
Perrott, D.R.
1982 Studies in the perception of auditory motion. Pp. 169-193 in R.W. Gatehouse, ed., *Localization of Sound: Theory and Applications.* Groton, Conn.: Amphora Press.
Perrott, D.R., and J. Tucker
1988 Minimum audible movement angle as a function of signal frequency and the velocity of the source. *Journal of the Acoustical Society of America* 83:1522-1527.
Perrott, D.R., K. Marlborough, P. Merrill, and T.Z. Strybel
1989 Minimum audible angle thresholds obtained under conditions in which the precedence effect is assumed to operate. *Journal of the Acoustical Society of America* 85:282-287.
Pickles, J.D.
1988 *An Introduction to the Physiology of Hearing,* 2nd ed. New York: Springer-Verlag.
Portnoff, M.R.
1980 Short-time Fourier analysis of sampled speech. *IEEE Transactions on Acoustics, Speech and Signal Processing* 29:55-69.
Richards, W.
1988 *Natural Computation.* Cambridge, Mass.: MIT Press.
Risset, J.C., and M.V. Mathews
1969 Analysis of musical instrument tones. *Physics Today* 22:23-40.
Roads, C., and J. Strawn
1988 *Foundations of Computer Music.* Cambridge, Mass.: MIT Press.
Scaletti, C.
1994 Sound synthesis algorithms for auditory data representations. In G. Kramer, ed., *Auditory Display: The Proceedings of ICAD'92, the First International Conference on Auditory Display.* Santa Fe Institute Studies in the Sciences of Complexity, Volume 18, Reading, Mass.: Addison-Wesley.
Scaletti, C., and A.B. Craig
1991 Using sound to extract meaning from complex data. Pp. 207-219 in *Proceedings of the SPIE (1459)*, San Jose, Calif.
Schottstaedt, B.
1977 The simulation of natural instrument tones using frequency modulation with a complex modulating wave. *Computer Music Journal* 1:46-50.
Schroeder, M.R.
1962 Natural sounding artificial reverberation. *Journal of the Audio Engineering Society* 10(3).
Searle, C.L., L.D. Braida, D.R. Cuddy, and M.F. Davis
1976 Binaural pinna disparity: Another auditory localization cue. *Journal of the Acoustical Society of America* 57:448-455.

Shaw, E.A.G.
1966 Earcanal pressure generated by circumaural and supraaural earphones. *Journal of the Acoustical Society of America* 39:471-479.

Shinn-Cunningham, B.G.
1994 Adaptation to Supernormal Auditory Localization Cues in an Auditory Virtual Environment. Unpublished doctoral dissertation, Department of Electrical Engineering, Massachusetts Institute of Technology.

Shinn-Cunningham, B.G., H. Lehnert, G. Kramer, E. Wenzel, and N.I. Durlach
1994 Auditory displays. In R. Gilkey and T. Anderson, eds., *Binaural and Spatial Hearing*. Hillsdale, N.J.: Lawrence Erlbaum.

Smith, J.O.
1992 Physical modeling using digital waveguides. *Computer Music Journal* 16(4):74-98.

Smith, S., R.D. Bergeron, and G.G. Grinstein
1990 Stereophonic and surface sound generation for exploratory data analysis. Pp. 125-132 in *Proceedings of CHI'90, ACM Conference on Human Factors in Computing Systems*, Seattle, Wash.

Stelmachowicz, P.G., K.A. Beauchaine, A. Kalberer, T. Langer, and W. Jesteadt
1988 The reliability of auditory thresholds in the 8 to 20 kHz range using a prototype audiometer. *Journal of the Acoustical Society of America* 83:1528-1535.

Strelow, E.R., and D.H. Warren
1985 Sensory substitution in blind children and neonates. Pp. 273-298 in D.H. Warren and E.R. Strelow, eds., *Electronic Spatial Sensing for the Blind*. Dordrecht, Netherlands: Martinus-Nijhoff.

Trahiotis, C., and L.R. Bernstein
1987 Some modern techniques and devices used to preserve and enhance the spatial qualities of sounds. Pp. 279-290 in W.A. Yost and G. Gourevitch, eds., *Directional Hearing*. New York: Springer-Verlag.

Van Veen, B.D., and R.L. Jenison
1991 Auditory space expansion via linear filtering. *Journal of the Acoustical Society of America* 90:231-240.

Vian, J., and J. Martin
1992 Binaural room acoustics simulation: Practical uses and applications. *Applied Acoustics* 36:293-306.

Warren, D.H., and E.R. Strelow
1984 Learning spatial dimensions with a visual sensory aid: Molyneaux revisited. *Perception* 13:331-350.
1985 Training the use of artificial spatial displays. In D.H. Warren and E.R. Strelow, eds., *Electronic Spatial Sensing for the Blind*. Dordrecht, Netherlands: Martinus-Nijhoff.

Warren, H., and R.R. Verbrugge
1984 Auditory perception of breaking and bouncing events: A case study in ecological acoustics. *Journal of Experimental Psychology: Human Perception and Performance* 10:704-712.

Wawrzynek, J.
1991 VLSI models for sound synthesis. Pp. 113-148 in G.D. Poli, A. Piccialli,

and C. Roads, eds., *Representations of Musical Signals.* Cambridge, Mass.: MIT Press.

Welch, R.B.

1978 *Perceptual Modification: Adapting to Altered Sensory Environments.* New York: Academic Press.

Wenzel, E.M.

1992 Localization in virtual acoustic displays. *Presence* 1:80-107.

1994 Spatial sound and sonification. In G. Kramer, ed., *Auditory Display: Sonification, Audification, and Auditory Interfaces.* Santa Fe Institute Studies in the Sciences of Complexity, Proceedings Vol. 18, Reading, Mass.: Addison-Wesley.

Wenzel, E.M., and S.H. Foster

1993 Perceptual consequences of interpolating head-related transfer functions during spatial synthesis. *Proceedings of the ASSP (IEEE) Workshop on Applications of Signal Processing to Audio & Acoustics,* New Paltz, N.Y., Oct. 17-20.

Wenzel, E.M., F.L. Wightman, and S.H. Foster

1988 A virtual display system for conveying three-dimensional acoustic information. Pp. 86-90 in *Proceedings of the Human Factors Society,* Anaheim, Calif.

Wenzel, E.M., P.K. Stone, S.S. Fisher, and S.H. Foster

1990 A system for three-dimensional acoustic "visualization" in a virtual environment workstation. Pp. 329-337 in *Proceedings of the IEEE Visualization '90 Conference,* San Francisco, Calif.

Wenzel, E.M., M. Arruda, D.J. Kistler, and F.L. Wightman

1993a Localization using non-individualized head-related transfer functions. *Journal of the Acoustical Society of America* 94:111-123.

Wenzel, E.M., W. Gaver, S.H. Foster, H. Levkowitz, and R. Powell

1993b Perceptual vs. hardware performance in advanced acoustic interface design. Pp. 363-366 in *Proceedings of INTERCHI'93, Conference on Human Factors in Computing Systems,* Amsterdam, Netherlands.

Wightman, F.L., and D.J. Kistler

1989a Headphone simulation of free-field listening I: Stimulus synthesis. *Journal of the Acoustical Society of America* 85:858-867.

1989b Headphone simulation of free-field listening II: Psychophysical validation. *Journal of the Acoustical Society of America* 85:868-878.

Wightman, F.L., D.J. Kistler, and M. Arruda

1992 Perceptual consequences of engineering compromises in synthesis of virtual auditory objects [Abstract]. *Journal of the Acoustical Society of America* 92:2332.

Woodhouse, J.

1992 Physical modeling of bowed strings. *Computer Music Journal* 16(4):43-56.

Yost, W.A.

1991 Auditory image perception and analysis: The basis for hearing. *Hearing Reseach* 56:8-18.

1994 *Fundamentals of Hearing,* 3rd ed. New York: Academic Press.

Yost, W.A., and G. Gourevitch, eds.
1987 *Directional Hearing*. New York: Springer-Verlag.

CHAPTER 4: HAPTIC INTERFACES

Adelstein, B.D., and M.J. Rosen
1992 Design and implementation of a force reflecting manipulandum for manual control research. Pp. 1-12 in M. Kazerooni, ed., *ASME 1992: Advances in Robotics* DSC Vol. 42.

Bach-y-Rita, P.
1982 Sensory substitution in rehabilitation. Pp. 361-383 in L. Illis, M. Sedgwick, and H. Granville, eds., *Rehabilitation of the Neurological Patient*. Oxford, England: Blackwell Scientific.

Bliss, J.C., and J.W. Hill
1971 Tactile Perception Studies Related to Teleoperator Systems. Stanford Research Institute, Menlo Park, Calif., Final Report, 20, NASA-CR-114346, June.

Bolanowski, S.J., Jr., G.A. Gescheider, R.T. Verrillo, and C.M. Checkosky
1988 Four channels mediate the mechanical aspects of touch. *Journal of the Acoustical Society of America* 84(5):1680-1694.

Brooks, F.P., M. Ouh-young, and J. Batter
1990 Project GROPE: Haptic displays for scientific visualization. *Computer Graphics* 24(4):177-185.

Brooks, T.L.
1990 Telerobotic response requirements. Pp. 113-120 in *Proceedings of the IEEE International Conference on Systems, Man and Cybernetics*, Los Angeles; also Report No. STX/ROB/90-03, STX Corporation, Lanham, Md.

Brooks, T.L., and A.K. Bejczy
1985 Hand Controllers for Teleoperation. Technical Report JPL Publication 85-11, Jet Propulsion Laboratory, Pasadena, Calif., March 1, 1985.

Burdea, G., J. Zhuang, E. Roskos, D. Silver, and N. Langrana
1992 A portable dextrous master with force feedback. *Presence* 1(1):18-28.

Burgess, P.R., and F.J. Clark
1969 Characteristics of knee joint receptors in the cat. *Journal of Physiology* (London) 203:317-335.

Clark, F.J., and K.W. Horch
1986 Kinesthesia. Chap. 13 in K.R. Boff, L. Kauffman, and J.P. Thomas, eds., *Handbook of Perception and Human Performance, Vol. 1: Sensory Processes and Perception*. New York: John Wiley.

Cybernet Systems Corporation
1992 *Per-Force: Programmable Environment Reality through Force*. Sales literature, Cybernet Systems Corporation.

Darian-Smith, I.
1984 The sense of touch: Performance and peripheral neural processes. Pp. 739-788 in *Handbook of Physiology—The Nervous System—III*. England: Oxford University Press.

Darian-Smith, I., I. Davidson, and K.O. Johnson
1980 Peripheral neural representation of spatial dimensions of a textured surface moving across the monkey's fingerpad. *Journal of Physiology* 309:135-146.

Durlach, N.I., L.A. Delhorne, A. Wong, W.Y. Ko, W.M. Rabinowitz, and J. Hollerbach
1989 Manual discrimination and identification of length by the finger-span method. *Perception and Psychophysics* 46(1):29-38.

Durlach, N.I., R.W. Pew, W.A. Aviles, P.A. DiZio, and D.L. Zeltzer
1992 Virtual Environment Technology for Training. Virtual Environment and Teleoperator Research Consortium (VETREC), Massachusetts Institute of Technology, BBN Report 7661, March 1992.

Edin, B.B.
1993 The Capacity of Hairy Skin Mechanoreceptors to Provide Information about Joint Configurations in Humans. 32nd Congress of the International Union of Physiological Sciences, Glasgow, Aug. 1-6.

Fasse, E.D.
1992 On the Use and Representation of Sensory Information of the Arm by Robots and Humans. Unpublished dissertation, Department of Mechanical Engineering, Massachusetts Institute of Technology.

Fasse, E.D., B.A. Kay, and N. Hogan
1990 Human haptic illusions in virtual object manipulation. *Proceedings of Annual Conference of IEEE Engineering in Medicine and Biology Society* 12(5):1917-1918.

Ferrel, W.R.
1980 The adequacy of stretch receptors in the cat knee joint for signalling joint angle throughout a full range of movement. *Journal of Physiology* (London) 299:85-99.

Ferrel, W.R., Gandevia, S.C. and D.I. McCloskey
1987 The role of joint receptors in human kinaesthesia when intramuscular receptors cannot contribute. *Journal of Physiology* 386:63-71.

Frisken-Gibson, S.F., P. Bach-Y-Rita, W.J. Tompkins, and J.G. Webster
1987 A 64-Solenoid, four-level fingertip search display for the blind. *IEEE Transactions on Biomedical Engineering* BME-34(12):963-965.

Goertz, R.
1964 Some work on manipulator systems at ANL: Past, present and a look at the future. In *Seminars on Remotely Operated Special Equipment*, Vol. 1.

Goodwin, G.M., D.I. McCloskey, and P.B.C. Matthews
1972 The contribution of muscle afferents to kinaesthesia shown by vibration induced illusions of movement and by the effects of paralysing the joint afferents. *Brain* 95:705-748.

Grigg, P., and B.J. Greenspan
1977 Response of primate joint afferent neurons to mechanical stimulation of knee joint. *Journal of Neurophysiology* 40:1-8.

Grigg, P., G.A. Finerman, and L.H. Riley
1973 Joint-position sense after total hip replacement. *Journal of Bone and Joint Surgery* 55A:1016-1025.

Hill, J.W.
1979 Study of Modeling and Evaluation of Remote Manipulation Tasks with Force Feedback. Final Report, JPL Contract 95-5170, March.

Hogan, N., B.A. Kay, E.D. Fasse, and F.A. Mussa-Ivaldi
1990 Haptic illusions: Experiments on human manipulation and perception of 'virtual objects.' Cold Spring Harbor Symposia on Quantitative Biology 55:925-931.

Honeywell Inc.
1989 Hand Controller Commonality Study. Technical Report, Honeywell Inc., Avionics Division, Clearwater, Fla., February 1989. Prepared for McDonnell Douglas Space Systems, Co., Huntington Beach, Calif.

Howe, R.D.
1992 A force-reflecting teleoperated hand system for the study of tactile sensing in precision manipulation. Pp. 1321-1326 in *IEEE International Conference on Robotics and Automation*, Nice, France.

Hunter, I.W., S. Lafontaine, P.M.F. Nielsen, P.J. Hunter, and J.M. Hollerbach
1990 Manipulation and dynamic mechanical testing of microscopic objects using a tele-micro-robot system. *IEEE Control Systems Magazine* 10(2):3-9.

Hunter, I., T.D. Doukoglou, S.R. Lafontaine, P.G. Charrette, L.A. Jones, M.A. Sagar, G.D. Mallinson, and P.J. Hunter
1994 A teleoperated microsurgical robot and associated virtual environment for eye surgery. *Presence* 2(4):265-280.

Iwata, H.
1990 Artificial reality with force-feedback: Development of desktop virtual space with compact master manipulator. *Computer Graphics* 24(4):165-170.

Iwata, H., T. Nakagawa, and T. Nakashima
1992 Force display for presentation of rigidity of virtual objects. *Journal of Robotics and Mechatronics* 4(1):39-42.

Jacobsen, S.C., E.K. Iversen, C.C. Davis, D.M. Potter, and T.W. McLain
1989 Design of a Multiple Degree of Freedom, Force Reflective Hand Master/Slave with a High Mobility Wrist. Paper presented at the Third Topical Meeting on Robotics and Remote Systems, March 13-16, Charleston, South Carolina.

Jacobus, H.N., A.J. Riggs, C.J. Jacobus, and Y. Weinstein
1992 Implementation issues for telerobotic handcontrollers: Human-Robot ergonomics. Pp. 284-314 in M. Rahimi and W. Karwowski, eds., *Human-Robot Interaction*. New York: Taylor & Francis.

Jau, B.M.
1992 Man-equivalent telepresence through four fingered human-like hand system. Pp. 843-848 in *IEEE International Conference on robotics and Automation*, Nice, France.

Johansson, R.S., and K.J. Cole
1992 Sensory-motor coordination during grasping and manipulative actions. *Current Opinion in Neurobiology* 2:815-823.

Johansson, R.S., and A.B. Vallbo
1979 Tactile sensibility in the human hand: Relative and absolute densities

of four types of mechanoreceptive units in glabrous skin. *Journal of Physiology* 286:283-300.

Johansson, R.S., and G. Westling

1984 Roles of glabrous skin receptors and sensorimotor memory in automatic control of precision grip when lifting rougher or more slippery objects. *Experimental Brain Research* 56:550-564.

Johnson, K.O., and S.S. Hsiao

1992 Neural mechanisms of tactual form and texture perception. *Annual Review of Neuroscience* 15:227-250.

Johnson, K.O., and J.R. Phillips

1981 Tactile spatial resolution - I. Two point discrimination, gap detection, grating resolution and letter recognition. *Journal of Neurophysiology* 46(6):1177-1191.

Jones, L.A.

1986 Perception of force and weight: Theory and research. *Psychological Bulletin* 100(1):29-42.

1989 Matching forces: Constant errors and differential thresholds. *Perception* 18(5):681-687.

Jones, L.A., and I.W. Hunter

1992 Human operator perception of mechanical variables and their effects on tracking performance. Pp. 49-53 in H. Kazerooni, ed., *Advances in Robotics, ASME DSC Vol. 42*.

Kaczmarek, K.A., and P. Bach-y-Rita

1993 Tactile displays. In W. Barfield and T. Furness III, eds., *Virtual Environments and Advanced Interface Design*. Oxford, England: Oxford University Press.

Knibestol, M., and A.B. Vallbo

1970 Single unit analysis of mechanoreceptor activity from the human glabrous skin. *Acta Physiologica Scandinavia* 80(6):178-195.

Lamb, G.D.

1983a Tactile discrimination of textured surfaces: Psychophysical performance measurements in humans. *Journal of Physiology* 338:551-565.

1983b Tactile discrimination of textured surfaces: Peripheral neural coding in the monkey. *Journal of Physiology* 338:567-587(a).

LaMotte, R.H., and M.A. Srinivasan

1991 Surface microgeometry: Tactile perception and neural encoding. In O. Franzen and J. Westman, eds., *Information Processing in the Somatosensory System*. Wenner-Gren International Symposium Series. London: Macmillan Press.

1993 Responses of cutaneous mechanoreceptors to the shape of objects applied to the primate fingerpad. *Acta Psychologica* 84:41-51.

LaMotte, R.H., and J. Whitehouse

1986 Tactile detection of a dot on a smooth surface. *Journal of Neurophysiology* 56:1109-1128.

Lederman, S.J., and M.M. Taylor

1972 Fingertip force, surface geometry, and the perception of roughness by active touch. *Perception Psychophysics* 12:401-408.

Loomis, J.M.
1979 An investigation of tactile hyperacuity. *Sensory Processes* 3:289-302.
Loomis, J.M., and S.J. Lederman
1986 Tactual perception. Chap. 31 in K.R. Boff, L. Kauffman, and J.P. Thomas, eds., *Handbook of Perception and Human Performance: Vol. I - Sensory Processes and Perception*. New York: John Wiley.
Massie, T.H., and J.K. Salisbury
1994 Probing Virtual Objects with the PHANToM Haptic Interface. Paper submitted to the ASME WAM, Nov. 6-11, 1994, Chicago, Illinois.
Matthews, P.B.C.
1982 Where does Sherrington's "muscular sense" originate? Muscles, joints or corollary discharges? *Annual Reviews of Neuroscience* 5:189-218.
1988 Proprioceptors and their contribution to somatosensory mapping: Complex messages require complex processing. *Canadian Journal of Physiological Pharmacology* 66:430-438.
McAffee, D.A., and P. Fiorini
1991 Hand Controller Design Requirements and Performance Issues in Telerobotics. Proceedings of International Conference on Advanced Robotics (ICAR), Pisa, June.
Meyer, K., H.L. Applewhite, and F.A. Biocca
1992 A survey of position trackers. *Presence* 1(2):173-200.
Minsky, M., M. Ouh-young, O. Steele, F.P. Brooks, Jr., and M. Behensky
1990 Feeling and seeing: Issues in force display. *Computer Graphics* 24(2):235-243.
Monkman, G.J.
1992 An electrorheological tactile display. *Presence* 1(2):219-228.
Morley, J.W., A.W. Goodwin, and I. Darian-Smith
1983 Tactile discrimination of gratings. *Experimental Brain Research* 49:291-299.
Mountcastle, V.B., and T.P.S. Powell
1959 Central nervous mechanisms subserving position sense and kinesthesis. *Bulletin of the Johns Hopkins Hospital* 105:173-200.
Mountcastle, V.B., R.H. LaMotte, and G. Carli
1972 Detection threshold for stimuli in humans and monkeys: Comparison with threshold events in mechanoreceptive afferent nerve fibers innervating monkey hand. *Journal of Neurophysiology* 35:122-136.
Noma, H., and H. Iwata
1993 Presentation of multiple dimensional data by 6.D.O.F force display. Pp. 1495-1500 in *IEEE/RSJ International Conference on Intelligent Robots and Systems*, Yokohama, Japan.
Pang, X.D., H.Z. Tan, and N.I. Durlach
1991 Manual discrimination of force using active finger motion. *Perception and Psychophysics* 49(6):531-540.
Phillips, J.R., and K.O. Johnson
1981a Tactile spatial resolution - II. Neural represntation of bars, edges and gratings in monkey primary afferents. *Journal of Neurophysiology* 46(6):1192-1203.
1981b Tactile spatial resolution - III. A continuum mechanics model of skin

predicting mechanoreceptor responses to bars, edges, and gratings. *Journal of Neurophysiology* 46(6):1204-1225.

Phillips, J.R., K.O. Johnson, and H.M. Browne
1983 A comparison of visual and two modes of tactual letter recognition. *Perception and Psychophysics* 34:243-249.

Phillips, J.R., K.O. Johnson, and S.S. Hsiao
1988 Spatial pattern representation and transformation in monkey somatosensory cortex. *Proceedings of the National Academy of Sciences (USA)* 85(6):1317-1321.

Poulton, E.C.
1974 *Tracking Skill and Manual Control.* New York: Academic Press.

Pubols, B.H.P., Jr.
1980 On-versus-off responses of raccoon glabrous skin rapidly adapting cutaneous mechanoreceptors. *Journal of Neurophysiology* 43:1558-1570.

Pubols, B.H.P., Jr., and L.M. Pubols
1976 Coding of mechanical stimulus velocity and indentation depth by squirrel monkey and raccoon skin mechanoreceptors. *Journal of Neurophysiology* 39:773-787.
1983 Tactile receptor discharge and mechanical properties of glabrous skin. *Fed. Proc.* 42:2528-2535.

Rabinowitz, W.M., A.J.M. Houtsma, N.I. Durlach, and L.A. Delhorne
1987 Multidimensional tactile displays: Identification of vibratory intensity, frequency, and contactor area. *Journal of the Acoustical Society of America* 82(4):1243-1252.

Reed, C.M., N.I. Durlach, and L.D. Braida
1982 Research on tactile communication: A review. *ASHA Monographs* 20.

Salcudean, S.E., N.M. Wong, and R.L. Hollis
1992 A force-reflecting teleoperation system with magnetically levitated master and wrist. Pp. 1420-1426 in *Proceedings of IEEE International Conference on Robotics and Automation*, Nice, France.

Sathian, K., A.W. Goodwin, K.T. John, and I. Darian-Smith
1989 Perceived roughness of a grating: Correlation with responses of mechanoreceptive afferents innervating the monkey's fingerpad. *Journal of Neuroscience* 9:1273-1279.

Schmult, B., and R. Jebens
1993 Application areas for a force-feedback joystick. Pp. 47-54 in H. Kazerooni, J.E. Colgate, and B.D. Adelstein, eds., *Advances in Robotics, Mechatronics and Haptic Interfaces*, ASME DSC Vol. 49.

Schneider, W.
1988 The tactile array stimulator. *Johns Hopkins API Technical Digest* 9(1):39-43.

Sheridan, T.B.
1992 *Telerobotics, Automation, and Supervisory Control.* Cambridge, Mass.: MIT Press.

Sherrick, C.E., and R.W. Cholewiak
1986 Cutaneous sensitivity. Ch. 12 in K.R. Boff, L. Kauffman, and J.P. Thomas, eds., *Handbook of Perception and Human Performance, Vol. 1: Sensory Processes and Perception.* New York: John Wiley.

Shimoga, K.B.
1992 Finger force and touch feedback issues in dexterous telemanipulation. In *Proceedings of NASA-CIRSSE International Conference on Intelligent Robotic Systems for Space Exploration*, Troy, N.Y.

Skoglund, S.
1956 Anatomical and physiological studies of knee joint innervation in the cat. *Acta Physiologica Scandinavica* 36(Suppl 124):1-101.

Srinivasan, M.A.
1989 Surface deflection of primate fingertip under line load. *Journal of Biomechanics* 22(4):343-349.

Srinivasan, M.A., and J.-S. Chen
1993 Human performance in controlling normal forces of contact with rigid objects. Pp. 119-125 in H. Kazerooni, J.E. Colgate, and B.D. Adelstein, eds., *Advances in Robotics, Mechatronics and Haptic Interfaces*, ASME DSC Vol. 49.

Srinivasan, M.A., and K. Dandekar
1992 Role of fingertip geometry in transmission of tactile mechanical signals. *Advances in Bioengineering* 22:569-572.

Srinivasan, M.A., and R.H. LaMotte
1991 Encoding of shape in the responses of cutaneous mechanoreceptors. Pp. 1323-1332 in O. Franzen and J. Westman, eds., *Information Processing in the Somatosensory System*. Wenner-Gren International Symposium Series. London: Macmillan Press.
1994 Tactual discrimination of softness. Manuscript submitted to *Journal of Neurophysiology*.

Srinivasan, M.A., J.M. Whitehouse, and R.H. LaMotte
1990 Tactile detection of slip: Surface microgeometry and peripheral neural codes. *Journal of Neuorophysiology* 63(6):1323-1332.

Sutherland, I.E.
1965 The ultimate display. *Proceedings of the IFIP Congress* 2:506-508.

Talbot, W.H., I. Darian-Smith, H.H. Kornhuber, and V.B. Mountcastle
1968 The sense of flutter-vibration: Comparison of the human capacity with response patterns of mechanoreceptive afferents from the monkey hand. *Journal of Neurophysiology* 31:301-334.

Tan, H.Z., X.D. Pang, and N.I. Durlach
1992 Manual resolution of length, force and compliance. Pp. 13-18 in H. Kazerooni, ed., *Advances in Robotics*, ASME DSC Vol. 42.

Tan, H.Z., N.I. Durlach, Y. Shao, and M. Wei
1993 Manual resolution of compliance when work and force cues are minimized. Pp. 99-104 in H. Kazerooni, J.E. Colgate, and B.D. Adelstein, eds., *Advances in in Robotics, Mechatronics and Haptic Interfaces*, ASME DSC Vol. 49.

Tan, H.Z., M.A. Srinivasan, B.S. Eberman, and B. Cheng
1994 Human factors for the design of force-reflecting haptic interfaces. *Proceedings of ASME Winter Annual Meeting* (in press).

TiNi Alloy Company
1990 Sales literature. Oakland, Calif.

Trimmer, W.S., K.J. Gabriel, and R. Mahadevan
1987 Silicon Electrostatic Motors. Pp. 857-860 in *Proceedings of Transducers '87*, Tokyo, Japan.

Valéry, P.
1938 Discours d'ouverture au Congrès de Chirurgie, Paris, N.R.F. Cited on P. 82 in R. Tubiana, ed., *The Hand, Vol. 1*. Philadelphia, Pa., W.B. Saunders Co., 1981.

Vallbo, A.B., and K.E. Hagbarth
1968 Activity from skin mechanoreceptors recorded percutaneously in awake human subjects. *Experimental Neurology* 21:270-289.

CHAPTER 5: POSITION TRACKING AND MAPPING

Agronin, M.L.
1987 The design of a nine-string six-degree-of-freedom force-feedback joystick for telemanipulation. Pp. 341-348 in *Proceedings of the NASA Workshop on Space Telerobotics*, Pasadena, Calif.

Andersson, R.L.
1993 A real experiment in virtual environments: A virtual batting cage. *Presence: Teleoperators and Virtual Environments* 2:16-33.

Applewhite, H.
1994 A new VR positioning method yielding pseudo-absolute location. *VRST '94: The ACM Symposium on Virtual Reality Systems and Technology*.

Atkeson, C.G., and J.M. Hollerbach
1985 Kinematic features of unrestrained vertical arm movements. *Journal of Neuroscience* 5:2318-2330.

Atkinson, W.D., K.E. Bond, G.L. Tribble, and K.R. Wilson
1977 Computing with feeling. *Computers and Graphics* 2:97-103.

Badler, N.I., M.J. Hollick, and J.P. Granieri
1993 Real-time control of a virtual human using minimal sensors. *Presence: Teleoperators and Virtual Environments* 2:82-86.

Bahill, A.T., A. Brockenbrough, and B.T. Troost
1981 Variability and development of a normative data base for saccadic eye movements. *Investigative Ophthalmology and Visual Science* 21:116-125.

Beraldin, J.-A., M. Rioux, F. Blais, L. Cournoyer, and J. Domey
1992 Registered intensity and range imaging at 10 mega-samples per second. *Optical Engineering* 31:88-94.

Beraldin, J.-A., S.F. El-Hakim, and L. Cournoyer
1993 Practical range camera calibration. *Videometrics II*. SPIE Vol. 2067.

Besl, P.J.
1988 Active, optical range imaging sensors. *Machine Vision and Applications* 1:127-152.

Blais, F., M. Rioux, and J. Domey
1991 Optical range image acquisition for the navigation of a mobile robot. Pp. 2574-2580 in *Proceedings of the IEEE International Conference Robotics and Automation*, Sacramento, Calif.

Brooks, T.L.
1990a *Telerobot Response Requirements.* Report No. STX/ROB/90-03. Lanham, Md.: STX Robotics.
1990b Telerobot response requirements. Pp. 113-120 in *Proceedings of the IEEE Conference on Systems, Man, and Cybernetics,* Los Angeles, Calif.

Canepa, G., J.M. Hollerbach, and A.J.M. Boelen
1994 Kinematic calibration by means of a triaxial accelerometer. *Proceedings of the IEEE International Conference on Robotics and Automation,* San Diego, Calif.

Chen, Y.D., J. Ni, and S.M. Wu
1993 Dynamic calibration and compensation of a 3D laser radar scanning system. *IEEE International Conference Robotics and Automation* 3:652-658.

Durlach, N.I., R.W. Pew, W.A. Aviles, P.A. DiZio, and D.L. Zeltzer
1992 *Virtual Environment Technology for Training.* BBN Report No. 7661. Cambridge, Mass.: BBN Systems and Technologies.

Fischer, P., R. Daniel, and K. Siva
1990 Specification and design of input devices for teleoperation. Pp. 540-545 in *Proceedings of the IEEE International Conference Robotics and Automation,* Cincinnati, Ohio.

Fohanno, T.
1982 Assessment of the mechanical performance of industrial robots. Pp. 349-358 in *Proceedings of the 12th International Symposium on Industrial Robots,* Paris, France.

Foxlin, E.
1993 *Inertial Head-Tracking.* Electrical Engineering and Computer Science, Massachusetts Institute of Technology.

Foxlin, E., and N. Durlach
1994 An inertial head-orientation tracker with automatic drift compensation for use with HMDs. Paper submitted to *VRST '94: The ACM Symposium on Virtual Reality Systems and Technology.*

Franklin, G.F., J.P. Powell, and M.L. Workman
1990 *Digital Control of Dynamic Systems.* Reading, Mass.: Addison-Wesley.

Friedmann, M., T. Starner, and A. Pentland
1992 Synchronization in virtual realities. *Presence: Teleoperators and Virtual Environments* 1:139-144.

Hebert, M., and T. Kanade
1989 3-D vision for outdoor navigation by an autonomous vehicle. Pp. 208-226 in M. Brady, ed., *Robotics Science.* Cambridge, Mass.: MIT Press.

Held, R., and N. Durlach
1987 *Telepresence, Time Delay, and Adaptation.* NASA Conference Publication 10023.

Hollerbach, J.M., L. Giugovaz, M. Buehler, and Y. Xu
1993 Screw axis measurement for kinematic calibration of the Sarcos Dextrous Arm. Pp. 1617-1621 in *Proceedings of the IEEE/RSJ International Conference on Intelligent Robots and Systems,* Yokohama, Japan.

Inchingolo, P., and M. Spanio
1985 On the identification and analysis of saccadic eye movements—A quantitative study of the processing procedures. *IEEE Transactions on Biomedical Engineering* BME-32:683-695.

Inoue, H.
1993 Vision based behavior: observation and control of robot behavior by real-time tracking vision. In preprints of the 6th International Symposium Robotics Research, Hidden Valley, Pa.

Iwata, H.
1991 Force display for virtual worlds. Pp. 111-116 in *Proceedings of the International Conference Artificial Reality and Tele-existence.*
1992 Force displays for walkthrough simulation. Pp. 481-486 in *Proceedings of the Second International Symposium on Measurement and Control in Robotics*, Tsukuba Science City, Japan.

Jiang, B.C., J.Y. Black, and R. Duraisamy
1988 A review of recent developments in robot metrology. *Journal of Manufacturing Systems* 7:339-357.

Jones, L.A., I.W. Hunter, S. Lafontaine, J.M. Hollerbach, and R. Kearney
1991 Wide-bandwidth measurements of human elbow joint mechanics. In *Proceedings of the Canadian Medical and Biological Engineering Conference* 17:135-136.

Kanade, T.
1993 Very fast 3D sensing hardware. In preprints of the 6th International Symposium Robotics Research, Hidden Valley, Pa.

Kang, S.B., and K. Ikeuchi
1993 Toward automatic robot instruction from perception—Recognizing a grasp from observation. *IEEE Transactions on Robotics and Automation* 9:432-443.

Kawamura, S., and K. Ito
1993 A new type of master robot for teleoperation using a radial wire drive system. Pp. 55-60 in *Proceedings of the IEEE/RSJ International Conference on Intelligent Robots and Systems,* Yokohama, Japan.

Kyle, S.A.
1993 Non-contact measurement for robot calibration. Pp. 79-100 in R. Bernhardt and S.L. Albright, eds., *Robot Calibration.* London: Chapman and Hall.

Lau, K., R. Hocken, and L. Haynes
1985 Robot performance measurements using automatic laser tracking techniques. *Robotics and Computer-Integrated Manufacturing* 2:227-236.

Liang, J., C. Shaw, and M. Green
1991 On temporal-spatial realism in the virtual reality environment. Pp. 19-25 in *Fourth Annual Symposium on User Interface Software and Technology.*

Maclean, S.G., M. Rioux, F. Blais, J. Grodski, P. Milgram, H.F.L. Pinkney, and B.A. Aikenhead
1990 Vision system development in a space simulation laboratory. *Close-Range Photogrammetry Meets Machine Vision* SPIE 1395:8-15.

Maeda, T., and S. Tachi
1992 Development of light-weight binocular head-mounted displays. Pp. 281-288 in *Proceedings of the Second International Symposium on Measurement and Control in Robotics*, Tsukuba Science City, Japan.

Meyer, K., H.L. Applewhite, and F.A. Biocca
1992 A survey of position trackers. *Presence: Teleoperators and Virtual Environments* 1:173-200.

Mooring, B.W., Z.S. Roth, and M.R. Driels
1991 *Fundamentals of Manipulator Calibration*. New York: Wiley Interscience.

Mulligan, I.J., A.K. Mackworth, and P.D. Lawrence
1989 A model-based vision system for manipulator position sensing. Pp. 186-193 in *Proceedings of the Workshop on Interpretation of 3D Scenes*, Austin, TX.

Neilson, P.D.
1972 Speed of response or bandwidth of voluntary system controlling elbow position in intact man. *Medical and Biological Engineering* 10:450-459.

Payannet, D., M.J. Aldon, and A. Liegeois
1985 Identification and compensation of mechanical errors for industrial robots. Pp. 857-864 in *Proceedings of the 15th International Symposium on Industrial Robots*, Tokyo, Japan.

Prenninger, J., M. Vincze, and H. Gander
1993a Contactless position and orientation measurement of robot end-effectors. *IEEE International Conference Robotics and Automation* 1:180-185.
1993b Measuring dynamic robot movements in 6 DOF and real time. Pp. 124-153 in R. Bernhardt and S.L. Albright, eds., *Robot Calibration*. London: Chapman and Hall.

Rioux, M.
1984 Laser range finder based on synchronized scanners. *Applied Optics* 23:3837-3844.

Rohling, R., and J.M. Hollerbach
1993 Calibrating the human hand for haptic interfaces. *Presence: Teleoperators and Virtual Environments* 2(4).

Rohling, R., J.M. Hollerbach, and S.C. Jacobsen
1993 Optimized fingertip mapping: A general algorithm for robotic hand teleoperation. *Presence: Teleoperators and Virtual Environments* 2(3).

Sato, M.
1991 Virtual work space for 3-dimensional modeling. Pp. 103-110 in *Proceedings of the International Conference Artificial Reality and Tele-existence*.

Soechting, J.F., and M. Flanders
1989 Sensorimotor representations for pointing to targets in three-dimensional space. *Journal of Neurophysiology* 62:582-594.

Soechting, J.F., and F. Lacquaniti
1988 Quantitative evaluation of the electromyographic responses to multidirectional load perturbations of the human arm. *Journal of Neurophysiology* 59:1296-1313.

Stone, H.W.
1987 *Kinematic Modeling, Identification, and Control of Robotic Manipulators.* Boston: Kluwer Academic Publishers.

Truong, D.M., and S.E. Feldon
1987 Sources of artifact in infrared recording of eye movement. *Investigative Ophthalmology and Visual Science* 28(6):1018-1022.

Ward, M., R. Azuma, R. Bennett, S. Gottschalk, and H. Fuchs
1992 A demonstrated optical tracker with scalable work area for head-mounted display system. Pp. 43-52 in *Proceedings of the 1992 Symposium Interactive 3-D Graphics,* Cambridge, Mass.

Wells, M.J., and M.W. Haas
1990 Head movement during simulated air-to-air engagements. In R.J. Lewandowski, ed., *Helmet-Mounted Displays II* SPIE 1290:16-29.

Xu, Y., I.W. Hunter, J.M. Hollerbach, and D.J. Bennett
1991 An airjet actuator system for identification of the human arm joint mechanical properties. *IEEE Transactions on Biomedical Engineering* 38:1111-1122.

Zhuang, H., B. Li, Z.S. Roth, and X. Xie
1992 Self-calibration and mirror center offset elimination of a multi-beam laser tracking system. *Robotics and Autonomous Systems* 9:255-269.

CHAPTER 6: WHOLE-BODY MOTION, MOTION SICKNESS, AND LOCOMOTION INTERFACES

Barany, R.
1906 Untersuchungen uber den vom vestibular appat des chres reflek torisch ausgelosten rythmischer nystagmus und seine begleiterscheinunger. *Mschr Chrenheilkd* 40:193-197.

Berthoz, A., B. Pavard, and L.R. Young
1975 Perception of linear horizontal self-motion induced by peripheral vision (linear vection): Basic characteristics and visual-vestibular interactions. *Experimental Brain Research* 23:471-489.

Biocca, F.
1992 Will simulation sickness slow down the diffusion of virtual environment technology? *Presence* 1(3):334-343.

Bles, W.
1981 Stepping around: Circular vection and Coriolis effects. In J. Long and A. Baddeley, eds., *Attention and Performance* IX. Hillsdale, N.J.: Lawrence Erlbaum.

Bles, W., and T.S. Kapteyn
1977 Circular vection and human posture. 1. Does the proprioceptive system play a role? *Agressologie* 18:325-328.

Bles, W., T.S. Kapteyn, T. Brandt, and F. Arnold
1980 The mechanism of physiological height vertigo. II. Posturography. *Acta Otolaryngologica* Stockholm, Sweden 89:534-540.

Brandt, T., W. Büchele, and F. Arnold
1977 Arthrokinetic nystagmus and ego-motion sensation. *Experimental Brain Research* 30:331-338.

Calkins, D.S., M.F. Reschke, R.S. Kennedy, and W.P. Dunlop
1987 Reliability of provocative tests of motion sickness susceptibility. *Aviation Space and Environmental Medicine* 58(9 Suppl):A50-54.

Coats, A.C., and S.H. Smith
1967 Body position and the intensity of caloric nystagmus. *Acta Otolaryngologica* 63:515-532.

Cohen, B., J.-I. Suzuki, and T. Raphan
1977 Role of the otolith organs in generation of horizontal nystagmus: Effects of selective labyrinthine lesions. *Brain Research* 276:159-164.

Cowings, P.S., K.H. Naifeh, and W.B. Toscano
1990 The stability of individual patterns of autonomic responses to motion sickness stimulation. *Aviation, Space, and Environmental Medicine* 61:5.

Crampton, G.
1990 *Motion and Space Sickness*. Boca Raton, Fla.: CRC Press.

Dichgans, J., and T. Brandt
1978 Visual-vestibular interaction: Effects on self-motion perception and postural control. Pp. 755-804 in *Handbook of Sensory Physiology*, Vol 8. New York: Springer.

DiZio, P., and J.R. Lackner
1991 Motion sickness susceptibility in parabolic flight and velocity storage activity. *Aviation, Space, and Environmental Medicine* 62:300-307.

Fukuda, T.
1975 Postural behavior in motion sickness. *Acta Otolaryngologica* (Stockholm) 330:9-14.

Goodwin, G.M., D.I. McCloskey, and P.B.C. Matthews
1972 Proprioceptive illusions induced by muscle vibration: Contribution by muscle spindles to perception? *Science* 175:1382-1384.

Graybiel, A.
1952 The oculogravic illusion. *American Medical Association Ophthalmology* 48:605-615.

Graybiel, A., and D.I. Hupp
1946 The oculo-gyral illusion, a form of apparent motion which may be observed following stimulation of the semicircular canals. *Journal of Aviation Medicine* 17:3-27.

Graybiel, A., and J. Knepton
1976 Sopite syndrome: a sometimes sole manifestation of motion sickness. *Aviation Space and Environmental Medicine* 47(8):873-882.

Graybiel, A., and J.R. Lackner
1980 Evaluation of the relationship between motion sickness symptomatology and blood pressure, heart rate, and body temperature. *Aviation, Space, and Environmental Medicine* 51: 211-214
1983 Motion sickness: Acquisition and retention of adaptation effects in three motion environments. *Aviation, Space, and Environmental Medicine* 54:307-311.

1987 Treatment of severe motion sickness with antimotion sickness drug injections. *Aviation, Space, and Environmental Medicine* 58:773-776.

Graybiel, A., C. Wood, E. Miller, and D. Cramer
1968 Diagnostic criteria for grading the severity of acute motion sickness. *Aerospace Medicine* 39:453-455.

Guignard, J.C., and M.E. McCauley
1990 The accelerative stimulus for motion sickness. Pp. 123-152 in G. Crampton, ed., *Motion and Space Sickness*. Boca Raton, Fla.: CRC Press.

Held, R., J. Dichgans, and J. Bauer
1975 Characteristics of moving visual areas influencing spatial orientation. *Vision Research* 15:357-365.

Howard, I.P.
1982 *Human Visual Orientation*. New York: John Wiley.

Howard, I.P., and W.B. Templeton
1966 *Human Spatial Orientation*. London: John Wiley.

Kennedy, R.S., A. Graybiel, R.C. McDonough, and F.D. Beckwith
1968 Symptomatology under storm conditions in the North Atlantic in control subjects and in persons with labyrinthine defects. *Acta Otolaryngologica* 66:533.

Kennedy, R.S., L.J. Hettinger, and M.G. Lilienthal
1990 Simulator sickness. Pp. 317-341 in G.H. Crampton, ed., *Motion and Space Sickness*. Florida: CRC Press.

Kennedy, R.S., N.E. Lane, M.G. Lilienthal, K.S. Berbaum, and L.J. Hettinger
1992 Profile analysis of simulator sickness symptoms: Application to virtual environment systems. *Presence* 1(3):295-301.

Lackner, J.R.
1977 Induction of illusory self-rotation and nystagmus by a rotating sound-field. *Aviation, Space, and Environmental Medicine* 48: 129-131.
1985 Human sensory-motor adaptation to the terrestrial force environment. Pp. 175-210 in D. Ingle, M. Jeannerod, and D. Lee, eds., *Brain Mechanisms and Spatial Vision*. Amsterdam: Nijhoff.
1988 Some proprioceptive influences on the perceptual representation of body shape and orientation. *Brain* 111:281-297.
1989 Human orientation, adaptation, and movement control. Pp. 29-50 in National Reseach Council, *Motion Sickness, Visual Displays, and Armored Vehicle Design*. Washington, D.C.: National Academy Press.
1990 Sensory-motor adaptation to high force levels in parabolic flight maneuvers. Pp. 527-548 in M. Jeannerod, ed., *Attention and Performance*. Hillsdale, N.J.: Lawrence Erlbaum.
1992a Spatial orientation in weightless environments. *Perception* 21: 803-812.
1992b Sense of body position in parabolic flight. *Annals of the New York Academy of Sciences* 656:329-339.
1992c Multimodal and motor influences on orientation: Implications for adapting to weightless and virtual environments. *Journal of Vestibular Research* 2:307-322.
1993 Orientation and movement in unusual force environments. *Psychological Science* 4(3):134-142.

Lackner, J.R., and P. DiZio
1984 Some efferent and somatosensory influences on body orientation and oculomotor control. Pp. 283-301 in L. Spillman and B.R. Wooten, eds., *Sensory Experience, Adaptation, and Perception.* Hillsdale, N.J.: Lawrence Erlbaum.
1988 Visual stimulation affects the perception of voluntary leg movements during walking. *Perception* 17:71-80.
1989 Altered sensory-motor control of the head as an etiological factor in space motion sickness. *Perceptual and Motor Skills* 68:784-786.
1992 Sensory-motor calibration processes constraining the perception of force and motion during locomotion. *Posture and Gait: Control Mechanisms* 1:93-104.
1993 Spatial stability, voluntary action and causal attribution during self-locomotion. *Journal of Vestibular Research* 3:15-23.
1994 Rapid adaptation to Coriolis force perturbations of arm movement trajectory. To appear in *Journal of Neurophysiology.*

Lackner, J.R., and A. Graybiel
1981 Illusions of postural, visual, and aircraft motion elicited by deep knee bends in the increased gravitoinertial force phase of parabolic flight. *Experimental Brain Research* 44:312-316.
1984a Elicitation of motion sickness by head movements in the microgravity phase of parabolic flight maneuvers. *Aviation, Space, and Environmental Medicine* 55:513-520.
1984b Influence of gravitoinertial force level on apparent magnitude of Coriolis cross-coupled angular accelerations and motion sickness. *NATO-AGARD Aerospace Medical Panel Symposium on Motion Sickness: Mechanisms, Prediction, Prevention and Treatment* AGARD-CP-372, 22:1-7.
1984c Perception of body weight and body mass at twice earth-gravity acceleration levels. *Brain* 107:133-144.
1985 Head movements elicit motion sickness during exposure to microgravity and macrogravity acceleration levels. Pp. 170-176 in M. Igarashi and F.O. Black, eds., *Proceedings of the VII International Symposium: Vestibular and Visual Control on Posture and Locomotor Equilibrium.* Basel, Switzerland: Krager.
1986 The effective intensity of Coriolis, cross-coupling stimulation is gravitoinertial force dependent: Implications for space motion sickness. *Aviation, Space, and Environmental Medicine* 57:229-235.
1987 Head movements in low and high gravitoinertial force environments elicit motion sickness: Implications for space motion sickness. *Aviation, Space, and Environmental Medicine* 58:A212-A217.

Lackner, J.R., and M.S. Levine
1979 Changes in apparent body orientation and sensory localiztion induced by vibration of postural muscles: Viratory myesthetic illusions. *Aviation, Space, and Environmental Medicine* 50:346-354.

Lackner, J.R., and D. Lobovits
1978 Incremental exposure facilitates adaptation to sensory rearrangement. *Aviation, Space, and Environmental Medicine* 49:362-364.

Lackner, J.R., and R.A. Teixeira
1977 Optokinetic motion sickness: Continuous head movements attenuate the visual induction of apparent self-rotation and symptoms of motion sickness. *Aviation, Space, and Environmental Medicine* 48:248-253.

Lackner, J.R., A. Graybiel, and P. DiZio
1991 Altered sensory-motor control of the body as an etiological factor in space motion sickness. *Aviation, Space, and Environmental Medicine* 62:765-771.

Lawson, B.D., F.A. Sunahara, and J.R. Lackner
1991 Physiological responses to visually induced motion sickness. *Society for Neuroscience Abstracts, 21st Annual Meeting* 17(1):317.

McCauley, M.E., ed.
1984 *Research Issues in Simulator Sickness: Proceedings of a Workshop.* Committee on Human Factors, National Research Council, Washington, D.C.: National Academy Press.

McCauley, M.E., and T.J. Sharkey
1992 Cybersickness: Perception of self-motion in virtual environments. *Presence* 1(3):311-318.

Miller, E.F., and A. Graybiel
1972 Semicircular canals as a primary etiological factor in motion sickness. *Aerospace Medicine* 43:1065-1074.

Minor, L.B., and J.M. Goldberg
1990 Influence of static head position on the horizontal nystagmus evoked by caloric, rotational, and optokinetic stimulation in the squirrel monkey. *Experimental Brain Research* 82:1-13.

Previc, F.H., R.V. Kenyon, and K.K. Gillingham
1991 The Effects of Dynamic Visual Roll on Postural and Manual Control and Self-motion Perception. Paper presented at the 2nd Annual Scientific Meeting of the Aerospace Medical Society, Cincinnati, Ohio.

Reason, J.T., and J.J. Brand
1975 *Motion Sickness.* New York: Academic Press.

Robinson, D.A.
1977 Linear addition of optokinetic and vestibular signals in the vestibular nucleus. *Experimental Brain Research* 30:447-450.

Stern, R.M., K.L. Koch, W.R. Stewart, and I.M. Lindbland
1987 Spectral analysis of tachygastria recorded during motion sickness. *Gastroenterology* 92:92-97.

Teixeira, R.A., and J.R. Lackner
1979 Optokinetic motion sickness: Attenuation of visually-induced apparent self-rotation by passive head movements. *Aviation, Space, and Environmental Medicine* 50:264-266.

Warren, W.H.
1993 Perception and control of ego-motion. In W. Epstein and J.J. Rogers, eds., *Handbook of Perception and Cognition, Vol. 5: Perception of Space and Motion.* Orlando, Fla.: Academic Press.

Welch, R.B., B. Bridgeman, S. Anand, and K.E. Browman
1993 Alternating prism exposure causes dual adaptation and generalization to a novel displacement. *Perception and Psychophysics* 54:195-204.

CHAPTER 7: OTHER INTERFACE COMPONENTS: SPEECH, PHYSIOLOGY, SMELL, AND TASTE

Bahl, L.R., F. Jelinek, and R. L. Mercer
1983 A maximum likelihood approach to continuous speech recognition. *IEEE Transactions on Pattern Analysis and Machine Intelligence, PAMI-5* 179-190.

Cacioppo, J.T., and R.E. Petty
1983 *Social Psychophysiology: A Sourcebook.* New York: Guilford.

Cacioppo, J.T., and L.G. Tassinary
1990 Psychophysiology and psychophysiological inference. In J.T. Cacioppo and L.G. Tassinary, eds., *Principles of Psychophysiology: Physical, Social and Inferential Elements.* New York: Cambridge University Press.

Calhoun, G., G. McMillan, and P. Morton
1993 Brain actuated control: A candidate enabling technology. Submitted to *IEEE transactions on Rehabilitation Engineering.*

Cater, J.P., D.C. Varner, S.D. Huffman, K. Miller, and E. Peterson
1994 Development of a Computer-Controlled Olfactory and Radiant Heat Delivery System for Virtual Environments. Project 05-9766, Southwest Research Institute, San Antonio, Tex.

Chow, Y.L., R. Schwartz, S. Roucos, O. Kimball, P.Price, G. Kubala, M. Dunham, M. Krasner, and J. Makhoul
1986 The role of word-dependent coarticulatory effects in a phoneme-based speech recognition system. *Proceedings of ICASSP* 86:1593-1596. Tokyo, Japan.

Coles, M.G.H., G. Gratton, T.R. Bashore, C.W. Eriksen, and E. Donchin
1985 A psychophysiological investigation of the continuous flow model of human information processing. *Journal of Experimental Psychology: Human Perception and Performance* 11:529-553.

Davis, S., and P. Mermelstein
1980 Comparison of parametric representations for monosyllabic word recognition in continuously spoken sentences. *IEEE Transactions on Acoustics, Speech and Signal Processing,* ASSP-28 357-366.

Dixon, N.R., and T. Martin, eds.
1979 *Automatic Speech and Speaker Recognition.* New York: IEEE.

Donchin, E., D. Karis, T. Bashores, M. Coles, and G. Gratton
1986 Cognitive psychophysiology and human information processing. In M.G.H. Coles, E. Donchin, and S. Porges, eds. *Psychophysiology: Systems, Processes, and Applications.* New York: Guilford Press.

Druckman, D., and J.I. Lacey, eds.
1989 *Brain and Cognition: Some New Technologies.* Committee on New Technologies in Cognitive Psychophysiology, Commission on Behavioral

and Social Sciences and Education, National Research Council. Washington, D.C.: National Academy Press.

Farry, K.A., and I.D. Walker
1993 Myoelectric teleoperation of a complex robotic hand. *IEEE International Conference on Robotics and Automation* 3:502-509.

Hartman, B.O.
1980 *Technical Evaluation Report of the Aerospace Medical Panel Working Group WG-08 on Evaluation of Methods to Assess Workload.* Report No. AGARD-AR-139. Neuilly-Sur-Seine, France: Advisory Group for Aerospace Research and Development.

Hassett, J.
1978 *A Primer of Psychophysiology.* San Francisco: W.H. Freeman.

Hiraiwa, A., N. Uchida, N. Sonehara, and K. Shimohara
1992 EMG pattern recognition by neural networks for prosthetic fingers control "cyber finger." Pp. 535-542 in *Proceedings of the Second International Symposium on Measurement and Control in Robotics,* Tsukuba Science City, Japan, Nov. 16-19.

Jacobsen, S.C., D.F. Knutti, R.T. Johnson, and H.H. Sears
1982 Development of the Utah Artificial Arm. *IEEE Transactions on Biomedical Engineering* BME29:249-269.

Junker, A.M., J.H. Schnurer, D.F. Ingle, and C.W. Downey
1988 *Loop-Closure of the Visual-Cortical Response (U).* Summary report number AAMRL-TR-88-014. Wright-Patterson Air Force Base, Ohio: Armstrong Aerospace Medical Research Laboratory.
1989 Brain actuated control of a roll axis tracking simulator. Pp. 714-717 in *Proceedings of the 1989 IEEE International Conference on Systems, Man, and Cybernetics.*

Karis, D., M. Fabiani, and E. Donchin
1984 P300 and memory: Individual differences in the von Restorff effect. *Cognitive Psychology* 16:177-216.

Klatt, D.H.
1987 A review of text-to-speech conversion of English. *Journal of the Acoustical Society of America* 82:737-793.

Kramer, A.
1993 Event Related Potentials and Mental Workload. Unpublished paper presented at the Workshop on Virtual Environment and Teleoperator Technology, February 26-27, 1993, to the Committee on Virtual Reality Research and Development, Commission on Behavioral and Social Sciences and Education, National Research Council, Washington, D.C.

Krantz, D., and S. Manuck
1984 Acute psychophysiological reactivity and risk of cardiovascular disease: A review and methodological critique. *Psychological Bulletin* 96:465-490.

Makhoul, J., and R. Schwartz
1994 State of the art in continuous speech recognition. Pp. 165-198 in *Voice Communication Between Humans and Machines.* Washington, D.C.: National Academy Press.

Martin, J.H., J.C.M. Brust, and S. Hilal
1991 Imaging the living brain. Pp. 309-324 in Eric R. Kandel, James H. Schwartz, and Thomas M. Jessell, eds., *Principles of Neural Science, Third Edition*. New York: Elsevier.

Meek, S.G., J.E. Wood, and S.C. Jacobsen
1990 Model-based, multi-muscle EMG control of upper-extremity prostheses. Pp. 360-376 in J.M. Winters and S.L-Y. Woo, eds., *Multiple Muscle Systems: Biomechanics and Movement Organization*. New York: Springer-Verlag.

Neville, H.J., M. Kutas, G. Chesney, and A.L. Schmidt
1986 Event-related brain potentials during initial encoding and recognition memory of congruous and incongruous words. *Journal of Memory and Language* 25:75-92.

Nguyen, L., R. Schwartz, F. Kubala, and P. Placeway
1993 Search algorithms for software-only real-time recognition with very large vocabularies. Pp. 91-95 in *Proceedings of the ARPA Human Language Technology Workshop*, Plainsboro, N.J., March. San Mateo, Calif.: Morgan-Kaufmann.

O'Shaughnessy, D.
1987 *Speech Communication, Human and Machine*. Reading, Mass.: Addison-Wesley.

Pallett, D., J. Fiscus, W. Fisher, J. Garafolo, B. Lund, and M. Pryzbocki
1994 1993 Benchmark tests for the ARPA spoken language program. *Proceedings of the ARPA Human Language Technology Workshop*, Plainsboro, N.J., March. San Mateo, Calif.: Morgan-Kaufmann.

Rabiner, L.R.
1983 Tutorial on isolated and connected word recognition. In H.W. Sdchiissler, ed., *Signal Processing II: Theories and Applications*. New York: Elsevier Science Publishers.

Schwartz, R., and S. Austin
1991 A comparison of several approximate algorithms for finding multiple (n-best) sentence hypotheses. Pp. 701-704 in *Proceedings of ICASSP*, Toronto, Canada.

Smith, J.J., and J.P. Kampine
1984 *Circulatory Physiology: The Essentials*. Baltimore, Md.: Williams and Wilkins.

Walter, W.G., R. Cooper, A.J. Aldridge, W.C. McCallum, and A.L. Winter
1964 Contingent negative variation: An electrical sign of sensorimotor association and expectancy in the human brain. *Nature* 203:380-384.

Warner, D., T. Anderson, and J. Johanson
1994 Bio-Cybernetics: A biologically responsive interactive interface—The next paradigm of human computer interaction. Pp. 237-241 in *Proceedings of Medicine and Virtual Reality*. San Diego, Calif.

Weber, M.A., and I.M. Drayer
1984 *Ambulatory blood pressure monitoring*. Dannstadt, Germany: Steinhopff-Verlag.

Wierwille, W.W.
1979 Physiological measures of aircrew mental workload. *Human Factors* 21(5):579-594.

Zeffiro, T.
1993 Brain Imaging Technology. Unpublished paper presented at the Workshop on Virtual Environment and Teleoperator Technology, February 26-27, 1993, to the Committee on Virtual Reality Research and Development, Commission on Behavioral and Social Sciences and Education, National Research Council, Washington, D.C.

CHAPTER 8: COMPUTER HARDWARE AND SOFTWARE FOR THE GENERATION OF VIRTUAL ENVIRONMENTS

Airey, J.M., J.H. Rohlf, and F.P. Brooks, Jr.
1990 Towards image realism with interactive update rates in complex virtual building environments. *Computer Graphics* 24(2):41.

Akeley, K.
1993 Reality Engine Graphics. Paper submitted to SIGGRAPH.

Badler, N.I., B.L. Webber, J.K. Kalita, and J. Esakov
1991 Animation from instructions. Pp. 51-93 in N.I. Badler, B.A. Barsky, and D. Zeltzer, eds., *Making Them Move: Mechanics, Control, and Animation of Articulated Figures*. San Mateo, Calif.: Morgan-Kaufmann.

Badler, N.I., C. Phillips, and B.L. Webber
1993 *Simulating Humans: Computer Graphics, Animation, and Control*. Oxford, England: Oxford University Press.

Baraff, D.
1989 Analytical methods for dynamic simulation of non-penetrating rigid bodies. *Computer Graphics* 23(3):223-232.

Baraff, D., and A. Witkin
1992 Dynamic simulation of non-penetrating flexible bodies. *Computer Graphics* 26:303-308. (Proceedings of SIGGRAPH 1992.)

Barzel, R., and A.H. Barr
1988 A modeling system based on dynamic constraints. *Computer Graphics* 22:179-188.

Baumgarte, J.
1972 Stabilization of constraints and integrals of motion in dynamical systems. *Computer Methods in Applied Mechanics and Engineering* 1:1-16.

Beckett, W., and N. Badler
1993 Integrated Behavioral Agent Architecture. *Proceedings of the Workshop on Computer Generated Forces and Behavior Representation*, Orlando, Fla.

Bhargava, H.K., and W.C. Branley, Jr.
1993 What would Agax have observed? Or, introducing imperfections in the belief systems of autonomous agents. *Proceedings of the Twenty-sixth Hawaii International Conference on System Sciences Vol. III*, January, Kauai, Hawaii. (Los Alamitos, Calif: IEEE Computer Society Press.)

Blanchard, C., S. Burgess, Y. Harvill, J. Lanier, A. Lasko, M. Oberman, and M. Teitel

1990 Reality built for two: A virtual reality toolkit. *Computer Graphics* 24(2):35-36.

Blattner, M., D. Sumikawa, and R. Greenberg

1989 Earcons and icons: Their structure and common design principles. *Human-Computer Interaction* 4(1). Hillsdale, N.J.: Lawrence Erlbaum.

Brooks, F.P., Jr.

1986 Walkthrough—A dynamic graphics system for simulating virtual buildings. Pages 9-21 in *Proceedings of the 1986 Workshop on Interactive 3D Graphics*, October 23-24.

Brooks, R.A.

1989 The whole iguana. Pp. 432-456 in M. Brady, ed., *Robotics Science*. Cambridge, Mass.: MIT Press.

Bryson, S.

1992a Paradigms for the shaping of surfaces in a virtual environment. *Proceedings of HICSS*, Kauai, Hawaii.

1992b Virtual space-time. In *Proceedings of Visualization 92*, Boston.

Bryson, S., and M. Gerald-Yamasaki

1992 The distributed virtual wind tunnel. In *Proceedings of Supercomputing*, Minneapolis, Minn.

Bryson, S., and C. Levit

1992 Virtual wind tunnel: An environment for the exploration of three dimensional unsteady flows. *IEEE Computer Graphics and Applications* 12(4):25-34.

Catmull, E., L. Carpenter, and R. Cook

1984 Private and public communication. Communication in reference to the number of polygons required to render reality, making certain assumptions about depth complexity and display resolution.

Celniker, G., and D. Gossard

1992 Deformable curve and surface finite elements for free-form shape design. *IEEE Computer Graphics and Applications* 25(4):257-66.

Ching, W., and N. Badler

1992 Fast motion planning for anthropometric figures with many degrees of freedom. In *IEEE International Conference on Robotics and Automation*, May.

Clark, J.H.

1976 Hierarchical geometric models for visible surface algorithms. *Communications of the ACM* 19(10):547-554.

Cohen, M.F.

1992 Interactive space-time control for animation. In *IEEE International Conference on Robotics and Automation* 26(2):293-303.

Conner, D.B., S.S. Snibbe, K.P. Herndon, D.C. Robbins, R.C. Zelenik, and A. van Dam

1992 Three-dimensional widgets. Pp. 197-208 in *Proceedings of 1992 SIGGRAPH Symposium on Interactive 3D Graphics*.

DeHaemer, M.J., Jr., and M.J. Zyda
1991 Simplification of objects rendered by polygonal approximations. *Computers and Graphics* 15(2):175-184.

de Mey, V., and S. Gibbs
1993 A multimedia testbed. In *Proceedings of ACM Multimedia*, Anaheim, Calif.

Ellis, S.R., M.K. Kaiser, and A.J. Grunwald, eds.
1991 *Pictorial Communication in Virtual and Real Environments*. London: Taylor and Francis.

Esakov, J., and N.I. Badler
1991 An architecture for high-level human task animation control. In P.A. Fishwick and R.B. Modjeski, eds., *Knowledge-Based Simulation*. New York: Springer-Verlag.

Fisher, S.S., M. McGreevy, J. Humphries, and W. Robinett
1986 Virtual environment display system. Pp. 77-87 of *Proceedings ACM Workshop on Interactive Graphics.*

Ford, L.
1985 Intelligent computer aided instruction. Pp. 106-126 in M.Y.a.A. Narayanan, ed., *Artificial Intelligence: Human Effects*. New York: Wiley Halsted Press.

Frisch, U., D. d'Homieres, B. Hasslacher, P. Lallemand, Y. Pomeau, and J.-P. Rivet
1987 Lattice gas hydrodynamics in two and three dimensions. *Complex Systems* 1:649-701.

Fuchs, H., J. Goldfeather, J. Hultquist, S. Spach, J. Austin, F. Brooks, J. Eyles, and J. Poulton
1985 Fast spheres, shadows, textures, transparencies, and image enhancements in pixel-planes. Pp. 111-120 in *SIGGRAPH '89* 23(3).

Fuchs, H., J. Poulton, J. Eyles, T. Greer, J. Goldfeather, D. Ellsworth, S. Molnar, G. Turk, B. Tebbs, and L. Israel
1989 Pixel-planes 5: A heterogeneous multiprocessor graphics system using processor-enhanced memories. Pp. 79-88 in *SIGGRAPH '89* 23 (3).

Fuchs, H., G. Bishop, K. Arthur, L. McMillan, R. Bajcsy, S.W. Lee, H. Farid, and T. Kanade
1994 Virtual space teleconferencing using a sea of cameras. *Robotic Applications in Telemedicine*, September.

Funkhouser, T.A., C.H. Sequin, and S. Teller
1992 Management of large amounts of data in interactive building walkthroughs. P. 11 of *Proceedings of the Computer Graphics Symposium on Interactive 3D Graphics.*

Geib, C.
1993 A consequence of incorporating intentions in means-end planning. AAAI Spring Symposium Series, *Foundations of Automatic Planning: The Classical Approach and Beyond*, Stanford, Calif.

Hahn, J.K.
1988 Realistic animation of rigid bodies. *Computer Graphics* 22:299-308.

Held, R., and N. Durlach
1991 Telepresence, time delay, and adaptation. In S.R. Ellis, M.K. Kaiser, and A.C. Grunwald, eds., *Pictorial Communication in Real and Virtual Environments.* Bristol, Penn.: Taylor and Francis.

Holloway, R., H. Fuchs, and W. Robinett
1992 Virtual-worlds research at the University of North Carolina at Chapel Hill. Course notes, Course #9: Implementation of Immersive Virtual Environments, SIGGRAPH '92, Chicago, July.

Jacoby, R.H.
1992 Using virtual menus in a virtual environment. *Proceedings of the Symposium on Electronic Imaging Science and Technology,* Vol. 1668. International Society for Optical Engineering/Society for Imaging Science and Technology.

Kramer, G.
1992 Some organizing principles for representing data with sound. In G. Kramer, ed., *Auditory Display: Proceedings of ICAD '92, First International Conference on Auditory Display.* SFI Studies in the Sciences of Complexity, Proceedings Volume 18. Reading, Mass.: Addison-Wesley.

Lin, M.C., and F.C. Canney
1992 *Efficient Collision Detection for Animation.* Proceedings of the Third Eurographics Workshop on Animation and Simulation. Cambridge, England: Eurographics Workshop on Animation and Simulation.

Maes, P.
1990 Situated agents can have goals. *Journal of Robotics and Autonomous Systems* 6(1&2):49-70.

Magnenat-Thalmann, N., and D. Thalmann, eds.
1990 *Synthetic Actors in Computer-Generated 3D Films.* Berlin: Springer-Verlag.
1991 Complex models for animating synthetic actors. *IEEE Computer Graphics and Applications* 11(5):32-44.

Maiocchi, R., and B. Pernici
1990 Directing an animated scene with autonomous actors. Pp. 41-60 in N. Magnenat-Thalmann and D. Thalmann, eds., *Computer Animation.* Tokyo, Japan: Springer-Verlag.

McKenna, M., and D. Zeltzer
1990 Dynamic simulation of autonomous legged locomotion. *Computer Graphics* 24(4):29-38.

Metaxas, D.
1992 Physics-Based Modeling of Nonrigid Objects for Vision and Graphics. Ph.D. thesis, Department of Computer Science, University of Toronto.
1993 Fast dynamic point-to-point constraint algorithm for deformable bodies. In *Proceedings of the 2nd International Conference on Discrete Element Methods,* Massachusetts Institute of Technology, Cambridge, Mass.

Metaxas, D., and D. Terzopoulos
1992a Dynamic deformation of solid primitives with constraints. *Computer Graphics* 26(2):309-312.

1992b Flexible multibody dynamics techniques for shape and nonrigid motion estimation and synthesis. Pp. 147-155 in *Proceedings of ASME Symposium on the Dynamics and Control of Flexible Multibody Systems*, Anaheim, Calif.

1993 Shape and nonrigid motional estimation through physics-based synthesis. *Pattern Analysis and Machine Intelligence* 15(6):580-591.

Miller, G.P.

1988 The motion dynamics of snakes and worms. *Computer Graphics* 22:169-178.

Molnar, S., J. Eyles, and J. Poulton

1992 PixelFlow: High-speed rendering using image composition. Pp. 231-240 in *SIGGRAPH '92* 26(2).

Moore, M., and J. Wilhelms

1988 Collision detection and response for computer animation. *Computer Graphics* 22:289-298.

Ohya, J., Y Kitamura, H. Takemura, F. Kishino, and N. Terashima

1993 Real-time reproduction of 3D human images in virtual space teleconferencing. Pp. 408-414 in *IEEE Virtual Reality Annual International Symposium*, September 18-22.

Pentland, A., and J. Williams

1989 Good vibrations: Modal dynamics for graphics and animation. *Computer Graphics* 23(3):215-222.

Platt, J., and A. Barr

1988 Constraint methods for flexible models. *Computer Graphics* 22:279-288.

Raibert, M.H., and J.K. Hodgkins

1992 Animation of dynamic legged locomotion. *Computer Graphics* 25(4)349-58. .

Ridsdale, G.

1990 Connectionist modelling of skill dynamics. *Journal of Visualization and Computer Animation* 1(2):66-72.

Rijpkema, H., and M. Girard

1991 Computer animation of knowledge-based human grasping. Pp. 339-348 in *Proceedings of ACM SIGGRAPH 91*, Las Vegas, Nev.

Robinett, W., and R. Holloway

1992 Implementation of flying, scaling and grabbing in virtual worlds. *Computer Graphics* 26(3):189.

Schroeder, P., and D. Zeltzer

1990 The virtual erector set: Dynamic simulation with linear recursive constraint propagation. *Computer Graphics* 24(2):23-32.

Schroeder, W.J., J.A. Zarge, and W.E. Lorensen

1992 Decimation of triangle meshes. *Computer Graphics* 26(2):65-69.

Sheridan, T.B., and W.R. Ferrell

1974 *Man Machine Systems*. Cambridge, Mass.: MIT Press.

Snibbe, S.S., K.P. Herndon, D.C. Robbins, D.B. Conner, and A. van Dam

1992 Using deformations to explore 3D widget design. Pp. 351-352 in *Proceedings of SIGGRAPH '92*, Chicago, Ill.

Szeliski, R., and D. Tonnesen
1992 Surface modeling with oriented particle systems. *Computer Graphics* 26(2):185-94.
Teller, S.J., and C.H. Sequin
1991 Visibility preprocessing for interactive walkthroughs. *Computer Graphics* 25(4):61-69.
Terzopoulos, D., and D. Metaxas
1991 Dynamic 3D models with local and global deformations: Deformable superquadrics. *IEEE Trans. Pattern Analysis and Machine Intelligence* 13(7):703-714.
Toffoli, T.
1983 Cellular automata as an alternative to differential equations in modelling physics. Interdisciplinary Workshop, Los Alamos, N.M. March. p. 117.
Turk, G.
1992 Retiling polygonal surfaces. *Computer Graphics* 26(2):55-64.
Van de Panne, M., E. Fiume, and Z. Vranesic
1990 Reusable motion synthesis using state-space controllers. *Computer Graphics* 24(4):225-234.
Webber, B., N. Badler, F.B. Baldwin, W. Becket, B. DiEugenio, C. Geib, M. Jung, L. Levison, M. Moore, and M. White
1993 Doing what you're told: Following task instructions in changing, but hospitable environments. Submitted to *AI Journal*.
Welch, W., and A. Witkin
1992 Variational surface modeling. *Computer Graphics* 26(2):157-166.
Wilson, K., M. Zyda, and D. Pratt
1992 NPSGDL: An object oriented graphics description language for virtual world appplication support. In *Proceedings of the Third Eurographics Workshop on Object-Oriented Graphics*, Champery, Switzerland.
Witkin, A., M. Gleicher, and W. Welch
1990 Interactive dynamics. *Computer Graphics* 2:11-24.
Witkin, A., and M. Kass
1988 Space-time constraints, computer graphics. *Computer Graphics* 22:159-168.
Witkin, A., and W. Welch
1990 Fast animation and control of non-rigid structures. *Computer Graphics* 24(4):243-252.
Zeltzer, D.
1983 Knowledge-based animation. Pp. 187-192 in *Proceedings ACM SIGGRAPH/SIGART Workshop on Motion*.
Zyda, M.J., J.G. Monahan, and D.R. Pratt
1992 NPSNET: Physically-based modeling enhancements to an object file format. Pp. 35-52 in Nadia Magnenat-Thalmann and Daniel Thalmann, eds., *Creating and Animating the Virtual World*. Tokyo, Japan: Springer-Verlag.

Zyda, M.J., K.P. Wilson, D.R. Pratt, J.G. Monahan, and J.S. Falby
1993a NPSOFF: An object description language for supporting virtual world construction. *Computers and Graphics* 17(4):457-464.

Zyda, M.J., D.R. Pratt, W.D. Osborne, and J.G. Monahan
1993b NPSNET: Real-time collision detection and response. *The Journal of Visualization and Computer Animation* 4(1):13-24.

CHAPTER 9: TELEROBOTICS

Advanced Research Projects Agency
1992 Proceedings of the Image Understanding Workshop. Defense Advanced Research Projects Agency, San Diego, Calif., January.

Ahmad, S.
1988 Issues in the design of multiprocessor robot control hardware. Pp. 127-337 in *Proceedings of the IEEE Workshop on Special Computer Architecture for Robotics*.

An, C.H., C.G. Atkeson, and J.M. Hollerbach
1988 *Model-Based Control of a Robot Manipulator*. Cambridge, Mass.: MIT Press.

Andary, J.F., and P.D. Spidaliere
1993 The development test flight of the flight telerobotic servicer: Design description and lessons learned. *IEEE Transactions on Robotics and Automation* 9:664-674.

Anderson, R.J., and M.W. Spong
1988 Bilateral control of teleoperators with time delay. In *Proceedings of the IEEE International Conference on Systems, Man, & Cybernetics* 1:131.
1989 Bilateral control of teleoperators with time delay. *IEEE Transactions on Automatic Control* 34:494-501.

Anderson, T., and M. Donath
1991 Animal behavior as a paradigm for developing robot autonomy. Pp. 145-168 in P. Maes, ed., *Design Autonomous Agents*. Cambridge, Mass.: MIT Press.

Andriot, C., and R. Fournier
1991 On the bilateral control of teleoperators with flexible joints and time delay by the passive approach. In *Proceedings of the 5th ICAR*, Pisa, Italy.

Arai, T., R. Stoughton, and J.P. Merlet
1992 Teleoperator assisted hybrid control for parallel link manipulator and its application to assembly task. Pp. 817-822 in *Proceedings of the Second International Symposium on Measurement and Control in Robotics*, Tsukuba Science City, Japan, November 16-19.

Arkin, R.C.
1992 Behavior-based robot navigation for extended domains. *Adaptive Behavior* 1(2):201-225.

Armstrong-Helouvry, B.
1991 *Control of Machines with Friction*. Boston, Mass.: Kluwer Academic.

Asada, H., and S.-K. Lim
1985 Design of torque sensors and torque feedback control for direct-drive arms. Pp. 277-284 in *ASME Winter Annual Meeting: Robotics and Manufacturing Automation*, PED-Vol. 15, Miami Beach, Fla. November 17-22.

Asada, H., and K. Youcef-Toumi
1987 *Direct Drive Robots: Theory and Practice.* Cambridge, Mass.: MIT Press.

Aviles, W.A., T.W. Hughes, H.R. Everett, A.Y. Umeda, S.W. Martin, A.H. Koyamatsu, M.R. Solorzano, R.T. Laird, and S.P. McArthur
1990 Issues in mobile robotics: The unmanned ground vehicle program teleoperated vehicle (TOV). Pp. 587-597 in *SPIE Vol. 1388, Mobile Robots V.*

Backes, P., S. Hayati, V. Hayward, and K. Tso
1989 The KALI multi-arm robot programming and control environment. Pp. 173-182 in *Proceedings of the NASA Conference on Space Telerobotics.* Pp. 173-182.

Badler, N.I., B.A. Barsky, and D. Zeltzer, eds.
1991 *Making Them Move: Mechanics, Control, and Animation of Articulated Figures.* San Mateo, Calif.: Morgan-Kaufmann.

Baron, S., G. Zacharias, R. Muralhidaran, and R. Lancraft
1980 PROCRU: A model for analyzing flight crew procedures in approach to landing. Pp. 71-76 in *Proceedings of 8th IFAC Congress*, Tokyo, Japan. (See also NASA Report CR-152397.)

Bederson, B.B., R.S. Wallace, and E.L. Schwartz
1992 Two miniature pan-tilt devices. Pp. 658-663 in *IEEE International Conference on Robotics and Automation*, Nice, France, May 10-15.

Bejczy, A.K., and M. Handlykken
1981 Experimental results with a six-degree-of-freedom force reflecting hand controller. In *Proceedings of the 17th Annual Conference on Manual Control*, Los Angeles, Calif.

Bejczy, A.K., and J.K. Salisbury
1980 Kinesthetic coupling between operator and remote manipulator. In *Proceedings: ASME International Computer Technology Conference* 1:197-211.
1983 Kinesthetic coupling for remote manipulators. *CIME (Computers in Mechanical Engineering)* 2:48-62.

Bejczy, A.K., and Z. Szakaly
1987 Universal computer control system (UCCS) for space telerobots. Pp. 318-324 in *Proceedings of the IEEE International Conference on Robotics and Automation.*

Bejczy, A.K., B. Hannaford, and Z. Szakaly
1988 Multi-mode manual control in telerobotics. In *Proceedings of Romany '88*, Udine, Italy, September 12-15.

Belanger, P.R.
1992 Estimation of angular velocity and acceleration from shaft encoder measurements. Pp. 585-592 in *IEEE International Conference on Robotics and Automation*, Nice, France, May 10-15.

Besl, P.J.
1988 Range Image Sensors. Technical Report GMR-6090. General Motors Research Laboratory, March.

Biggers, K.B., S.C. Jacobsen, and C.C. Davis
1989 Linear analysis of a force reflective teleoperator. In *Proceedings of NASA Conference on Space Telerobotics*. Pasadena, Calif., January 31.

Book, W.J., and D.P. Hannema
1980 Master-slave manipulator performance for various dynamic characteristics and positioning task parameters. *IEEE Transactions on Systems, Man, and Cybernetics* 10:764-771.

Brenan, C.J.H., T.D. Doukoglou, I.W. Hunter, and S. Lafontaine
1993 Characterization and use of a novel position for microposition control of a linear motor. *Review of Scientific Instruments* 63:349-356.

Brooks, R.
1986 A robust layer control systems for a mobile robot. *IEEE Journal of Robotics and Automation* 2(1):14-23.
1990 A robot that walks: Emergent behaviors from a carefully evolved network. Pp. 29-39 in P.H. Winston and S.A. Shellard, eds., *Artificial Intelligence at MIT*. Cambridge, Mass.: MIT Press.
1991 Intelligence without representation. *Artificial Intelligence* 47(1-3):139-159.

Brooks, T.L., and I. Ince
1992 Operator vision aids for telerobotic assembly and servicing in space. Pp. 886-891 in *Proceedings of the IEEE International Conference on Robotics and Automation*, Nice, France, May 10-15.

Brown, R.H., S.C. Schneider, and M.G. Mulligan
1992 Analysis of algorithms for velocity estimation from discrete position versus time data. *IEEE Transactions on Industrial Electronics* 39:11-19.

Brunner, B., G. Hirzinger, K. Landzettel, and J. Heindl
1993 Multisensory shared autonomy and tele-sensor-programming—Key issues in the space robot technology experiment ROTEX. Pp. 2123-2131 in *Proceedings of the IEEE/RSJ International Conference on Intelligent Robots and Systems*, Yokohama, July 26-30.

Buehler, M., L. L. Whitcomb, F. Levin, and D. Koditscheck
1989 A distributed message passing computation and I/O enginge for real-time motion control. Pp. 478-483 in *Proceedings of the American Control Conference*.

Busby Associates, Inc.
1990 *Undersea Vehicles Directory - 1990-1991*. Arlington, Va.: Busby Assoc., Inc.

Charette, P.G., I.W. Hunter, and C.J.H. Brenan
1993 A complete high performance heterodyne interferometer displacement transducer for microactuator control. *Review of Scientific Instruments* 63:241-248.

Chen, J.B., R.S. Fearing, B.S. Armstrong, and J.W. Burdick
1986 NYMPH: A multiprocessor for manipulation applications. Pp. 1721-

1736 in *Proceedings of the IEEE International Conference on Robotics and Automation.*

Cheng, C.-C.
1991 Predictor Displays: Theory, Development and Application to Towed Submersibles. Doctoral thesis, Massachusetts Institute of Technology, June.

Chu, Y.Y., and W.B. Rouse
1979 Adaptive allocation of decision making responsibility between human and computer in multi-task situations. *IEEE Transactions on Systems, Man and Cybernetics* 9(12):769-788.

Clark, D.
1989 HIC: An operating system for hierarchical servo loops. Pp. 1004-1009 in *Proceedings of the IEEE International Conference on Robotics and Automation.*

Colgate, J.E.
1991 Power and Impedance Scaling in Bilateral Manipulation. Pp. 2292-2297 in *Proceedings of IEEE International Conference on Robotics and Automation,* Sacramento, Calif.

Colgate, J.E., and N. Hogan
1988 Robust control of dynamically interacting systems. *International Journal of Control* 48:65-88.

Connell, J.H.
1990 *Minimalist Mobile Robotics.* San Diego: Academic Press.

Conway, L., R. Volz, and M. Walker
1987 Teleautonomous systems: Methods and architectures for intermingling autonomous and telerobotic technology. Pp. 1121-1130 in *Proceedings 1987 IEEE International Conference on Robotics and Automation,* March 31-April 3, Raleigh, N.C.

Cutkosky, M., and J.M. Hyde
1993 Manipulation control with dynamic tactile sensing. In *6th International Symposium on Robotics Research,* Hidden Valley, Pa., Oct. 2-5.

Dario, P.
1989 Tactile sensing for robots: Present and future. Pp. 133-146 in O. Khatib, J.J. Craig, and T. Lozano-Perez, eds., *The Robotics Review* 1. Cambridge, Mass.: MIT Press.
1992 Microbiotics: Shifting robotics technology toward a different scale world. Pp. 343-358 in O. Khatib, J.J. Craig, and T. Lozano-Perez, eds., *The Robotics Review* 2. Cambridge, Mass.: MIT Press.

Dario, P., and D. De Rossi
1985 Tactile sensors and the gripping challenge. *IEEE Spectrum* 22(8):46-52.

Dario, P., R. Valleggi, M. Pardini, and A. Sabatini
1991 A miniature device for medical intracavity intervention. Pp. 171-175 in *Proceedings of the IEEE Workshop on Micro Electro Mechanical Systems.* Nara, Japan, Jan. 30-Feb. 2.

Das, H.
1989 Kinematic Control and Visual Display of Redundant Teleoperators. Ph.D. Thesis, Massachusetts Institute of Technology.

Das, H., H. Zak, W.S. Kim, A.K. Bejczy, and P.S. Schenker
1992 Operator performance with alternative manual control modes in teleoperation. *Presence* 1:201-218.

De Rossi, D., G. Canepa, F. Germagnoli, G. Magenes, A. Caiti, and T. Parisini
1993 Skin-like tactile sensor arrays for contact stress field extraction. *Materials Science and Engineering C: Biomimetic Materials, Sensors, and Systems* 1(1):23.

Desai, R.S., B. Wilcox, and R. Bedard
1992 Robotic vehicle overview. *Unmanned Systems* 10(4): 16-21.

deSilva, C.W.
1989 *Control Sensors and Actuators.* Englewood Cliffs, N.J.: Prentice-Hall.

Engelberger, J.L.
1989 *Robotics in Service.* Cambridge, Mass.: MIT Press.

Espiau, B.
1987 The advanced teleoperation project. Pp. 19-27 in A. Morecki, G. Bianchi, and K. Kedzior, eds., *RoManSy 6: Proceedings of the 6th CISM-IFToMM Symposium on Theory and Practice of Robots and Manipulators.* Cambridge, Mass.: MIT Press.

Everett, H.R., G.A. Gilbreath, T. Tran, and J.M. Nieusma
1990 Modeling the Environment of a Mobile Security Robot. Technical Document 1835, Naval Ocean Systems Center, San Diego, Calif.

Everett, H.R., E.H. Stitz, and D.E. DeMuth
1992 *Survey of a Collision Avoidance and Ranging Sensors for Mobile Robots.* Naval Command, Control and Ocean Surveillance Center, RDT&E Division. TR 1194, December.

Everett, H.R., G.A. Gilbreath, R.T. Laird, and T.A. Heath-Pastore
1993 Multiple Robot Host Architecture: Mobile Detection, Assessment, and Response System (MDARS). Technical Note 1710, Revision 1. NCCOSC RDT&E Division. San Diego, Calif.

Ferrell, W.R.
1965 Remote Manipulation with Transmission Delay. Pp. 24-32 in *IEEE Transactions on Human Factors in Electronics*, September.

Feynman, R.P.
1960 There's plenty of room at the bottom: An invitation to open up a new field of physics. *Engineering and Science* 23(Feb.):22-26.

Flash, T., and N. Hogan
1985 The coordination of arm movement: An experimentally confirmed mathematical model. *Journal of Neuroscience* 5(July):1688-1703

Flatau, C.R.
1973 The manipulator as a means of extending our dextrous capabilities to larger and smaller scales. Pp. 47-50 in *Proceedings of the 21st Conference on Remote Systems Technology.*

Fletcher, B.E., and J.L. Fuqua
1991 Advanced Security Vehicle (ASV) Major Bid and Proposal, Naval Ocean System Center. TR 1421. San Diego, Calif., April.

Freund, J.F.
1986 *New Technologies in U.S. Navy Undersea Vehicles.* Naval Sea Systems Command, Washington, D.C.

Fukuda, T., K. Tanie, and T. Mitsuka
1987 A new method of master slave type teleoperation for a micro-manipulator. In *Proceedings of the IEEE Micro Robots and Teleoperators Workshop,* Hyannis, Mass., November.

Funda, J., T.S. Lindsay, and R.P. Paul
1992 Teleprogramming: Toward delay-invariant remote manipulation. *Presence: Teleoperators and Virtual Environments* 1:29-44.

Goertz, R.C.
1964 Manipulator systems development at ANL. Pp. 117-136 in *Proceedings of the 12th Conference on Remote Systems Technology,* ANS, November.

Goertz, R.C., and W.M. Thompson
1954 Electronically controlled manipulator. *Nucleonics* (November):46-47.

Goldenberg, A.A., and D. Bastas
1990 On the bilateral control of force reflecting teleoperation. In *Proceedings IFAC '90.* August.

Gopinath, G., and S. Thakkar
1990 CHAOS: Why one cannot have only an operating system for real-time applications. *Computer* (June):60-69.

Gordon, A.
1993 Underwater laser line scan technology. Pp. 164-170 in *Underwater Intervention 93: The 11th Annual ROV Conference and Exposition,* Marine Technology Society, New Orleans, La.

Gosselin, C.M., and E. Lavoie
1993 On the kinetic design of spherical three-degree-of-freedom parallel manipulators. *International Journal of Robotics Research* 12:394-402.

Graves, S., J. Mollenhauer, and B.D. Morgan
1994 Dynamic session management for telerobotic control and simulation. To appear in the *Proceedings of the IEEE International Conference on Robotics and Automation.*

Griffin, K.A., and J.W. Murphy
1992 *Low Cost GPS/INS System for UAV Navigation.* AUVS-92. Huntsville. June.

Handlykken, M., and T. Turner
1980 Control system analysis and synthesis for a six degree-of-freedom universal force-reflecting hand controller. Pp. 1197-1205 in *Proceedings of the IEEE Conference on Decision and Control.*

Hannaford, B.
1989 A design framework for teleoperators with kinesthetic feedback. *IEEE Transactions on Robotics and Automation* 5:426-434.
1990 Scaling, impedance, and power flows in force reflecting teleoperation. *Robotics Research: Proceedings of ASME Winter Annual Meeting* 26:229-232.
1991 Kinesthetic feedback techniques in teleoperated systems. In C.

Leondes, ed., *Advances in Control and Dynamic Systems.* San Diego, Calif.: Academic Press.

Hannaford, B., and W.S. Kim

1989 Force reflection, shared control, and time delay in telemanipulation. In *Proceedings of the IEEE International Conference on Systems, Man, and Cybernetics.* Cambridge, Mass., November.

Hannaford, B., and Z. Szakaly

1992 Force-feedback cursor control. *NASA Tech Briefs* 16(5):item 21.

Hannaford, B., L. Wood, B. Guggisberg, D. McAffee, and H. Zak

1989 *Performance Evaluation of a Six-Axis Generalized Force-Reflecting Teleoperator.* JPL Publication 89-18, Pasadena, Calif., June 15.

Hannaford, B., L. Wood, D. McAffee, and H. Zak

1991 Performance evaluation of a six-axis generalized force reflecting teleoperator. In *IEEE Transactions on Systems, Man, and Cybernetics* 21:620-633.

Hashimoto, M., Y. Kiyosawa, H. Hirabayashi, and R.P. Paul

1991 A joint torque sensing technique for robots with harmonic drives. Pp. 1034-1039 in *Proceedings of the IEEE International Conference on Robotics and Automation*, Sacramento, Calif., April 9-11.

Hashimoto, T., T.B. Sheridan, and M.V. Noyes

1986 Effects of predicted information in teleoperation through a time delay. *Japanese Journal of Ergonomics* 22(2).

Hatamura, Y., and H. Morishita

1990 Direct coupling system between nanometer world and human world. Pp. 203-208 in *Proceedings of the IEEE Workshop on Micro Electro Mechanical Systems*, Napa Valley, Calif., Feb. 11-14.

Hayati, S., T. Lee, K. Tso, P. Backes, and J. Lloyd

1990 A testbed for a unified teleoperated-autonomous dual-arm robotic system. Pp. 1090-1095 in *Proceedings of the IEEE International Conference on Robotics and Automation.* Cincinnati, Ohio, May 13-18.

Hayati, S., H. Seraji, W.S. Kim, and J. Balarum

1992 Remote surface inspection for space applications. *AIAA 92-1017.* AIAA Aerospace Design Conference, Irvine, Calif., Feb. 3-6.

Hayward, V., and R.P. Paul

1984 Introduction to RCCL: a robot control library. Pp. 283-294 in *Proceedings of the IEEE International Conference on Robotics and Automation.*

Hayward, V., N. Chafye, X. Chen, and B. Duplat

1993 Kinematic decoupling in mechanisms and application to a passive hand controller design. *Journal of Robotic Systems* 10(5).

Heath, T.A.

1992 Human-machine interface development for teleoperated vehicles: Improved attitude awareness with gravity references sensors. Technical report in process by the Naval Command, Control and Ocean Surveillance Center.

Hebert, M., T. Kanade, and I. Kweon

1988 3-D Vision Techniques for Autonomous Vehicles. Technical report CMU-RI-TR-88-12. Carnegie Mellon University.

Held, R.M., and N.I. Durlach
1992 Telepresence. *Presence: Teleoperators and Virtual Environments* 1(1):109-112.

Henrion, W., et al.
1990 Wide dynamic range direct digital accelerometer. Pp. 153-157 in *IEEE Solid-State Sensor and Actuator Workshop*, Hilton Head.

Hill, J.W.
1979 Study of modeling and evaluation of remote manipulation tasks with force feed-back. Final Report 5,6 JPL Contract 95-5718, Stanford Research Institute.

Hirose, M., and K. Yokoyama
1992 VR application for transmission of synthetic sensation. Pp. 145-154 in *Proceedings of the Second International Conference on Artificial Reality and Tele-Existence*, Tokyo, Japan.

Hirose, S., and K. Yoneda
1989 Robotic sensors with photodetecting technology. Pp. 271-287 in *Proceedings of the International Symposium on Industrial Robots*, Tokyo, Japan.
1990 Development of optical 6-axial force sensor and its signal calibration considering non-linear interference. Pp. 46-53 in *Proceedings of the IEEE International Conference on Robotics and Automation*, Cincinnati, Ohio.

Hirose, M., K. Hirota, and K. Yokoyama
1992 A study on the transmission of synthetic sensation. Pp. 487-492 in *Proceedings of the Second International Symposium on Measurement and Control in Robotics*, Tsukuba Science City, Japan.

Hirzinger, G.
1987 The space and telerobotic concepts of the DFVLR ROTEX. Pp. 443-449 in *Proceedings of the IEEE International Conference on Robotics and Automation*, Raleigh, N.C.

Hirzinger, G., and J. Dietrich
1986 Multisensory robots and sensor based path generation. Pp. 1992-2001 in *Proceedings of the IEEE International Conference on Robotics and Automation*, San Francisco, Calif.

Hirzinger, G., B. Brunner, J. Dietrich, and J. Heindl
1993 Sensor-based space robotics ROTEX and its telerobotic features. *IEEE Transactions on Robotics and Automation* 9: 649-663.

Hogan, N.
1985a Impedance control: An approach to manipulation: Part I—Theory. *Journal of Dynamic Systems, Measurement, and Control* 107:1-7.
1985b Impedance control: An approach to manipulation: Part II—Implementation. *Journal of Dynamic Systems, Measurement, and Control* 107:8-16.
1985c Impedance control: An approach to manipulation: Part III— Applications. *Journal of Dynamic Systems, Measurement, and Control* 107:17-24.

Hollerbach, J.M.
1982 A recursive formulation of manipulator dynamics and comparative

study of dynamic formulations and complexity. Pp. 73-87 in Brady et al., eds, *Robot Motion*, Cambridge, Mass.: MIT Press.

1985 Optimum kinetic design for a seven degree of freedom manipulator. Pp. 215-222 in H. Hanafusa and H. Inoue, eds., *Robotics Research: The Second International Symposium*. Cambridge, Mass.: MIT Press.

1987 Robot hands and tactile sensing. Pp. 317-342 in W.E.L. Grimson and R.S. Patil, eds., *AI in the 1980's and Beyond*. Cambridge, Mass.: MIT Press.

Hollerbach, J.M., I.W. Hunter, and J. Ballantyne

1992 A comparative analysis of actuator technologies for robotics. Pp. 299-342 in O. Khatib, J.J. Craig, and T. Lozano-Perez, eds., *The Robotics Review 2*. Cambridge, Mass.: MIT Press.

Hollerbach, J.M., J. Lang, E. Vaaler, I. Garabieta, R. Sepe, S. Umans, and I.W. Hunter

1993 The McGill/MIT direct drive motor project. In *Proceedings of the IEEE International Conference on Robotics and Automation*, Atlanta, Georgia.

Hollis, R.L., S. Salcudean, and D.W. Abraham

1990 Toward a telenanorobotic manipulation system with atomic scale force feedback and motion resolution. Pp. 115-119 in *Proceedings of the IEEE Workshop on Micro Electro Mechanical Systems*, Napa Valley, Calif.

Hong, J., and X. Tan

1989 Calibrating a VPL DataGrove for teleoperating the Utah/MIT hand. Pp. 1752-1757 in *Proceedings of the IEEE International Conference on Robotics and Automation*, Scottsdale, Ariz.

Howe, R.D., and M.R. Cutkosky

1992 Touch sensing for robotic manipulation and recognition. Pp. 55-112 in O. Khatib, J.J. Craig, and T. Lozano-Perez, eds., *The Robotics Review 2*. Cambridge, Mass.: MIT Press.

Hunter, I.W., S. Lafontaine, P.M.F. Nielsen, P.J. Hunter, and J.M. Hollerbach

1990 Manipulation and dynamic mechanical testing of microscopic objects using a tele-microbot system. *IEEE Control Systems Magazine* 10(2):3-9.

Hunter, I.W., S. Lafontaine, J.M. Hollerbach, and P.J. Hunter

1991 Fast reversible NiTi fibers for use in microbotics. Pp. 166-170 in *Proceedings of the IEEE Workshop on Micro Electro Mechanical Systems*, Nara, Japan.

Ikeuchi, K.

1993 Assembly Plan from Observation. In *6th International Symposium on Robotics Research*, Hidden Valley, Pa.

Ikuta, K., M. Tsukamoto, and S. Hirose

1988 Shape memory alloy servo actuator system with electric resistance feedback and application for active endoscope. Pp. 427-430 in *Proceedings of the IEEE International Conference on Robotics and Automation*, Philadelphia, Pa.

Izaguirre, A., M. Hashimoto, R.P. Paul, and V. Hayward

1992 A new computational structure for real-time dynamics. *International Journal of Robotics Research* 11(4):346-362.

Jacobsen, S.C., E.K. Iversen, D.F. Knutti, R.T. Johnson, and K.B. Biggers
1986 Design of the Utah/MIT dextrous hand. Pp. 1520-1532 in *Proceedings of the IEEE International Conference on Robotics and Automation*, San Francisco, Calif.

Jacobsen, S.C., R.H. Price, J.E. Wood, T.H. Rytting, and M. Rafaelof
1989 The wobble motor: Design, fabrication, and testing of an eccentric-motion electrostatic microactuator. Pp. 1536-1546 in *Proceedings of the IEEE International Conference on Robotics and Automation*, Scottsdale, Ariz.

Jacobsen, S.C., E.K. Iverson, C.C. Davis, D.M. Potter, and T.W. McLain
1990a Design of a multiple degree of freedom, force reflective hand master/slave with a high mobility wrist. In *3rd Topical Meeting on Robotics and Remote Systems*, Charleston, S.C.

Jacobsen, S.C., F.M. Smith, E.K. Iverson, and D.K. Backman
1990b High performance, high dexterity, force reflective teleoperator. In *Proceedings of the 38th Conference on Remote Systems Technology*, Washington, D.C.

Jacobsen, S.C., F.M. Smith, D.K. Backman, and E.K. Iverson
1991 High performance, high dexterity, force reflective teleoperator II. In *ANS Topical Meeting on Robotics and Remote Systems*, Albuquerque, N.M.

Jansen, J.F., and J.N. Herndon
1990 Design of a telerobotic controller with joint torque sensors. Pp. 1109-1115 in *Proceedings of the IEEE International Conference on Robotics and Automation*, Cincinnati, Ohio.

Jansen, J.F., and R.L. Kress
1991 Control of a teleoperator system with redundancy based on passivity conditions. Pp. 478-484 in *Proceedings of the IEEE International Conference on Robotics and Automation*, Sacramento, Calif.

Jau, B.M.
1992 Man-equivalent telepresence through four-fingered human-like hand system. Pp. 843-848 in *IEEE International Conference on Robotics and Automation*, Nice, France.

Jones, J.L., and A.M. Flynn
1993 *Mobile Robots: Inspiration to Implementation.* Wellesley, Mass.: A.K. Peters.

Jones, L.A., and I.W. Hunter
1990 Influence of the mechanical properties of a manipulandum on human operator dynamics: 1. Elastic stiffness. *Biological Cybernetics* 62:299-307.
1992 Human operator perception of mechanical variables and their effects on tracking performance. In *ASME Winter Annual Meeting: Issues in the Development of Kinesthetic Displays for Teleoperation and Virtual Environments*, Anaheim, Calif.
1993 Influence of the mechanical properties of a manipulandum on human operator dynamics: 2. Viscosity. *Biological Cybernetics* (in press).

Juliussen, E.
1994 Which low-end workstation? *IEEE Spectrum* 31(4):51-59.

Kan, E.P., J. Lee, and M. Junod
1990 An integrated FRHC-PUMA-EE teleautonomous system. Pp. 397-404 in M. Jamshidi and M. Saif, eds., *Robotics and Manufacturing* (vol.3). New York: ASME Press.

Kanade, T.
1992 CMU image understanding program. Pp. 51-61 in *DARPA Image Understanding Workshop*, San Diego, Calif.

Kazerooni, H., and J. Guo
1993 Human extenders. *Journal of Dynamic Systems, Measurements, and Control* 115(2B):281-290.

Kazerooni, H., T.I. Tsay, and K. Hollerbach
1993 A controller design framework for telerobotic systems. *IEEE Transactions on Control Systems Technology* 1:50-62.

Kelley, A.J., and S.E. Salcudean
1993 MagicMouse: Tactile and kinesthetic feedback in the human-computer interface using an electromagnetically actuated input/output device. *IEEE Transactions on Robotics and Automation* (in press).

Khoshzaban, M., F. Sassani, and P.D. Lawrence
1992 Autonomous kinematic calibration of industrial hydraulic manipulators. In M. Jamshidi, R. Lumia, J. Mullins, and M. Shahinpoor, eds., *Robotics and Manufacturing* 4:577-584. New York: ASME Press.

Khosla, P.K.
1988 Some experimental results on model-based control schemes. In *Proceedings of the IEEE International Conference on Robotics and Automation* 3:1380-1385. Philadelphia, Pa.

Kim, W.S., F. Tendick, S.R. Ellis, and L.W. Stark
1987 A comparison of position and rate control for telemanipulations with consideration of manipulator system dynamics. *IEEE Journal of Robotics and Automation* 3:426-436.

Kim, W.S., B. Hannaford, and A.K. Bejczy
1992 Force-reflection and shared complaint control in operating telemanipulators with time delay. In *IEEE Transactions on Robotics and Automation* 8:176-185.

Kim, W.S., K.S. Tso, and S. Hayati
1993 An operator interface design for a telerobotic inspection system. *AIAA 93-1160*. AIAA Aerospace Design Conference, Irvine, Calif.

Kruth-Microwave Electronics Company
1989 Final Report Phase I. Naval Sea Systems Command NSSC Contract No. N00024-88-C-5155. Hanover, Md.

Kvasnica, M.
1992 Six-component force-torque sensing by means of one square CCD or PSD element. Pp. 213-219 in *Proceedings of the 2nd International Symposium on Measurement and Control in Robotics*, Tsukuba Science City, Japan.

Laird, R.T., H.R. Everett, and G.A. Gilbreath
1993 A host architecture for multiple robot control. In *American Nuclear*

Society 5th Topical Meeting on Robotics and Remote Handling, Knoxville, Tenn.

Landsberger, S., and T.B. Sheridan
1985 A new design for parallel link manipulators. In *Proceedings of the IEEE International Conference on Cybernetics and Society*, Tuscon, Ariz.

Lathrop, R. H.
1985 Parallelism in manipulator dynamics. *International Journal of Robotics Research* 4(2):346-362. Cambridge, Mass.: MIT Press.

Latombe, J.-C.
1991 *Robot Motion Planning*. Norwell, Mass.: Kluwer Academic.

Laughlin, D.R., A.A. Ardaman, and H.R. Sebesta
1992 Inertial angular rate sensors: Theory and applications. *Sensors* 9(10):20-24.

Lawn, C.A., and B. Hannaford
1993 Performance testing of passive communication and control in teleoperation with time delay. *IEEE International Conference on Robotics and Automation* 3:776-783.

Lee, T.S., S. Hayati, K.S. Tso, and P.G. Backes
1990 Dual-arm voice-controlled force-reflecting teleoperation system. Pp. 419-426 in M. Jamshidi and M. Saif, eds., *Robotics and Manufacturing* (vol. 3). New York: ASME Press.

Long, G.L., and C.L. Collins
1992 A pantograph linkage parallel platform master hand controller for force reflection. Pp. 390-395 in *IEEE International Conference on Robotics and Automation*, Nice, France.

Luh, J.Y.S., M.W. Walker, and R.P.C. Paul
1980 Online computational scheme for mechanical manipulators. *Journal of Dynamic Systems, Measurement, and Control* 102:69-70.

Luh, J.Y.S., W.D. Fisher, and R.P.C. Paul
1983 Joint torque control by a direct feedback for industrial robots. Pp. 153-161 in *IEEE Transactions on Automatic Control* AC-28.

Ma, D., J.M. Hollerbach, and Y. Xu
1994 Gravity based autonomous calibration for robot manipulators. In *Proceedings of the IEEE International Conference on Robotics and Automation*.

Machida, K., Y. Toda, T. Iwata, M. Kawachi, and T. Nakamura
1988 Development of a graphic simulator augmented teleoperator system for space applications. Pp. 358-364 in *Proceedings of 1988 AIAA Conference on Guidance, Navigation, and Control*.

Mack, B., S. McClure, and R. Ravindran
1991 A ground testbed for evaluating concepts for the special purpose dextrous manipulator. Pp. 884-889 in *Proceedings of the IEEE International Conference on Robotics and Automation*, Sacramento, Calif.

Maekawa, H., K. Tanie, K. Komoriya, M. Kaneko, K. Horiguchi, and T. Sugawara
1992 Development of a finger-shaped tectile sensor and its evaluation by active touch. Pp. 1327-1334 in *IEEE International Conference on Robotics and Automation*, Nice, France.

Mason, M.T.
1981 Compliance and force control for computer controlled manipulators. Pp. 418-432 in *IEEE Transactions on Systems, Man, and Cybernetics*, Vol. SMC-11.
1989 Robotic manipulation: Mechanics and planning. Pp. 262-290 in M. Brady, ed., *Robotics Science*. Cambridge, Mass.: MIT Press.
McCammon, I.D., and S.C. Jacobsen
1990 Tactile sensing and control for the Utah/MIT hand. Pp. 239-266 in S.T. Venkataraman and T. Iberall, eds., *Dextrous Robot Hands*. New York: Springer-Verlag.
McGovern, D.W.
1990 Experiences and Results in Teleoperation of Land Vehicles. Sandia National Laboratory Report SAND90-0299.UC-515, April.
McLean, G.F., B. Prescott, and R. Podhorodeski
1994 Teleoperated system performance evaluation. *IEEE Transactions on Systems, Man, and Cybernetics* 24:796-804.
Mettala, E.G.
1992 CMU Image Understanding Program. Pp. 159-171 in *DARPA Image Understanding Workshop*.
Milgram, P., S. Zhai, and D. Drascic
1993 Applications of augmented reality for human-robot communication. Pp. 1467-1472 in *Proceedings of the IEEE/RSJ International Conference on Intelligent Robots and Systems*, Yokohama, Japan, July 26-30.
Millitech Corporation
1989 Modular Millimeter Wave FMCW Sensor. Final Report for Naval Sea Systems Command NSSC Contract No. N00024-88-C-5144. South Deerfield, Mass.
Mitsuishi, M., S. Warisawa, Y. Hatamura, T. Nagao, and B. Kramer
1992 Trial of a remote reality-based manufacturing system in Japan operated from the United States. Pp. 1481-1498 in *Japan-USA International Workshop on Flexible Automation*.
Morinaga, W.S., and R.T. Hoffman
1991 Advanced Tethered Vehicle. IEEE 07803-0202-8/91/0000-1268.
Nakamura, Y., and H. Hanafusa
1986 Inverse kinematic solutions with singularity robustness for robot manipulator control. *Journal of Dynamic Systems, Measurement, and Control* 108:163-171.
Nakamura, Y., T. Yoshikawa, and I. Futamata
1988 Design and signal processing of six-axis force sensors. Pp. 75-82 in R. Bolles and B. Roth, eds., *Robotics Research: The Fourth International Symposium*. Cambridge, Mass.: MIT Press.
Narasimham, S., D.M. Siegel, and J.M. Hollerback
1988 Condor: A revised architecture for controlling the Utah-MIT hand. Pp. 446-449 in *Proceedings of the IEEE International Conference on Robotics and Automation*.
Newman, W.S., and J.J. Patel
1991 Experiments in torque control of the Adept One robot. Pp. 1867-1872

in *Proceedings of the IEEE International Conference on Robotics and Automation*, Sacramento, Calif.

Nicholls, H.R., and M.H. Lee
1989 A survey of robot tactile sensing technology. *International Journal of Robotics Research* 8(3):3-30.

Nielsen, P., I. Hunter, S. Lafontaine, and S. Martel
1989 A parallel computation and control computer for microrobotics. Pp. 472-477 in *Proceedings of the American Control Conference.*

Niemeyer, G., and J.J. Slotine
1991 Stable adaptive teleoperation. *IEEE Journal of Oceanic Engineering* 16:152-162.

Niino, R., S. Egawa, and T. Higuchi
1993 High-power and high-efficiency electrostatic actuator. Pp. 236-241 in *Proceedings of the IEEE Workshop on Micro Electro Mechanical Systems,* Fort Lauderdale, Fla.

Novak, J.L., and J.T. Feddema
1992 A capacitance-based proximity sensor for whole arm obstacle avoidance. Pp. 1307-1314 in *IEEE International Conference on Robotics and Automation*, Nice, France.

Noyes, M.V.
1984 Superposition of Graphics on Low Bit-rate Video as an Aid to Teleoperation. Master's thesis, Massachusetts Institute of Technology.

Okada, T.
1982 Development of an optical distance sensor for robots. *International Journal of Robotics Research* 1(4):3-14.

Okada, T., and U. Rembold
1991 Proximity sensor using a spiral-shaped light-emitting mechanism. *IEEE Transactions Robotics and Automation* 7:798-805.

Ollendorf, S.
1990 Robotics for space applications. Pp. 413-418 in M. Jamshidi and M. Saif, eds., *Robotics and Manufacturing*, Vol. 3. New York: ASME Press.

Ouh-young, M., M. Pique, J. Hughes, N. Srinivasan, and F.P. Brooks, Jr.
1988 Using a manipulator for force display in molecular docking. Pp. 1824-1829 in *Proceeding of the IEEE International Conference on Robotics and Automation.* Philadelphia, Pa.

Oyama, E., N. Tsunemoto, S. Tachi, and Y. Inoue
1993 Remote manipulation using virtual environment. *Presence: Teleoperators and Virtual Environments* 2:112-124.

Paden, B., and S. Sastry
1988 Optimal kinematic design of 6R manipulators. *International Journal of Robotics Research* 7(2):43-61.

Pao, L., and T.H. Speeter
1989 Transformation of human hand positions for robotic hand control. Pp. 1758-1763 in *Proceedings of the IEEE International Conference on Robotics and Automation,* Scottsdale, Ariz.

Papanikolopoulos, N.P., and P.K. Khosla
1992 Shared and traded telerobotic visual control. Pp. 878-885 in *IEEE International Conference on Robotics and Automation*, Nice, France.

Park, J.H.
1991 Supervisory Control of Robot Manipulators for Gross Motions. Ph.D. thesis, Massachusetts Institute of Technology.

Parker, N.R., S.E. Salcudean, P.D. Lawrence
1993 Application of Force Feedback to Heavy Duty Machines. Proceedings of *IEEE International Conference on Robotics and Automation* 3:375-381.

Partaatmdja, O., B. Benhabib, A. Sun, and A.A. Goldenberg
1992 An electrooptic orientation sensor for robotics. *IEEE Transactions Robotics and Automation* 8:111-119.

Paul, H.A., B. Mittlestadt, W.L. Bargar, B. Musits, R.H. Taylor, P.Kazanzides, B.Williamson, and W. Hanson
1992 A surgical robot for total hip replacement. Pp. 606-611 in *IEEE International Conference on Robotics and Automation*, Nice, France.

Pennington, J.E.
1986 Space telerobotics: A few more hurdles. Pp. 813-816 in *Proceedings on the IEEE International Conference on Robotics and Automation*, San Francisco.

Pepper, R.L., and J.D. Hightower
1984 Research Issues in Teleoperator Systems. A paper presented at the 28th Annual Human Factors Society Meeting, San Antonio, Texas.

Pfeffer, L., O. Khatib, and J. Hake
1989 Joint torque sensory feedback in the control of a PUMA manipulator. *IEEE Transactions Robotics and Automation* 5:418-425.

Pugh, A., ed.
1986 *Robot Sensors, Volume 2—Tactile and Non-Vision*. New York: Springer-Verlag.

Raibert, M.H.
1990 Pp. 149-179 in Winston and Shellard, eds., *Legged Robots, Artificial Intelligence at MIT*. Cambridge, Mass.: MIT Press.

Raibert, M.H., and J.J. Craig
1981 Hybrid position/force control of manipulators. *ASME Journal of Dynamic Systems, Measurement, and Control* 102:126-133.

Raju, G.J.
1988 Operator Adjustable Impedance in Bilateral Remote Manipulation. Ph.D. dissertation, Department of Mechanical Engineering, Massachusetts Institute of Technology.

Ramstein, C., and V. Hayward
1994 The Pantograph: A large workspace haptic device for a multimodal human-computer interaction. *Computer-Human Interaction CHI'94, ACM*, Boston, April 24-28.

Rasmussen, D.
1992 A natural visual interface for precision telerobot control. Pp. 170-179 in *SPIE Vol. 1833 Telemanipulator Technology*, Boston, November.

Rizzi, A., L. Whitcomb, and D. Koditschek
1992 Distributed real-time control of a spatial robot juggler. Pp. 12-24. *Computer* May.
Rohling, R.N., and J.M. Hollerbach
1993 Optimized fingertip mapping for teleoperation of dextrous robot hands. In *Proceedings of the IEEE International Conference on Robotics and Automation*, Atlanta, Ga.
Rosheim, M.E.
1990 Design of an omnidirectional arm. Pp. 2162-2167 in *Proceedings of the IEEE International Conference of Robotics and Automation*, Cincinnati, Ohio.
Russell, M.
1983 Odex 1: The first functionoid. *Robotics Age* 5(5):12-18.
Salcudean, S.E., N.M. Wong, and R.L. Hollis
1992 A force-deflecting teleoperation system with magnetically levitated master and wrist. Pp. 1420-1426 in *IEEE International Conference on Robotics and Automation*, Nice, France.
Salisbury, J.K.
1985 Kinematic and force analysis of articulated hands. Pp. 2-167 in M.T. Mason and J.K. Salisbury, eds., *Robot Hands and the Mechanics of Manipulation*. Cambridge, Mass.: MIT Press.
Salisbury, K.
1980 Active stiffness control of a manipulator in Cartesian coordinates. In *Conference on Decision and Control, 19th Annual Symposium on Adaptive Processes*, Albuquerque, N.Mex.
Salisbury, K., B. Eberman, M. Levin, and W. Townsend
1990 The design and control of an experimental whole-arm manipulator. Pp. 233-242 in H. Miura and S. Arimoto, eds., *Robotics Research: The 5th International Symposium*. Cambridge, Mass.: MIT Press.
Sato, T., and S. Hirai
1988 MEISTER: A model enhanced intelligent and skillful teleoperational robot system. Pp. 155-162 in R. Bolles and B. Roth, eds., *Robotics Research: The 4th International Symposium*. Cambridge, Mass.: MIT Press.
Sayers, C., and R. Paul
1993 Synthetic fixturing. Pp. 37-46 in *Advances in Robotics, Mechatronics and Haptic Interfaces, DSC-49*, ASME Winter Annual Meeting, New Orleans, Nov. 28-Dec. 3.
Sheridan, T.B.
1960 Human metacontrol. In *Proceedings of Annual Conference on Manual Control*, Wright Patterson Air Force Base, Ohio.
1970 On how often the supervisor should sample. *IEEE Transactions on Systems Science and Cybernetics* 6(2):140-145.
1989 Telerobotics. *Automatica* 25(4):487-507.
1992a *Telerobotics, Automation, and Human Supervisory Control*. Cambridge, Mass.: MIT Press.
1992b Musings on telepresence and virtual presence. *Presence: Teleoperators and Virtual Environments* 1(1):120-125.

1992c Defining our terms. *Presence: Teleoperators and Virtual Environments* 1:272-274.

1993 Space teleoperation through time delay: Review and prognosis. *IEEE Transactions on Robotics and Automation* 9:592-606.

Sheridan, T.B., and R.T. Hennessy, eds.

1984 *Research and Modeling of Supervisory Control Behavior.* National Research Council, Commission on Behavioral and Social Sciences and Education, Committee on Human Factors. Washington, D.C.: National Academy Press.

Sheridan, T.B., G.J. Raju, F.T. Buzan, W. Yared, and J. Park

1989 Adjustable impedance, force feedback and command language aids for telerobotics. Parts 1-4 in *Proceedings of the NASA Conference on Space Telerobotics,* 8-part MIT progress report. Cambridge, Mass.: MIT Press.

Smurlo, R.P., and H.R. Everett

1992 Intelligent security assessment for a mobile robot. In *Proceedings of Sensors Exposition.* Chicago, Ill.

Song, S., and K.J. Waldron

1989 *Machines That Walk.* Cambridge, Mass.: MIT Press.

Spain, E.H.

1992 Effects of Camera Aiming Technique and Display Type on Unmanned Ground Vehicle Performance. Final Report to the Department of Defense Unmanned Ground Vehicle Joint Program Office.

Speeter, T.H.

1992 Transforming human hand motion for telemanipulation. *Presence: Teleoperators and Virtual Environments* 1:63-79.

Spooner, M.G., and C.H. Weaver

1955 An Analysis and Analog-computer Study of a Force Reflecting Positional Servomanipulator. Technical Report Reprint No. 264. University of Wisconsin Engineering Station.

Starr, G.P., and C.W. Wilson

1992 Design of a torque controller for the Adept-2 robot. *Journal of Intelligent and Robotic Systems* 6:183-201.

Stewart, D.B., D.E. Schmitz, and P.K. Khosla

1989 CHIMERA II: A real-time multiprocessing environment for sensor-based roto control. Pp. 265-271 in *Proceedings of the IEEE International Symposium on Intelligent Control.*

Strassberg, Y., A.A. Goldenberg, and J.K. Mills

1992 A new control scheme for bilateral teleoperating systems: Lyapunov stability analysis. Pp. 837-852 in *Proceedings of the IEEE International Conference on Robotics and Automation.*

Swartz, B.A.

1993 Diver and ROV deployable laser range gate underwater imaging systems, Underwater Intervention 93. Pp. 193-197 in the *11th Annual ROV Conference and Exposition,* Marine Technology Society, New Orleans, La.

Tachi, S.

1991 Toward virtual existence in real and/or virtual worlds. Pp. 85-94 in

Proceedings of the International Conference on Artificial Reality and Tele-existence.

Tachi, S., H. Arai, and T. Maeda

1989 Development of an anthropomorphic tele-existence slave robot. Pp. 385-390 in *Proceedings of the International Conference on Advanced Mechatronics*, Tokyo, Japan.

1990a Tele-existence master slave system for remote manipulation. Pp. 343-348 in *Proceedings of the IEEE International Workshop on Intelligent Robots and Systems.*

1990b Tele-existence master slave system for remote manipulation (II). Pp. 85-90 in *Proceedings of the 29th IEEE Conference on Decision and Control.*

Tendick, F., R.W. Jennings, G. Tharp, and L. Stark

1993 Sensing and manipulation problems in endoscopic surgery: Experiment, analysis, and observation. *Presence: Teleoperators and Virtual Environments* 2:66-81.

Thorpe, C.E.

1990 *Vision and Navigation: The Carnegie Mellon Navlab.* Norwell, Mass.: Kluwer Academic Publishers.

Todd, D.J.

1986 *Fundamentals of Robot Technology.* New York: John Wiley.

Trimmer, W.

1989 Micromechanical systems. Pp. 1.1-1.32 in *3rd Toyota Conference on Integrated Micro Motion Systems*, Tokyo, Japan.

Tsach, U., T.B. Sheridan, and A. Buharali

1983 Failure detection and location in process control: Integrating a new model-based tachnique with other methods. In *Proceedings of American Control Conference*, San Francisco.

Tsai, L.-W., and A.P. Morgan

1985 Solving the kinematics of the most general six- and five-dgree-of-freedom manipulators by continuation methods. *ASME Journal of Mechanisms, Transmissions, and Automation in Design* 107:189-200.

Uhrich, R.W., and J.M. Walton

1993 AUSS-Navy Advanced Unmanned Search System. *Sea Technology* (February).

VanZandt, T.R., T.W. Kenny, and W.J. Kaiser

1992 Novel position sensor technologies for microaccelerometers. *SPIE* 1694:165-171.

Vertut, J., and P. Coiffet

1985a *Teleoperation and Robotics: Evolution and Development.* London: Kogan Page.

1985b *Teleoperation and Robotics: Applications and Technology.* London: Kogan Page.

Vijaykumar, R., M.J. Tsai, and K.J. Waldron

1985 Geometric optimization of manipulator structures for working volume and dexterity. Pp. 228-237 in *Proceedings of the IEEE Conference on Robotics and Automation*, St. Louis, Missouri.

Vischer, D., and O. Khatib
1990 Design and development of torque-controlled joints. Pp. 271-286 in V. Hayward and O. Khatib, eds., *Experimental Robotics 1—The First International Symposium.* New York: Springer-Verlag.
Waldron, J.J., and R.B. McGee
1986 The adaptive suspension vehicle. Pp. 7-12 in *IEEE Control Systems Magazine*
Waldron, K.J., and K.H. Hunt
1991 Series parallel dualities in actively coordinated mechanisms. *International Journal of Robotics Research* 10.
Wallace, R.S., and D.G. Taylor
1991 Low-torque-ripple switched reluctance motors for direct-drive robotics. *IEEE Transactions on Robotics and Automation* 7:733-742.
Walton, J., M. Cook, and R. Urich
1993 Advanced unmanned search systems. Pp. 243-249 in *Underwater Intervention 93: the 11th Annual ROV Conference and Exposition,* Marine Technology Society, New Orleans, La.
Wampler, C.W.
1986 Manipulator inverse kinematic solutions based on vector formulations and damped least-square methods. *IEEE Transactions on Systems, Man, and Cybernetics* SMC-16:93-101.
Wen, J.T.
1988 Controller synthesis for infinite dimensional systems based on a passivity approach. Pp. 724-729 in *Proceedings of the 27th Conference on Decision and Control,* Austin, Tex.
Whitney, D.
1969 Resolved motion rate control of manipulators and human prosthesis. *IEEE Transactions on Man-Machine Systems* MMS-10:47-53.
1985 Historical perspectives and state of the art in robot force control. Pp. 262-268 in *Proceedings of the IEEE Conference on Robotics and Automation.*
Wilcox, B.H.
1992 Robotic vehicles for planetary exploration. In C. Weisbin, ed., *Robotic Systems to Augment Man's Capability in Space.* A special issue of *International Journal of Applied Intelligence* 2. Norwell, Mass.: Kluwer Academic Publishers.
Wittenburg, R.C.
1987 Automobile collision avoidance systems capitalizes on bean steerable antenna. *Electronic Engineering Times* 2(February).
Wu, E.C., J.C. Hwang, and J.T. Chlakek
1993 Fault-tolerant joint development for the space shuttle remote manipulator system: Analysis and experiment. *IEEE Transactions on Robotics and Automation* 9:675-684.
Yokokohji, Y., and T. Yoshikawa
1992 Bilateral control of master-slave manipulators for ideal kinesthetic coupling—Formulation and experiment. Pp. 849-885 in *Proceedings of the IEEE International Conference on Robotics and Automation.*

Yoshichida, F., and K. Michitaka
1993 2400-MFlops reconfigurable parallel VLSI processor for robot control. Pp. 149-154 in *Proceedings of the IEEE International Conference on Robotics and Automation.*

Yoshikawa, T., and T. Miyazaki
1989 A six-axis force sensor with three-dimensional cross-shape structure. Pp. 249-255 in *Proceedings of the IEEE International Conference on Robotics and Automation*, Scottsdale, Ariz.

Yun, W., and R.T. Howe
1992 Recent developments in silicon microaccelerometers. *Sensors* 9(10).

Yun, W., et al.
1992 Surface micromachined, digitally force-balanced accelerometer with integrated CMOS detection circuitry. Pp. 126-131 in *IEEE Solid-State Sensor Actuator Workshop*, Hilton Head, S.C.

Zai, L.C., L.F. Durfee, D.G. Manzer, J.P. Karadis, M.P. Mastro, and L.W. Landerman
1992 Control of a Hummingbird Minipositioner with a multi-transputer MARC Controller. Pp. 543-541 in *Proceedings of the IEEE International Conference on Robotics and Automation.*

Zhai, S., and P. Milgram
1993a Human performance evaluation of manipulation schemes in virtual environments. *Proceedings of the IEEE Virtual Reality Annual International Symposium*, Seattle, Sept.
1993b Human performance evaluation of isometric and elastic rate controllers in a 6 DOF tracking task. *Proceedings of the SPIE, Vol. 2057: Telemanipulator Technology* Boston, Mass. Sept. 7-10.

CHAPTER 10: NETWORKING AND COMMUNICATIONS

Ballart, R., and Y.-C. Ching
1989 SONET: Now it's the standard optical network. *IEEE Communications Magazine* 29(3):8-15.

Bryson, S., and C. Levit
1992 Virtual wind tunnel: An environment for the exploration of three dimensional unsteady flows. *IEEE Computer Graphics and Applications* 12(4):25-34.

Cattlett, C.E.
1992 Balancing resources. *IEEE Spectrum* 29(9):48-55.

Cavanaugh, J.D., and T.J. Salo
1992 Internetworking with ATM WANS. Research report, Minnesota Supercomputer Center, Inc.

Gelernter, D.
1991 *Mirror Worlds: or the Day Software Puts the Universe in a Shoebox...How It will Happen and What It Will Mean.* New York: Oxford University Press.

Habib, I.W., and T.N. Saadawi
1991 Controlling flow and avoiding congestion in broadband networks. *IEEE Communications Magazine* 29(10):46-53.
Hsing, R.T., C.-T. Chen, and J. Bellisio
1993 Video communications and services in the copper loop. *IEEE Communications Magazine* 31(1):62-68.
Institute for Simulation and Training
1991 Protocol data units for entity information and entity interaction in a distributed interactive simulation. Military Standard (DRAFT). Document number IIST-PD-90-2. Orlando, Fla.
Johnson, J.T.
1992 Turning the clock ahead on tomorrow's network. *Data Communications* 21(12):43-62.
Miller, D.C., A.R. Pope, and R.M. Waters
1989 Long-Haul Networking of Simulators. BBN Systems and Technologies Corporation research paper.
National Research Council
1994 *Realizing the Information Future: The Internet and Beyond.* NRENAISSANCE Committee; Computer Science and Telecommunications Board; Commission on Physical Sciences, Mathematical, and Applications, Washington, D.C.: National Academy Press.
Pausch, R.
1991 Virtual reality on five dollars a day. Pp. 265-270 in *Proceedings of CHI '91*. New Orleans: ACM Press.
Pope, A.
1989 The SIMNET Network and Protocols. BBN Report No. 7102. Cambridge, Mass.: BBN Systems and Technologies.
Presence
1994 Networked Virtual Environments and Teleoperation. Special issue.
Reddy, R.
1992 Advanced Distributed Simulation Concept Briefing. ADS Concept Brief.
Rudin, H., and R. Williamson
1989 Faster, more efficient streamlined protocols. *IEEE Communications Magazine* 27(6):10-12.
Thorpe, J.
1987 The new technology of large scale simulator networking: Implications for mastering the art of warfighting. In *Proceedings of the Ninth Interservice Industry Training Systems Conference.*

CHAPTER 11: EVALUATION OF SYNTHETIC ENVIRONMENT SYSTEMS

Meister, D.
1985 *Behavioral Analysis and Measurement Methods.* New York: John Wiley.

CHAPTER 12: SPECIFIC APPLICATIONS OF SE SYSTEMS

Adam, J.A., ed.
1993 Virtual reality is for real. *IEEE Spectrum* (October):22-29.

Airey, J.M., J.H. Rohlf, and F. P. Brooks, Jr.
1990 Towards image realism with interactive update rates in complex virtual building environments. *Computer Graphics* 24(2):41.

Anderson, J.R.
1993 *Cognitive Tutors: Lessons Learned*. Technical report. Pittsburgh, Pa.: Carnegie Mellon University.

Badler, N.I., C. Phillips, and B.L. Webber
1993 *Simulating Humans: Computer Graphics, Animation, and Control*. Cambridge, England: Cambridge University Press.

Bailey, R.W., A.L. Imbmbo, and K.A. Zucker
1991 Establishment of a laparoscopic cholecystectomy training program. *The American Surgeon* 57(4):231-236.

Bajura, M., H. Fuchs, and R. Ohbuchi
1992 Merging virtual objects with the real world: Seeing ultrasound images. *Computer Graphics* 26(2):203-210.

Bricken, M., and C.M. Byrne
1992 Summer Students in Virtual Reality: A Pilot Study on Educational Applications of Virtual Reality Technology. Technical Report from Human Interface Technology Laboratory of the Washington Technology Center, University of Washington.

Brooks, F.P., Jr., M. Ouh-young, J.J. Blatter, and P.J. Kilpatrick
1990 Project GROPE—Haptic displays for scientific visualization. *Computer Graphics: Proceedings of SIGGRAPH '90* 24(4).

Brown, J.S., A. Collins, and P. Duguid
1989 Situated cognition and the culture of learning. *Educational Researcher* 18(1):32-42.

Bruckman, A.
1992-93 Identity Workshop: Emergent Social and Psychological Phenomena in Text-Based Virtual Reality. In *Proceedings of INET 93*. Available via anonymous ftp from media.mit.edu in pub/MediaMOO/Papers/identity-workshop.{ps,rtf}

Brunner, B., G. Hirzinger, K. Landzettel, and J. Heindl
1993 Multisensory shared autonomy and tele-sensor-programming—Key issues in the space robot Technology Experiment ROTEX. Pp. 2123-2131 in *Proceedings of the IEEE/RSJ International Conference on Intelligent Robots and Systems*, Yokohama, Japan

Bryson, S.
1992 Virtual spacetime: An environment for the visualization of curved spacetimes via geodesic flows. In *Proceedings of the IEEE Visualization T92*, Boston.

Bryson, S., and M. Gerald-Yamasaki
1992 The distributed virtual wind tunnel. In *Proceedings of Supercomputing '92*, Minneapolis, Minn.

Bryson, S., and C. Levit
1992 The virtual wind tunnel: An environment for the exploration of three dimensional unsteady flows. *Computer Graphics and Applications*, July
Carey, S.
1987 *Conceptual Change in Childhood.* Cambridge, Mass.: MIT Press.
Caudell, T.P., and D.W. Mizell
1992 Augmented reality: An application of heads-up display technology to manual manufacturing processes. *IEEE* (January):659-669.
Chapman, R.L., J.L. Kennedy, A. Newell, and W.C. Biel
1959 The Systems Research Laboratory's air defense experiments. *Management Science* 5(3):251-269.
Cognition and Technology Group at Vanderbilt
1993 Anchored instruction and situated cognition revisited. *Educational Technology* 22(3):52-70.
Craig, L.C.
1980 Office automation at Texas Instruments, Incorporated. Pp. 202-214 in M.L. Moss, ed., *Telecommunications and Productivity.* Reading, Mass.: Addison-Wesley.
Cruz-Neira, C., D.J. Sandin, T.A. DeFanti, R.V. Kenyon, and J.C. Hart
1992 The CAVE: Audio visual experience automatic virtual environment. *Communications of the ACM* 35(6):65-72.
Cruz-Neira, C., D.J. Sandin, and T.A. DeFanti
1993a Surround-screen projection-based virtual reality: The design and implementation of the CAVE. *Computer Graphics: Proceedings of SIGGRAPH '93.* August.
Cruz-Neira, C., J. Leigh, C. Barnes, S. Cohen, S. Das, R. Englemann, R. Hudson, M. Papka, L. Siegel, C. Vasilakis, D.J. Sandin, and T.A. DeFanti
1993b Scientists in wonderland: A report on visualization applications in the CAVE virtual reality environment. In *Proceedings of the IEEE Symposium on Research Frontiers in Virtual Reality*, October.
Delp, S., P. Loan, M. Hoy, F. Zajac, E. Topp, and J. Rosen
1990 An interactive graphics-based model of the lower extremity to study orthopaedic surgical procedures. *IEEE Transactions on Biomedical Engineering* 37(8 August):757-767.
Denenberg, V.H.
1954 The Training Effectiveness of a Tank Hull Trainer. HumRRO Technical Report No. 3., Washington, D.C.
Ellis, S.R., ed.
1993 *Pictorial Communication in Virtual and Real Environments.* London: Taylor and Francis.
Fanning, T., and B. Raphael
1986 Computer teleconferencing: Experience at Hewlett-Packard. *Proceedings of a Conference on Computer-Supported Cooperative Work.* New York: The Association for Computing Machinery.
Fisher, S.
1993 Current Status of VR and Entertainment. Presentation to the Commit-

tee on Virtual Reality Research and Development. Woods Hole, Mass., August.

Fitts, P.M., L. Schipper, J.S. Kidd, M. Shelly, and C. Kraft
1958 Some concepts and methods for the conduct of system research in a laboratory setting. In G. Finch and F. Cameron, eds., *Air Force Human Engineering, Personnel and Training Research.* Washington D.C.: National Academy of Sciences.

Fogle, R.F.
1992 The use of teleoperators in hostile environment applications. Pp. 61-66 in *IEEE International Conference on Robotics and Automation*, Nice, France.

Fordyce, S.W.
1974 *NASA Experiences in Telecommunications as a Substitute for Transportation.* Washington, D.C.: National Aeronautics and Space Administration. April.

Fuerzeig, W.
1992 Visualization tools for model-based inquiry. In *Proceedings of the Conference on Technology Assessment.* UCLA Center for Technology Assessment, Los Angeles, Calif.

Fujiwara, S., R. Kanehara, T. Okada, and T. Sanemori
1993 An articulated multi-vehicle robot for inspection and testing of pipeline interiors. Pp. 509-516 in *Proceedings of the IEEE/RSJ International Conference on Intelligent Robots and Systems*, Yokohama, Japan.

Fukuda, T., H. Hosokai, and M. Otsuka
1987 Autonomous pipeline inspection and maintenance robot with inch worm mobile mechanism. Pp. 539-544 in *Proceedings of the IEEE International Conference on Robotics and Automation.* Raleigh, N.C.

Glenn, A., and L. Ehman
1987 *Computer-Based Education in Social Studies.* Social Studies Development Center, Indiana University and ERIC Clearinghouse for Social Studies/Social Science Education.

Goldenberg, A.A., J. Wiercienski, P. Kuzan, C. Szymczyk, R.G. Fenton, and B. Shaver
1992 A remote manipulator for forestry operations. Pp. 2792-2795 in *IEEE International Conference on Robotics and Automation*, Nice, France.

Greenleaf, W.J.
1994 DataGlove and DataSuit: Virtual reality technology applied to the measurement of human movement. Pp. 63-69 in *Proceedings of Medicine and Virtual Reality.* San Diego, Calif.

Guetzkow, H., ed.
1962 *Simulation in the Social Sciences.* Englewood Cliffs, N.J.: Prentice-Hall.

Haber, R.N.
1986 Flight simulation. *Scientific American* 255(1):96-103.

Hall, D.M., and W.K. Walsh
1993 The changing face of textile research. *Textile World* (September):90-100.

Harel, I., and S. Papert, eds.
1991 *Constructionism.* Norwood, N.J.: Ablex Publishing Corp.

Harrigan, R.W.
1993 Automating the operation of robots in hazardous environments. Pp. 1211-1219 in *Proceedings of the IEEE/RSJ International Conference on Intelligent Robots and Systems*, Yokohama, Japan.

Hattinger, G., et al., eds.
1990 Ars Electronica 1990 Band II: Virtuelle Welten (in German and English). Linzer Veranstaltungsgesellschaft mbH. Linz, Austria: Brucknerhaus.

Hays, R.T., and M.J. Singer
1989 *Simulation Fidelity in Training System Design*. New York: Springer-Verlag.

Hirose, S.
1987 Wall climbing vehicle using internally balanced magnetic unit, RoManSy 6. Pp. 420-427 in A. Morecki, G. Bianchi, and K. Kedzior, eds., *Proceedings of the 6th CISM-IFToMM Symposium on Theory and Practice of Robots and Manipulators*. Cambridge, Mass.: MIT Press.

Hirzinger, G., B. Brunner, J. Dietrich and J. Heindl
1993 Sensor-based space robotics-ROTEX and its telerobotic features. *IEEE Transactions on Robotics and Automation* 9:649-663.

Hodges, L.F., J. Bolter, E. Mynatt, W. Ribarsky, and R. van Teylingen
1993 Virtual environments research at the Georgia Tech GVU Center. *Presence* 2(3):234-243.

Hunter, I.W., L.A. Jones, M.A. Sagar, T.D. Doukoglou, S.R. Lafontaine, P.G. Charette, G.D. Mallinson, and P.J. Hunter
1993 A Teleoperated Microsurgical Robot and Associated Virtual Environment for Eye Surgery. Technical Report. Montreal: McGill University.

Kibbee, J.M., C.J. Craft, and B. Nanus
1961 *Management Games*. New York: Reinhold Publishing Corp.

Kijima, R., K. Shirakawa, M. Hirose, and K. Nihei
1994 Virtual sand box: Development of an application of virtual environments for clinical medicine. *Presence* 3(1): 45-59.

Kim, W.S., and A.K. Bejczy
1993 Demonstration of a high-fidelity predictive/preview display technique for telerobotic servicing in space. *IEEE Transactions on Robotics and Automation* 9:698-702.

Kim, W.S., P.S. Schenker, A.K. Bejczy, and S. Hayati, S.
1993 Advanced graphics interfaces for telerobotic servicing and inspection. Pp. 303-309 in *Proceedings of the IEEE/RSJ International Conference on Intelligent Robots and Systems*, Yokohama, Japan.

Kozak, J.J., P.A. Hancock, E. Arthur, and S. Chrysler
1993 Transfer of training from virtual reality. *Ergonomics* 36:777-784.

Krishnaswamy, G.M., and A.K. Elshennawy
1992 Concurrent engineering deployment: An enhanced "customer product" approach. *Computers and Industrial Engineering* 23(1-4):503-506.

Kwitowski, A.J., W.D. Mayercheck, and A.L. Brautigam
1992 Teleoperation for continuous miners and haulage equipment. *IEEE Transactions on Industry Applications* 28:1118-1126.

Leonardo
1994 Virtual Reality: Venus return or vanishing point. *Leonardo* 27:277-302.

Levin, H., et al.
1984 Cost Effectiveness of Four Educational Interventions. IFG Report No. 84-A11. Stanford University.

Lineham, T.E., ed.
1993 *Computer Graphics Visual Proceedings*. New York: Association for Computing Machinery.

Loeffler, L.E., and T. Anderson, eds.
1994 *The Virtual Reality Casebook*. New York: Van Nostrand Reinhold.

Loftin, R.B., B. Lee, S. Mueller, and R. Way
1991 An intelligent tutoring system for physics problem solving. In *Proceedings of Physics Computer '91*, San Jose, Calif.

Mann, R.W.
1985 Computer aided surgery. In *RESNA 8th Annual Conference*, Memphis, Tenn.

Maxis Corp.
1991 SimAnt, SimEarth, SimCity: Software Products and Reference Manuals. Orinda, Calif.

McCormick, B., T.A. DeFanti, and M.D. Brown, eds.
1987 Visualization in scientific computing. *Computer Graphics* 21(6).

McLane, R.C., and J.D. Wolf
1965 *Final Research Report on the Experimental Evaluation of Symbolic and Pictorial Displays for Submarine Control*. St. Paul, Minn.: Honeywell, Inc.

Merickel, M.L.
1990 The creative technologies project: Will training in 2D/3D graphics enhance kids' cognitive skills? *T.H.E. Journal* 18(5):55-58.

Mizell, D.
1993 Virtual Reality Research at Boeing. Presentation to the National Research Council's Committee on Virtual Reality Research and Development.

Moser, M.A., ed.
1991 *Virtual Seminar on the Bioapparatus*. Banff, Alberta: Banff Centre for the Arts.

Munakata, T., S. Murakami, Y. Matsumoto, S. Nakagaki, T. Honda, K. Shibanuma, S. Kakudate, K. Oka, T. Terakado, and M. Kondoh
1993 Manipulator for in-vessel remote maintenance of fusion experimental reactor. Pp. 1220-1224 in *Proceedings of the IEEE/RSJ International Conference on Intelligent Robots and Systems*, Yokohama, Japan.

National Council of Teachers of Mathematics
1989 *Curriculum and Evaluation Standards for School Mathematics*. National Council of Teachers of Mathematics. Reston, Virginia.

Nishi, A., Y. Wakasugi, and K. Watanabe
1986 Design of a robot capable of moving on a vertical wall. *Advanced Robotics* 1(1):33-46.

Office of Technology Assessment
1988 *Power On! New Tools for Teaching and Learning.* OTA-SET-379. Washington, D.C.: Government Printing Office.

Okada, T., and T. Kanade
1987 A three-wheeled self-adjusting vehicle in a pipe, FERRET-1. *International Journal of Robotics Research* 6(4):60-75.

Orlansky, J., and J. String
1977 *Cost-Effectiveness of Flight Simulators for Military Training,* Vol. 1. Arlington, Va.: The Institute for Defense Analyses.

Palmiter, S., and J. Elkerton
1993 Animated demonstrations for learning procedural computer-based tasks. *Human-Computer Interaction* 8:193-216.

Pantelidis, V.S.
1993 *Virtual Reality and Education: A Bibliography.* Available via anonymous ftp from u.washington.edu in the directory /pub/user-supported/VirtualReality/misc/papers under the filename Pantelids-VR-Education-Bibl.txt.

Piaget, J., and B. Inhelder
1967 *The Child's Conception of Space.* New York: Norton.

Pimentel, K., and B. Blau
1994 System architecture issues related to multiple user VR systems: Teaching your system to share. *IEEE Computer Graphics and Applications,* January.

Potemkin, E., P. Astafurov, A. Osipov, M. Malenkov, V. Mishkenyuk, and P. Sologub
1992 Remote-controlled robots for repair and recovery in the zones of high radiation levels. Pp. 80-82 in *IEEE International Conference on Robotics and Automation,* Nice, France.

Regian, W.E., and W.I. Shebilske
1992 Virtual reality: An instructional medium for visual-spatial tasks. *Journal of Communication* 42(4):136-149.

Ritchie, M.L., and F.A. Muckler
1954 Retroaction as a function of discrimination and motor variables. *Journal of Experimental Psychology* 48(6):409-415.

Rolfe, A.
1992 The problem of maintenance of the Joint European Torus Tokamak fusion nuclear experiment. *Spar Journal of Engineering and Technology* 1:7-13.

Rosen J., A. Lasko-Harvill, and R. Satava
1994 Virtual reality and surgery. In G. Burdea, ed., *Computer Integrated Surgery.* Cambridge, Mass.: MIT Press.

Sanders, J.H., and F.J. Tedesco
1993 Telemedicine: Bringing medical care to isolated communities. *Journal of the Medical Association of Georgia* (May):237-241.

Satava, R.

1993a Surgery 2001: A technologic framework for the future. *Surgical Endoscopy* 7:111-113.

1993b Virtual reality surgical simulator: The first steps. *Surgical Endoscopy* 7:203-205.

Sauder, B.J., P.D. Lawrence, and U. Wallensteiner

1992 Coordinated control systems applied to the forest industry. Pp. 52-55 in *Proceedings of the Forest Sector 2000*, ISTC.

Schmidt, R.A., and D.E. Young

1987 Transfer of movement control in motor skill learning. Pp. 47-59 in S.M. Cormier and J.D. Hagman, eds., *Transfer of Learning: Contemporary Research and Applications*. San Diego: Academic Press.

Schug, M.C., and H.S. Kepner, Jr.

1984 Choosing computer simulations in social studies. *The Social Studies* 75(Sept/Oct):211-215.

Secretary's Commission on Achieving Necessary Skills

1991 *What Work Requires of Schools: A SCANS Report for America 2000*. Secretary's Commission on Achieving Necessary Skills, U.S. Department of Labor.

Sheridan, T.B.

1993a Cooperation among Humans and among Humans and Machines. Talk delivered at Woods Hole, Mass., August.

1993b Space teleoperation through time delay: Review and prognosis. *IEEE Transactions on Robotics and Automation* 9:592-606.

Sheridan, T.B., and W.R. Ferrill

1974 *Man Machine Systems*. Cambridge, Mass.: MIT Press.

Shirazi, E.

1991 *Telecommuting: Moving the Work to the Workers*. Washington, D.C.: U.S. Department of Transportation.

Sinaiko, H.W., and T.G. Belden

1965 The indelicate experiment. Pp. 343-348 in Spiegel, J., and D.E. Walker, eds., *Second Congress on the Information System Sciences*. Washington, D.C.: Spartan Books.

Song, D., and M.L. Norman

1993 Cosmic explorer: A virtual reality environment for exploring cosmic data. In *Proceedings of IEEE Symposium on Research Frontiers in Virtual Reality*, October.

Steward, A.

1993 Virtual Reality Applications to the Textile Industry. Presentation to the National Research Council's Committee on Virtual Reality Research and Development. Woods Hole, Mass., August.

Stix, G.

1992 Reach out: Touch is added to virtual reality simulations. *Scientific American* 264(2):134.

Stone, H.W., and G. Edmonds

1992 HAZBOT: A hazardous materials emergency response mobile robot.

Pp. 67-73 in *IEEE International Conference on Robotics and Automation*, Nice, France.

Taylor, E.R., W. Robinett, V.L. Chi, F.P. Brooks, Jr., W.V. Wright, R.S. Williams, and E.S. Snyder

1993 The nanomanipulator: A virtual reality interface for a scanning tunneling microscope. *Proceedings of Siggraph 1993*. Anaheim, Calif.

Taylor, T.

1980 Telecommunications and public safety. Pp. 280-288 in M.L. Moss, ed., *Telecommunications and Productivity*. Reading, Mass.: Addison-Wesley.

Thorndike, E.L.

1903 *Educational Psychology*. New York: Lemcke and Buechner.

Thorpe, J.

1993 Synthetic Environments Strategic Plan. Draft 3B. Defense Advanced Research Projects Agency, Alexandria, Va.

Tinker, P.

1993 The Rockwell International Virtual Reality Laboratory. Rockwell International Corporation, Thousand Oaks, Calif.

Trivedi, M.M., and C.X. Chen

1993 Developing telerobotic systems using virtual reality concepts. Pp. 352-359 in *Proceedings of the IEEE/RSJ International Conference on Intelligent Robots and Systems*, Yokohama, Japan.

Tyre, T.

1989 Live T.V. broadcasts from ocean floor bring new depth to science education. *T.H.E. Journal* August:42-46.

Ullman, N.

1993 High-tech connection between schools and science expeditions enlivens classes. *Wall Street Journal* March 17:B1 and B10.

VPL Research, Inc.

1989 *Virtual Reality at Texpo '89*. Redwood City, Calif.: VPL Research, Inc.

1991 Brain Tour. Videotape for Lederle, Inc., demonstration. May.

Wallensteiner, U., P. Stager, and P. Lawrence

1988 A human factors evaluation of teleoperated hand controllers. Pp. 291-296 in *International Symposium on Teleoperation and Control*.

Warner, D., T. Anderson, and J. Johanson

1994 Bio-Cybernetics: A biologically responsive interactive interface—The next paradigm of human computer interaction. Pp. 237-241 in *Proceedings of Medicine and Virtual Reality*. San Diego, Calif.

Weghorst, S., J. Prothero and T. Furness

1994 Virtual images in the treatment of Parkinson's disease akinesia. *Conference Proceedings: Virtual Reality Meets Medicine II*. San Diego, Calif.

Whitcomb, L.L., and D.R. Yoerger

1993 A new distributed real-time control system for the JASON underwater robot. Pp. 368-374 in *Proceedings of the IEEE/RSJ International Conference on Intelligent Robots and Systems*, Yokohama, Japan.

White, B.

1984 Designing computer games to help physics students understand Newton's laws of motion. *Cognition and Instruction* 1(1):69-108.

William, A., C.S. Cooper, and R.J. Fisher
1993 Predictors of laparoscopic complications after formal training in laparoscopic surgery. *Journal of the American Medical Association* 270(22):2689-2692.

Williams, A.C., and R.E. Flexman
1949 *An Evaluation of the Link SNJ Operational Trainer as an Aid on Contact Flight Training*. Port Washington, N.Y.: Naval Special Devices Center.

Yam, P.
1993 Surreal science: Virtual reality finds a place in the classroom. *Scientific American* 268 (February):103ff.

APPENDIXES

A

Biographical Sketches

NATHANIEL DURLACH (Chair) is a senior scientist in the Department of Electrical Engineering and Computer Science at the Massachusetts Institute of Technology and has been co-director of the Sensory Communication Group in the Research Laboratory of Electronics there for over 20 years. He has also been a visiting scientist in the Biomedical Engineering Department of Boston University for five years. He received an M.A. degree from Columbia University in mathematics and took courses at Harvard University in psychology and biology. He is the author (or coauthor) of numerous book chapters and refereed articles in such journals as *Perception and Psychophysics* and the *Journal of the Acoustical Society of America*; he continues to review articles, proposals, and research programs in the field of psychophysics; and he has received the silver medal award for outstanding work in psychoacoustics by the Acoustical Society of America. Recently, his research interests have focused on teleoperator and virtual environment systems, with special emphasis on the human-machine interfaces used in such systems. He is cofounder and director of the MIT Virtual Environment and Teleoperator Research Consortium, as well as cofounder and managing editor of the new MIT Press journal *Presence: Teleoperators and Virtual Environments*.

STEVE BRYSON is an employee of Computer Sciences Corporation working under contract for the Applied Research Office of the Numerical Aerodynamics Simulation Systems Division at the NASA-Ames Research Center. His current research involves the application of virtual reality techniques for scientific visualization, of which the virtual wind tunnel is the main focus. He began work in the virtual reality field in 1984 at VPL Research, working on a graphics-based programming environment using the prototype DataGlove for input; later he was involved in work on the DataGlove model II. He joined the VIEW lab at the NASA-Ames Research Center in 1987, where he was involved in integrating the various input-output and graphics systems into a virtual environment. This included research in software architectures for virtual reality systems and human factors. In 1991, he was cochair of the IEEE Symposium on Research Frontiers in Virtual Reality and is program cochair of the IEEE Virtual Reality Annual International Symposium.

NORMAN HACKERMAN is chairman of the Scientific Advisory Board at the Robert A. Welch Foundation. He is president emeritus of Rice University, where he was president and a professor of chemistry for 15 years. Prior to that, he had a long and distinguished career at the University of Texas at Austin, where he served as president and held various positions, including director of the Corrosion Research Laboratory and on state boards and on committees focusing on various aspects of research and education. He was technical editor and then editor of the *Journal of the Electrochemical Society* from 1950 to 1990. He has served on a dozen National Academy of Science/National Research Council panels and committees and is a past chairman of the Board on Energy Studies and of the Commission on Physical Sciences, Mathematics, and Applications. He is the author or coauthor of more than 200 publications. He has a Ph.D. in chemistry from Johns Hopkins University.

JOHN M. HOLLERBACH is professor of computer science at the University of Utah. From 1989 to 1994 (including his time of membership on the committee) he was the natural sciences and engineering/Canadian Institute for Advanced Research professor of robotics at McGill University. From 1982 to 1989 he was on the faculty of the Department of Brain and Cognitive Sciences and a member of the Artificial Intelligence Laboratory at the Massachusetts Institute of Technology. He received a Ph.D. in computer science from MIT in 1978. He was a member of the Administrative Committee of the IEEE Robotics and Automation Society from 1989 to 1993. He is a technical editor of the *IEEE Transactions on Robotics and Automation*, treasurer of the *IEEE/ASME Journal of Microelectromechanical Systems*, and a senior editor of *Presence*.

JAMES R. LACKNER is Riklis professor of physiology and director of the Ashton Graybiel Spatial Orientation Laboratory at Brandeis University. He received B.Sc. and Ph.D. degrees from the Massachusetts Institute of

Technology. He has served on the National Research Council's Committee on Vision from 1986 to 1992 and the Committee on Space Biology and Medicine from 1990 to the present. From 1986 to 1990 he was provost of Brandeis University. He is the section editor for spatial orientation of the *Journal of Vestibular Research* and is on the editorial board of *Presence.* His research interests concern human movement and orientation control in unusual sensory and force environments.

HERBERT S. LIN is senior staff officer for the Computer Science and Telecommunications Board. Previously he was a professional staff member for the House Armed Services Committee under Representative Les Aspin working on strategic modernization and arms control issues. He has also worked as a researcher, instructor, and visiting scholar at the Massachusetts Institute of Technology, Cornell University, and the University of Washington. He received a Ph.D. in physics from the Massachusetts Institute of Technology.

ANNE S. MAVOR is study director for the Committee on Virtual Reality Research and Development and for the Committee on Human Factors. Her previous work as a National Research Council senior staff officer has included a study of modeling cost and performance of military enlistment, a review of federally sponsored education research activities, and a study to evaluate performance appraisal for merit pay. For the past 25 years her work has concentrated on human factors, cognitive psychology, and information system design. Prior to joining the NRC she worked for the Essex Corporation, a human factors research firm, and served as a consultant to the College Board. She has an M.S. in experimental psychology from Purdue University.

J. MICHAEL MOSHELL is associate professor of computer science at the University of Central Florida. He received a Ph.D. in computer science from Ohio State University in 1975 and spent the next 10 years at the University of Tennessee. He currently serves as chief scientist of the Visual Systems Laboratory in the Institute for Simulation and Training of the University of Central Florida. He is active in the Association for Computing Machinery and is an associate editor for the journal *Presence.* His research interests concern the application of simulation and virtual environments to education and training.

RANDY PAUSCH is an associate professor of computer science at the University of Virginia. He received a B.S. in computer science from Brown University and a Ph.D. in computer science from Carnegie Mellon University. He is a National Science Foundation presidential young investigator and a Lilly Foundation teaching fellow. His primary interests are human-computer interaction and undergraduate education.

RICHARD W. PEW is a psychologist with Bolt Beranek and Newman, Inc., where he is principal scientist and manager of the Cognitive Sciences and Systems Department. From 1960 to 1974 he was at the University of Michigan, where he received a Ph.D. and was on the faculty for 11 years. He was the first chairman of the National Research Council's Committee on Human Factors. Throughout his career he has been concerned with issues related to human performance and system design, ranging from the design of a control panel for an advanced music synthesizer to the impact of the introduction of automation into aircraft cockpits.

WARREN ROBINETT is a designer of interactive computer graphics software and hardware and president and founder of Virtual Reality Games, Inc., a developer of virtual reality video games for the home market. In 1978, he designed the Atari video game Adventure, the first graphical adventure game. In 1980, he was cofounder and chief software engineer at The Learning Company, a publisher of educational software. There he designed Rocky's Boots, a computer game that teaches digital logic design to 11-year-old children. Rocky's Boots won software of the year awards from three magazines in 1983. In 1986 Robinett worked as a research scientist at the NASA-Ames Research Center, where he designed the software for the Virtual Environment Workstation, NASA's pioneering virtual reality project. From 1989 to 1992 at the University of North Carolina, he directed the virtual reality and nanomanipulator projects. He is an associate editor for the journal *Presence*.

JOSEPH ROSEN is an associate professor of plastic and reconstructive surgery at Dartmouth-Hitchcock Medical Center and chief of the section on plastic and reconstructive surgery at the Veterans Administration Medical Center in White River Junction, Vermont. He is also adjunct professor of engineering at the Thayer School of Engineering at Dartmouth College. His clinical specialty is complex microsurgical reconstructions of congenital and acquired problems in plastic and reconstructive surgery. He operates on and reconstructs all parts of the body and interacts with surgeons in every specialty. His research interests include bionics, human-machine interfaces, nerve repair and evaluation, artificial nerve grafts, transplantation of limbs, and computer simulations of complex surgical procedures.

MANDAYAM A. SRINIVASAN is a principal research scientist in the Department of Mechanical Engineering at the Massachusetts Institute of Technology and a member of the Sensory Communication Group in the

Research Laboratory of Electronics at MIT. After receiving degrees in civil and aeronautical engineering in India, he received a Ph.D. in mechanical engineering from Yale University. Before moving to MIT in 1987, he was a member of the research faculty in the Department of Anesthesiology at the Yale University School of Medicine. His research interests include all aspects of human sensorimotor interactions with environments, especially in the context of human and machine haptics. He is a founding member of the MIT Virtual Environment and Teleoperator Consortium and is on the editorial board of the journal *Presence*.

JAMES J. THOMAS is a technology manager at the Applied Physics Center at Battelle Pacific Northwest Laboratories in Richland, Washington. He specializes in the research, design, and implementation of innovative information systems technology. Prior to working at Battelle, he worked at General Motors Research Laboratories. He has been named among the Top 100 Scientific Innovators by *Science Digest* and among the Top 100 Innovators in Science and Technology by *Research and Development*. In addition, he was awarded the Federal Laboratories Consortium technology transfer award for innovation in transferring research technology to industry and universities. He publishes, has been editor of scientific journals, gives invited talks, teaches as an adjunct associate professor for Washington State University, lectures for professional short courses, and is actively involved as a motivator and leader in computer graphics and human interface technology at the professional society SIGGRAPH. He was cochair for the SIGGRAPH annual conference in 1987, vice chair for SIGGRAPH from 1987 to 1989, and chair of ACM SIGGRAPH from 1989 to 1992.

ANDRIES VAN DAM is L. Herbert Ballou university professor and professor of computer science at Brown University. He has been on Brown's faculty since 1965 and was one of the founders of the Department of Computer Science and its first chairman, from 1979 to 1985. His research has concerned computer graphics, text-processing and hypermedia systems, and workstations. He has been working for more than 25 years on systems for creating and reading electronic books, based on high-resolution interactive graphics systems, for use in teaching and research. He received a Ph.D. from the University of Pennsylvania. *Introduction to Computer Graphics*, coauthored with J. Foley, S. Feiner, J. Hughes, and R. Phillips, was published in 1993. He has also authored or coauthored numerous other books and papers.

ELIZABETH WENZEL is director of the Spatial Auditory Displays Lab in the Aerospace Human Factors Research Division at the NASA-Ames Research Center, directing development of real-time display technology and conducting basic and applied research in auditory perception and localization in three-dimensional virtual acoustic displays. She is an associate editor of the journal *Presence* and has published a number of articles and spoken at many conferences on the topics of virtual acoustic environments. She received a Ph.D. in cognitive psychology with an emphasis in psychoacoustics from the University of California, Berkeley. From 1985 to 1986 she was a National Research Council postdoctoral research associate at NASA-Ames, working on the auditory display of information for aviation systems.

ANDREW WITKIN is a professor of computer science at Carnegie Mellon University. Previously he was director of the perception and graphics groups at Schlumberger's Palo Alto Research Lab. He received a B.A. in psychology from Columbia College and a Ph.D. in psychology from the Massachusetts Institute of Technology. He has published extensively in the areas of computer vision and computer graphics. He serves as an associate editor for *ACM Transactions on Graphics*, has served on numerous conference program committees, and is a fellow of the American Association for Artificial Intelligence. His awards include best paper prizes at the National Conference on Artifical Intelligence and the International Joint Conference on Artificial Intelligence; the grand prix for animation at the 1987 Parigraph competition in Paris, France; and the grand prix for computer graphics at the Prix Ars Electronica 1992 in Linz, Austria.

EUGENE WONG is the pro vice chancellor for research and development at the Hong Kong University of Science and Technology; during his time of service on the committee he was at the University of California, Berkeley. He assumed his current position upon his recent retirement as professor of electrical engineering and computer science at the University of California, where he had a 32-year career as a faculty member, most recently as department chairman. In January 1993, he completed a three-year stint as associate director of the White House Office of Science and Technology Policy. His research interests have spanned a wide range of topics, two principal ones being stochastic processes and their applications and database management systems. He was one of the architects of INGRES, a pioneering relational database management system, and contributed to the development of query processing, database semantics, and distributed systems. He has been a consultant to a number of major corporations and was a founder of the INGRES Corporation. He received B.S., A.M., and Ph.D. degrees in electrical engineering from Princeton University.

MICHAEL ZYDA is a professor in the Department of Computer Science at the Naval Postgraduate School. He is also the academic associate and associate chair for academic affairs in that department. His main focus in research is in the area of computer graphics, specifically the development of large-scale, networked, three-dimensional virtual environments and visual simulation systems. He is the senior editor for virtual environments for the quarterly *Presence*; for that journal, he has coedited special issues on Pacific Rim virtual reality and telepresence.

B

Contributors

Many individuals contributed to the committee's thinking and its drafting of various sections of the report by serving as presenters, consultants, and advisers. The list below acknowledges these contributors and their affiliations.

Bishnu Atal, American Telephone & Telegraph
Kurt Akeley, Silicon Graphics, Inc.
Walter Aviles, Massachusetts Institute of Technology
Norman Badler, University of Pennsylvania
Paul Barham, Naval Postgraduate School
Klaus Biggers, University of Utah
William Bricken, University of Washington
Martin Buehler, McGill University
Matt Conway, University of Virginia
Hari Das, Jet Propulsion Laboratory
Thomas DeFanti, University of Illinois-Chicago
Paul DiZio, Brandeis University
John Falby, Naval Postgraduate School
Scott Foster, Crystal River Engineering
Eric Foxlin, Massachusetts Institute of Technology
Blake Hannaford, University of Washington
Vincent Hayward, McGill University
Eric Howlett, Leep Systems
Ian Hunter, McGill University
Charles Hutchenson, Dartmouth College

Bonnie John, Carnegie Mellon University
Kenneth Kaplan, Harvard University
Kristen M. Kelleher, Naval Postgraduate School
Thomas Knight, Massachusetts Institute of Technology
Arthur Kramer, University of Illinois
Ronald Kruk, CAE Electronics
Jaron Lanier, Consultant
Jeng-Feng Lee, Massachusetts Institute of Technology
Ming C. Lin, Naval Postgraduate School
Michael R. Macedonia, Naval Postgraduate School
John Makhoul, BBN Laboratories, Inc.
David Mizell, Boeing
Joshua Mogal, Silicon Graphics, Inc.
Steve Molnar, University of North Carolina
Thomas Piantanida, SRI International
David Pratt, Naval Postgraduate School
Richard Satava, Advanced Research Projects Agency
Barbara Shinn-Cunningham, Massachusetts Institute of Technology
Kenneth Stevens, Massachusetts Institute of Technology
Alan Steward, Apparel CIM Center
Susumu Tachi, University of Tokyo
Thomas Wiegand, Massachusetts Institute of Technology
James Winget, Silicon Graphics, Inc.
Thomas Zeffiro, National Institutes of Health
David Zeltzer, Massachusetts Institute of Technology

Advisers

Scott Fisher, Telepresence
Richard Held, Massachusetts Institute of Technology
Thomas Sheridan, Massachusetts Institute of Technology
George Zweig, Los Alamos National Laboratory

Consultant

Harold Van Cott, National Research Council

Liaison

Lawrence W. Stark, Committee on Human Factors

Sponsors

Bernard Corona, Human Research and Engineering Directorate, Army Research Laboratory

Brenda Thein, Human Research and Engineering Directorate, Army Research Laboratory
John Tangney, Air Force Office of Scientific Research
Col. William Strickland, Human Resource Directorate, Armstrong Laboratory, Brooks AFB
Hendrick Ruck, Human Resource Directorate, Armstrong Laboratory, Brooks AFB
Kenneth Boff, Human Engineering Division, Crew Systems Directorate, Armstrong Laboratory, Wright Patterson AFB
Robert Eggleston, Human Engineering Division, Crew Systems Directorate, Armstrong Laboratory, Wright-Patterson AFB
Claire Gordon, U.S. Army Natick R&D Center
James Sampson, U.S. Army Natick R&D Center
James Jenkins, National Aeronautics and Space Administration
Y. T. Chien, National Science Foundation
John Hestenes, National Science Foundation
George Cotter, National Security Agency
Norman Glick, National Security Agency
Sharon Stanfield, Sandia National Laboratory

Index

A

D

I

N

O

P

R

S

T

W

X